AF832216

Subcellular Biochemistry

Volume 107

Series Editor

Tapas K. Kundu, Transcription and Disease Laboratory, JNCASR, Bangalore, Karnataka, India

Advisory Editors

J. Robin Harris, Institute of Molecular Physiology, University of Mainz, Mainz, Germany

Andreas Holzenburg, University of Texas Rio Grande Valley, Harlingen, USA

Viktor I. Korolchuk, Biosciences Institute, Faculty of Medical Sciences, Newcastle University, Newcastle upon Tyne, UK

Victor Bolanos-Garcia, Department of Biological and Medical Sciences, Oxford Brookes University, Oxford, UK

Jon Marles-Wright, School of Natural and Environmental Sciences, Newcastle University, Newcastle upon Tyne, UK

The book series SUBCELLULAR BIOCHEMISTRY is a renowned and well recognized forum for disseminating advances of emerging topics in Cell Biology and related subjects. All volumes are edited by established scientists and the individual chapters are written by experts on the relevant topic. The individual chapters of each volume are fully citable and indexed in Medline/Pubmed to ensure maximum visibility of the work.

Viktor I. Korolchuk • J. Robin Harris
Editors

Biochemistry and Cell Biology of Ageing: Part V, Anti-Ageing Interventions

Springer

Editors
Viktor I. Korolchuk
Biosciences Institute, Faculty of
Medical Sciences
Newcastle University
Newcastle upon Tyne, UK

J. Robin Harris
Institute of Molecular Physiology
University of Mainz
Mainz, Rheinland-Pfalz, Germany

ISSN 0306-0225 ISSN 2542-8810 (electronic)
Subcellular Biochemistry
ISBN 978-3-031-66767-1 ISBN 978-3-031-66768-8 (eBook)
https://doi.org/10.1007/978-3-031-66768-8

© The Editor(s) (if applicable) and The Author(s), under exclusive license to Springer Nature Switzerland AG 2024
This work is subject to copyright. All rights are solely and exclusively licensed by the Publisher, whether the whole or part of the material is concerned, specifically the rights of translation, reprinting, reuse of illustrations, recitation, broadcasting, reproduction on microfilms or in any other physical way, and transmission or information storage and retrieval, electronic adaptation, computer software, or by similar or dissimilar methodology now known or hereafter developed.
The use of general descriptive names, registered names, trademarks, service marks, etc. in this publication does not imply, even in the absence of a specific statement, that such names are exempt from the relevant protective laws and regulations and therefore free for general use.
The publisher, the authors and the editors are safe to assume that the advice and information in this book are believed to be true and accurate at the date of publication. Neither the publisher nor the authors or the editors give a warranty, expressed or implied, with respect to the material contained herein or for any errors or omissions that may have been made. The publisher remains neutral with regard to jurisdictional claims in published maps and institutional affiliations.

This Springer imprint is published by the registered company Springer Nature Switzerland AG
The registered company address is: Gewerbestrasse 11, 6330 Cham, Switzerland

If disposing of this product, please recycle the paper.

Preface

This volume is the fifth book in the *Biochemistry and Cell Biology of Ageing* sub-series, following four earlier instalments published by Springer within the *Subcellular Biochemistry* series. The previous books provided a comprehensive exploration of the basic and clinical aspects of ageing biology, offering answers to fundamental questions such as why and how we age. In this book, however, we shift the focus to the current options available for delaying the ageing process and promoting healthy lifespan extension.

To achieve this, we have invited experts in the field to share their perspectives on emerging anti-ageing interventions. Although many of these interventions are still in the pre-clinical phase, our contributors provide in-depth discussions of evidence from laboratory models and, where possible, link this to data emerging from human trials. The topics covered include interventions targeting telomerase, proteostasis, metabolic pathways, cellular senescence, and stem cells (see Chapter list). Additionally, several chapters explore how lifestyle choices, such as dietary interventions—including caloric restriction, vitamins, minerals, and supplements—can help prevent age-related diseases. These insights will be particularly valuable for those who are not yet ready to engage with more extreme longevity regimens.

For those new to the field, this book serves as an accessible entry point, complementing the content of the earlier volumes in the series (https://www.springer.com/series/6515/). It should be viewed as a part of a broader educational resource that, together with the preceding books, offers a comprehensive foundation in ageing research.

The field of ageing research has experienced an unprecedented surge in both fundamental and applied research in recent years. This growth is driven by the increasing recognition of ageing as a critical area of study by funding bodies, the urgent need to address the challenges posed by the rapidly growing elderly population, and the rise of a new "longevity" industry focused on combating ageing itself. The influx of funding into both academic and commercial sectors offers hope that the concepts presented in this book series will soon translate into real-world interventions, improving the quality of life for the ageing population.

These are exciting times for the field, and we look forward to incorporating new topics and revising existing ones in future volumes commissioned by Springer Nature. The knowledge presented in this book, together with the earlier volumes, provides a solid educational foundation for undergraduate biomedical science students, medical students, postgraduate researchers, clinicians, and academics. Furthermore, it will be of great interest to the general public, offering practical advice on promoting health and well-being through lifestyle choices. With both e-book and e-chapter availability, we aim to ensure broad accessibility to a wide range of readers.

Newcastle upon Tyne, UK — Viktor I. Korolchuk
Mainz, Germany — J. Robin Harris

Contents

About the Editors

Viktor I. Korolchuk is a Professor of Cell Biology at Newcastle University, UK. His scientific interests lie in the area of intracellular protein trafficking and degradation pathways and their relevance to human health and ageing. The current focus of research in his laboratory is the key pro-longevity pathway called autophagy (literally self-eating) which is essential for cellular quality control and homeostasis.

J. Robin Harris is an Honorary Professor of the University of Mainz, who specialized in macromolecular electron microscopy. He has been the Series Editor of the *Subcellular Biochemistry* Book Series for many years and his broad scientific interests are reflected in the diversity of content of the Series.

Chapter 1
Telomerase-Mediated Anti-Ageing Interventions

Phoebe L. Dunn, Dhenugen Logeswaran, and Julian J. -L. Chen

Abstract The ageing process involves a gradual decline of chromosome integrity throughout an organism's lifespan. Telomeres are protective DNA-protein complexes that cap the ends of linear chromosomes in eukaryotic organisms. Telomeric DNA consists of long stretches of short "TTAGGG" repeats that are conserved across most eukaryotes including humans. Telomeres shorten progressively with each round of DNA replication due to the inability of conventional DNA polymerase to completely replicate the chromosome ends, known as the "end-replication problem". The telomerase enzyme counteracts the telomeric DNA loss by *de novo* addition of telomeric repeats onto chromosomal ends. Germline and stem cells maintain significant levels of telomerase activity to maintain telomere length and can divide almost indefinitely. However, the differentiation of stem cells accompanies the inactivation of telomerase gene expression, resulting in the progressive shortening of telomeres in somatic cells over successive divisions. Critically short telomeres elicit and sustain a persistent DNA damage response leading to permanent growth arrest of cells known as cellular senescence, a hallmark of cellular ageing. The accumulation of senescent cells in tissues and organs contributes to organismal ageing. Thus, the prevention of telomere shortening is a promising means to delay or even reverse cellular ageing. In this chapter, we summarize potential anti-ageing interventions that mitigate telomere shortening through increasing telomerase level or activity and discuss these strategies' risks, benefits, and future outlooks.

Keywords DNA replication · Hayflick limit · End replication problem · Cell senescence · Telomere shortening · Telomerase activation · Ribonucleoprotein

P. L. Dunn
School of Life Sciences, Arizona State University, Tempe, Arizona, USA
e-mail: pldunn@asu.edu

D. Logeswaran · J. J. undefined.-L. Chen (✉)
School of Molecular Sciences, Arizona State University, Tempe, Arizona, USA
e-mail: dlogeswa@asu.edu; jlchen@asu.edu

© The Author(s), under exclusive license to Springer Nature Switzerland AG 2024

V. I. Korolchuk, J. R. Harris (eds.), *Biochemistry and Cell Biology of Ageing: Part V, Anti-Ageing Interventions*, Subcellular Biochemistry 107,
https://doi.org/10.1007/978-3-031-66768-8_1

The Hayflick Limit

We age because our cells and tissues cannot forever renew themselves. Pioneering experiments by Hayflick and Moorhead demonstrated that cells in culture divide a finite number of times before undergoing growth arrest, termed the Hayflick limit (Hayflick 1965; Hayflick and Moorhead 1961). This irreversible growth arrest accompanies cellular morphological and physiological changes known as cellular senescence (Huang et al. 2022). However, the molecular events underlying the Hayflick limit remained unknown until a decade later when Alexey Olovnikov and James Watson independently proposed the seminal hypothesis that conventional DNA polymerase is unable to completely replicate the ends of linear DNA, known as the "end-replication problem". Conventional DNA polymerase synthesizes DNA in the 5′-to-3′ direction as incoming nucleotides are added to the 3′-hydroxyl group of the growing DNA strand or RNA primer (Fig. 1.1). The DNA synthesis on the leading strand results in a blunt end that is later processed by 5′-to-3′ exonucleases Apollo and Exo1 to generate a 3′ single-stranded DNA (ssDNA) overhang (Makarov et al. 1997). Therefore, after each round of DNA replication, the processed leading

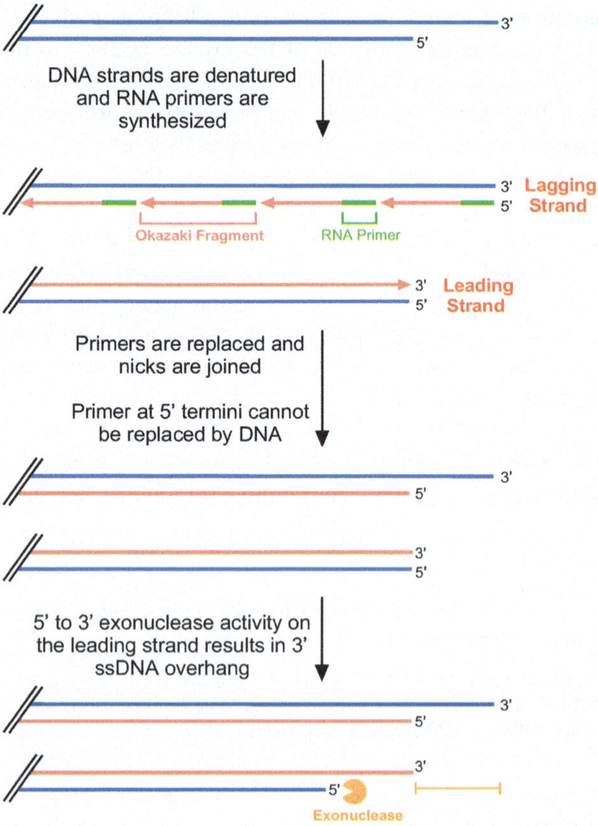

Fig. 1.1 The end-replication problem results in DNA shortening. The ends of eukaryotic chromosome harbour 3′-overhang tails. Conventional DNA polymerases synthesize DNA in the 5′-to-3′ direction and thus replicate the two parental DNA strands with different mechanisms, i.e. continuous leading vs discontinuous lagging strand synthesis. The leading strand synthesis results in a blunt end which is then processed by exonucleases Apollo and Exo1 to regenerate the 3′-overhang tail. The processed leading strand thus becomes shorter than the lagging strand

strand of the newly synthesized DNA becomes slightly shorter than the lagging strand or the parental DNA (Sfeir et al. 2005). Over numerous cell divisions, the chromosome ends gradually shorten to a critical length, imposing a limit on the replicative capacity of cells, i.e., the Hayflick limit.

Telomerase Confers Cellular Immortality

Telomeres are nucleoprotein complexes that protect the ends of linear chromosomes from end-to-end fusions or from being recognized as damaged DNA. In humans and most animals, telomeric DNA comprises short, tandem repeats e.g. TTAGGG in humans (Fig. 1.2). The length of human telomeric DNA ranges from 10 to 15 kb with a terminal 3′ ssDNA overhang (Makarov et al. 1997; Moyzis et al. 1988). The telomeric DNA is continuously decorated by a group of six proteins known as the Shelterin complex in mammals (de Lange 2018). Three proteins, TRF1, TRF2, and POT1, directly bind telomeric DNA and serve as a scaffold for binding RAP1, TIN2, and TPP1. The Shelterin complex not only protects the telomeric DNA from being recognized as damaged DNA but also recruits the telomerase enzyme to counteract the shortening of telomeric DNA resulting from the end-replication problem.

The telomerase enzyme was first discovered in the ciliate *Tetrahymena thermophila* by Carol Greider and Elizabeth Blackburn in the mid-1980s (Greider and Blackburn 1985). It synthesizes new TTAGGG DNA repeats at chromosome ends to replenish the telomeric DNA lost during DNA replications (Fig. 1.2). The extended G-rich strand of telomeric DNA then binds the CTC1-STN1-TEN1 (CST) protein complex to recruit the DNA polymerase α for fill-in DNA synthesis of the C-rich strand (Miyake et al. 2009; F. Wang et al. 2012). The three complexes, Shelterin, telomerase, and CST, function in a cooperative way to maintain the telomere length. The expression of telomerase is repressed in differentiated somatic cells but highly active in germline and stem cells, which prevents telomere shortening and confers infinite cell division, i.e., cellular immortality (Harley et al. 1990). Moreover, reactivation of telomerase expression is a critical step of tumorigenesis, enabling the formation of immortal cancer cells.

Telomerase Is a Multi-Component RNA-Protein Complex

Telomerase is a large ribonucleoprotein complex comprised of two core subunits essential for catalysis (Fig. 1.3). This catalytic core includes the telomerase reverse transcriptase (TERT) protein and the non-coding telomerase RNA (TR). The TERT protein is the catalytic component that harbours the active site for DNA polymerization, while the TR component provides a short RNA template for DNA repeat synthesis and contains structural elements facilitating processive repeat addition. In

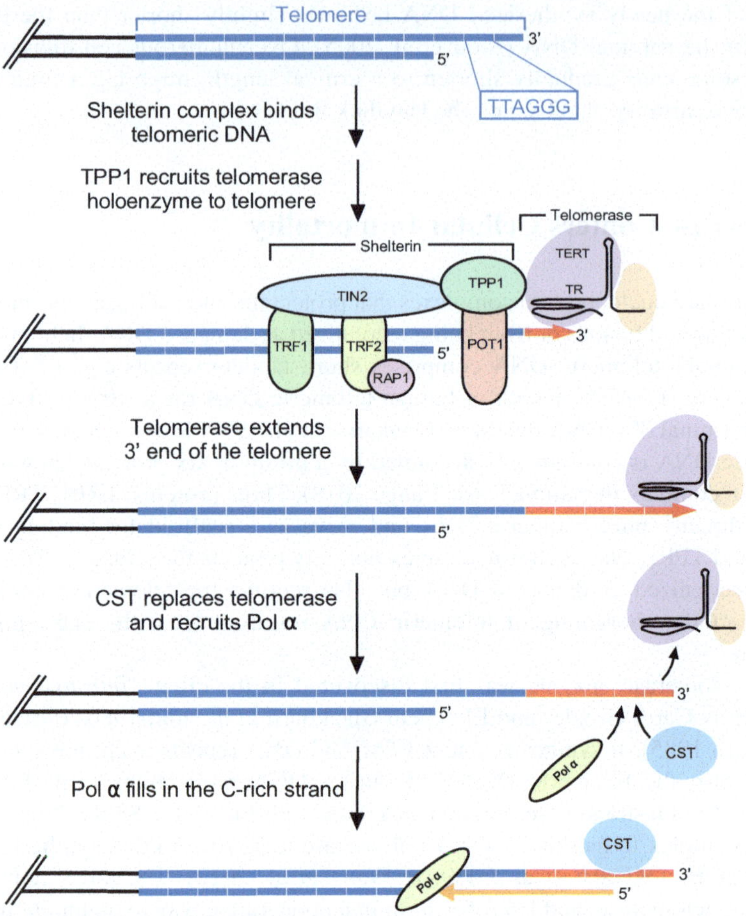

Fig. 1.2 Telomerase as a solution to the end-replication problem. The telomeric regions of linear chromosomes are composed of hexanucleotide repeats (shown as blue rectangles) of TTAGGG. The Shelterin complex binds along the telomeric DNA and is composed of 6 different proteins. Three proteins bind to telomeric DNA: TRF1 (light green) and TRF2 (yellow) bind to dsDNA, while POT1 (red) binds to the ssDNA overhang. RAP1 (pink) binds to TRF2. TIN2 (blue) interacts with both TRF1 and TRF2 and serves as a bridge to TPP1 (green), which binds POT1 and plays a role in recruiting other proteins to the telomere, including telomerase. Once telomerase (purple) is brought to the telomere through its interaction with TPP1, it is free to synthesize telomeric repeats and extend the telomere. Telomerase (purple) ameliorates the effect of the end replication problem by extending the 3′ overhang of the telomeric DNA through *de novo* addition of telomeric repeats (red). The CST complex (blue) interacts with telomere-binding proteins to recruit DNA Polymerase α (Pol α, yellow) for synthesizing the C-rich strand (orange), resulting in lengthening of the telomeric DNA

addition to the template, the TR component contains two structural domains, pseudoknot and CR4/5, essential for assembly with TERT protein and required for the enzymatic activity of telomerase (Fig. 1.3). The telomerase holoenzyme also

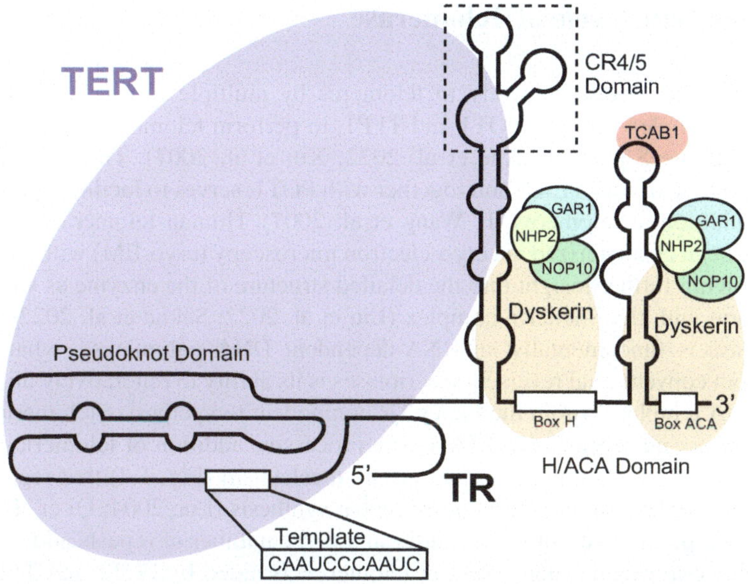

Fig. 1.3 Composition of Telomerase ribonucleoprotein enzyme. The two core components of telomerase are the protein component (TERT, shown in purple) and the RNA component (TR, shown in black). TERT binds to the 5′ region of TR containing the template sequence, forming the catalytic domain of telomerase. At the 3′ end of TR, two H/ACA ribonucleoprotein (RNP) complexes form and play a role in the biogenesis of the RNA. These complexes are each composed of four proteins: dyskerin (orange), NHP2 (yellow), GAR1 (blue) and NOP10 (green). A conserved motif CAB box in the distal loop of the H/ACA domain is bound by protein TCAB1 that regulates the localization of hTR to the Cajal body in the nucleoplasm

contains several accessory proteins in addition to the core components. These accessory components are clade-specific, with high diversity across eukaryotic kingdoms (Podlevsky and Chen 2016). In humans and other vertebrates, two sets of the dyskerin complex that includes four proteins, dyskerin, NHP2, NOP10, and GAR1, bind to the 3′ distal H/ACA domain of the TR, forming the box H/ACA snoRNP lobe of the human telomerase complex (J. L. Chen et al. 2000; Nguyen et al. 2018). A structural element called the CAB box in the H/ACA domain binds to an additional accessory protein called TCAB1 for localization of the TR to the Cajal body in the nucleoplasm for holoenzyme assembly. These telomerase accessory proteins are crucial for *in vivo* biogenesis of the telomerase holoenzyme and are potential targets for interventions to enhance telomerase levels *in vivo*.

The Catalytic Cycle of Telomerase

Telomerase is recruited actively to telomeres by multiple factors, including the Shelterin complex proteins POT1 and TPP1, to perform telomeric DNA synthesis during the late S-phase (Sekne et al. 2022; Xin et al. 2007). The TPP1 protein directly interacts with TERT, and together with POT1, serves to facilitate processive repeat addition at telomeres (F. Wang et al. 2007). Human telomerase has most recently been characterized via cryo-electron microscopy (cryo-EM) with TPP1 and POT1, giving further insight into the detailed structure of the enzyme as well as its interaction with the Shelterin complex (Liu et al. 2022; Sekne et al. 2022). While telomerase is fundamentally an RNA-dependent DNA polymerase, what sets it apart from conventional reverse transcriptases is its ability to reiteratively utilize the short RNA template within the larger TR component to synthesize telomeric DNA repeats of six nucleotides GGTTAG. The processive addition of telomeric repeats relies on a unique "template translocation" mechanism (Fig. 1.4) that regenerates the RNA template after each round of repeat synthesis (Lue 2004; Qi et al. 2012). The *in vivo* processivity of repeat addition, i.e. the number of repeats added during each telomerase recruitment event at telomeres, is aided by POT1 and TPP1 that prevent premature dissociation of telomerase from the chromosome ends during template translocation. Similarly, the *in vitro* primer extension assays of telomerase activity produce a characteristic ladder banding pattern with 6-nt increments owing to the product release during template translocation (Greider 1991). While the detailed mechanism governing template translocation requires further investigation, advances in telomerase structural biology have provided crucial pieces of the puzzle toward a complete understanding of this telomerase-specific property. For instance, the TR template-DNA substrate duplex is maintained at a constant 4 bp duplex with concomitant formation and unwinding of a new base pair with each nucleotide addition at opposing ends of the duplex (Ghanim et al. 2021; He et al. 2021). Furthermore, crucial amino acids from TERT are likely responsible for "sensing" the sequence-dependent pause signal that slows down the incorporation of the first dG nucleotide during each catalytic cycle (Fig. 1.4) (Y. Chen et al. 2018; Wan et al. 2021). The first dG incorporation as a rate-limiting step of the telomerase catalytic cycle explains previous findings that high dGTP concentration significantly increases telomerase repeat addition activity and processivity *in vitro* (Maine et al. 1999). The detailed structures of telomerase and the deep understanding of telomerase catalytic mechanisms provide rich resources for developing therapeutics targeting telomerase.

Telomere Biology Disorders

Highly proliferative or immortal cells rely on telomerase function to confer their capacity to divide. As such, telomerase activity is readily detectable in highly proliferative cells including germline and stem cells (Meyerson et al. 1997), which

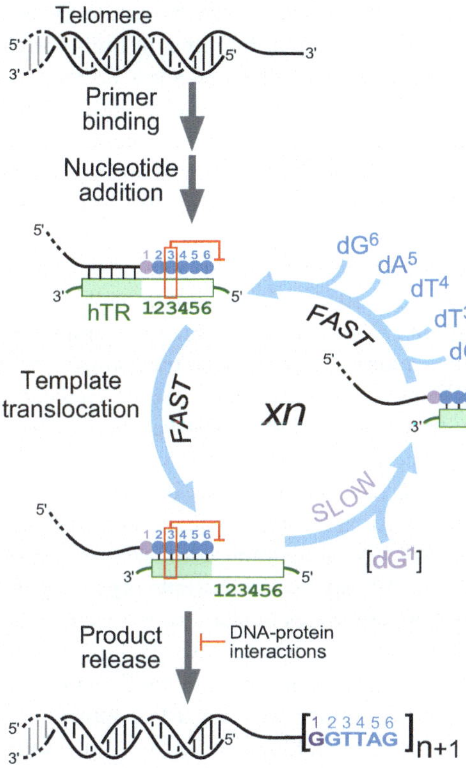

Fig. 1.4 Telomerase catalytic cycle. (Adapted from (Y. Chen et al. 2018)). Telomerase adds six nucleotides, dG^1, dG^2, dT^3, dT^4, dA^5, and dG^6, to the 3′-end of telomere, (violet and blue circles) and then pauses at the end of the template. A template translocation event regenerates the template by realigning the template relative to the 3′-end of the DNA primer. Progression to the next catalytic cycle is impeded by the pause signal from the DNA/RNA hybrid, causing a slow dG^1 residue incorporation (violet). Slow incorporation of the dG^1 residue promotes product release, which is counteracted by DNA-protein interactions through the TERT anchor sites and telomerase accessory POT1-TPP1 protein complex. Successful dG^1 residue incorporation is proceeded by the rapid incorporation of five other residues, dG^2, dT^3, dT^4, dA^5, and dG^6 (blue), completing one catalytic cycle. The number of repeats added corresponds to the number of catalytic cycles completed before the complete disassociation of the telomeric DNA from the telomerase enzyme. Increasing the dGTP concentration increases the rate of the slow dG^1 residue incorporation, which increases telomerase repeat addition processivity and rate

largely preserve telomere lengths despite undergoing an extremely large number of cell divisions. Genetic mutations affecting the functions of the core components and accessory proteins of telomerase can result in telomerase deficiency and accelerated telomere shortening. This telomerase deficiency and abnormally short telomeres are associated with premature ageing diseases such as dyskeratosis congenita (DC), idiopathic pulmonary fibrosis (IPF), and aplastic anaemia (AA), known as telomere biology disorders (Revy et al. 2023; Stanley and Armanios 2015). Numerous mutations in genes associated with telomere maintenance have been characterized in patients with DC, the first reported short telomere syndrome (Mason and Bessler 2011). DC patients exhibit a myriad of abnormalities in highly proliferative cells and tissues, including hematopoietic stem cells, and show premature ageing symptoms such as aberrant skin pigmentation, nail dystrophy, and leukoplakia. The two other short telomere syndromes, IPF and AA, share genetic and clinical overlap with DC but show distinct symptoms, including lung scarring and insufficient production of new blood cells by the bone marrow, respectively (Alder and Armanios 2022; Calado and Young 2008).

Human somatic cells lack telomerase expression and thus experience significant telomere attrition following each cell division. Somatic cells are mortal and can undergo only a finite number of cell divisions. Telomerase repression and telomere attrition in somatic cells act as a tumour suppression mechanism that limits cellular proliferative capacity. Incidentally, in more than 85% of cancers, expression of TERT is reactivated at the transcription level to drive oncogenesis and maintain telomere lengths for unlimited cell growth (Shay 2016).

Although mouse models that have homozygous deletions for telomerase core components reproduce normally for at least 4–5 generations with no visible ageing phenotypes (Blasco et al. 1997), late-generation telomerase-negative mice exhibit fitness defects, including shorter life spans and multiple organ degeneration. While the presence of dramatically long telomeres in laboratory mice may explain the lack of degenerative phenotypes in early-generation mice lacking telomerase activity, shorter life-spans, even in first-generation telomerase knockout mice, suggests non-telomere related function/s of telomerase with regard to longevity (Geserick and Blasco 2006). Interestingly, restoring telomerase activity in telomerase-negative mice improves many fitness-associated indicators, establishing telomerase as a bona fide anti-ageing target (Jaskelioff et al. 2011).

While laboratory mice have provided crucial insight into the relationship between telomerase and ageing in a mammalian system, fish models have been developed to better understand the role of telomerase in tissue regeneration and wound healing. For example, zebrafish, a popular vertebrate model, exhibits remarkable regenerative abilities, including repair of various organs following injury. This capability is telomerase dependent as fish lacking active TERT protein fail to regenerate damaged heart tissue (Bednarek et al. 2015). This requirement for active telomerase in tissue rejuvenation seems to be consistent in mammalian systems as well (Aix et al. 2016; Lin et al. 2018).

The Delicate Balance of Telomere Length

There is a clear correlation between telomere length and the age of the organism (Fig. 1.4). Telomere length is viewed as a "mitotic clock" that counts down as cells divide and expires as cells reach the Hayflick limit and are unable to divide further (Harley et al. 1992). Mortal somatic cells from newborns have longer telomeres on average and undergo more cell division or live longer than those from adults. The aged somatic cells from older adults suffer from the accumulation of critically short telomeres, which activate DNA damage response pathways, leading to cellular senescence (d'Adda di Fagagna et al. 2003). The accumulation of senescent cells in organs is thought to drive organismal ageing. As the primary mechanism responsible for maintaining telomeres, telomerase is vital to the organismal ageing process.

At the organismal level, it is less clear if telomere length has a direct causative effect on organismal ageing or can be used to predict lifespan in different species (Daniali et al. 2013; Simons 2015). For example, humans have longer life spans than mice but have telomere lengths ranging from 5 to 15 kb, which is significantly shorter than telomere length of >50 kb in mice. Studies indicate the rates at which telomeres shorten, rather than their initial lengths, may play a more crucial role in the lifespan of organisms (Vera et al. 2012). The telomere shortening rates in humans and mice seem notably distinct at approximately 70 bp per year and 7000 bp per year, respectively. This was suggested to be true for a number of vertebrates (Whittemore et al. 2019b). Thus, strategies that modulate telomere shortening rates could be explored as potential anti-ageing interventions.

It is assumed that with a constant rate of telomere attrition, individuals with initially long telomeres may have a longer life span than individuals with initially short telomeres. However, abnormally long telomeres were shown to correlate with an increased risk of malignancies (DeBoy et al. 2023; Schmutz et al. 2020). In fact, as a tumour suppression mechanism, telomere shortening destines old cells to senescence, which prevents indefinite division and tumorigenesis (Maciejowski and de Lange 2017). The number of critically short telomeres rather than the average telomere length is critical for determining cell division capacity in this context (Hemann et al. 2001). In the absence of telomerase, somatic cells with long telomeres will be prone to be tumorigenic until the depletion of telomeres (Taboski et al. 2012).

Telomerase-Mediated Anti-Ageing Interventions

Telomere attrition is a pivotal hallmark of ageing. Interventions to mitigate or prevent telomere degradation have been explored as an anti-ageing strategy. In this context, we discuss potential interventions that aim to increase the expression levels or the intrinsic activity of telomerase to avert telomere degradation and alleviate ageing symptoms.

Increasing hTERT Levels

Although most telomerase holoenzyme components exhibit constitutive expression, hTERT is downregulated by transcriptional repression during somatic differentiation. A combination of epigenetic modifications and chromatin remodelling in the promoter region is responsible for strong hTERT repression (Lewis and Tollefsbol

2016). It has been shown that hTERT expression is sufficient to reinstate telomerase activity and rescue telomere shortening in telomerase-negative somatic cells (Bodnar et al. 1998). Additionally, in human Hutchinson–Gilford progeria syndrome cells, the transient introduction of hTERT was sufficient to increase telomerase activity and lengthen telomeres for up to 3 days and increase cell proliferation for up to 3 months after transfections without conferring immortalization of the cells (Li et al. 2019). Thus, reactivation and increasing the expression of hTERT is an attractive anti-ageing modality, although the risk of cancer as a result of telomerase activation needs to be investigated further.

Multiple hTERT reactivation strategies have been explored in cell and animal models, including gene therapy with viral vectors, cell therapy, hormone therapy, and activation with small molecules. Strategies to increase hTERT expression can generally be classified into two categories: (1) strategies that introduce a new copy of the gene under the control of a stronger promoter and (2) strategies that seek to reactivate or increase the expression of endogenous TERT gene. For the former strategy, transgenic manipulations using both viral and transposable elements as delivery mechanisms have been successful. Adeno-associated viral (AAV) vectors are a mature group of gene-delivery vehicles that have been investigated heavily for decades to treat hereditary diseases through replacement, silencing, or addition of genes and, more recently, gene editing (D. Wang et al. 2019). AAVs have been used to deliver TERT cDNA under the control of the cytomegalovirus (CMV) promoter to mice, resulting in telomere elongation and the reduction of age-related phenotypes in various tissues without increased cancer incidence (Bernardes de Jesus et al. 2012; Munoz-Lorente et al. 2018; Povedano et al. 2018). Furthermore, AAV9-TERT treatment of mice that model aplastic anemia, myocardial infarction, and pulmonary fibrosis demonstrated therapeutic effects, including improved hematopoietic stem cell and bone marrow function (Bar et al. 2014, 2016; Povedano et al. 2018). Similarly, the improvement of neurodegenerative phenotypes has been observed in mouse models of neurodegeneration following telomerase activation via AAV9-TERT introduction (Whittemore et al. 2019a).

Due to their efficient transduction into most tissues compared to AAVs, adenoviral vectors (AVVs) have also been used to deliver TERT into aged rabbit models with improvements observed in ischemic wound healing (Mogford et al. 2006). Transgenic TERT delivered via non-viral methods have also shown longevity benefits in vertebrates. For example, delivery of a TERT transgene cassette that inserts via transposition in zebrafish homozygous TERT knockout^{-} lines showed that tissue-specific expression of telomerase in the gut mitigated telomere shortening and significantly extended the lifespan of transgenic fish (El Mai et al. 2023). Cell therapy targeting hTERT has also been applied in pre-clinical settings by transplanting cells transduced with the hTERT gene into animal models. The introduction of hTERT-transduced endothelial progenitor cells (EPCs) in rats resulted in tissues with reduced apoptosis, increased renal function, and the reduction of fibrosis (Shuai et al. 2012). Following transplantation of mouse mesenchymal cells genetically engineered to overexpress telomerase in hindlimb ischemic mouse models, improved tissue repair and blood vessel rejuvenation were observed (Madonna et al. 2013).

In contrast to introducing a new copy of the hTERT gene or hTERT-activated cells into the organism, many hTERT modulating strategies attempt to reactivate the existing copy of hTERT by increasing the gene's transcription. Minimizing the introduction of extraneous factors, including engineered hTERT transgenes, may improve safety. These strategies include treatment with hormones that affect the activity or expression of hTERT. Hormones, specifically estrogen, have been shown to directly regulate and promote the transcription of hTERT by interacting with estrogen response elements (EREs) in the hTERT promoter (Kyo et al. 1999). Androgens have also been shown to increase telomerase activity and TERT mRNA, presumably by the same activation mechanism (Misiti et al. 2000). Danazol, an androgenic hormone, has already been used in clinical studies to treat patients with short telomere syndromes (Townsley et al. 2016). Hypoxia-inducible factor 1 (HIF1) has also been linked to telomerase activation at the transcriptional level by binding hypoxia response elements (HREs) found in the hTERT promoter (Nishi et al. 2004). Hyperbaric oxygen therapy has been implicated in increasing telomere length via the activation of HIF1, and therefore may be an interesting route for increasing hTERT transcription (Sunkari et al. 2015; Hachmo et al. 2020). However, hyperbaric oxygen therapy as a means to increase telomerase activity still needs to be thoroughly investigated.

Small molecule activations, both natural and synthetic, have also been explored to increase the transcription of hTERT, although their mechanisms need to be thoroughly investigated. TA-65 is a commercially available compound marketed as a telomerase activator. TA-65 supplementation appears to play its role in telomerase activation through the mitogen-activated protein kinase (MAPK) pathway, as treatment of mice with TA-65 has been shown to result in increased levels of mRNA from MAPK-related transcription factors (Bernardes de Jesus et al. 2012). The MAP kinase is suggested to phosphorylate certain E26 transformation–specific (ETS) transcription factors in response to growth factor signalling. The activated ETS can then migrate to the nucleus and activate the hTERT promoter, thereby transcriptionally activating hTERT (Dwyer et al. 2007).

Another small molecule, called TERT activator compound (TAC), has also been shown to increase TERT transcription in mice models and human cell lines (Shim et al. 2024). TAC can reverse the epigenetic silencing of the TERT promoter region, thereby increasing transcription of the TERT gene. TAC appears to do this through the MEK/ERK/AP-1 pathway. Promisingly, mice were able to take TAC for extended periods without any adverse effects or importantly, without carcinogenesis. Naturally aged mice also demonstrated improved neuroinflammation, neurogenesis, and cognitive function. However, the potential toxicities and long-term effects, including cancer incidence, of TAC still need to be explored further in humans.

In addition to promoter regulation, hTERT mRNA is post-transcriptionally regulated by alternative splicing. The hTERT gene contains 16 exons, which can be alternatively spliced to yield several different isoforms (Cong et al. 1999; Wick et al. 1999). One alternative splicing event in somatic cells results in the hTERT mRNA missing the second exon and being targeted for non-sense mediated decay, therefore silencing hTERT (Penev et al. 2021). This study also identified that, in pluripotent stem cells and cancer cells with active telomerase, the second exon of hTERT is retained with the help of SON, a nuclear speckle protein. A complete understanding of hTERT regulation by alternative splicing opens a new area to seek potential therapeutic targets for anti-ageing and telomere biology disease treatments, including approaches to facilitate retention of the exon 2 to produce functional hTERT mRNA.

In addition to its reverse transcriptase activity, TERT has been shown to have non-canonical functions that may improve ageing phenotypes. Catalytically inactive TERT can still promote proliferation through transcriptional regulation of the Wnt and Myc pathways (Choi et al. 2008). Therefore, an unintended but possibly beneficial side effect of increasing hTERT levels could be the activation of these pathways. For example, it has been shown that in the absence of TR or using a catalytically inactive TERT, spermatogonial stem cells with high levels of TERT can outcompete those with lower levels (Hasegawa et al. 2024). The involvement of TERT in stimulating stem cell competition seems to be via the activation of the Myc pathway.

Increasing hTR Accumulation

Unlike hTERT, hTR is present at steady-state levels in somatic cells (Nakamura et al. 1997). It appears to be crucial to maintain a constant level of hTR, as patients who are haploinsufficient for hTR have short telomeres, low telomerase activity, and develop autosomal dominant dyskeratosis congentia, a bone marrow failure syndrome (Goldman et al. 2005). The low, but still constant, level of hTR in somatic cells relative to immortal cells may make it a more attractive candidate for telomerase-related anti-ageing strategies (Avilion et al. 1996). Increasing hTR levels, either by promoting transcription or stabilizing the RNA, has the potential to increase the activity of telomerase specifically in stem cells, where both hTERT and hTR are present and telomerase is active, as opposed to reactivating telomerase in somatic cells. Transcription and maturation of hTR are thought to be the rate-limiting steps of telomerase holoenzyme assembly and are consequently tightly regulated (Greider 2006). Therefore, carefully increasing hTR accumulation may increase the assembly of functional telomerase RNPs and subsequent extension of telomeres.

Targeting hTR expression can be achieved at either the transcriptional level or during the RNA's maturation. Transcription of hTR is mainly driven by an RNA Polymerase II promoter and then processed via a biogenesis pathway distinct from mRNAs and snRNAs (Feng et al. 1995; Fu and Collins 2003). Promoter analysis of hTR revealed consensus sites for binding transcription factors including those involved in haematopoiesis. (Zhao et al. 1998). Additionally, at least one methylation site within hTR has been identified along with a 3′ proximal m6A site proposed which suggests that detailed studies are needed to unravel epigenetic regulation of hTR (Han et al. 2020; H. Tang et al. 2018). Regulation of both hTR and hTERT biosynthesis is proposed to be integrated likely to maintain productive RNP assembly. Transcription of both the hTR and hTERT genes appears to be regulated by cell signalling pathways, including the mitogen-activated protein kinase (MAPK) pathway, and regulatory proteins like transcriptional co-repressor C-terminal binding protein (CtBP) and hypoxia-inducible factor 1 (HIF-1) (Cairney and Keith 2008). Understanding how the transcription of hTR is regulated helps reveal potential therapeutic targets for telomere biology disease and anti-ageing treatments aimed at maintaining the steady state levels of hTR in cells (Batista et al. 2022).

The maturation pathway is the other crucial determinant of cellular hTR levels. The hTR biogenesis pathway involves numerous proteins important for the proper processing, stabilization, and localization of the RNA into its mature form. Several structural motifs within hTR can be bound by proteins or protein complexes, including the H/ACA box, Cajal body box (CAB box), and biogenesis-promoting box (BIO box) (Qin and Autexier 2021). The proteins that bind these motifs have the potential to be modulated in an effort to alter the accumulation of hTR. For example, mutations in the H/ACA domain of hTR can prevent hTR accumulation and can lead to a form of dyskeratosis congenita where the assembly of the pre-RNP is heavily impacted (Mitchell et al. 1999; Trahan and Dragon 2009). The 5′ end of hTR also contains guanosine (G) tracts that form G-quadruplexes which can be recognized by the helicase DHX36 (Sexton and Collins 2011). Disruptions in the formation of the G-quadruplexes can result in decreased levels of hTR, implicating DHX36 to play a role in hTR accumulation and could be a target for anti-ageing therapies. In addition to promoting the maturation of hTR, preventing hTR degradation could also be a viable strategy to increase hTR accumulation. The degradation pathway of hTR has been previously targeted in potential treatments for short telomere diseases. PAPD5 is a non-canonical poly (A) polymerase that destabilizes TR by essentially tagging it for degradation (Nagpal and Agarwal 2020). Mutations in poly(A)ribonuclease (PARN), a ribonuclease involved in 3′ hTR processing into its mature form following the addition of a poly(A) tail by PAPD5, can result in accumulation of hTR precursors that are degraded (Roake et al. 2019). A small molecule inhibitor of PAPD5, BCH001, was identified through high-throughput screening and has been introduced to cells with a mutated PARN, resulting in increased accumulation of mature hTR and resumed telomerase activity (Nagpal et al. 2020).

While the box H/ACA module protects the 3′ end of hTR, a 2, 2, 7-trimethylguanosine (TMG) cap is found on the 5′ end and is important for the retention of hTR in the nucleus (L. Chen et al. 2020). This is done by trimethylguanosine synthase 1 (TGS1), which post-transcriptionally hypermethylates the 5′ cap into a trimethylguanosine (TMG) cap on hTR, as well as other non-coding RNAs, including snRNAs and snoRNAs (L. Chen et al. 2020; Mouaikel et al. 2002). It has been shown that TGS1 has a negative effect on the accumulation of hTR, where TGS1 down-regulation results in an increased amount of mature hTR (L. Chen et al. 2020). This implies that TGS1 downregulation could be a potential target in the treatment of short telomere diseases and in anti-ageing therapies, although changing its expression could affect other RNAs processed by TGS1. Proteins that play a role in the maturation and stability of hTR and assembly of the RNP could also be targeted to increase the accumulation of both in the cell, including dyskerin, NHP2, NOP10, NAF1, and GAR1 (Egan and Collins 2012). However, targeting these proteins needs to be carefully considered given their roles in the maturation of other cellular RNAs. Regardless, targeting hTR transcription and biogenesis in anti-ageing therapies and in short telomere disease treatments still needs further investigation.

Cell reprogramming is the process of transforming somatic cells into pluripotent cells (known as induced pluripotent cells) by changing the expression of certain transcription factors. This strategy has been presented to alleviate symptoms caused by the telomere biology disease, aplastic anaemia. hTR-haploinsufficient cells have been reprogrammed into cells with higher differential potential that maintain telomerase activity (X. Tang et al. 2023). It has been shown that reprogramming human somatic cells into pluripotent cells can result in the lengthening of the telomeres (Suhr et al. 2009). Cell reprogramming has not been extensively researched in aplastic anaemia, but the strategy provides a different approach worthy of pursuit towards the treatment of short telomere diseases and anti-ageing interventions.

Increasing Telomerase Activity and Processivity

Most telomere-mediated anti-ageing therapies have targeted increasing the levels of the telomerase core components, hTERT and hTR. However, another approach is to increase the activity of existing and assembled telomerase holoenzymes. One molecule that appears to be activator of telomerase activity is thymidine (dT) nucleotides (Mannherz and Agarwal 2023). While the synthesis of "TTAGGG" repeats requires dT as a substrate, the authors engineered the telomerase enzyme to synthesize a dT-free repeat sequence. The engineered enzyme retained the telomerase activation properties of dT, showing this activation mechanism is substrate-independent and suggests an allosteric dT binding site in telomerase. Importantly, dT supplementation demonstrated telomere restoration in patient-derived induced pluripotent stem cell models, suggesting a favourable therapeutic application for dT. Although further evaluation is needed for its dosage and safety *in vivo*, given dT is a natural monomer of DNA, it is likely to be less toxic at reasonable concentrations than most compounds evaluated for telomerase activation. Other small molecule activators may also have an allosteric effect of telomerase, but their mechanisms have not been explored thoroughly.

Telomere and Telomerase Associated Proteins

While the telomerase core is composed of hTR and hTERT, two sets of heterotetrameric proteins bind to hTR to form the telomerase holoenzyme. Accessory proteins within the holoenzyme can play a role in the assembly, localization, stabilization, and recruitment of the holoenzyme to the telomeres and could be targeted as a means of increasing the assembly of the telomerase RNP.

Telomere-binding proteins also could be used as targets for anti-ageing therapeutics. Telomeres are bound by Shelterin complexes involved in protecting the telomeric DNA and regulating telomerase recruitment. Mutations in one member of the Shelterin complex, POT1, have been associated with telomere biology diseases, indicating its presence is needed for normal telomere function (Revy et al. 2023). Another protein in the Shelterin complex, TRF1, has been previously identified as a potential target for anti-ageing therapies. Overexpression of TRF1 using adeno-associated vectors (AAVs) in mice resulted in fewer short telomeres and improved ageing biomarkers (Derevyanko et al. 2017). The protein klotho may be an anti-ageing treatment target involving both POT1 and TRF1. Mutated or reduced levels of klotho have been correlated with premature ageing phenotypes (Kuro-o et al. 1997). More specifically, klotho deficiency is correlated with decreased telomerase activity as well as decreased expression of POT1 and increased expression of TRF1 (Ullah and Sun 2019). Overexpression of klotho in ageing transgenic mice has led to decreased ageing phenotypes and improved kidney functions, which demonstrates the potential of klotho in anti-ageing treatments (Oishi et al. 2021).

Concluding Remarks

Telomeres play a vital role in maintaining genomic integrity. The length and structure of telomeres are crucial for protecting chromosome ends and preventing cell senescence that contributes to organismal ageing. While interventions that reduce telomere shortening rate or increase telomere length are expected to prevent or reverse cellular ageing, overextension of telomeres is also not desired as it may contribute to familial haematopoiesis diseases (DeBoy et al. 2023). Thus, any effective anti-ageing interventions targeting telomere or telomerase must prioritise strict regulation to maintain telomere length within the optimal range.

References

Aix E, Gutierrez-Gutierrez O, Sanchez-Ferrer C, Aguado T, Flores I (2016) Postnatal telomere dysfunction induces cardiomyocyte cell-cycle arrest through p21 activation. J Cell Biol 213(5):571–583. https://doi.org/10.1083/jcb.201510091

Alder JK, Armanios M (2022) Telomere-mediated lung disease. Physiol Rev 102(4):1703–1720. https://doi.org/10.1152/physrev.00046.2021

Avilion AA, Piatyszek MA, Gupta J, Shay JW, Bacchetti S, Greider CW (1996) Human telomerase RNA and telomerase activity in immortal cell lines and tumor tissues. Cancer Res 56(3):645–650

Bar C, Bernardes de Jesus B, Serrano R, Tejera A, Ayuso E, Jimenez V et al (2014) Telomerase expression confers cardioprotection in the adult mouse heart after acute myocardial infarction. Nat Commun 5. https://doi.org/10.1038/ncomms6863

Bar C, Povedano JM, Serrano R, Benitez-Buelga C, Popkes M, Formentini I et al (2016) Telomerase gene therapy rescues telomere length, bone marrow aplasia, and survival in mice with aplastic anemia. Blood 127(14):1770–1779. https://doi.org/10.1182/blood-2015-08-667485

Batista LFZ, Dokal I, Parker R (2022) Telomere biology disorders: time for moving towards the clinic? Trends Mol Med 28(10):882–891. https://doi.org/10.1016/j.molmed.2022.08.001

Bednarek D, Gonzalez-Rosa JM, Guzman-Martinez G, Gutierrez-Gutierrez O, Aguado T, Sanchez-Ferrer C et al (2015) Telomerase is essential for Zebrafish Heart Regeneration. Cell Rep 12(10):1691–1703. https://doi.org/10.1016/j.celrep.2015.07.064

Bernardes de Jesus B, Vera E, Schneeberger K, Tejera AM, Ayuso E, Bosch F, Blasco MA (2012) Telomerase gene therapy in adult and old mice delays aging and increases longevity without increasing cancer. EMBO Mol Med 4(8):691–704. https://doi.org/10.1002/emmm.201200245

Blasco MA, Lee HW, Hande MP, Samper E, Lansdorp PM, DePinho RA, Greider CW (1997) Telomere shortening and tumor formation by mouse cells lacking telomerase RNA. Cell 91(1):25–34. https://doi.org/10.1016/s0092-8674(01)80006-4

Bodnar AG, Ouellette M, Frolkis M, Holt SE, Chiu CP, Morin GB et al (1998) Extension of life-span by introduction of telomerase into normal human cells. Science 279(5349):349–352. https://doi.org/10.1126/science.279.5349.349

Cairney CJ, Keith WN (2008) Telomerase redefined: integrated regulation of hTR and hTERT for telomere maintenance and telomerase activity. Biochimie 90:13–23

Calado RT, Young NS (2008) Telomere maintenance and human bone marrow failure. Blood 111(9):4446–4455. https://doi.org/10.1182/blood-2007-08-019729

Chen JL, Blasco MA, Greider CW (2000) Secondary structure of vertebrate telomerase RNA. Cell 100(5):503–514. https://doi.org/10.1016/s0092-8674(00)80687-x

Chen Y, Podlevsky JD, Logeswaran D, Chen JJ (2018) A single nucleotide incorporation step limits human telomerase repeat addition activity. EMBO J 37(6). https://doi.org/10.15252/embj.201797953

Chen L, Roake CM, Galati A, Bavasso F, Micheli E, Saggio I et al (2020) Loss of human TGS1 Hypermethylase promotes increased telomerase RNA and telomere elongation. Cell Rep 30(5):1358–1372.e1355. https://doi.org/10.1016/j.celrep.2020.01.004

Choi J, Southworth LK, Sarin KY, Venteicher AS, Ma W, Chang W, Cheung P, Jun S, Artandi MK, Shah N, Kim SK, Artandi SE (2008) TERT promotes epithelial proliferation through transcriptional control of a Myc- and Wnt-related developmental program. PLoS Genetics 4(1):e10. https://doi.org/10.1371/journal.pgen.0040010

Cong YS, Wen J, Bacchetti S (1999) The human telomerase catalytic subunit hTERT: organization of the gene and characterization of the promoter. Hum Mol Genet 8(1):137–142. https://doi.org/10.1093/hmg/8.1.137

d'Adda di Fagagna F, Reaper PM, Clay-Farrace L, Fiegler H, Carr P, Von Zglinicki T et al (2003) A DNA damage checkpoint response in telomere-initiated senescence. Nature 426(6963):194–198. https://doi.org/10.1038/nature02118

Daniali L, Benetos A, Susser E, Kark JD, Labat C, Kimura M et al (2013) Telomeres shorten at equivalent rates in somatic tissues of adults. Nat Commun 4:1597. https://doi.org/10.1038/ncomms2602

de Lange T (2018) Shelterin-mediated telomere protection. Annu Rev Genet 52:223–247. https://doi.org/10.1146/annurev-genet-032918-021921

DeBoy EA, Tassia MG, Schratz KE, Yan SM, Cosner ZL, McNally EJ et al (2023) Familial clonal hematopoiesis in a long telomere syndrome. N Engl J Med 388(26):2422–2433. https://doi.org/10.1056/NEJMoa2300503

Derevyanko A, Whittemore K, Schneider RP, Jimenez V, Bosch F, Blasco MA (2017) Gene therapy with the TRF1 telomere gene rescues decreased TRF1 levels with aging and prolongs mouse health span. Aging Cell 16:1353–1368

Dwyer J, Li H, Xu D, Liu JP (2007) Transcriptional regulation of telomerase activity: roles of the the Ets transcription factor family. Ann N Y Acad Sci 1114(1):36–47

Egan ED, Collins K (2012) Biogenesis of telomerase ribonucleoproteins. RNA 18

El Mai M, Bird M, Allouche A, Targen S, Serifoglu N, Lopes-Bastos B et al (2023) Gut-specific telomerase expression counteracts systemic aging in telomerase-deficient zebrafish. Nat Aging 3(5):567–584. https://doi.org/10.1038/s43587-023-00401-5

Feng J, Funk WD, Wang S, Weinrich SL, Avilion RC, Chiu C et al (1995) The RNA of component human telomerase. Science 269:1236–1241. https://doi.org/10.1126/science.7544491

Fu D, Collins K (2003) Distinct biogenesis pathways for human telomerase RNA and H/ACA small nucleolar RNAs. Mol Cell 11:1361–1372

Geserick C, Blasco MA (2006) Novel roles for telomerase in aging. Mech Ageing Dev 127(6):579–583. https://doi.org/10.1016/j.mad.2006.01.017

Ghanim GE, Fountain AJ, van Roon AM, Rangan R, Das R, Collins K, Nguyen THD (2021) Structure of human telomerase holoenzyme with bound telomeric DNA. Nature 593(7859):449–453. https://doi.org/10.1038/s41586-021-03415-4

Goldman F, Bouarich R, Kulkarni S, Freeman S, Du H-Y, Harrington L et al (2005) The effect of TERC haploinsufficiency on the inheritance of telomere length. Proc Natl Acad Sci 102(47):17119–17124

Greider CW (1991) Telomerase is processive. Mol Cell Biol 11(9):4572–4580. https://doi.org/10.1128/mcb.11.9.4572-4580.1991

Greider CW (2006) Telomerase RNA levels limit the telomere length equilibrium. Cold Spring Harb Symp Quant Biol 71:225–229. https://doi.org/10.1101/sqb.2006.71.063

Greider CW, Blackburn EH (1985) Identification of a specific telomere terminal transferase activity in Tetrahymena extracts. Cell 43(2 Pt 1):405–413. https://doi.org/10.1016/0092-8674(85)90170-9

Hachmo Y, Hadanny A, Abu Hamed R, Daniel-Kotovsky M, Catalogna M, Fishlev G et al (2020) Hyperbaric oxygen therapy increases telomere length and decreases immunosenescence in isolated blood cells: a prospective trial. Aging (Albany NY) 12(22):22445–22456. https://doi.org/10.18632/aging.202188

Han S, Zhao BS, Myers SA, Carr SA, He C, Ting AY (2020) RNA-protein interaction mapping via MS2- or Cas13-based APEX targeting. Proc Natl Acad Sci U S A 117(36):22068–22079. https://doi.org/10.1073/pnas.2006617117

Harley CB, Futcher AB, Greider CW (1990) Telomeres shorten during ageing of human fibroblasts. Nature 345(6274):458–460. https://doi.org/10.1038/345458a0

Harley CB, Vaziri H, Counter CM, Allsopp RC (1992) The telomere hypothesis of cellular aging. Exp Gerontol 27(4):375–382. https://doi.org/10.1016/0531-5565(92)90068-b

Hasegawa K, Zhao Y, Garbuzov A, Corces MR, Neuhöfer P, Gillespie VM, Cheung P, Belk JA, Huang YH, Wei Y, Chen L, Chang HY, Artandi SE (2024). Clonal inactivation of Tert impairs stem cell competition. Nature 632:201–208

Hayflick L (1965) The limited in vitro lifetime of human diploid cell strains. Exp Cell Res 37:614–636. https://doi.org/10.1016/0014-4827(65)90211-9

Hayflick L, Moorhead PS (1961) The serial cultivation of human diploid cell strains. Exp Cell Res 25:585–621. https://doi.org/10.1016/0014-4827(61)90192-6

He Y, Wang Y, Liu B, Helmling C, Susac L, Cheng R et al (2021) Structures of telomerase at several steps of telomere repeat synthesis. Nature 593(7859):454–459. https://doi.org/10.1038/s41586-021-03529-9

Hemann MT, Strong MA, Hao LY, Greider CW (2001) The shortest telomere, not average telomere length, is critical for cell viability and chromosome stability. Cell 107(1):67–77. https://doi.org/10.1016/s0092-8674(01)00504-9

Huang W, Hickson LJ, Eirin A, Kirkland JL, Lerman LO (2022) Cellular senescence: the good, the bad and the unknown. Nat Rev Nephrol 18(10):611–627. https://doi.org/10.1038/s41581-022-00601-z

Jaskelioff M, Muller FL, Paik JH, Thomas E, Jiang S, Adams AC et al (2011) Telomerase reactivation reverses tissue degeneration in aged telomerase-deficient mice. Nature 469(7328):102–106. https://doi.org/10.1038/nature09603

Kuro-o M, Matsumura Y, Aizawa H, Kawaguchi H, Suga T, Utsugi T et al (1997) Mutation of the mouse klotho gene leads to a syndrome resembling ageing. Nature 390:45–51

Kyo S, Takakura M, Kanaya T, Zhuo W, Fujimoto K, Nishio Y et al (1999) Estrogen activates telomerase. Cancer Res 59:5917–5921

Lewis KA, Tollefsbol TO (2016) Regulation of the telomerase reverse transcriptase subunit through epigenetic mechanisms. Front Genet 7:83. https://doi.org/10.3389/fgene.2016.00083

Li Y, Zhou G, Bruno IG, Zhang N, Sho S, Tedone E et al (2019) Transient introduction of human telomerase mRNA improves hallmarks of progeria cells. Aging Cell 18(4):e12979. https://doi.org/10.1111/acel.12979

Lin S, Nascimento EM, Gajera CR, Chen L, Neuhofer P, Garbuzov A et al (2018) Distributed hepatocytes expressing telomerase repopulate the liver in homeostasis and injury. Nature 556(7700):244–248. https://doi.org/10.1038/s41586-018-0004-7

Liu B, He Y, Wang Y, Song H, Zhou ZH, Feigon J (2022) Structure of active human telomerase with telomere shelterin protein TPP1. Nature 604(7906):578–583. https://doi.org/10.1038/s41586-022-04582-8

Lue NF (2004) Adding to the ends: what makes telomerase processive and how important is it? Bioessays 26(9):955–962. https://doi.org/10.1002/bies.20093

Maciejowski J, de Lange T (2017) Telomeres in cancer: tumour suppression and genome instability. Nat Rev Mol Cell Biol 18(3):175–186. https://doi.org/10.1038/nrm.2016.171

Madonna R, Taylor DA, Geng YJ, De Caterina R, Shelat H, Perin EC, Willerson JT (2013) Transplantation of mesenchymal cells rejuvenated by the overexpression of telomerase and myocardin promotes revascularization and tissue repair in a murine model of hindlimb ischemia. Circ Res 113(7):902–914. https://doi.org/10.1161/CIRCRESAHA.113.301690

Maine IP, Chen SF, Windle B (1999) Effect of dGTP concentration on human and CHO telomerase. Biochemistry 38(46):15325–15332. https://doi.org/10.1021/bi991596+

Makarov VL, Hirose Y, Langmore JP (1997) Long G tails at both ends of human chromosomes suggest a C strand degradation mechanism for telomere shortening. Cell 88(5):657–666. https://doi.org/10.1016/s0092-8674(00)81908-x

Mannherz W, Agarwal S (2023) Thymidine nucleotide metabolism controls human telomere length. Nat Genet 55:568–580. https://doi.org/10.1038/s41588-023-01339-5

Mason PJ, Bessler M (2011) The genetics of dyskeratosis congenita. Cancer Genet 204(12):635–645. https://doi.org/10.1016/j.cancergen.2011.11.002

Meyerson M, Counter CM, Eaton EN, Ellisen LW, Steiner P, Caddle SD et al (1997) hEST2, the putative human telomerase catalytic subunit gene, is up-regulated in tumor cells and during immortalization. Cell 90(4):785–795. https://doi.org/10.1016/s0092-8674(00)80538-3

Misiti S, Nanni S, Fontemaggi G, Cong YS, Wen J, Hirte HW et al (2000) Induction of hTERT expression and telomerase activity by estrogens in human ovary epithelium cells. Mol Cell Biol 20(11):3764–3771. https://doi.org/10.1128/MCB.20.11.3764-3771.2000

Mitchell JR, Cheng J, Collins K (1999) A box H/ACA small nucleolar RNA-like domain at the human telomerase RNA 3′ end. Mol Cell Biol 19(1):567–576

Miyake Y, Nakamura M, Nabetani A, Shimamura S, Tamura M, Yonehara S et al (2009) RPA-like mammalian Ctc1-Stn1-Ten1 complex binds to single-stranded DNA and protects telomeres independently of the Pot1 pathway. Mol Cell 36(2):193–206. https://doi.org/10.1016/j.molcel.2009.08.009

Mogford JE, Liu WR, Reid R, Chiu CP, Said H, Chen SJ et al (2006) Adenoviral human telomerase reverse transcriptase dramatically improves ischemic wound healing without detrimental immune response in an aged rabbit model. Hum Gene Ther 17(6):651–660. https://doi.org/10.1089/hum.2006.17.651

Mouaikel J, Verheggen C, Bertrand E, Tazi J, Bordonné R (2002) Hypermethylation of the cap structure of both yeast snRNAs and snoRNAs requires a conserved methyltransferase that is localized to the nucleolus. Mol Cell 9(4):891–901. https://doi.org/10.1016/s1097-2765(02)00484-7

Moyzis RK, Buckingham JM, Cram LS, Dani M, Deaven LL, Jones MD et al (1988) A highly conserved repetitive DNA sequence, (TTAGGG)n, present at the telomeres of human chromosomes. Proc Natl Acad Sci U S A 85(18):6622–6626. https://doi.org/10.1073/pnas.85.18.6622

Munoz-Lorente MA, Martinez P, Tejera A, Whittemore K, Moises-Silva AC, Bosch F, Blasco MA (2018) AAV9-mediated telomerase activation does not accelerate tumorigenesis in the context of oncogenic K-Ras-induced lung cancer. PLoS Genet 14(8):e1007562. https://doi.org/10.1371/journal.pgen.1007562

Nagpal N, Agarwal S (2020) Telomerase RNA processing: implications for human health and disease. Stem Cells 38:1532–1543
Nagpal N, Wang J, Zeng J, Lo E, Moon DH, Luk K et al (2020) Small molecule PAPD5 inhibitors restore telomerase activity in patient stem cells. Cell Stem Cell 26(6):896–909
Nakamura TM, Morin GB, Chapman KB, Weinrich SL, Andrews WH, Lingner J et al (1997) Telomerase catalytic subunit homologs from fission yeast and human. Science 277:955–959
Nguyen THD, Tam J, Wu RA, Greber BJ, Toso D, Nogales E, Collins K (2018) Cryo-EM structure of substrate-bound human telomerase holoenzyme. Nature 557(7704):190–195. https://doi.org/10.1038/s41586-018-0062-x
Nishi H, Nakada T, Kyo S, Inoue M, Shay JW, Isaka K (2004) Hypoxia-inducible factor 1 mediates upregulation of telomerase (hTERT). Mol Cell Biol 24(13):6076–6083. https://doi.org/10.1128/mcb.24.13.6076-6083.2004
Oishi H, Doi S, Nakashima A, Ike T, Maeoka Y, Sasaki K et al (2021) Klotho overexpression protects against renal aging along with suppression of transforming growth factor-b1 signaling pathways. Am J Physiol Renal Physiol 321:F799–F811
Penev A, Bazley A, Shen M, Boeke JD, Savage SA, Sfeir A (2021) Alternative splicing is a developmental switch for hTERT expression. Mol Cell 81(11):2349–2360.e2346. https://doi.org/10.1016/j.molcel.2021.03.033
Podlevsky JD, Chen JJ (2016) Evolutionary perspectives of telomerase RNA structure and function. RNA Biol 13(8):720–732. https://doi.org/10.1080/15476286.2016.1205768
Povedano JM, Martinez P, Serrano R, Tejera A, Gomez-Lopez G, Bobadilla M et al (2018) Therapeutic effects of telomerase in mice with pulmonary fibrosis induced by damage to the lungs and short telomeres. Elife 7. https://doi.org/10.7554/eLife.31299
Qi X, Xie M, Brown AF, Bley CJ, Podlevsky JD, Chen JJ (2012) RNA/DNA hybrid binding affinity determines telomerase template-translocation efficiency. EMBO J 31(1):150–161. https://doi.org/10.1038/emboj.2011.363
Qin J, Autexier C (2021) Regulation of human telomerase RNA biogenesis and localization. RNA Biol 18(305–315). https://doi.org/10.1080/15476286.2020.1809196
Revy P, Kannengiesser C, Bertuch AA (2023) Genetics of human telomere biology disorders. Nat Rev Genet 24:86–108
Roake CM, Chen L, Chakravarthy AL, Ferrell JE, Raffa GD, Artandi SE (2019) Disruption of telomerase RNA maturation kinetics precipitates disease. Mol Cell 74(4):688–700. e683
Schmutz I, Mensenkamp AR, Takai KK, Haadsma M, Spruijt L, de Voer RM et al (2020) TINF2 is a haploinsufficient tumor suppressor that limits telomere length. Elife 9. https://doi.org/10.7554/eLife.61235
Sekne Z, Ghanim GE, van Roon AM, Nguyen THD (2022) Structural basis of human telomerase recruitment by TPP1-POT1. Science 375(6585):1173–1176. https://doi.org/10.1126/science.abn6840
Sexton AN, Collins K (2011) The 5′ guanosine tracts of human telomerase RNA are recognized by the G-quadruplex binding domain of the RNA helicase DHX36 and function to increase RNA accumulation. Mol Cell Biol 31(4):736–743. https://doi.org/10.1128/MCB.01033-10
Sfeir AJ, Chai W, Shay JW, Wright WE (2005) Telomere-end processing the terminal nucleotides of human chromosomes. Mol Cell 18(1):131–138. https://doi.org/10.1016/j.molcel.2005.02.035
Shay JW (2016) Role of telomeres and telomerase in aging and cancer. Cancer Discov 6(6):584–593. https://doi.org/10.1158/2159-8290.CD-16-0062
Shim HS, Iaconelli J, Shang X, Li J, Lan ZD, Jiang S, Nutsch K, Beyer BA, Lairson LL, Boutin AT, Bollong MJ, Schultz PG, DePinho RA (2024) TERT activation targets DNA methylation and multiple aging hallmarks. Cell 187(15):4030–4042. https://doi.org/10.1016/j.cell.2024.05.048
Shuai L, Li X, He Q, Dang X, Chen H, Zhou P et al (2012) Angiogenic effect of endothelial progenitor cells transfected with telomerase reverse transcriptase on peritubular microvessel in five out of six subtotal nephrectomy rats. Ren Fail 34(10):1270–1280. https://doi.org/10.3109/0886022X.2012.723592
Simons MJ (2015) Questioning causal involvement of telomeres in aging. Ageing Res Rev 24(Pt B):191–196. https://doi.org/10.1016/j.arr.2015.08.002

Stanley SE, Armanios M (2015) The short and long telomere syndromes: paired paradigms for molecular medicine. Curr Opin Genet Dev 33:1–9. https://doi.org/10.1016/j.gde.2015.06.004

Suhr ST, Chang EA, Rodriguez RM, Wang K, Ross PJ, Beyhan Z et al (2009) Telomere dynamics in human cells reprogrammed to pluripotency. PLoS One 4(12):e8124. https://doi.org/10.1371/journal.pone.0008124

Sunkari VG, Lind F, Botusan IR, Kashif A, Liu ZJ, Ylä-Herttuala S et al (2015) Hyperbaric oxygen therapy activates hypoxia-inducible factor 1 (HIF-1), which contributes to improved wound healing in diabetic mice. Wound Repair Regen 23(1):98–103

Taboski MA, Sealey DC, Dorrens J, Tayade C, Betts DH, Harrington L (2012) Long telomeres bypass the requirement for telomere maintenance in human tumorigenesis. Cell Rep 1(2):91–98. https://doi.org/10.1016/j.celrep.2011.12.004

Tang H, Wang H, Cheng X, Fan X, Yang F, Zhang M et al (2018) HuR regulates telomerase activity through TERC methylation. Nat Commun 9(1):2213. https://doi.org/10.1038/s41467-018-04617-7

Tang X, Xu R, Wang Y, Chen K, Cui S (2023) TERC haploid cell reprogramming: a novel therapeutic strategy for aplastic anemia. Mol Med 29(1):94. https://doi.org/10.1186/s10020-023-00691-w

Townsley DM, Dumitriu B, Liu D, Biancotto A, Weinstein B, Chen C et al (2016) Danazol treatment for telomere diseases. N Engl J Med 374(20):1922–1931. https://doi.org/10.1056/NEJMoa1515319

Trahan C, Dragon F (2009) Dyskeratosis congenita mutations in the H/ACA domain of human telomerase RNA affect its assembly into a pre-RNP. RNA 15(2):235–243

Ullah M, Sun Z (2019) Klotho deficiency accelerates stem cells aging by impairing telomerase activity. Gerontol Soc Am 74(9):1396–1407

Vera E, Bernardes de Jesus B, Foronda M, Flores JM, Blasco MA (2012) The rate of increase of short telomeres predicts longevity in mammals. Cell Rep 2(4):732–737. https://doi.org/10.1016/j.celrep.2012.08.023

Wan F, Ding Y, Zhang Y, Wu Z, Li S, Yang L et al (2021) Zipper head mechanism of telomere synthesis by human telomerase. Cell Res 31(12):1275–1290. https://doi.org/10.1038/s41422-021-00586-7

Wang F, Podell ER, Zaug AJ, Yang Y, Baciu P, Cech TR, Lei M (2007) The POT1-TPP1 telomere complex is a telomerase processivity factor. Nature 445(7127):506–510. https://doi.org/10.1038/nature05454

Wang F, Stewart JA, Kasbek C, Zhao Y, Wright WE, Price CM (2012) Human CST has independent functions during telomere duplex replication and C-strand fill-in. Cell Rep 2(5):1096–1103. https://doi.org/10.1016/j.celrep.2012.10.007

Wang D, Tai PWL, Gao G (2019) Adeno-associated virus vector as a platform for gene therapy delivery. Nat Rev Drug Discov 18(5):358–378. https://doi.org/10.1038/s41573-019-0012-9

Whittemore K, Derevyanko A, Martinez P, Serrano R, Pumarola M, Bosch F, Blasco MA (2019a) Telomerase gene therapy ameliorates the effects of neurodegeneration associated to short telomeres in mice. Aging 11(10):2916–2948. https://doi.org/10.18632/aging.101982

Whittemore K, Vera E, Martinez-Nevado E, Sanpera C, Blasco MA (2019b) Telomere shortening rate predicts species life span. Proc Natl Acad Sci U S A 116(30):15122–15127. https://doi.org/10.1073/pnas.1902452116

Wick M, Zubov D, Hagen G (1999) Genomic organization and promoter characterization of the gene encoding the human telomerase reverse transcriptase (hTERT). Gene 232(1):97–106. https://doi.org/10.1016/S0378-1119(99)00108-0

Xin H, Liu D, Wan M, Safari A, Kim H, Sun W et al (2007) TPP1 is a homologue of ciliate TEBP-beta and interacts with POT1 to recruit telomerase. Nature 445(7127):559–562. https://doi.org/10.1038/nature05469

Zhao J, Hoare SF, McFarlane R, Muir S, Parkinson EK, Black DM, Keith WN (1998) Cloning and characterization of human and mouse telomerase RNA gene promoter sequences. Oncogene 16:1345–1350

Chapter 2
Proteasome Activators and Ageing: Restoring Proteostasis Using Small Molecules

Arun Upadhyay and Vibhuti Joshi

Abstract Ageing is an inevitable phenomenon that remains under control of a plethora of signalling pathways and regulatory mechanisms. Slowing of cellular homeostasis and repair pathways, declining genomic and proteomic integrity, and deficient stress regulatory machinery may cause accumulating damage triggering initiation of pathways leading to ageing-associated changes. Multiple genetic studies in small laboratory organisms focused on the manipulation of proteasomal activities have shown promising results in delaying the age-related decline and improving the lifespan. In addition, a number of studies indicate a prominent role of small molecule-based proteasome activators showing positive results in ameliorating the stress conditions, protecting degenerating neurons, restoring cognitive functions, and extending life span of organisms. In this chapter, we provide a brief overview of the multi-enzyme proteasome complex, its structure, subunit composition and variety of cellular functions. We also highlight the strategies applied in the past to modulate the protein degradation efficiency of proteasome and their impact on rebalancing the proteostasis defects. Finally, we provide a descriptive account of proteasome activation mechanisms and small molecule-based strategies to improve the overall organismal health and delay the development of age-associated pathologies.

Keywords Ageing · Proteasome · Ubiquitin-proteasome system · Small molecules Lifespan

A. Upadhyay (✉)
Department of Neurology, Northwestern University Feinberg School of Medicine, Chicago, IL, USA

Department of Bioscience and Biomedical Engineering, Indian Institute of Technology Bhilai, Chhattisgarh, India
e-mail: arun@iitbhilai.ac.in

V. Joshi (✉)
Department of Biotechnology, Bennett University, Greater Noida, Uttar Pradesh, India
e-mail: Vibhuti.Joshi@bennett.edu.in

© The Author(s), under exclusive license to Springer Nature Switzerland AG 2024

V. I. Korolchuk, J. R. Harris (eds.), *Biochemistry and Cell Biology of Ageing: Part V, Anti-Ageing Interventions*, Subcellular Biochemistry 107,
https://doi.org/10.1007/978-3-031-66768-8_2

Introduction

An organism's lifecycle involves development, growth, and ageing. Our understanding of ageing is particularly poor in comparison to the other two. We primarily define ageing in terms of accumulating damage, reduced homeostasis mechanisms, and a slowdown in defence mechanisms. Ageing is a tightly regulated multidimensional retardation of molecular and cellular maintenance. These changes could be observed at multiple levels. Genomic integrity is lost, epigenetic regulation is distorted, mitochondria are deregulated, intracellular communication is lost, and nutrient-sensing and metabolism are flawed (Vermeij et al. 2016). In parallel, the efficiencies of various protein quality control (PQC) pathways are massively declined. Notably, the first step towards ensuring the quality of newly synthesised proteins is chaperone-mediated folding of the polypeptide chains (Saibil 2013). However, some proteins escape from these quality control mechanisms, raising the need for the dedicated machinery which removes them from cells before they start forming amyloids. An altered protein metabolism, orchestrated by inefficient chaperoning and slowed proteasomal degradation, may result in altered global protein turnover.

Loss of proteostasis is one obvious hallmark of ageing. The proteostasis network of the cells and organisms includes protein synthesis, maintenance, and degradation machinery (Amanullah et al. 2017). While protein synthesis is a primary function of cytosolic ribosomes, maintenance is supervised and performed by multiple families of molecular chaperones. On the other hand, protein degradation is shared among proteasomal and lysosomal pathways. More specifically, lysosomal degradation mechanisms, including autophagy, tend to degrade bulk cellular debris, including proteins or aggregates of proteins (Joshi et al. 2020; Nixon 2013). Previous studies have shown that upregulated autophagy may help reduce the cytosolic protein aggregates (Sarkar et al. 2007; Upadhyay et al. 2018). On the other hand, proteasomal degradation includes targeted degradation of linearized proteins/polypeptide chains following target-specific ubiquitination of substrate proteins. The degradation of proteins is time-dependent and regulated based on the turnover rates and lifetimes of each cellular protein.

Cross-species studies indicate that the turnover rates of proteins are negatively correlated with an organism's lifespan. In simpler terms, protein degradation is slower in organisms with greater lifespans; and the turnover rates diminish in long-lived animals without affecting proteostasis (Swovick et al. 2021). Contrarily, animals with shorter lifespan relish higher metabolic rates by replenishing the older, damaged proteins (Santra et al. 2019). The inability in doing so may affect the stability and function of cellular proteins. Multiple concurrent cellular processes may affect protein turnover and in turn, the half-life of proteins. However, a granular description of protein turnover is challenging due to technological limitations and the lack of appropriate models. A limited number of previous studies have comprehended the abundances of proteins across various ageing time points and have elucidated that pharmacological manipulation of proteasome machinery may help delay age-associated changes.

The proteome integrity is compromised during ageing as the burden of accumulated proteins builds up. These changes are also accompanied by oxidative and proteotoxic stress followed by declining efficiency of lysosomal and proteasomal protein degradation pathways (Chen et al. 2011; Taylor and Dillin 2011). Among the two major degradation systems, the roles of proteasomal machinery in ageing-associated pathways are explored in detail. For example, we can understand the importance of proteasomal machinery in cellular proteostasis through extensive studies on neurodegenerative diseases, which established that the lack of efficiency of the ubiquitin-proteasome system (UPS) may affect the cellular efficiency to get rid of toxic protein aggregates (Labbadia and Morimoto 2015; Sundaria et al. 2021). Ageing is one of the biggest risk factors for many diseases, especially neurodegeneration. The manifestation of multiple neurodegenerative diseases is directly or indirectly linked with ageing-associated changes. Loss of cellular proteostasis is one major hallmark of most of these late-age conditions (Lopez-Otin et al. 2013; Schmauck-Medina et al. 2022).

Proteasome being a crucial player in this network has been studied rigorously over the past few decades for its roles in affecting age-associated lack of protein quality control. Inefficient proteasomes may lead to a global proteotoxic overburden and a proteome-wide decline in the controlled protein turnover (Chen et al. 2011; Mishra et al. 2018). As discussed, the regulation of protein turnover is an important although less appreciated denominator of most pathological conditions, especially proteinopathies (Amanullah et al. 2017; Upadhyay et al. 2017). Proteinopathy is a general term used for the conditions and diseases manifested because of perturbed proteostasis leading to the generation of proteotoxic stress and disturbed cellular signalling pathways and metabolism. Genetic and pharmacological manipulation of proteasomal activities remained at the centre of many such studies performed in different animal models (Cattaneo et al. 2023; Gami and Wolkow 2006; Yang et al. 2023). An upregulation of chymotrypsin-like and trypsin-like activities of proteasome has been reported using genetic manipulation in different cellular models. A similar upregulation in proteasomal degradation could be achieved by phenolic components of bee pollen (Graikou et al. 2011).

Although none of the animal models truly represent human ageing, most of these studies indicated that increasing the proteolytic activity of proteasome can reduce the proteotoxic burden and may enhance life expectancy in both disease and healthy conditions (Upadhyay 2021). Since genetic manipulation remains a challenging task, pharmacological modulation of distinct 26S proteasome activities has been actively investigated (Njomen and Tepe 2019; Thibaudeau and Smith 2019). Naturally-derived small molecules provide highly robust tools to modulate the proteolytic activities of the proteasome, thus have been utilised to understand the complex relationship of the proteasome, proteostasis, and ageing (Badhwar and Upadhyay 2020; Myeku and Duff 2018). In this chapter, we will provide an overview of small molecules that can increase the proteolytic efficacy of proteasome and rescue some of the age-associated or pathological changes.

Why Proteasome?

Proteasomal degradation of proteins is a highly regulated process that is orchestrated by multiple protein players, including enzymes and other accessory components. The most important of these is the large multiprotein complex called 26S proteasome. Studies suggest that the stability and turnover of more than 20% of cellular proteins are regulated by proteasomes (Baugh et al. 2009). To do so, proteasomes form a large proportion, i.e. up to 20 μg/mg of total soluble cellular proteins (Dahlmann 2007; Kuehn et al. 1986). Although proteasomes are primarily responsible for the degradation of ubiquitinated proteins, there is evidence of degrading non-ubiquitinated proteins by this protease complex. Oxidized cellular proteins are the primary targets of proteasomal degradation in a ubiquitin- and ATP-independent manner (Demasi et al. 2013). Multiple proteins have been found to be involved in this proteasome-mediated degradation mechanism. For example, a proteasome transcriptional activator Rpn4 has been shown to be degraded despite mutations or the absence of lysine residues (Ju and Xie 2004). Interestingly, Rpn4 is also targeted through ubiquitin-dependent degradation via a different N-terminal degradation signal. Furthermore, ornithine decarboxylase and thymidylate synthase are cellular enzymes degraded exclusively by proteasomes in a ubiquitin-independent manner (Bercovich et al. 1989; Pena et al. 2006).

Multiple genetic studies have previously shown that modulation of proteasome activities may help improving varying pathological conditions. Multiple myeloma is the first example of a condition that has been shown to be targeted by proteasome inhibitors. Bortezomib is the first FDA-approved drug of this class; with, carfilzomib and ixazomib are the other two added up later to this list (Buac et al. 2013). Not only multiple myeloma but many other cancers, like mantle cell lymphoma have been targeted by these drugs (Mishra et al. 2018). While inhibition of proteasomes may help to target cancer cells by initiating stress response, activation of this proteolytic machinery, on the other hand, may help preserve cellular proteostasis. Therefore, proteasome activators have been proposed to treat conditions like protein misfolding, and neurodegeneration. Although proteasome activators have been proven to be beneficial in delaying the deteriorating changes in many neurodegenerative disorders, their clinical application is still under investigation (Dal Vechio et al. 2014; Joshi et al. 2019). On the other hand, what has been shown more convincingly is an increased lifespan of several organisms and models when the activity of the UPS is restored or upregulated (Chondrogianni et al. 2015b; Krahn et al. 2017).

The Ubiquitin-Proteasome System

The proteasomes are substrate-degrading complexes embedded within a wider regulatory network called the ubiquitin-proteasome system, which also consists of substrate-recruiting machinery. The recruitment of substrates is orchestrated by the

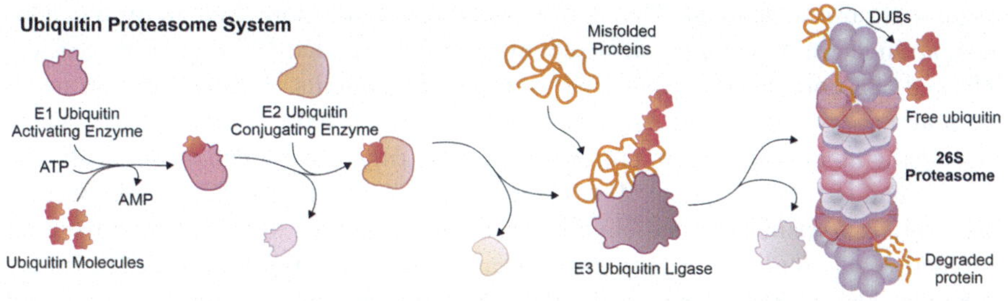

Fig. 2.1 A brief overview of the ubiquitin-proteasome system: The UPS machinery consists of a battery of enzymes to target a substrate protein for proteasomal degradation. The very first step in the cascade of enzymatic attachment of ubiquitin to the target protein is the activation of a small ~8.5 kDa ubiquitin protein in an ATP dependent manner by an enzyme called E1 (a ubiquitin-activating enzyme). In the next step, the activated ubiquitin molecule is attached to the next protein called the E2 ubiquitin-conjugating enzyme. This enzyme continues the reaction by transferring the ubiquitin to the next protein in the cascade, called the E3 ubiquitin ligating enzyme. The E3 ligases can bind to the E2 enzymes and transfer the ubiquitin to the substrate directly. In some instances, they bind the ubiquitin first and then attach to the substrate in two consecutive steps. In the last step, before transferring the substrate protein to the 26S proteasome for degradation, the ubiquitin proteins are removed from the proteins and linearized polypeptide enters the 20S tunnel for the proteolytic cleavage

sequential action of multiple enzymes. These reactions attach a small 76 amino acid long protein ubiquitin (Ub) to a substrate protein, during the process termed ubiquitination. Ubiquitination is a type of posttranslational modification (PTM), often considered as a death signal for proteins. Ubiquitin activating (E1) enzyme initiates the ATP-dependent activation of Ub as a first step followed by the transfer of the activated molecule to a conjugating enzyme (E2). The next step in this cascade of reactions is the attachment of activated ubiquitin to a target substrate using an intermediate ligating (E3) enzyme with very high specificity toward its substrates (Ciechanover 1998; Hershko and Ciechanover 1998). The ubiquitinated substrates are redirected towards proteasomal machinery, where the final but highly important deubiquitination reaction is catalysed by a number of enzymes called deubiquitinases (DUBs). Some of the proteasomal subunits may also have DUB activity (discussed in the next section). Our proteome consists of two E1s, approximately 35–40 E2s, more than 600 E3 ligases, while the number of DUBs is 80 (Mayor et al. 2016). All these proteins make an extensive arsenal of tools that orchestrate targeted degradation and hence the turnover rates of cellular proteins (see Fig. 2.1).

Proteasomes: The Trash Bins of the Cells

Proteasomes are multisubunit protein complexes that grind cellular proteins into smaller polypeptides (Demartino and Gillette 2007; Rechsteiner and Hill 2005). The proteasomal protein degradation could be both ubiquitination-dependent and ubiquitination-independent. This complex is made up of more than 30 protein subunits, arranged in two independent subunit complexes called 19S regulatory particle (RP) and 20S core particle (CP). The twenty-eight subunits of 20S CP are arranged into four stacked heptameric rings: two alpha (α) rings on the outer and two beta (β) rings on the inner sides. As mentioned, both these rings have seven distinct subunits (α1- α7, and (β1- β7, respectively). Three of the β-subunits- β1, β2, and β5 possess caspase-like, trypsin-like, and chymotrypsin-like activities to cleave the incoming unfolded polypeptides in a long barrel-shaped structure (Bard et al. 2018; Finley 2009; Groll et al. 2000; Jager et al. 1999). Notably, the thymus contains additional β5t subunits, while immune-proteasome in antigen-presenting cells (APCs) have additional variants of the three proteolytic subunits, referred as β1i, β2i, and β5i. While inner β-rings cleave the substrates by three successive enzymatic activities, the N-termini of outer α-ring subunits are responsible for regulating the entry of proteins through the proteolytic cavity of CP. When docked with 19S RP, a complete 26S proteasome holoenzyme is formed (for details see Fig. 2.2).

The 19S (also called PA700) complex is composed of 19 subunits that are arranged in two subcomplexes called base and lid (Bajorek and Glickman 2004). The base consists of six Rpt1-6 subunits arranged in the form of two superimposed rings (a pore). These subunits have AAA+ ATPase activities, thus helping unfold and translocate the proteins through the pore (Bar-Nun and Glickman 2012). Three additional non-ATPase scaffolding subunits Rpn1, -2, and -13 facilitate the functioning of RP. Nine other (Rpn3, Rpn5-9, and Rpn11-12) subunits form a lid that sits on top of the base and regulates anchoring, substrate recognition, and deubiquitination steps before entry and degradation. Notably, Rpn10 is a ubiquitin receptor protein and Rpn11 is a metalloprotease and contains deubiquitinase activity, which is required to remove the ubiquitin chains from the substrates before their degradation. Regulation of CP can also be regulated by means and mechanisms other than those facilitated by 19S RP. For example, PA200 or PA28 (an 11S complex) are alternative activators that bind to CP and regulate the protein degradation in an ATP- and ubiquitin-independent manner. A HEAT repeat containing Blm10 protein is a yeast ortholog of PA200 that forms functional Blm-CP or Blm10-CP-RP complexes to provide opening for substrate proteins into the proteolytic chamber (Schmidt et al. 2005).

The regulation of proteasomes is a highly complex process. Several PTMs have been shown to regulate the activities of proteasomal subunits (Kors et al. 2019; Tsimokha et al. 2020). S-glutathionylation of α-ring subunits at cysteine residues is one such PTM that regulates opening of the 20S proteolytic core (Demasi et al. 2003; Silva et al. 2008). O-GlcNAc transferase (OGT) mediated O-GlcNAc modification of 19s ATPase subunit Rpt2 leads to reversible inhibition of proteasomal activity (Zhang et al. 2003). Similarly, cyclic AMP-dependent phosphorylation of proteasome subunits (Rpt6, PSMD11) mediated by protein kinase A (PKA) increases the efficiency of the proteasome in clearing the misfolded proteins (Lokireddy et al. 2015; Zhang et al. 2007). In yeast, Rpn4 is an upstream transcriptional regulator that controls activation of all proteasomal subunits under stress conditions, for example, unfolded protein response, oxidative stress, heat shock or

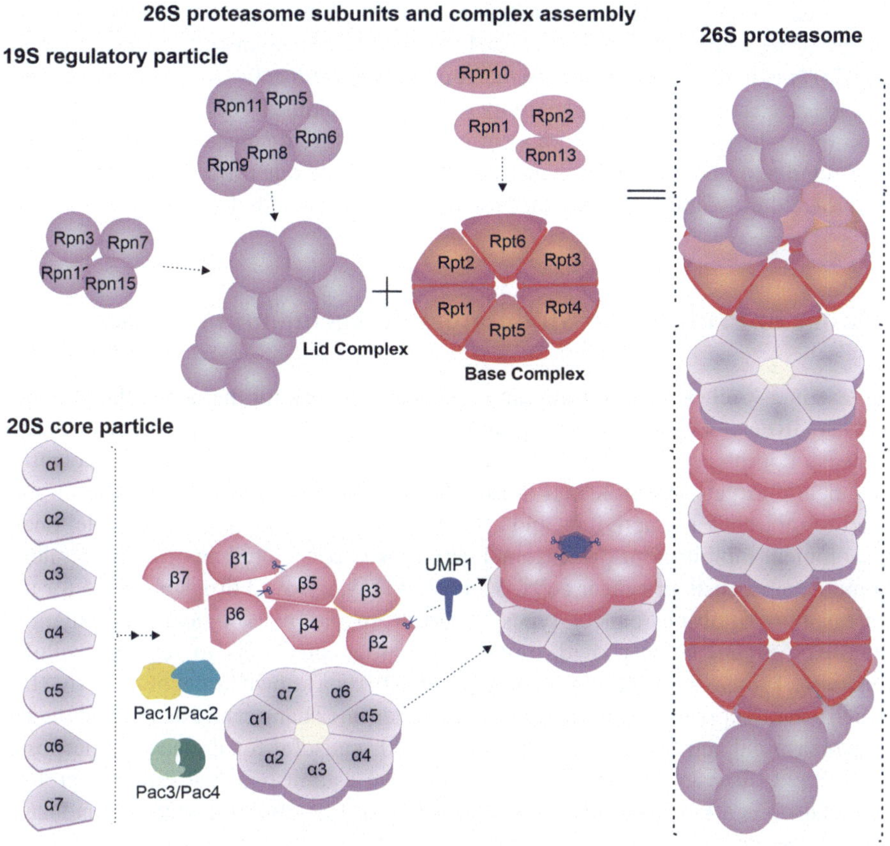

Fig. 2.2 Subunit composition of 26S proteasome: The multi-proteolytic machinery of the proteasome majorly consists of two subcomplexes: the 20S proteolytic core complex and the 19S regulatory complex. Both these complexes are made up of several components or subunits with distinct but highly cohesive functions. The top panel represents all major subunits forming the 19S complex. Apart from providing scaffolds, the subunits of the 19S complex provide ATPase and deubiquitinase activities. These subunits are important for regulating the entrance of a polypeptide chain into the proteolytic core complex. As shown at the bottom of the figure, the core 20S is a complex of four heptameric rings, two outer alpha, and two inner beta rings. Three subunits β1, β2 and β5 have caspase-like, trypsin-like, and chymotrypsin-like proteolytic activities, which help degrade the long polypeptide into smaller peptides

protein aggregation, etc. (Shirozu et al. 2015; Xie and Varshavsky 2001). In plants, NAC53-NAC78 is a heterodimeric transcriptional factor that regulates the expression of proteasomal subunits, other UPS enzymes, and molecular chaperones (Gladman et al. 2016). In a similar manner, nuclear respiratory factor 1 (Nrf1) and Nrf2 regulate the expression of proteasomal subunits in mammals (Pickering et al. 2013; Radhakrishnan et al. 2010). Another protein skinhead-1 (SKN-1), an Nrf2 orthologue in worms, when overexpressed increases the lifespan of *Caenorhabditis elegans* (Chondrogianni et al. 2015a). Another study in *C. elegans* showed that

Rpn-6 when upregulated is sufficient to restore the proteostasis defects and increase longevity in worms (Vilchez et al. 2012b). The mammalian orthologue of Rpn6, PSMD11, is highly expressed in human embryonic stem cells (hESCs) and contributes to increased assembly and activity of 26S proteasome (Vilchez et al. 2012a). In addition to the synthesis of component proteins, the architecture of proteasome is a well-regulated process. More than 10 molecular chaperone proteins are involved in the molecular assembly of the 26S proteasome (Tomko and Hochstrasser 2013).

History of Proteasome Modulators in Ageing and Diseases

Proteasomes orchestrate an upstream regulatory network for managing the turnover and half-lives of a significant proportion of cellular proteins. Therefore, these multi-protease complexes and other UPS components are considered highly potential therapeutic targets. As indicated in Fig. 2.3, a number of strategies have been utilized to modulate the activities and efficiency of proteasomes. For example, pharmacological inhibition of proteasomal activities is considered an effective strategy to induce apoptosis in cancer cells. Multiple FDA-approved drugs are used for the treatment of multiple cancer types. Apart from pharmacological manipulation, microRNAs (miRNAs) provide important endogenous tools to regulate cellular functions. Proteasome maturation protein (POMP) is a known target of miRNA-101, which can disrupt proteasome assembly and function (Zhang et al. 2015). Modulation of proteasomal activity in both directions could be useful for therapeutic purposes in different diseases. A plethora of studies have established that efficient protein degradation is a key limiting factor to lifespan. In fact, multiple studies conducted in rat, mice, and marmoset brain areas (cortex, striatum, substantia nigra, and spinal cord) confirmed an age-dependent decrease in proteasome activity (Keller et al. 2000; Zeng et al. 2005). Therefore, an upregulated proteasome function may help reduce proteotoxic stress by coordinated removal of proteins with increased efficiency, thus contributing to extended lifespan. Unlike inhibition which exaggerates the proteotoxic load and activates cell death pathways in cancer cells, the enhanced proteasomal clearance provides an opportunity for cells to ameliorate proteotoxic damage and restore lost proteostatic balance. This may help in the survival of degenerating neurons or other cells and delay multiple age-associated deteriorative changes. In this section, we are providing an updated account of research done so far with small molecules as proteasome activators.

1. **Natural molecules as proteasome activators**

   Natural compounds are a vast category of chemical compounds obtained from natural sources like fungi, bacteria, plants, or animals. These compounds are well known for their therapeutic potential and modulation of many biological activities (Atanasov et al. 2021). One of these biological processes is proteasome activation. As mentioned in earlier sections, the proteasome is the key cellular machinery to degrade misfolded or abnormal proteins (Dahlmann 2016; Kudriaeva and Belogurov 2019). In many disease conditions, these misfolded

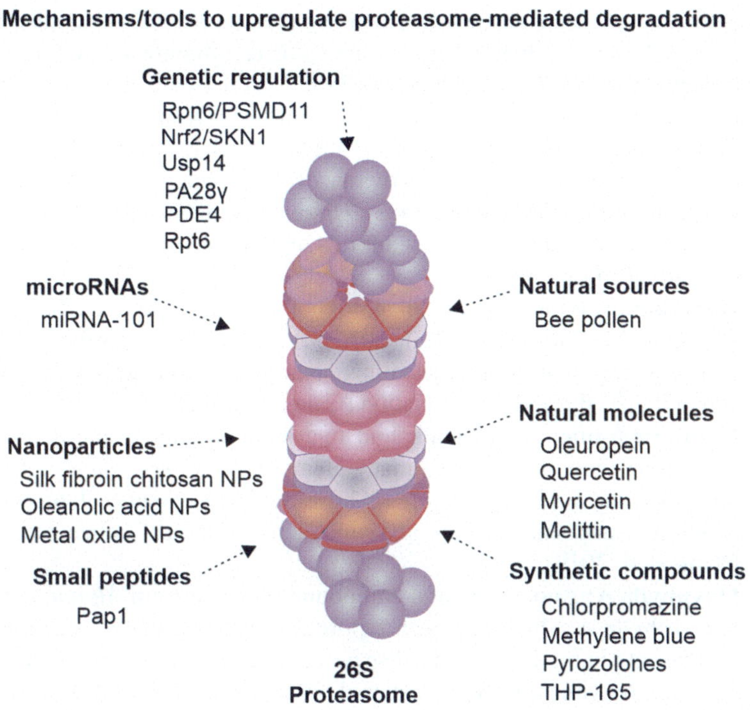

Fig. 2.3 Multiple ways of activating proteasome activities: Several genetic and molecular methods have been applied to study the effect of enhancing the proteasomal activities in health, disease, and ageing conditions to understand its downstream impact on these conditions. Multiple proteasome activators have shown neuroprotective effects in neurodegenerating brains. Apart from benefiting stress regulation in multiple ways, studies have also confirmed that increasing the efficiency of proteasomes in small organisms may help increase their lifespan

proteins accumulate and cause the death of cells (Moreno-Gonzalez and Soto 2011; Soto and Pritzkow 2018). Elimination of misfolded proteins before aggregation is only possible by activation of proteasomes to their maximum capacity. There are several natural compounds studied for proteasome activation; here, we are providing a brief account of some of the most-studied proteasome activators.

(a) **Oleuropein:** The natural compound obtained from the *Olea europea* leaf extract, olive oil, and olives is known as oleuropein. It has been reported that oleuropein works as a 20S stimulator, to activate proteasome (Trader et al. 2017). Crude extract of the 20S and 26S proteasomes was tested with oleuropein for their activation, and elevated degradation of proteins has been observed under oxidative stress. Oleuropein activates all three proteasome activities in human embryonic fibroblasts (Katsiki et al. 2007). Oleuropein is also known to activate proteasome 20S β5 subunits, as observed in the neonatal piglet model (El Demerdash et al. 2021).

(b) **Betulinic Acid (BA):** Triterpene derivative betulinic acid, commonly obtained from the bark of *Betula pubescens*, also works as a proteasome activator via chymotrypsin-like activity (Lou et al. 2021). Unlike oleuropein, BA plays a dual role as an activator or inhibitor of the proteasome (Huang et al. 2007). BA enhances all three proteasome activities, including β1, β2, and β5, as observed with HeLa cell lines (Kim et al. 2020).

(c) **Ursolic acid**: Another triterpenoid ursolic acid is structurally similar to betulinic acid and acts as an activator of the chymotrypsin-like activity of the proteasome (Wang et al. 2022). Ursolic acid is the natural derivative of corosolic acid and is widely used for its antitumor activity (Kim et al. 2014; Qi et al. 2021).

(d) **Harmine:** Harmine is the alkaloid obtained from the medicinal plant *Peganum harmala* (Patel et al. 2012). A mouse study performed by Cia et al. reported that harmine also increases proteasome activity by all three subunits (β1, β2, and β5) activation (Cai et al. 2019).

(e) **Canthine-6-one:** Another alkaloid that showed proteasome activation is canthine-6-one, isolated from many plants such as *Aerva lanata*, *Ailanthus altissima*, and *Simaba ferruginea* (Showalter 2013). Yuan et al. reported that canthine-6-one enhances proteasome activity by β5 subunit activation (Yuan et al. 2019).

(f) **Oxyphylla A:** Another natural compound obtained from *Alpinia oxyphylla* is oxyphylla A which has been reported to activate the β5 subunit of the proteasome in the Parkinson's disease cellular model to promote the degradation of α-synuclein (Sonninen et al. 2020; Zhou et al. 2020).

(g) **Sulforaphane:** Isothiocyanate isolated from broccoli (*Brassica oleracea*), sulforaphane is also known to increase proteasome activity by enhancing chymotrypsin activity (Gan et al. 2010). The elevated activity of proteasomes in β1, β2, and β5 subunits has been observed after the treatment with sulforaphane in the Huntington's disease mouse model (Liu et al. 2014).

(h) **Zerumbone:** A sesquiterpene obtained from the plant *Zingiber zerumbet*. Ohnishi et al. reported that pre-treatment of Hepa1c1c7 cells with zerumbone showed an increase in proteolytic activity by activation of the β5 subunit of the proteasome (Ohnishi et al. 2013).

(i) **Fatty acids:** Initially, it was observed that the activity of the plant's 20S proteasome gets enhanced by the treatment with fatty acids (Watanabe and Yamada 1996). Further, Dahlmann et al. and others found that fatty acids from an animal source increase the activity of the β1 proteasome subunit (Dahlmann et al. 1985; Vigouroux et al. 2003).

(j) **Phospholipids:** Animal fatty acids like cardiolipin (diphosphatidylglycerol) showed an increase in chymotrypsin-like and caspase-like activities of the proteasome (Ruiz de Mena et al. 1993). Another phospholipid that increases the proteasome β5 subunit is lysophosphatidylinositol (LPI), commonly obtained from sea urchin sperm (Matsumura and Aketa 1991).

(k) **Sulfatides:** Chymotrypsin-like activity of proteasome has also been enhanced by galactosylceramide sulfate (SM4) and lactosylceramide sulfate (SM3) (Ohkubo et al. 1991).

(l) **Heparin:** Heparin obtained from an animal sources selectively activates the trypsin-like activity of the proteasome (Yukawa et al. 1991).

(m) **PA28, PA700, and PA200:** These are also known as cellular proteasome activators. Overexpression of these proteins leads to enhanced degradation by the proteasome (Rechsteiner et al. 2000).
(n) **Kinases:** Proteasome can be activated by phosphorylation by multiple different kinases such as Dual Specificity Tyrosine Phosphorylation Regulated Kinase 2 (DYRK2) and Calcium/Calmodulin Dependent Protein Kinase II Alpha (CaMKIIα); these kinases were tested in a mouse model and cell lines and found to enhance proteasome activity by phosphorylation (Djakovic et al. 2009; Ranek et al. 2013).
(o) **PafE:** Proteasome accessory factor E (PafE, Rv3780) was also found to activate proteasome, especially for the *Mycobacterium tuberculosis* (Mtb) proteasome. Its function is very similar to eukaryotic proteasome activators PA26 and PA28 (Bai et al. 2016).
(p) **18α-glycyrrhetinic acid:** Recently, it has been discovered that 18α-glycyrrhetinic acid, when fed to *Caenorhabditis elegans*, showed enhanced proteasome activity, and extended their lifespan. Similar effects were also observed with mammalian and murine cell cultures (Papaevgeniou et al. 2016).

2. **Synthetic molecules as proteasome activators**

The small synthetic molecules as proteasome inhibitors have been known for many years, but researchers also found proteasome activators (Dahlmann et al. 1993). Small molecules can affect proteasome function either directly or indirectly through the 26S proteasome.

(a) **Sodium dodecyl sulfate (SDS):** It has been reported long back that a low concentration of SDS increases the chymotrypsin-like activity of the proteasome reversibly (Shibatani and Ward 1995; Tanaka et al. 1989). SDS might cause partial denaturation of the 20S subunit to induce gate opening, but the actual mechanism of proteasome activation by SDS is not yet known (George and Tepe 2021).
(b) **TCH-165:** It has been observed that TCH-165, generated by imidazoline scaffold, also acts as a mild activator of proteasome and enhances degradation of multiple proteins like tau, α-synuclein, etc., in cell culture (Njomen et al. 2018, 2022).
(c) **AM-404 and MK-866:** Trader et al. found that two small molecules AM-404 and MK-866 have great potential as 20S stimulators. Both molecules have been tested in multiple cell culture experiments and found to enhance the degradation of α-synuclein (Coleman et al. 2019; Trader et al. 2017).
(d) **Dihydroquinazolines:** Similar to other synthetic molecules, it also enhances the 20S proteasome activity and increases the degradation of proteins associated with neurodegeneration (Fiolek et al. 2021b).
(e) **Pyrazolones:** In a murine model of Amyotrophic Lateral Sclerosis (ALS), treatment of small molecule pyrazolones showed a protective effect by activating the 26S proteasome subunit (Trippier et al. 2014). Further research reported that pyrazolones bind the α-ring surfaces and modulate gate dynamics to activate proteasome. Aminopyrine and nifenazone are the two pyrazo-

lones studied for their therapeutic potential for Alzheimer's disease by enhancing proteasome activity (Santoro et al. 2020).

(f) **Chlorpromazines:** Screening of NIH clinical collection and Prestwick libraries, Jones et al. reported a 20% increase in chymotrypsin-like activity of the proteasome (Jones et al. 2017).

(g) **Methylene Blue:** Alzheimer's mouse model treated with methylene blue reported reduced Aβ level and further analysis observed an increase in chymotrypsin and trypsin-like activity of the proteasome in the brain of a model mouse (Medina et al. 2011).

(h) **Fluspirilene and Acylfluspirilene:** Fluspirilene and its synthetic analogues are also reported to increase 20S proteasome activity and restore its activity when impaired by intrinsically disordered proteins (Fiolek et al. 2021a).

3. **Other small molecules**

Similar to small molecules, peptides may provide important tools to regulate the activity of proteasomes. Proteasome-activating peptide 1 (Pap1) upregulates proteasomal degradation by assisting the gate opening and thus reduces oxidative stress in mouse fibroblasts and human neuroblastoma cells (Dal Vechio et al. 2014). Inhibition of a DUB enzyme USP14 provides an indirect method of upregulating the proteasome activity and thus reduces proteotoxic stress by clearing aberrant proteins (Lee et al. 2010). IU1, a selective inhibitor of Usp14 (a DUB), accelerates the degradation of aggregator proteins via the proteasomal pathway (Lee et al. 2010). Targeting PKA by small molecules (e.g. rolipram) also leads to an increase in the intrinsic proteasomal activity and confers neuroprotection in a Tg4510, a tauopathy mouse model (Lokireddy et al. 2015; Myeku et al. 2016). Table 2.1 provides a list of known small molecule-based activators of 26S proteasome.

Table 2.1 A summary of known small molecule-based proteasome activators

Proteasome activator	Target	Effect	Model system	Reference
Natural compounds				
Oleuropein	β1, β2, β5	Delayed senescence	IMR90, WI38	(Katsiki et al. 2007)
Quercetin	β5	Delayed senescence	HFL-1 cells	(Chondrogianni et al. 2010)
Myricetin	E6-AP	Aggregate clearance	Cos-7 cells	(Joshi et al. 2019)
Curcumin	β5	Heat shock response	Human keratinocytes	(Ali and Rattan 2006)
t-BHQ	Nrf2 β5	Pluripotency	hESC, iPSC	(Jang et al. 2014)
Sulphorafane	β1, β2, β5	Htt degradation	Mouse	(Liu et al. 2014)
18α-glycyrrhetinic acid	Nrf2, β5	Lifespan extension	HFL-1 cells	(Kapeta et al. 2010)
Melittin	β5	Reduced neuroinflammation	hSOD1G93A mice	(Yang et al. 2011)

(continued)

Table 2.1 (continued)

Proteasome activator	Target	Effect	Model system	Reference
n-butylidenephthalide	Rpn6, β5	Reduced neuronal death	C. elegans	(Fu et al. 2014a)
Betulinic acid	β5	Reduced level of HIF-1 alpha	HeLa cells	(Kim et al. 2020)
Canthin-6-one	β5	α-Synuclein degradation	HEK293 cells	(Yuan et al. 2019)
Acetylcorynoline	Rpn5	Increased lifespan	C. elegans	(Fu et al. 2014b)
Fatty acids	β5	Increased muscle proteasome activities	Rats	(Vigouroux et al. 2003)
Harmine	β1, β2, β5	α-Synuclein degradation	Transgenic mouse line (M83)	(Cai et al. 2019)
Heparin	β2	Activates trypsin like activity	In vitro assay	(Yukawa et al. 1991)
Lysophospholipids	β5	Activate sperm proteasome	Sea urchin	(Matsumura and Aketa 1991)
Oxyphylla A	β5	Promote α-syn degradation	PC12 stable cell lines	(Zhou et al. 2020)
Sulfatides	β5	Chymotrypsin like activity increase	In vitro assay	(Ohkubo et al. 1991)
Zerumbone	β5	Increases intracellular proteolysis	Hepa1c1c7 cells	(Ohnishi et al. 2013)
Other small molecules				
PAP-1	β5	Gate opening	Neuroblastoma cells	(Dal Vechio et al. 2014)
Pramipexole	β5	Neuroprotection	C57BL/6 mice	(Li et al. 2010)
Methylene blue	β2, β5	Aβ clearance	3xTg-AD mice	(Medina et al. 2011)
Pyrazolones	β5	Neuroprotection	PC12 cells	(Trippier et al. 2014)
Chlorpromazine	β1, β2, β5	Tau, α-syn clearance	HEK293T, U87-MG cells	(Jones et al. 2017)
TCH-165	β1, β2, β5	Selective IDP degradation	HEK293T, U87-MG cells	(Njomen et al. 2018)
Fluspirilene	β1, β2, β5	Prevent accumulation of IDP	HEK293T cells	(Fiolek et al. 2021a)
Dihydroquinazolines	β1, β2, β5	Degrade α-syn	In vitro assay	(Fiolek et al. 2021b)
SDS	β1, β2, β5	Increased peptidase activity	In vitro assay	(Shibatani and Ward 1995)
MK-866	20S CP	Degrade α-syn	HEK-293T	(Trader et al. 2017)
AM-404	20S CP	Degrade α-syn	SH-SY5Y and HEK293T cells	(Coleman et al. 2019)

Future Directions

Human ageing is a highly complex process spanning several decades. Efficient protein turnover machinery is a prerequisite for healthy ageing. Numerous studies addressed the protein turnover dynamics, including turnover of proteasome subunits, functioning of 19S and 20S complexes, and implications of modulating these activities on ageing-associated pathways in animal models. However, in comparison to human ageing, the most easily available and highly used animal models have significantly shorter lifespans ranging between a few days to months. Therefore, the major challenge in understanding this highly complex process is finding a suitable model system and designing relevant experiments that can recapitulate molecular and cellular pathways similar to what the human body presents. It is evident from numerous studies that cellular proteostasis ensures the maintenance and availability of appropriately folded three-dimensional states of proteins. Notably, the UPS is one arm of this cellular quality control mechanism. Proteasomal degradation is one crucial step in the regulation of global protein turnover dynamics. It is accompanied by multiple other accessory pathways that monitor the prevention of misfolded proteins, their aggregation, and refolding or cellular removal of aggregation-prone proteins. Several strategies for modulating some of these pathways have been harnessed for therapeutic benefits in multiple diseases. Enhancing proteasomal degradation is proposed and has been experimentally shown to extend lifespan in experimental models. However, due to our shortfall in the current understanding of ageing pathways and the unavailability of suitable models to study ageing, we are still far from utilising these tools for the extension of the human lifespan.

References

Ali RE, Rattan SI (2006) Curcumin's biphasic hormetic response on proteasome activity and heat-shock protein synthesis in human keratinocytes. Ann N Y Acad Sci 1067:394–399. https://doi.org/10.1196/annals.1354.056

Amanullah A, Upadhyay A, Joshi V, Mishra R, Jana NR, Mishra A (2017) Progressing neurobiological strategies against proteostasis failure: challenges in neurodegeneration. Prog Neurobiol 159:1–38. https://doi.org/10.1016/j.pneurobio.2017.08.005

Atanasov AG, Zotchev SB, Dirsch VM, International Natural Product Sciences Taskforce, Supuran CT (2021) Natural products in drug discovery: advances and opportunities. Nat Rev Drug Discov 20:200–216. https://doi.org/10.1038/s41573-020-00114-z

Badhwar R, Upadhyay A (2020) Modulation of cellular protein quality control pathways using small natural molecules. Bioactive natural products in drug discovery, pp 609–641

Bai L, Hu K, Wang T, Jastrab JB, Darwin KH, Li H (2016) Structural analysis of the dodecameric proteasome activator PafE in mycobacterium tuberculosis. Proc Natl Acad Sci U S A 113:E1983–E1992. https://doi.org/10.1073/pnas.1512094113

Bajorek M, Glickman MH (2004) Keepers at the final gates: regulatory complexes and gating of the proteasome channel. Cell Mol Life Sci 61:1579–1588. https://doi.org/10.1007/s00018-004-4131-y

Bar-Nun S, Glickman MH (2012) Proteasomal AAA-ATPases: structure and function. Biochim Biophys Acta 1823:67–82. https://doi.org/10.1016/j.bbamcr.2011.07.009

Bard JAM, Goodall EA, Greene ER, Jonsson E, Dong KC, Martin A (2018) Structure and function of the 26S proteasome. Annu Rev Biochem 87:697–724. https://doi.org/10.1146/annurev-biochem-062917-011931

Baugh JM, Viktorova EG, Pilipenko EV (2009) Proteasomes can degrade a significant proportion of cellular proteins independent of ubiquitination. J Mol Biol 386:814–827. https://doi.org/10.1016/j.jmb.2008.12.081

Bercovich Z, Rosenberg-Hasson Y, Ciechanover A, Kahana C (1989) Degradation of ornithine decarboxylase in reticulocyte lysate is ATP-dependent but ubiquitin-independent. J Biol Chem 264:15949–15952

Buac D, Shen M, Schmitt S, Kona FR, Deshmukh R, Zhang Z, Neslund-Dudas C, Mitra B, Dou QP (2013) From bortezomib to other inhibitors of the proteasome and beyond. Curr Pharm Des 19:4025–4038. https://doi.org/10.2174/1381612811319220012

Cai CZ, Zhou HF, Yuan NN, Wu MY, Lee SM, Ren JY, Su HX, Lu JJ, Chen XP, Li M et al (2019) Natural alkaloid harmine promotes degradation of alpha-synuclein via PKA-mediated ubiquitin-proteasome system activation. Phytomedicine 61:152842. https://doi.org/10.1016/j.phymed.2019.152842

Cattaneo M, Beltrami AP, Thomas AC, Spinetti G, Alvino V, Avolio E, Veneziano C, Rolle IG, Sponga S, Sangalli E et al (2023) The longevity-associated BPIFB4 gene supports cardiac function and vascularization in aging cardiomyopathy. Cardiovasc Res. https://doi.org/10.1093/cvr/cvad008

Chen B, Retzlaff M, Roos T, Frydman J (2011) Cellular strategies of protein quality control. Cold Spring Harb Perspect Biol 3:a004374. https://doi.org/10.1101/cshperspect.a004374

Chondrogianni N, Georgila K, Kourtis N, Tavernarakis N, Gonos ES (2015a) 20S proteasome activation promotes life span extension and resistance to proteotoxicity in Caenorhabditis elegans. FASEB J 29:611–622. https://doi.org/10.1096/fj.14-252189

Chondrogianni N, Kapeta S, Chinou I, Vassilatou K, Papassideri I, Gonos ES (2010) Anti-ageing and rejuvenating effects of quercetin. Exp Gerontol 45:763–771. https://doi.org/10.1016/j.exger.2010.07.001

Chondrogianni N, Voutetakis K, Kapetanou M, Delitsikou V, Papaevgeniou N, Sakellari M, Lefaki M, Filippopoulou K, Gonos ES (2015b) Proteasome activation: an innovative promising approach for delaying aging and retarding age-related diseases. Ageing Res Rev 23:37–55. https://doi.org/10.1016/j.arr.2014.12.003

Ciechanover A (1998) The ubiquitin-proteasome pathway: on protein death and cell life. EMBO J 17:7151–7160. https://doi.org/10.1093/emboj/17.24.7151

Coleman RA, Muli CS, Zhao Y, Bhardwaj A, Newhouse TR, Trader DJ (2019) Analysis of chain length, substitution patterns, and unsaturation of AM-404 derivatives as 20S proteasome stimulators. Bioorg Med Chem Lett 29:420–423. https://doi.org/10.1016/j.bmcl.2018.12.030

Dahlmann B (2007) Role of proteasomes in disease. BMC Biochem 8(Suppl 1):S3. https://doi.org/10.1186/1471-2091-8-s1-s3

Dahlmann B (2016) Mammalian proteasome subtypes: their diversity in structure and function. Arch Biochem Biophys 591:132–140. https://doi.org/10.1016/j.abb.2015.12.012

Dahlmann B, Becher B, Sobek A, Ehlers C, Kopp F, Kuehn L (1993) In vitro activation of the 20S proteasome. Enzyme Protein 47:274–284. https://doi.org/10.1159/000468685

Dahlmann B, Rutschmann M, Kuehn L, Reinauer H (1985) Activation of the multicatalytic proteinase from rat skeletal muscle by fatty acids or sodium dodecyl sulphate. Biochem J 228:171–177. https://doi.org/10.1042/bj2280171

Dal Vechio FH, Cerqueira F, Augusto O, Lopes R, Demasi M (2014) Peptides that activate the 20S proteasome by gate opening increased oxidized protein removal and reduced protein aggregation. Free Radic Biol Med 67:304–313. https://doi.org/10.1016/j.freeradbiomed.2013.11.017

Demartino GN, Gillette TG (2007) Proteasomes: machines for all reasons. Cell 129:659–662. https://doi.org/10.1016/j.cell.2007.05.007

Demasi M, Netto LE, Silva GM, Hand A, de Oliveira CL, Bicev RN, Gozzo F, Barros MH, Leme JM, Ohara E (2013) Redox regulation of the proteasome via S-glutathionylation. Redox Biol 2:44–51. https://doi.org/10.1016/j.redox.2013.12.003

Demasi M, Silva GM, Netto LE (2003) 20 S proteasome from Saccharomyces cerevisiae is responsive to redox modifications and is S-glutathionylated. J Biol Chem 278:679–685. https://doi.org/10.1074/jbc.M209282200

Djakovic SN, Schwarz LA, Barylko B, DeMartino GN, Patrick GN (2009) Regulation of the proteasome by neuronal activity and calcium/calmodulin-dependent protein kinase II. J Biol Chem 284:26655–26665. https://doi.org/10.1074/jbc.M109.021956

El Demerdash N, Chen MW, O'Brien CE, Adams S, Kulikowicz E, Martin LJ, Lee JK (2021) Oleuropein activates neonatal neocortical proteasomes, but proteasome gene targeting by AAV9 is variable in a clinically relevant piglet model of brain hypoxia-ischemia and hypothermia. Cells 10. https://doi.org/10.3390/cells10082120

Finley D (2009) Recognition and processing of ubiquitin-protein conjugates by the proteasome. Annu Rev Biochem 78:477–513. https://doi.org/10.1146/annurev.biochem.78.081507.101607

Fiolek TJ, Keel KL, Tepe JJ (2021a) Fluspirilene analogs activate the 20S proteasome and overcome proteasome impairment by intrinsically disordered protein oligomers. ACS Chem Nerosci 12:1438–1448. https://doi.org/10.1021/acschemneuro.1c00099

Fiolek TJ, Magyar CL, Wall TJ, Davies SB, Campbell MV, Savich CJ, Tepe JJ, Mosey RA (2021b) Dihydroquinazolines enhance 20S proteasome activity and induce degradation of alpha-synuclein, an intrinsically disordered protein associated with neurodegeneration. Bioorg Med Chem Lett 36:127821. https://doi.org/10.1016/j.bmcl.2021.127821

Fu RH, Harn HJ, Liu SP, Chen CS, Chang WL, Chen YM, Huang JE, Li RJ, Tsai SY, Hung HS et al (2014a) n-butylidenephthalide protects against dopaminergic neuron degeneration and alpha-synuclein accumulation in Caenorhabditis elegans models of Parkinson's disease. PLoS One 9:e85305. https://doi.org/10.1371/journal.pone.0085305

Fu RH, Wang YC, Chen CS, Tsai RT, Liu SP, Chang WL, Lin HL, Lu CH, Wei JR, Wang ZW et al (2014b) Acetylcorynoline attenuates dopaminergic neuron degeneration and alpha-synuclein aggregation in animal models of Parkinson's disease. Neuropharmacology 82:108–120. https://doi.org/10.1016/j.neuropharm.2013.08.007

Gami MS, Wolkow CA (2006) Studies of Caenorhabditis elegans DAF-2/insulin signaling reveal targets for pharmacological manipulation of lifespan. Aging Cell 5:31–37. https://doi.org/10.1111/j.1474-9726.2006.00188.x

Gan N, Wu YC, Brunet M, Garrido C, Chung FL, Dai C, Mi L (2010) Sulforaphane activates heat shock response and enhances proteasome activity through up-regulation of Hsp27. J Biol Chem 285:35528–35536. https://doi.org/10.1074/jbc.M110.152686

George DE, Tepe JJ (2021) Advances in proteasome enhancement by small molecules. Biomol Ther 11. https://doi.org/10.3390/biom11121789

Gladman NP, Marshall RS, Lee KH, Vierstra RD (2016) The proteasome stress regulon is controlled by a pair of NAC transcription factors in arabidopsis. Plant Cell 28:1279–1296. https://doi.org/10.1105/tpc.15.01022

Graikou K, Kapeta S, Aligiannis N, Sotiroudis G, Chondrogianni N, Gonos E, Chinou I (2011) Chemical analysis of Greek pollen – antioxidant, antimicrobial and proteasome activation properties. Chem Cent J 5:33. https://doi.org/10.1186/1752-153X-5-33

Groll M, Bajorek M, Kohler A, Moroder L, Rubin DM, Huber R, Glickman MH, Finley D (2000) A gated channel into the proteasome core particle. Nat Struct Biol 7:1062–1067. https://doi.org/10.1038/80992

Hershko A, Ciechanover A (1998) The ubiquitin system. Annu Rev Biochem 67:425–479. https://doi.org/10.1146/annurev.biochem.67.1.425

Huang L, Ho P, Chen CH (2007) Activation and inhibition of the proteasome by betulinic acid and its derivatives. FEBS Lett 581:4955–4959. https://doi.org/10.1016/j.febslet.2007.09.031

Jager S, Groll M, Huber R, Wolf DH, Heinemeyer W (1999) Proteasome beta-type subunits: unequal roles of propeptides in core particle maturation and a hierarchy of active site function. J Mol Biol 291:997–1013. https://doi.org/10.1006/jmbi.1999.2995

Jang J, Wang Y, Kim HS, Lalli MA, Kosik KS (2014) Nrf2, a regulator of the proteasome, controls self-renewal and pluripotency in human embryonic stem cells. Stem Cells 32:2616–2625. https://doi.org/10.1002/stem.1764

Jones CL, Njomen E, Sjogren B, Dexheimer TS, Tepe JJ (2017) Small molecule enhancement of 20S proteasome activity targets intrinsically disordered proteins. ACS Chem Biol 12:2240–2247. https://doi.org/10.1021/acschembio.7b00489

Joshi V, Mishra R, Upadhyay A, Amanullah A, Poluri KM, Singh S, Kumar A, Mishra A (2019) Polyphenolic flavonoid (Myricetin) upregulated proteasomal degradation mechanisms: eliminates neurodegenerative proteins aggregation. J Cell Physiol 234:20900–20914. https://doi.org/10.1002/jcp.28695

Joshi V, Upadhyay A, Prajapati VK, Mishra A (2020) How autophagy can restore proteostasis defects in multiple diseases? Med Res Rev 40:1385–1439. https://doi.org/10.1002/med.21662

Ju D, Xie Y (2004) Proteasomal degradation of RPN4 via two distinct mechanisms, ubiquitin-dependent and -independent. J Biol Chem 279:23851–23854. https://doi.org/10.1074/jbc.C400111200

Kapeta S, Chondrogianni N, Gonos ES (2010) Nuclear erythroid factor 2-mediated proteasome activation delays senescence in human fibroblasts. J Biol Chem 285:8171–8184. https://doi.org/10.1074/jbc.M109.031575

Katsiki M, Chondrogianni N, Chinou I, Rivett AJ, Gonos ES (2007) The olive constituent oleuropein exhibits proteasome stimulatory properties in vitro and confers life span extension of human embryonic fibroblasts. Rejuvenation Res 10:157–172. https://doi.org/10.1089/rej.2006.0513

Keller JN, Huang FF, Markesbery WR (2000) Decreased levels of proteasome activity and proteasome expression in aging spinal cord. Neuroscience 98:149–156. https://doi.org/10.1016/s0306-4522(00)00067-1

Kim HJ, Cho HS, Ban HS, Nakamura H (2020) Suppression of HIF-1alpha accumulation by betulinic acid through proteasome activation in hypoxic cervical cancer. Biochem Biophys Res Commun 523:726–732. https://doi.org/10.1016/j.bbrc.2020.01.031

Kim JH, Kim YH, Song GY, Kim DE, Jeong YJ, Liu KH, Chung YH, Oh S (2014) Ursolic acid and its natural derivative corosolic acid suppress the proliferation of APC-mutated colon cancer cells through promotion of beta-catenin degradation. Food Chem Toxicol 67:87–95. https://doi.org/10.1016/j.fct.2014.02.019

Kors S, Geijtenbeek K, Reits E, Schipper-Krom S (2019) Regulation of proteasome activity by (post-) transcriptional mechanisms. Front Mol Biosci 6:48

Krahn JH, Kaschani F, Kaiser M (2017) Turning-ON proteasomes. Cell Chem Biol 24:653–655. https://doi.org/10.1016/j.chembiol.2017.06.005

Kudriaeva AA, Belogurov AA (2019) Proteasome: a nanomachinery of creative destruction. Biochemistry (Mosc) 84:S159–S192. https://doi.org/10.1134/S0006297919140104

Kuehn L, Dahlmann B, Reinauer H (1986) Tissue distribution of the multicatalytic proteinase in the rat: an immunological and enzymic study. Cienc Biol (Portugal) 11:101–107

Labbadia J, Morimoto RI (2015) The biology of proteostasis in aging and disease. Annu Rev Biochem 84:435–464. https://doi.org/10.1146/annurev-biochem-060614-033955

Lee BH, Lee MJ, Park S, Oh DC, Elsasser S, Chen PC, Gartner C, Dimova N, Hanna J, Gygi SP et al (2010) Enhancement of proteasome activity by a small-molecule inhibitor of USP14. Nature 467:179–184. https://doi.org/10.1038/nature09299

Li C, Guo Y, Xie W, Li X, Janokovic J, Le W (2010) Neuroprotection of pramipexole in UPS impairment induced animal model of Parkinson's disease. Neurochem Res 35:1546–1556. https://doi.org/10.1007/s11064-010-0214-3

Liu Y, Hettinger CL, Zhang D, Rezvani K, Wang X, Wang H (2014) Sulforaphane enhances proteasomal and autophagic activities in mice and is a potential therapeutic reagent for Huntington's disease. J Neurochem 129:539–547. https://doi.org/10.1111/jnc.12647

Lokireddy S, Kukushkin NV, Goldberg AL (2015) cAMP-induced phosphorylation of 26S proteasomes on Rpn6/PSMD11 enhances their activity and the degradation of misfolded proteins. Proc Natl Acad Sci U S A 112:E7176–E7185. https://doi.org/10.1073/pnas.1522332112

Lopez-Otin C, Blasco MA, Partridge L, Serrano M, Kroemer G (2013) The hallmarks of aging. Cell 153:1194–1217. https://doi.org/10.1016/j.cell.2013.05.039

Lou H, Li H, Zhang S, Lu H, Chen Q (2021) A review on preparation of Betulinic acid and its biological activities. Molecules 26. https://doi.org/10.3390/molecules26185583

Matsumura K, Aketa K (1991) Activation of proteasome in sea urchin sperm by Lysophosphatidylinositol and by sperm lipids: (proteasome/sea urchin/sperm/acrosome reaction/lysophospholipid). Dev Growth Differ 33:259–266. https://doi.org/10.1111/j.1440-169X.1991.00259.x

Mayor T, Sharon M, Glickman MH (2016) Tuning the proteasome to brighten the end of the journey. Am J Physiol Cell Physiol 311:C793–C804. https://doi.org/10.1152/ajpcell.00198.2016

Medina DX, Caccamo A, Oddo S (2011) Methylene blue reduces abeta levels and rescues early cognitive deficit by increasing proteasome activity. Brain Pathol 21:140–149. https://doi.org/10.1111/j.1750-3639.2010.00430.x

Mishra R, Upadhyay A, Prajapati VK, Mishra A (2018) Proteasome-mediated proteostasis: novel medicinal and pharmacological strategies for diseases. Med Res Rev 38:1916–1973. https://doi.org/10.1002/med.21502

Moreno-Gonzalez I, Soto C (2011) Misfolded protein aggregates: mechanisms, structures and potential for disease transmission. Semin Cell Dev Biol 22:482–487. https://doi.org/10.1016/j.semcdb.2011.04.002

Myeku N, Clelland CL, Emrani S, Kukushkin NV, Yu WH, Goldberg AL, Duff KE (2016) Tau-driven 26S proteasome impairment and cognitive dysfunction can be prevented early in disease by activating cAMP-PKA signaling. Nat Med 22:46–53. https://doi.org/10.1038/nm.4011

Myeku N, Duff KE (2018) Targeting the 26S proteasome to protect against Proteotoxic diseases. Trends Mol Med 24:18–29. https://doi.org/10.1016/j.molmed.2017.11.006

Nixon RA (2013) The role of autophagy in neurodegenerative disease. Nat Med 19:983–997. https://doi.org/10.1038/nm.3232

Njomen E, Osmulski PA, Jones CL, Gaczynska M, Tepe JJ (2018) Small molecule modulation of proteasome assembly. Biochemistry 57:4214–4224. https://doi.org/10.1021/acs.biochem.8b00579

Njomen E, Tepe JJ (2019) Proteasome activation as a new therapeutic approach to target proteotoxic disorders. J Med Chem 62:6469–6481. https://doi.org/10.1021/acs.jmedchem.9b00101

Njomen E, Vanecek A, Lansdell TA, Yang YT, Schall PZ, Harris CM, Bernard MP, Isaac D, Alkharabsheh O, Al-Janadi A et al (2022) Small molecule 20S proteasome enhancer regulates MYC protein stability and exhibits antitumor activity in multiple myeloma. Biomedicine 10. https://doi.org/10.3390/biomedicines10050938

Ohkubo I, Gasa S, Namikawa C, Makita A, Sasaki M (1991) Human erythrocyte multicatalytic proteinase: activation and binding to sulfated galacto- and lactosylceramides. Biochem Biophys Res Commun 174:1133–1140. https://doi.org/10.1016/0006-291x(91)91538-n

Ohnishi K, Nakahata E, Irie K, Murakami A (2013) Zerumbone, an electrophilic sesquiterpene, induces cellular proteo-stress leading to activation of ubiquitin-proteasome system and autophagy. Biochem Biophys Res Commun 430:616–622. https://doi.org/10.1016/j.bbrc.2012.11.104

Papaevgeniou N, Sakellari M, Jha S, Tavernarakis N, Holmberg CI, Gonos ES, Chondrogianni N (2016) 18alpha-glycyrrhetinic acid proteasome activator decelerates aging and Alzheimer's disease progression in Caenorhabditis elegans and neuronal cultures. Antioxid Redox Signal 25:855–869. https://doi.org/10.1089/ars.2015.6494

Patel K, Gadewar M, Tripathi R, Prasad SK, Patel DK (2012) A review on medicinal importance, pharmacological activity and bioanalytical aspects of beta-carboline alkaloid "Harmine". Asian Pac J Trop Biomed 2:660–664. https://doi.org/10.1016/S2221-1691(12)60116-6

Pena MM, Xing YY, Koli S, Berger FG (2006) Role of N-terminal residues in the ubiquitin-independent degradation of human thymidylate synthase. Biochem J 394:355–363. https://doi.org/10.1042/BJ20051479

Pickering AM, Staab TA, Tower J, Sieburth D, Davies KJ (2013) A conserved role for the 20S proteasome and Nrf2 transcription factor in oxidative stress adaptation in mammals, Caenorhabditis elegans and Drosophila melanogaster. J Exp Biol 216:543–553. https://doi.org/10.1242/jeb.074757

Qi Z, Yang G, Deng T, Wang J, Zhou H, Popov SA, Shults EE, Wang C (2021) Design and linkage optimization of ursane-thalidomide-based PROTACs and identification of their targeted-degradation properties to MDM2 protein. Bioorg Chem 111:104901. https://doi.org/10.1016/j.bioorg.2021.104901

Radhakrishnan SK, Lee CS, Young P, Beskow A, Chan JY, Deshaies RJ (2010) Transcription factor Nrf1 mediates the proteasome recovery pathway after proteasome inhibition in mammalian cells. Mol Cell 38:17–28. https://doi.org/10.1016/j.molcel.2010.02.029

Ranek MJ, Terpstra EJ, Li J, Kass DA, Wang X (2013) Protein kinase g positively regulates proteasome-mediated degradation of misfolded proteins. Circulation 128:365–376. https://doi.org/10.1161/CIRCULATIONAHA.113.001971

Rechsteiner M, Hill CP (2005) Mobilizing the proteolytic machine: cell biological roles of proteasome activators and inhibitors. Trends Cell Biol 15:27–33. https://doi.org/10.1016/j.tcb.2004.11.003

Rechsteiner M, Realini C, Ustrell V (2000) The proteasome activator 11 S REG (PA28) and class I antigen presentation. Biochem J 345(Pt 1):1–15

Ruiz de Mena I, Mahillo E, Arribas J, Castano JG (1993) Kinetic mechanism of activation by cardiolipin (diphosphatidylglycerol) of the rat liver multicatalytic proteinase. Biochem J 296(Pt 1): 93–97. https://doi.org/10.1042/bj2960093

Saibil H (2013) Chaperone machines for protein folding, unfolding and disaggregation. Nat Rev Mol Cell Biol 14:630–642. https://doi.org/10.1038/nrm3658

Santoro AM, Lanza V, Bellia F, Sbardella D, Tundo GR, Cannizzo A, Grasso G, Arizzi M, Nicoletti VG, Alcaro S et al (2020) Pyrazolones activate the proteasome by gating mechanisms and protect neuronal cells from beta-amyloid toxicity. ChemMedChem 15:302–316. https://doi.org/10.1002/cmdc.201900612

Santra M, Dill KA, de Graff AMR (2019) Proteostasis collapse is a driver of cell aging and death. Proc Natl Acad Sci U S A 116:22173–22178. https://doi.org/10.1073/pnas.1906592116

Sarkar S, Davies JE, Huang Z, Tunnacliffe A, Rubinsztein DC (2007) Trehalose, a novel mTOR-independent autophagy enhancer, accelerates the clearance of mutant huntingtin and alpha-synuclein. J Biol Chem 282:5641–5652. https://doi.org/10.1074/jbc.M609532200

Schmauck-Medina T, Moliere A, Lautrup S, Zhang J, Chlopicki S, Madsen HB, Cao S, Soendenbroe C, Mansell E, Vestergaard MB et al (2022) New hallmarks of ageing: a 2022 Copenhagen ageing meeting summary. Aging (Albany NY) 14:6829–6839. https://doi.org/10.18632/aging.204248

Schmidt M, Haas W, Crosas B, Santamaria PG, Gygi SP, Walz T, Finley D (2005) The HEAT repeat protein Blm10 regulates the yeast proteasome by capping the core particle. Nat Struct Mol Biol 12:294–303. https://doi.org/10.1038/nsmb914

Shibatani T, Ward WF (1995) Sodium dodecyl sulfate (SDS) activation of the 20S proteasome in rat liver. Arch Biochem Biophys 321:160–166. https://doi.org/10.1006/abbi.1995.1381

Shirozu R, Yashiroda H, Murata S (2015) Identification of minimum Rpn4-responsive elements in genes related to proteasome functions. FEBS Lett 589:933–940. https://doi.org/10.1016/j.febslet.2015.02.025

Showalter HD (2013) Progress in the synthesis of canthine alkaloids and ring-truncated congeners. J Nat Prod 76:455–467. https://doi.org/10.1021/np300753z

Silva GM, Netto LE, Discola KF, Piassa-Filho GM, Pimenta DC, Barcena JA, Demasi M (2008) Role of glutaredoxin 2 and cytosolic thioredoxins in cysteinyl-based redox modification of the 20S proteasome. FEBS J 275:2942–2955. https://doi.org/10.1111/j.1742-4658.2008.06441.x

Sonninen TM, Goldsteins G, Laham-Karam N, Koistinaho J, Lehtonen S (2020) Proteostasis disturbances and inflammation in neurodegenerative diseases. Cells 9. https://doi.org/10.3390/cells9102183

Soto C, Pritzkow S (2018) Protein misfolding, aggregation, and conformational strains in neurodegenerative diseases. Nat Neurosci 21:1332–1340. https://doi.org/10.1038/s41593-018-0235-9

Sundaria N, Upadhyay A, Prasad A, Prajapati VK, Poluri KM, Mishra A (2021) Neurodegeneration & imperfect ageing: technological limitations and challenges? Mech Ageing Dev 200:111574. https://doi.org/10.1016/j.mad.2021.111574

Swovick K, Firsanov D, Welle KA, Hryhorenko JR, Wise JP Sr, George C, Sformo TL, Seluanov A, Gorbunova V, Ghaemmaghami S (2021) Interspecies differences in proteome turnover kinetics are correlated with life spans and energetic demands. Mol Cell Proteomics 20:100041. https://doi.org/10.1074/mcp.RA120.002301

Tanaka K, Yoshimura T, Ichihara A (1989) Role of substrate in reversible activation of proteasomes (multi-protease complexes) by sodium dodecyl sulfate. J Biochem 106:495–500. https://doi.org/10.1093/oxfordjournals.jbchem.a122880

Taylor RC, Dillin A (2011) Aging as an event of proteostasis collapse. Cold Spring Harb Perspect Biol 3. https://doi.org/10.1101/cshperspect.a004440

Thibaudeau TA, Smith DM (2019) A practical review of proteasome pharmacology. Pharmacol Rev 71:170–197. https://doi.org/10.1124/pr.117.015370

Tomko RJ Jr, Hochstrasser M (2013) Molecular architecture and assembly of the eukaryotic proteasome. Annu Rev Biochem 82:415–445. https://doi.org/10.1146/annurev-biochem-060410-150257

Trader DJ, Simanski S, Dickson P, Kodadek T (2017) Establishment of a suite of assays that support the discovery of proteasome stimulators. Biochim Biophys Acta Gen Subj 1861:892–899. https://doi.org/10.1016/j.bbagen.2017.01.003

Trippier PC, Zhao KT, Fox SG, Schiefer IT, Benmohamed R, Moran J, Kirsch DR, Morimoto RI, Silverman RB (2014) Proteasome activation is a mechanism for pyrazolone small molecules displaying therapeutic potential in amyotrophic lateral sclerosis. ACS Chem Nerosci 5:823–829. https://doi.org/10.1021/cn500147v

Tsimokha AS, Artamonova TO, Diakonov EE, Khodorkovskii MA, Tomilin AN (2020) Post-translational modifications of extracellular proteasome. Molecules 25. https://doi.org/10.3390/molecules25153504

Upadhyay A (2021) Natural compounds in the regulation of proteostatic pathways: an invincible artillery against stress, ageing, and diseases. Acta Pharm Sin B 11:2995–3014. https://doi.org/10.1016/j.apsb.2021.01.006

Upadhyay A, Amanullah A, Mishra R, Kumar A, Mishra A (2018) Lanosterol suppresses the aggregation and cytotoxicity of misfolded proteins linked with neurodegenerative diseases. Mol Neurobiol 55:1169–1182. https://doi.org/10.1007/s12035-016-0377-2

Upadhyay A, Joshi V, Amanullah A, Mishra R, Arora N, Prasad A, Mishra A (2017) E3 ubiquitin ligases neurobiological mechanisms: development to degeneration. Front Mol Neurosci 10:151

Vermeij WP, Hoeijmakers JH, Pothof J (2016) Genome integrity in aging: human syndromes, mouse models, and therapeutic options. Annu Rev Pharmacol Toxicol 56:427–445. https://doi.org/10.1146/annurev-pharmtox-010814-124316

Vigouroux S, Farout L, Clavel S, Briand Y, Briand M (2003) Increased muscle proteasome activities in rats fed a polyunsaturated fatty acid supplemented diet. Int J Biochem Cell Biol 35:749–755. https://doi.org/10.1016/s1357-2725(02)00389-8

Vilchez D, Boyer L, Morantte I, Lutz M, Merkwirth C, Joyce D, Spencer B, Page L, Masliah E, Berggren WT et al (2012a) Increased proteasome activity in human embryonic stem cells is regulated by PSMD11. Nature 489:304–308. https://doi.org/10.1038/nature11468

Vilchez D, Morantte I, Liu Z, Douglas PM, Merkwirth C, Rodrigues AP, Manning G, Dillin A (2012b) RPN-6 determines C. elegans longevity under proteotoxic stress conditions. Nature 489:263–268. https://doi.org/10.1038/nature11315

Wang N, Wang E, Wang R, Muhammad F, Li T, Yue J, Zhou Y, Zhi D, Li H (2022) Ursolic acid ameliorates amyloid beta-induced pathological symptoms in Caenorhabditis elegans by activating the proteasome. Neurotoxicology 88:231–240. https://doi.org/10.1016/j.neuro.2021.12.004

Watanabe N, Yamada S (1996) Activation of 20S proteasomes from spinach leaves by fatty acids. Plant Cell Physiol 37:147–151. https://doi.org/10.1093/oxfordjournals.pcp.a028925

Xie Y, Varshavsky A (2001) RPN4 is a ligand, substrate, and transcriptional regulator of the 26S proteasome: a negative feedback circuit. Proc Natl Acad Sci U S A 98:3056–3061. https://doi.org/10.1073/pnas.071022298

Yang EJ, Kim SH, Yang SC, Lee SM, Choi SM (2011) Melittin restores proteasome function in an animal model of ALS. J Neuroinflammation 8:69. https://doi.org/10.1186/1742-2094-8-69

Yang JH, Hayano M, Griffin PT, Amorim JA, Bonkowski MS, Apostolides JK, Salfati EL, Blanchette M, Munding EM, Bhakta M et al (2023) Loss of epigenetic information as a cause of mammalian aging. Cell 186:305–326 e327. https://doi.org/10.1016/j.cell.2022.12.027

Yuan NN, Cai CZ, Wu MY, Zhu Q, Su H, Li M, Ren J, Tan JQ, Lu JH (2019) Canthin-6-one accelerates alpha-Synuclein degradation by enhancing UPS activity: drug target identification by CRISPR-Cas9 whole genome-wide screening technology. Front Pharmacol 10:16. https://doi.org/10.3389/fphar.2019.00016

Yukawa M, Sakon M, Kambayashi J, Shiba E, Kawasaki T, Ariyoshi H, Mori T (1991) Proteasome and its novel endogeneous activator in human platelets. Biochem Biophys Res Commun 178:256–262. https://doi.org/10.1016/0006-291x(91)91807-o

Zeng BY, Medhurst AD, Jackson M, Rose S, Jenner P (2005) Proteasomal activity in brain differs between species and brain regions and changes with age. Mech Ageing Dev 126:760–766. https://doi.org/10.1016/j.mad.2005.01.008

Zhang F, Hu Y, Huang P, Toleman CA, Paterson AJ, Kudlow JE (2007) Proteasome function is regulated by cyclic AMP-dependent protein kinase through phosphorylation of Rpt6. J Biol Chem 282:22460–22471. https://doi.org/10.1074/jbc.M702439200

Zhang F, Su K, Yang X, Bowe DB, Paterson AJ, Kudlow JE (2003) O-GlcNAc modification is an endogenous inhibitor of the proteasome. Cell 115:715–725. https://doi.org/10.1016/s0092-8674(03)00974-7

Zhang X, Schulz R, Edmunds S, Kruger E, Markert E, Gaedcke J, Cormet-Boyaka E, Ghadimi M, Beissbarth T, Levine AJ et al (2015) MicroRNA-101 suppresses tumor cell proliferation by acting as an endogenous proteasome inhibitor via targeting the proteasome assembly factor POMP. Mol Cell 59:243–257. https://doi.org/10.1016/j.molcel.2015.05.036

Zhou H, Li S, Li C, Yang X, Li H, Zhong H, Lu JH, Lee SM (2020) Oxyphylla a promotes degradation of alpha-Synuclein for neuroprotection via activation of immunoproteasome. Aging Dis 11:559–574. 10.14336/AD.2019.0612

Chapter 3
Chaperone Activators

Siarhei A. Dabravolski

Abstract Ageing is a complex yet universal and inevitable degenerative process that results in a decline in the cellular capacity for repair and adaptation to external stresses. Therefore, maintaining the appropriate balance of the cellular proteome is crucial. In addition to the ubiquitin-proteasome and autophagy-lysosomal systems, molecular chaperones play a vital role in a sophisticated protein quality control system. Chaperones are responsible for the correct protein assembly, folding, and translocation of other proteins when cells are subjected to various stresses. The equilibrium of chaperones is pivotal for maintaining health and longevity, as a deficiency in their function and quantity can contribute to the development of various diseases and accelerate the ageing processes. Conversely, their overexpression has been associated with tumour growth and progression. In this work, we discuss recent research focused on the application of various natural and artificial substances, as well as physical and nutritional stresses, to activate molecular chaperones and prolong both life- and healthspan. Furthermore, we emphasise the significance of autophagy, apoptosis, mTOR and inflammation signalling pathways in chaperone-mediated extension of life- and healthspan.

Keywords Ageing · Chaperone · Longevity · Apoptosis · Proteostasis · Ubiquitin-proteasome system · Autophagy-lysosomal system

Introduction

The normal functioning of the Eukaryotic organism is based on the effective response to various external and internal factors mediated through thousands of different proteins. The ageing process is accompanied by a noticeable, gradual and progressive decline of the organism functionality on the cellular, sub-cellular and

S. A. Dabravolski (✉)
Department of Biotechnology Engineering, ORT Braude College, Karmiel, Israel

© The Author(s), under exclusive license to Springer Nature Switzerland AG 2024

V. I. Korolchuk, J. R. Harris (eds.), *Biochemistry and Cell Biology of Ageing: Part V, Anti-Ageing Interventions*, Subcellular Biochemistry 107,
https://doi.org/10.1007/978-3-031-66768-8_3

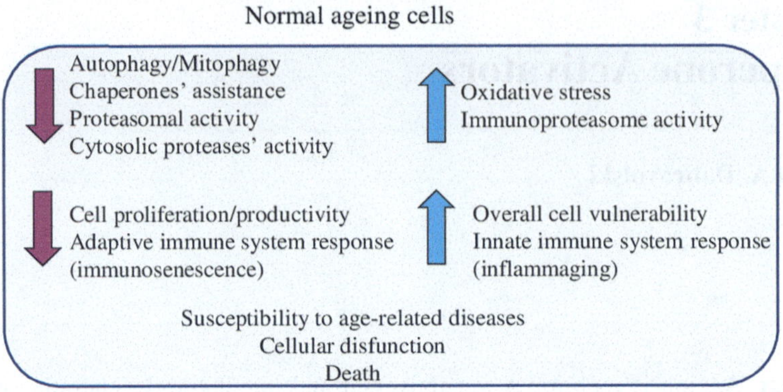

Fig. 3.1 Schematic representation of the ageing-associated changes in proteostasis of normal cells. Ageing is associated with a gradual decline in functionality of cellular structures and components, which subsequently resulted in adverse consequences leading to the aging-related diseases. The magenta arrows indicate a decline in the listed process, while blue arrows suggest an increase

molecular levels, which makes it more susceptible to diseases and stresses, and, eventually, resulting in death (Ruano 2021). This decline affects also the cellular proteome, particularly disrupting every stage of the protein biosynthesis and maturation processes, resulting in the gradual accumulation of the damaged, misfolded, toxic and/or unnecessary proteins that impair cell functionality in general (Chen et al. 2023). The dynamic system of production of functional proteins, including protein synthesis, folding, maintenance and various post-translational modifications, as well as the degradation of malfunctioning (misfolded, toxic and/or aggregated) proteins, is known as the proteostasis network. However, similar to other cellular proteins, the cellular quality control proteins are also subject to ageing and a gradual decline in functionality, leading to the dysregulation of the proteostasis network (Witkowski et al. 2021).

Loss of proteostasis, along with genomic instability, epigenetic alterations, deregulated nutrient-sensing, cellular senescence, chronic inflammation, telomere attrition, disabled macro-autophagy, stem cell exhaustion, altered intercellular communication, mitochondrial dysfunction, and dysbiosis, is considered one of the 12 major hallmarks of ageing (López-Otín et al. 2023). Impaired proteostasis ultimately disrupts the biosynthesis and maintenance of functionally active proteins, leading to cellular dysfunction and eventual demise (Fig. 3.1). The dysregulation of the proteostasis network is particularly prominent in specific cell types characterised by periods of quiescence (immune cells like macrophages and T lymphocytes) and arrested proliferation (such as cardiac muscle and neurons) (Witkowski et al. 2021; Mittelbrunn and Kroemer 2021).

Protein homeostasis is achieved through sophisticated protein quality control (PQC) mechanisms, encompassing three major components: molecular chaperones, ubiquitin-proteasome and autophagy-lysosomal systems (Fig. 3.2). There mechanisms regulate protein folding and eliminate misfolded, malfunctional, toxic and aggregated proteins, thereby facilitating cellular adaptation to dynamic stress

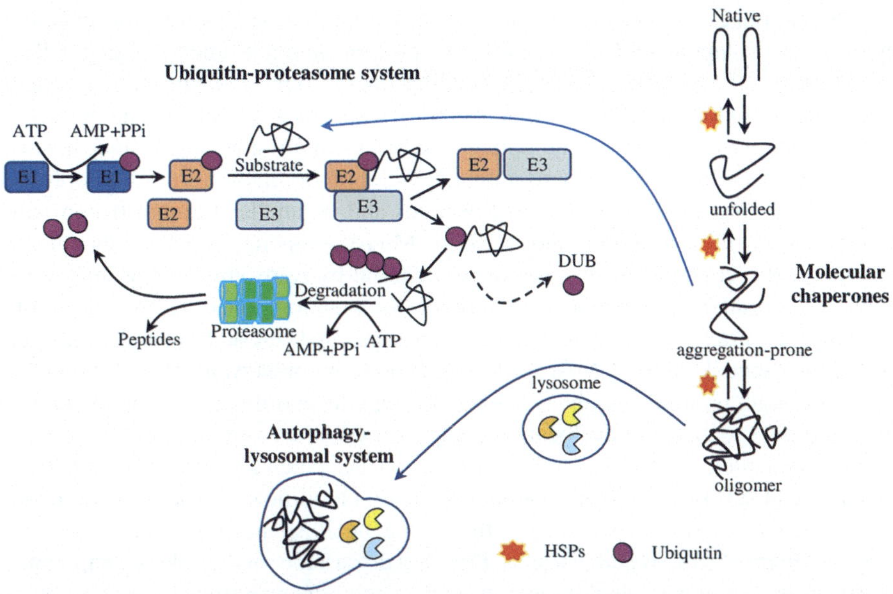

Fig. 3.2 Overview of the protein quality control mechanisms. Protein homeostasis network includes proteolytic (ubiquitin-proteasome and autophagy-lysosomal) and repair (molecular chaperones) systems. Chaperones promote the correct folding of damaged and misfolded proteins to acquire its functional conformation. Alternatively, aggregation-prone proteins can be degraded by ubiquitin-proteasome system, when target proteins are marked with ubiquitin and recognised by the proteasome. The deubiquitylating (DUB) enzymes can remove ubiquitin from one protein and use it on another damaged protein. Defective proteins and malfunctional organelles can be recycled by the autophagy-lysosomal system via fusion with lysosomes, which are filled with over 60 different hydrolytic enzymes to degrade proteins, lipids, carbohydrates, and nucleic acids

conditions (Hoppe and Cohen 2020). However, cellular proteolytic mechanisms (ubiquitin-proteasome and autophagy-lysosome systems) also experience functional decline during ageing and are frequently associated with neurodegenerative diseases such as Alzheimer's and Parkinson's diseases (Sjödin et al. 2019; Watanabe et al. 2020).

The ubiquitin-proteasome system (UPS) serves as the primary intracellular pathway for protein degradation and turnover, accounting for approximately 80% of cellular proteins involved in essential cellular processes (such as gene transcription and translation, cell cycle progression, antigen presentation and apoptosis). Withing the UPS, the ubiquitin-conjugating enzyme (E2) is transporting ubiquitin (consisting of 76 highly conserved amino acids) from the ubiquitin-activating enzyme (E1) to the ubiquitin-ligase enzyme (E3) and substrate, thereby marking it for degradation. Ultimately, substrates tagged with polyubiquitin chains are recognised and degraded by the proteasome (Liu et al. 2020). However, the process of ubiquitylation is dynamic and reversible, as deubiquitylating enzymes could remove ubiquitin molecules from one protein and re-used to mark another target protein (Hwang et al. 2022).

The autophagy-lysosomal system is another important mechanism for recycling cytoplasmic components (such as defective proteins and malfunctional organelles) to maintain cellular homeostasis. Depending on the type of damage, source and abundance of cytoplasmic material delivered to lysosomes, autophagy can be classified as chaperone-mediated autophagy (CMA), micro-autophagy and macro-autophagy (Li et al. 2018). Macro-autophagy, the predominant form of autophagy, involved the degradation of damaged proteins and organelles through fusion with lysosomes (Griffey and Yamamoto 2022). Micro-autophagy, on the other hand, entails the direct engulfment of the target material by lysosomes and primary contributes to organelle maintenance, membrane composition, transition to logarithmic growth phase after starvation-induced growth arrest, and cell survival under nitrogen starvation (Schuck 2020). CMA, a selective form of autophagy, involves the translocation of target proteins into the lysosome through the coordinated action of chaperones and the specialised translocation complexes (Bourdenx et al. 2021).

In this chapter, our focus is on various approaches (such as application of natural and artificial substances, physical stresses, and diet modifications) used to modulate the levels of chaperones, extending both life- and healthspan, while enhancing resistance to diverse stresses and diseases. Detailed discussion on the role of chaperones in cancer and ageing-related disease, as well as neurodegenerative diseases (such as Alzheimer's and Parkinson's) can be found in recent comprehensive reviews (Chaplot et al. 2020; Tittelmeier et al. 2020; Wang et al. 2021a; Frankowska et al. 2022; Chen et al. 2023). Interested readers are encouraged to refer to these publications for further information.

Molecular Chaperones

Molecular chaperones are defined as proteins capable of interacting with, stabilise, or help the other protein to acquire its functional conformation. Additionally, chaperones are also involved in oligomeric assembling, refolding of stress-modified/damaged proteins and protein trafficking. Molecular chaperones are highly conserved and ubiquitous proteins, encompassing structurally diverse families that play a crucial role in protein folding and preventing protein aggregation, thereby maintaining cellular protein homeostasis (Kawagoe et al. 2022).

The best-known chaperone family members are heat-shock proteins (HSPs), which are regulated by the heat shock transcription factors (HSFs) in response to various stresses or at the specific developmental stage and account for approximately 5–10% of total proteins in most cells. Further chaperone families are divided into several classes according to their molecular weight and functional properties (HSP40 or J-proteins, HSP60 or chaperonins, HSP70, HSP90, HSP100 or Clp proteins, and small HSPs (sHSP)) (Hartl et al. 2011). These chaperone classes rely on repeated cycles of ATP binding and hydrolysis to assist in protein folding. However, exists also a class of small (with a molecular mass between 8 and 28 kDa), ATP-independent HSPs known as holdases that prevent protein aggregation (Mitra et al. 2022). In general, chaperones are known for their low substrate specificity and weak binding to their clients (Arhar et al. 2021).

The chaperones of the sHSP family prevent proteins aggregation and participate in diverse cellular processes (suppress ROS production and demonstrate anti-inflammatory and immunomodulatory properties) (Reinle et al. 2022). Chaperonins (or HSP60s) have been extensively studied in eukaryotic cells, where they interact with some newly synthesised proteins in the cytosol (such as tubulins and actin), play an important role in neuroinflammation, cancer and viral infection progression (Wyżewski et al. 2018; Tang et al. 2022). Members of the HSP70 and HSP90 family play the most prominent role in heat-shock stress response, various signalling pathways, and also have been shown to take part in various cellular (such as apoptosis, ferroptosis vesicle-mediated transport, or targeted protein degradation) and pathological (such as cancer development, neurodegenerative and neuroinflammatory diseases and viral infections) processes (Lackie et al. 2017; Genest et al. 2019; Morán Luengo et al. 2019).

Especially, chaperone system plays the crucial role in the maturation of proteins of the major histocompatibility class I and class II (MHC I and MHC II). In the case of MHC I, chaperones are involved on the stages of their synthesis, maturation, folding, intracellular trafficking, antigenic cargo optimisation, release of the ready molecule to the cell surface, and binding with their cognate receptors on cytotoxic CD8+ T cells (Thomas and Tampé 2019). For MHC II proteins, chaperones are involved in the maturation and transport of MHC II molecules in the antigen-presenting cells (Dijkstra and Yamaguchi 2019). Another important role was characterised in detail for organelle-specific chaperones, which control the folding of the membrane and lumen localised proteins. For example, ER chaperones are involved in the unfolded protein response (UPR), which is activated to restore the protein folding balance in ER lumen (Read and Schröder 2021).

Chaperone Activators for Longevity and Health Ageing

During recent years, researchers have found numerous ways to activate chaperones (beyond the heat-shock itself) in different model species and achieve beneficial effects (increase life- and healthspan, resistance to stresses and toxins) (summarised in Table 3.1). Further in this section we review recent results of the application of various ways (administration of natural and synthetic compounds, diet modifications and physical treatments) to activate chaperones in several model organisms (*Caenorhabditis elegans*, *Drosophila melanogaster*, yeast, rats, and mice) and resulted in life- and healthspan extension, increased resistance to physical stresses and diseases.

Table 3.1 Chaperone activators with life- and healthspan expanding effects

Species	Chaperone activator	Effect on HSPs	Positive effects	Negative effects	Reference
C. elegans	Xyloketal B	↑HSP70	Alleviated age-related decline in physical performance; enhanced heat-stress resistance; reduced $A\beta_{3\text{-}42}$-associated paralysis	–	(Zhou et al. 2018)
	Chlorogenic acid, 1,5-dicaffeoylquinic acid and 1,3-dicaffeoylquinic acid extracted from *Lonicera japonica*	↑HSP16.2	Enhanced the mean lifespan by 21.87%, improved age-associated physiological functions; up-regulated SOD3 and MTL1, and decreased ROS levels and fat accumulation; enhanced oxidative and heat stress resistance; reduced $A\beta_{3\text{-}42}$-associated paralysis	–	(Yang et al. 2018)
	Extract of *Cuscuta chinensis*	↑HSP12.6 and HSP16.1	Enhanced physical performance, heat stress resistance and extended the mean lifespan by 24%	–	(Sayed et al. 2021)
	Metolazone	↑^{mt}HSP6; ↑HSP60 and HSP70	Extended lifespan by activating ^{mt}UPR	–	(Ito et al. 2021)
	Tanespimycin and monorden (HSP90 inhibitors)	↑HSF1	Extended lifespan	–	(Fuentealba et al. 2019) (Janssens et al. 2019)

Drosophila melanogaster	*Phyllanthus emblica*	↑HSP27	Enhanced SOD activity and reduced ROS accumulation and lipid peroxidation in aged flies	–	(Dwivedi and Lakhotia 2016)
	Aronia melanocarpa	↑HSP68 and HSP20	Extended the mean lifespan by 18% and improved locomotor activity in aged flies; decreased MDA accumulation, increased expression of *SOD*, *CAT*, *GPX* and *Jafrac1*	–	(Jo and Imm 2017)
	Curcumin	↑HSP70 and HSP83; Under heat stress – ↓HSP70 and HSP83	Extended mean lifespan under various heat stresses by 8.7–16.4%; reduced MDA levels and increased expression of *SOD*, *CAT* and *PHGPx*	–	(Chen et al. 2018a)
	Withaferin A	↓HSP68 and HSP83	Extended the median lifespan by 7.7–11.1%, enhanced oxidative and heat stress resistance; affected expression of *PrxV*, *Gadd45* and *Ku80* genes in gender-specific way	–	(Koval et al. 2021)
	Phosphatidyl-ethanolamine-binding proteins	↑HSP26	Extended the median and median lifespan, reduced age-related locomotor decline;	–	(Känel et al. 2022)
	Prenatal hyperbaric normoxia	↑HSP70	Extended the lifespan and enhanced oxidative and heat stress resistance; increased expression of *CAT*, *GS*, *Jafrac1* and *SOD*	–	(Yu et al. 2016)
	Alpha-ketoglutarate	↑HSP22 and HSP70	Extended lifespan by 8.54%, enhanced physical performance and heat stress resistance; up-regulated AMPKα and down-regulated mTOR signalling pathways	Reduced fecundity	(Su et al. 2019)
	Isoflavone puerarin	↓HSP70	Extended lifespan, improved physical performance and resistance to oxidative and starvation stresses; up-regulated PSMB5 and Sirtuin-1, CAT, SOD1, Shaker, Methuselah, AMPK and autophagy-related genes, while down-regulated PARP-1 and reduced TOR phosphorylation.	Reduced fecundity	(Kang et al. 2023)
	Tryptamide	↑HSF-1, HSP68 and HSP70s	Extended lifespan and reduced age-related locomotor decline, up-regulated Sirtuin-1	–	(Kanno et al. 2022)
	High-fat diet	↑HSP83	Increased CAT and SOD activities, ACSL1 and ACeCS1 levels; decreased glucose level and increased *MPK2* expression	Increased body weight, reduced mobility, survival and lifespan; increased ROS production and level of triglycerides;	(Trindade de Paula et al. 2016)

(continued)

Table 3.1 (continued)

Species	Chaperone activator	Effect on HSPs	Positive effects	Negative effects	Reference
S. cerevisiae	Calorie restriction	Ids2, a co-chaperone of HSP82	Extended lifespan; Ids2 is activated by PP2C	Ids2 is inactivated by PKA upon glucose intake, thus preventing interaction with HSP90 and leading to lifespan shortening	(Chen et al. 2018b)
Rats	Resveratrol	↑HSP70	Alleviated heat-stress-mediated liver injury; decreased levels of MDA and TNFα, increased levels of GSH and GPx in liver; decreased serum levels of ALP and AST, increased levels of caspase 3; down-regulated NF-κB in aged rats	–	(Khafaga et al. 2019)
Ercc1$^{-/\Delta}$ mice	Geldanamycin, tanespimycin and alvespimycin (HSP90 inhibitors)	↑HSF1	Down-regulated the anti-apoptotic PI3K/AKT pathway; reduced age-related symptoms, improved overall body condition, thus indicating healthspan extension		(Fuhrmann-Stroissnigg et al. 2017)

Chaperone Activation in Caenorhabditis elegans

The nematode *C. elegans* is one of the main model organisms for fundamental research in ageing and genetics. Because of the remarkable similarities in terms of generic background, many genes associated with human diseases and ageing process are phylogenetically close to those in *C. elegans*, thus, making the nematode a suitable model to study development of human disease and to acquire novel longevity-associated genes for further follow-up (Sutphin et al. 2017). There are some advantages for *C elegans* as a model system: small body size, low cost, rapid generation time, high reproductive rate, transparency, ease of cultivation, long-term cryopreservation, invariant cell number, and development. Also, application of gene knock-down RNAi technology on *C. elegans* is specific, rapid, effective and easy to perform with the nematode model via feeding, microinjection or soaking the nematodes in dsRNA (Conte et al. 2015). CRISPR/Cas9 system, an advance method for genome editing and production of targeted mutations, was also successfully used on *C. elegans* (Dickinson and Goldstein 2016).

Because lifespan is a genetically regulated trait, several long- and short-lived *C. elegans* mutants have been isolated. For example, *daf-2* and *age-1* mutants exhibit significant lifespan extension. DAF-2 is a homologue of the insulin/insulin-like growth factor 1 receptor, which acts through a conserved Phosphoinositide 3-Kinase Alpha (PI3K)/Akt pathway and inhibits the activity of an FOXO family transcription factor DAF-16. Similarly, *age-1* encodes a homologue of mammalian PI3K, thus, also depends on DAF-16 (Bao et al. 2020). Multiple recent studies have found that nematode longevity can be increased through feeding with certain substances (natural and chemically synthesised), which could influence expression of life-prolonging genes and lead to an extended lifespan (Dilberger et al. 2021; McIntyre et al. 2022; Schmitt and Eckert 2022). Further, we discuss recent papers where such external stimuli have affected chaperone and resulted in extended lifespan.

Xyloketal B, a marine ketone compound isolated from mangrove fungus *Xylaria sp.* and demonstrated unique antioxidant activity and protective effects, which suggested its high therapeutic potential (Gong et al. 2022). As it was shown, several xyloketal derivatives were beneficial for *C. elegans'* lifespan and healthspan. Compound 15, a benzo-1, 3-oxazine xyloketal derivative, protected worms in heat-stress test and attenuated ageing-related decline in pumping and bending assays. Interestingly, compound 15 increased expression of *HSP70*, while its beneficial effects were abolished in *hsf-1* mutant. *In silico* simulations suggested that xyloketal derivatives interacted with the DNA-binding domain of HSF-1, thus strengthening its interaction with target DNA. Furthermore, compound 15 effectively reduced paralysis of the *C. elegans* model of Alzheimer's disease (strain CL4176 expressing human $A\beta_{3\text{-}42}$) (Zhou et al. 2018).

Similarly, ethanol extract of traditional Chinese medical plant *Lonicera japonica* (Thunb.) enhanced the mean lifespan of *C.elegans* by over 21.87% and improved age-associated physiological functions. Treatment with *L.japonica* extract up-regulated expression of stress-inducible genes: *hsp-16.2*, antioxidant superoxide dismutase-3 (*sod-3*) and metal detoxification metallothionein-1 (*mtl-1*). Furthermore, lifespan extension and health benefits were abolished in *daf-2*, *daf-16*, *hsf-1* and *mev-1* mutants. Also, treatment with *L.japonica* extract decreased ROS levels and fat accumulation, enhancing worm's survival under oxidative and thermal stresses. Additionally, the paralysis in CL4176 strain was delayed by both native *L. japonica* extract and combination of its three main components—chlorogenic acid, 1,5-dicaffeoylquinic acid and 1,3-dicaffeoylquinic acid (Yang et al. 2018). Extract of another traditional Chinese medical plant, *Cuscuta chinensis* (Lam.), enhanced physical parameters (the swimming behaviour and the pharyngeal pumping rate), increased heat stress resistance and increased the mean lifespan by 24%. These beneficial effects were accompanied by increased expression of *hsp-16.1* and *hsp-12.6* genes (Sayed et al. 2021).

Metolazone, a diuretic drug used to reduce the swelling and fluid retention caused by heart failure or kidney disease, was demonstrated to prolong *C. elegans'* lifespan by activating mitochondrial unfolded protein response. Metolazone treatment specifically up-regulated mitochondrial chaperone *hsp-6*, while ER-specific chaperone *hsp-4* was not affected. These results were also confirmed in HeLa cells, where Metolazone up-regulated the expression of *mthsp70* and *hsp60*, but did not change the expression of ER chaperones. Interestingly, Metolazone did not extend lifespan in *atfs-1* and *ubl-5* mutants, which encode orthologous for human genes activating transcription factor 4 (ATF4) and ubiquitin like 5 (UBL5), two crucial UPR^{mt} activators. Furthermore, Metolazone required functional *nkcc-1* gene (orthologous of human solute carrier family 12, sodium:potassium:chloride symporter, which is involved in homeostatic process and inorganic ion transmembrane transport), however its connection to UPR^{mt} and longevity is currently unknown and require further investigation (Ito et al. 2021).

Inhibitors of chaperones (especially HSP90 and HSP70) have demonstrated a beneficial effect in the treatment of various types of malignancies (Albakova et al. 2022). Unexpectedly, some inhibitors showed the pro-longevity effect in anti-ageing screenings. The well-known HSP90 inhibitor tanespimycin, a derivative of the antibiotic geldanamycin, which is used to treat certain types of cancer, especially kidney tumours and called anti-tumour antibiotic, was shown to extend lifespan of *C. elegans*. Thus, treatment with tanespimycin activated HSF-1 in a dose-dependent way with no signs of toxicity. Interestingly, tanespimycin failed to extend lifespan in *hsp90* (RNAi mutant with shortened lifespan), suggesting that tanespimycin acted in an *hsp-90* dependent manner, while severe HSP90 depletion was harmful (Fuentealba et al. 2019). Similarly, these results were later confirmed in another research, where tanespimycin and monorden (a natural fungal macrocyclic antibiotic) extended *C. elegans* lifespan in HSP90-dependent way (Janssens et al. 2019).

Chaperone Activation in *Drosophila melanogaster*

Drosophila melanogaster is a classical model organism for research in genetics, physiology, microbial pathogenesis, evolution and other areas. Also, it is widely used to study molecular mechanisms of ageing (Piper and Partridge 2018; Clancy et al. 2023) and human diseases characterised by accelerated ageing, such as, for example, Werner syndrome (Cassidy et al. 2019). Further, we discuss recent papers where external stimuli have been demonstrated to affect chaperone levels and improved D. melanogaster lifespan.

A popular herbal formula, Amalaki Rasayana (AR), which is based on Phyllanthus emblica L. fruits and some other components, is traditionally used as anti-ageing supplementation (Swain et al. 2012). On the molecular level, AR feeding enhanced SOD activity and reduced ROS accumulation and lipid peroxidation in aged flies. Interestingly, the expression levels of *shp70* and *hsp83* genes were not affected by AR feeding either in control or in heat-shock conditions, while hsp27 levels were elevated under both feeding conditions (Dwivedi and Lakhotia 2016). The extract of another plant—aronia (AE) (Aronia melanocarpa Michx.), effectively extended the mean D. melanogaster lifespan by 18% and improved locomotor activity at advanced age. On the molecular level, such longevity extension was accompanied by decreased accumulation of MDA (by 49.3%), increased levels of antioxidant enzymes (SOD, CAT and glutathione peroxidase (GPX) and expression of stress resistance genes (*hsp68*, *lethal(2)-essential-for-life (l(2)efl* or *hsp20*), and *thioredoxin peroxidase 1* (*Jafrac1*) (Jo and Imm 2017).

A common spice curcumin (derived from plant Curcuma longa L.) is known for a wide range of beneficial properties (anti-inflammatory, antioxidative and many other, including also anti-ageing) (Zia et al. 2021). Flies supplemented with curcumin demonstrated an extended mean lifespan under various heat stresses (8.7–16.4%), reduced MDA levels and increased expression of antioxidants (SOD, CAT and Phospholipid-hydroperoxide glutathione peroxidase). Interestingly, under normal temperature conditions, curcumin supplementation increased hsp70 and hsp83 expression, while under heat-shock conditions curcumin decreased their expression. These results suggested that the effect of curcumin on the lifespan extension is linked to enhanced thermal tolerance (Chen et al. 2018a). Similarly, Withaferin A (WA), a steroidal lactone, derived from plant Iochroma arborescens (Schltdl.), increased flies' resistance to hyperthermia and oxidative stress, and prolong the median lifespan by 7.7–11.1%. Accordingly, WA treatment improved the intestinal barrier permeability in older flies and affected the expression of antioxidant defence (*Peroxiredoxin V*, *prxV*), recognition of DNA damage (*Growth arrest and DNA damage-inducible 45*, *gadd45*), heat shock proteins (*hsp68* and *hsp83*), and repair of double-strand breaks (*Ku80*) genes in gender-specific way. Therefore, in male flies WA treatment decreased expression levels of *hsp68* and *hsp83*, and increased expression levels of *gadd45* and *prxV* genes, while in female flies the expression levels of *gadd45*, *hsp68* and *hsp83* genes were decreased (Koval et al. 2021).

Phosphatidylethanolamine-binding proteins (PEBPs) are highly conserved family of signalling proteins involved in regulation of various cellular processes (such as apoptosis, differentiation and cell proliferation) and shown to improve cells vitality, growth and productivity even when expressed in heterologous systems (Kronenberg et al. 2021). Recently, it was also shown that PEBPs can affect ageing-related processes. Accordingly, expression of *Nicotiana tabacum* (L.) PEBP protein (NtFT4) in *D. melanogaster* reduced age-related locomotor decline and extend lifespan (both mean and median). Similar results were obtained with overexpression of the *D. melanogaster* orthologous *CG7054* gene, while its RNAi suppression reduced longevity. *NtFT4* expression modulated various genes related to longevity, stress response and protein ubiquitination and phosphorylation. Specifically, NtFT4 interacted with HSP26 and stabilised its expression in older flies, thus supporting proteome maintenance system (Känel et al. 2022).

Interestingly, mild physical stress also can be beneficial for organism survival, induce cytoprotective and defence mechanism (Oh et al. 2011). Thus, prenatal hyperbaric normoxia treatment (2 atm absolute pressure with 10% O_2) enhanced *D. melanogaster* lifespan, motor performance and resistance to oxidative and heat stresses, with no disruptive effect on developmental rate or adult body weight. These effects were accompanied by increased expression of *hsp70*, *cat*, *glutathione synthase* (*GS*), *Jafrac 1* and *sod*. These data suggested that mild physical stress may act as a hormetic factor to improve longevity and healthspan in Drosophila (Yu et al. 2016).

Alpha-ketoglutarate (AKG), one of the key metabolites of the tricarboxylic acid cycle, is crucial for energy metabolism, protein synthesis and other vital processes. Number of studies suggested that AKG is also an important regulator for lifespan and healthspan (Naeini et al. 2023). Recently, the molecular mechanisms of the dietary AKG-mediated anti-ageing effect were elucidated. Dietary supplementation of flies with 5μM AKG extended their lifespan (up to 8.54%), enhanced vertical climbing ability and resistance to heat stress. However, the flies' reproductive performance was reduced, and the resistance to oxidative stress and tolerance to starvation were not improved. Also, AKG administration up-regulated the expression of *hsp22* and *hsp70* genes and AMPKα signalling pathway (***Cryptochrome-1***, *FoxO, Hepatocyte nuclear factor 4, p300, Sirtuin 1*) and down-regulated mTOR pathway-related genes (*Histone deacetylase 4, PI3K, TORC, polar granule component (PGC)*, and *Sterol regulatory element binding protein* (*SREBP*). Furthermore, AKG supplementation reduced the ATP/ADP ratio and increased the autophagy rate. These data suggested that one of the common diet ingredients may regulate nutrient-sensing and energy pathways to regulate cell metabolism and lifespan in response to various stresses (Su et al. 2019).

Similarly, supplementation with plant-derived isoflavone puerarin significantly extended the lifespan of *D. melanogaster* by upregulating the expression of *sirtuin-1* and *proteasome subunit beta 5* (*PSMB5*). Additionally, *CAT*, *SOD1*, *Shaker* (potassium ion channel) and *Methuselah* (G-protein-coupled receptor) genes were up-regulated, while *HSP70* and *poly [ADP-ribose] polymerase* (*PARP-1*) were down-regulated. Furthermore, increased level of adenosine 5′-monophosphate (AMP)-activated protein kinase (AMPK) resulted in increased ATP content. Also, autophagy related genes (such as *autophagy-associated gene 1* (*ATG1*), *ATG5* and *ATG8b*) were up-regulated, while phosphorylation of TOR protein was decreased. Physical performance (climbing), resistance to starvation and oxidation were improved, while fecundity was reduced and accompanied by the decreased levels of proteins involved in gametogenesis (such as bag of marbles (Bam) and transforming growth factor beta type II receptor (Punt)) (Kang et al. 2023).

Cacao is another dietary product, which was shown to extend life in both *C. elegans* and *D. melanogaster* (Bahadorani and Hilliker 2008; Munasinghe et al. 2021). Recent research has identified that the active lifespan-prolonging component of cacao (*Theobroma cacao* (L.) is a fatty acid tryptamide. On the molecular level, tryptamide increased expression of *Sirtuin 1* and *HSF-1*, which subsequently, up-regulated different HSP proteins (such as HSP68 and many members of HSP70 family). Additionally, to lifespan extension, flies fed with tryptamide were less affected with age-related muscle weakness (accessed via hill-climbing test) (Kanno et al. 2022).

Interestingly, the consumption of a high-fat diet (HFD), a common unhealthy habit, was shown to negatively affect the normal metabolism and lifespan of flies. Seven-day supplementation resulted in increased body weight (5–14%), reduced survival (28–35%) and maximum lifespan. Also, flies fed on HFD demonstrated reduced mobility (negative geotaxis test), increased ROS production (47%) and levels of triglycerides. However, the activities of CAT and SOD were increased, while the level of glucose was decreased. In parallel, the levels of lipid metabolism-associated genes (*Acyl-CoA Synthetase Long Chain Family Member 1* (*ACSL1*) and *Acyl-CoA Synthetase Short Chain Family Member 1* (*ACeCS1*) were increased. Furthermore, HFD induced a higher expression of *HSP83* and *MAP kinase activated protein-kinase-2* (*MPK2*) mRNA. These data demonstrated that HFD administration may negatively affect metabolism, signalling pathways, and might be involved in fly's lifespan shortening (Trindade de Paula et al. 2016).

Chaperone Activation in Other Model Systems

Saccharomyces cerevisiae is an important model system for studying the biochemistry and molecular biology of mammals, including humans. A huge part of our knowledge on the various signalling pathways involved in the cell cycle, resistance to toxins, cellular growth and death has been obtained in this model system. One of the major advantages of the yeast as a model system is a set of well-developed technologies for high throughput screenings, which allows to identify and study genes and pathway associated with various cellular processes and diseases, including also ageing (Mirisola and Longo 2022). Mice and rats with expected lifespan up to 4 years are also considered as short-lived species. Despite the long history of laboratory experiments using mice as a prime mammalian model of human ageing, very little is known about ageing-related functional changes in mice. The recent rise of modern omics and gene editing technologies allow to better characterise ageing processes in mice and wider introduce it as a model organism for ageing research (Holtze et al. 2021; Yanai and Endo 2021).

Interesting pathway, linking nutrients sensing, proteostasis and lifespan control, was recently described in yeast (*Saccharomyces cerevisiae*). Among proteins, identified in quantitative mass spectrometry screen under calorie restriction (a well-known method to extend lifespan), Ids2, a protein with glycosyltransferase activity involved in chaperone-mediated protein folding, was activated by phosphatase 2C (PP2C) and inactivated by the glucose-sensing PKA (Regulatory subunit of the cyclic AMP-dependent protein kinase) upon glucose intake. Therefore, Ids2 acted as a co-chaperone to form a complex with Hsp82, and phosphorylation hamper its association with HSP90 and resulted in lifespan shortening. These results suggested that glucose sensing mechanism may orchestrate protein folding and facilitate lifespan extension (Chen et al. 2018b).

Recently, a popular antioxidant resveratrol (RES) was shown to effectively alleviate heat-stress-mediated liver injury in both young and old rats. Young and old rats, supplemented with resveratrol, had decreased levels of MDA and TNFα, while increased levels of reduced glutathione and glutathione peroxidase in liver. Also, RES treated rats had decreased serum levels of alkaline phosphatase (ALP) and aspartate transaminase (AST), and increased levels of caspase 3 and HSP70, when compared to the corresponding heat-stressed rats. Interestingly, Nuclear factor kappa-light-chain-enhancer of activated B cells (NF-κB), the key player regulating many inflammatory genes, was down-regulated only in aged rats. These data suggested that RES, a dietary polyphenols-based supplementation, exerted hepatoprotective effect in the young and old rats via its anti-inflammatory, antioxidant, and anti-apoptotic activities. Taking into account defined difference in activities on young and old rats, it could be a promising treatment specifically for elder group (Khafaga et al. 2019).

Screening of the library of autophagy-regulating compounds on rapidly senescing primary *Ercc1*$^{-/-}$ murine embryonic fibroblasts has identified well-known HSP90 inhibitors geldanamycin and 17-AAG (tanespimycin) as promising anti-ageing drugs. Further testing of the geldanamycin-derived HSP90 inhibitor 17-DMAG (alvespimycin) confirmed that it selectively kills senescent cells by down-regulating the anti-apoptotic PI3K/AKT pathway, thus leading to apoptosis of senescent cells. Application of 17-DMAG on *Ercc1*$^{-/\Delta}$ mice (a model system of a human progeroid (premature ageing) syndrome) improved overall body condition and reduced age-related symptoms (such as ataxia, tremor, gait disorder, loss of forelimb grip strength, kyphosis, coat condition and dystonia), thus indicating a health span extension (Fuhrmann-Stroissnigg et al. 2017).

Future Perspectives and Conclusion

In this review, we discuss the application of various natural and chemically engineered substances to protect the stability and functional properties of the proteome with an aim of improving life- and healthspan through modulation of molecular chaperone levels. Ageing is driven by different types of stress including oxidative, inflammatory, and genotoxic, therefore, life- and healthspan extension could be achieved through the reduction of stresses, remediation of the stress-induced damage, increased clearance of aged cells or activation of pro-survival pathways. As discussed in this chapter, chaperones can greatly facilitate these processes. Besides chaperone-modulating properties, various natural compounds mostly combine antioxidant and/or anti-inflammatory properties, which also increase stress resistance and adaptability. On the other side, complex artificial chemicals (such as HSP90 inhibitors) acted in a more sophisticated way, affecting apoptosis, autophagy and mTOR mechanisms. Also, it is important to note that most HSP90 inhibitors have no effect on lifespan extension, suggesting that only some mechanisms by which autophagy and mTOR regulated are important for the modulation of life- and healthspan.

Interestingly, in some aspects, senescent cells are very similar to cancer cells, which are known to utilise chaperone-stabilising properties on proteins of anti-apoptotic and pro-survival pathways, therefore avoiding clearance. Also, many HSPs have been shown to be overexpressed in many tapes of cancer and associated with a poor prognosis for patients (Wang et al. 2021b) (Fig. 3.3). For example, HSP90 client protein AKT, the key regulators of the anti-apoptotic PI3K/AKT pathway, is up-regulated in senescent sells. Also, in several types of cancer cells, AKT suppressed apoptosis induced by osmotic and oxidative stress, chemotherapeutics, and irradiation. However, application of HSP90 inhibitors disrupted normal AKT functioning and down-regulated this survival pathway, which resulted in apoptosis and cells clearance (Astle et al. 2012). Similar results were demonstrated for several plant-derived flavonoids, which targeted the survival pathway and could be

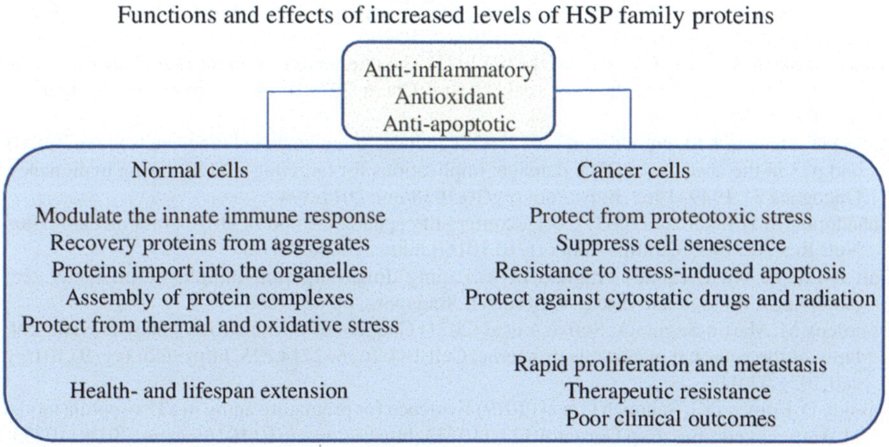

Fig. 3.3 The role of increased levels of HSP family proteins in normal and cancer cells. HSP family proteins provide anti-inflammatory, antioxidant and anti-apoptotic effects. In normal cells activation of HSPs showed beneficial effects, such as increased health- and lifespan. In cancer cells, HSP family proteins are overexpressed and cause several negative effects, which resulted in rapid proliferation and metastasis, resistance to anti-cancer therapeutics and poor clinical outcomes

considered as effective senolytics (Zhu et al. 2015). Also, HSPs can promote tumour angiogenesis through stabilisation of vascular growth factor (VEGF) and nitric oxide synthase (eNOS) (Tae et al. 2017), or stimulate epithelial-mesenchymal transition through down-regulation of E-cadherin, thus promoting metastasis and chemoresistance (Xue et al. 2022).

Recent analysis of 216 eukaryotic genomes suggested a strong correlation between the relative chaperone network size, proteome complexity and longevity in Metazoa (Draceni and Pechmann 2019). However, further analysis of the chaperone's expression in various organs of young and old mice suggested that the expression of individual chaperones does not change consistently during ageing. Surprisingly, ageing was associated with the abolished coordination among chaperones in a tissue-specific way (Soltanmohammadi et al. 2021). These results emphasise the importance of chaperones as a crucial target not only for ageing and ageing-associated degenerative disease but also for cancer initiation and progression. However, it also accentuates that markers beyond levels of individual HSP/target expression should be considered when exploring ageing and associated pathologies. While reviewed methods to activate chaperones and extend life- and healthspan have been linked to several pathways (such as inflammation, mTOR and autophagy), the exact molecular pathway connecting chaperones and longevity modulation is still unknown and requires further investigation.

References

Albakova Z, Mangasarova Y, Albakov A, Gorenkova L (2022) HSP70 and HSP90 in cancer: cytosolic, endoplasmic reticulum and mitochondrial chaperones of tumorigenesis. Front Oncol 12:829520. https://doi.org/10.3389/fonc.2022.829520

Arhar T, Shkedi A, Nadel CM, Gestwicki JE (2021) The interactions of molecular chaperones with client proteins: why are they so weak? J Biol Chem 297:101282. https://doi.org/10.1016/j.jbc.2021.101282

Astle MV, Hannan KM, Ng PY et al (2012) AKT induces senescence in human cells via mTORC1 and p53 in the absence of DNA damage: implications for targeting mTOR during malignancy. Oncogene 31:1949–1962. https://doi.org/10.1038/onc.2011.394

Bahadorani S, Hilliker AJ (2008) Cocoa confers life span extension in Drosophila melanogaster. Nutr Res 28:377–382. https://doi.org/10.1016/j.nutres.2008.03.018

Bao J, Liu B, Wu C (2020) Progress of anti-aging drugs targeting autophagy. In: Le W (ed) Autophagy: biology and diseases. Springer, Singapore, pp 681–688

Bourdenx M, Martín-Segura A, Scrivo A et al (2021) Chaperone-mediated autophagy prevents collapse of the neuronal metastable proteome. Cell 184:2696–2714.e25. https://doi.org/10.1016/j.cell.2021.03.048

Cassidy D, Epiney DG, Salameh C et al (2019) Evidence for premature aging in a Drosophila model of Werner syndrome. Exp Gerontol 127:110733. https://doi.org/10.1016/j.exger.2019.110733

Chaplot K, Jarvela TS, Lindberg I (2020) Secreted chaperones in neurodegeneration. Front Aging Neurosci 12:268. https://doi.org/10.3389/fnagi.2020.00268

Chen Y, Liu X, Jiang C et al (2018a) Curcumin supplementation increases survival and lifespan in *Drosophila* under heat stress conditions. Biofactors 44:577–587. https://doi.org/10.1002/biof.1454

Chen Y-C, Jiang P-H, Chen H-M et al (2018b) Glucose intake hampers PKA-regulated HSP90 chaperone activity. eLife 7:e39925. https://doi.org/10.7554/eLife.39925

Chen X-Q, Shen T, Fang S-J et al (2023) Protein homeostasis in aging and cancer. Front Cell Dev Biol 11:1143532. https://doi.org/10.3389/fcell.2023.1143532

Clancy D, Chtarbanova S, Broughton S (2023) Editorial: model organisms in aging research: Drosophila melanogaster. Front Aging 3:1118299. https://doi.org/10.3389/fragi.2022.1118299

Conte D, MacNeil LT, Walhout AJM, Mello CC (2015) RNA interference in *Caenorhabditis elegans*. CP Mol Biol 109. https://doi.org/10.1002/0471142727.mb2603s109

Dickinson DJ, Goldstein B (2016) CRISPR-based methods for *Caenorhabditis elegans* genome engineering. Genetics 202:885–901. https://doi.org/10.1534/genetics.115.182162

Dijkstra JM, Yamaguchi T (2019) Ancient features of the MHC class II presentation pathway, and a model for the possible origin of MHC molecules. Immunogenetics 71:233–249. https://doi.org/10.1007/s00251-018-1090-2

Dilberger B, Weppler S, Eckert GP (2021) Phenolic acid metabolites of polyphenols act as inductors for hormesis in C. elegans. Mech Ageing Dev 198:111518. https://doi.org/10.1016/j.mad.2021.111518

Draceni Y, Pechmann S (2019) Pervasive convergent evolution and extreme phenotypes define chaperone requirements of protein homeostasis. Proc Natl Acad Sci U S A 116:20009–20014. https://doi.org/10.1073/pnas.1904611116

Dwivedi V, Lakhotia SC (2016) Ayurvedic Amalaki Rasayana promotes improved stress tolerance and thus has anti-aging effects in Drosophila melanogaster. J Biosci 41:697–711. https://doi.org/10.1007/s12038-016-9641-x

Frankowska N, Lisowska K, Witkowski JM (2022) Proteolysis dysfunction in the process of aging and age-related diseases. Front Aging 3:927630. https://doi.org/10.3389/fragi.2022.927630

Fuentealba M, Dönertaş HM, Williams R et al (2019) Using the drug-protein interactome to identify anti-ageing compounds for humans. PLoS Comput Biol 15:e1006639. https://doi.org/10.1371/journal.pcbi.1006639

Fuhrmann-Stroissnigg H, Ling YY, Zhao J et al (2017) Identification of HSP90 inhibitors as a novel class of senolytics. Nat Commun 8:422. https://doi.org/10.1038/s41467-017-00314-z

Genest O, Wickner S, Doyle SM (2019) Hsp90 and Hsp70 chaperones: collaborators in protein remodeling. J Biol Chem 294:2109–2120. https://doi.org/10.1074/jbc.REV118.002806

Gong H, Bandura J, Wang G-L et al (2022) Xyloketal B: a marine compound with medicinal potential. Pharmacol Ther 230:107963. https://doi.org/10.1016/j.pharmthera.2021.107963

Griffey CJ, Yamamoto A (2022) Macroautophagy in CNS health and disease. Nat Rev Neurosci 23:411–427. https://doi.org/10.1038/s41583-022-00588-3

Hartl FU, Bracher A, Hayer-Hartl M (2011) Molecular chaperones in protein folding and proteostasis. Nature 475:324–332. https://doi.org/10.1038/nature10317

Holtze S, Gorshkova E, Braude S et al (2021) Alternative animal models of aging research. Front Mol Biosci 8:660959. https://doi.org/10.3389/fmolb.2021.660959

Hoppe T, Cohen E (2020) Organismal protein homeostasis mechanisms. Genetics 215:889–901. https://doi.org/10.1534/genetics.120.301283

Hwang J-T, Lee A, Kho C (2022) Ubiquitin and ubiquitin-like proteins in cancer, neurodegenerative disorders, and heart diseases. IJMS 23:5053. https://doi.org/10.3390/ijms23095053

Ito A, Zhao Q, Tanaka Y et al (2021) Metolazone upregulates mitochondrial chaperones and extends lifespan in Caenorhabditis elegans. Biogerontology 22:119–131. https://doi.org/10.1007/s10522-020-09907-6

Janssens GE, Lin X-X, Millan-Ariño L et al (2019) Transcriptomics-based screening identifies pharmacological inhibition of Hsp90 as a means to defer aging. Cell Rep 27:467–480.e6. https://doi.org/10.1016/j.celrep.2019.03.044

Jo AR, Imm J-Y (2017) Effects of aronia extract on lifespan and age-related oxidative stress in Drosophila melanogaster. Food Sci Biotechnol 26:1399–1406. https://doi.org/10.1007/s10068-017-0180-5

Känel P, Noll GA, Schroedter K et al (2022) The tobacco phosphatidylethanolamine-binding protein NtFT4 increases the lifespan of Drosophila melanogaster by interacting with the proteostasis network. Aging 14:2989–3029. https://doi.org/10.18632/aging.204005

Kang A-W, Sun C, Li H-T et al (2023) Puerarin extends the lifespan of *Drosophila melanogaster* by activating autophagy. Food Funct 14:2149–2161. https://doi.org/10.1039/D2FO02800J

Kanno K, Kayashima Y, Tamura K et al (2022) Fatty acid tryptamide from cacao elongates Drosophila melanogaster lifespan with sirtuin-dependent heat shock protein expression. Sci Rep 12:12080. https://doi.org/10.1038/s41598-022-16471-1

Kawagoe S, Ishimori K, Saio T (2022) Structural and kinetic views of molecular chaperones in multidomain protein folding. IJMS 23:2485. https://doi.org/10.3390/ijms23052485

Khafaga AF, Noreldin AE, Taha AE (2019) The adaptogenic anti-ageing potential of resveratrol against heat stress-mediated liver injury in aged rats: role of HSP70 and NF-kB signalling. J Therm Biol 83:8–21. https://doi.org/10.1016/j.jtherbio.2019.04.012

Koval L, Zemskaya N, Aliper A et al (2021) Evaluation of the geroprotective effects of withaferin A in Drosophila melanogaster. Aging 13:1817–1841. https://doi.org/10.18632/aging.202572

Kronenberg J, Schrödter K, Noll GA et al (2021) The tobacco phosphatidylethanolamine-binding protein NtFT4 simultaneously improves vitality, growth, and protein yield in human cells. Biotechnol Bioeng 118:3770–3786. https://doi.org/10.1002/bit.27853

Lackie RE, Maciejewski A, Ostapchenko VG et al (2017) The Hsp70/Hsp90 chaperone machinery in neurodegenerative diseases. Front Neurosci 11:254. https://doi.org/10.3389/fnins.2017.00254

Li J, Zhang D, Wiersma M, Brundel BJJM (2018) Role of autophagy in proteostasis: friend and foe in cardiac diseases. Cells 7:279. https://doi.org/10.3390/cells7120279

Liu W, Tang X, Qi X et al (2020) The ubiquitin conjugating enzyme: an important ubiquitin transfer platform in ubiquitin-proteasome system. IJMS 21:2894. https://doi.org/10.3390/ijms21082894

López-Otín C, Blasco MA, Partridge L et al (2023) Hallmarks of aging: an expanding universe. Cell 186:243–278. https://doi.org/10.1016/j.cell.2022.11.001

McIntyre RL, Liu YJ, Hu M et al (2022) Pharmaceutical and nutraceutical activation of FOXO3 for healthy longevity. Ageing Res Rev 78:101621. https://doi.org/10.1016/j.arr.2022.101621

Mirisola MG, Longo VD (2022) Yeast chronological lifespan: longevity regulatory genes and mechanisms. Cells 11:1714. https://doi.org/10.3390/cells11101714

Mitra R, Wu K, Lee C, Bardwell JCA (2022) ATP-independent chaperones. Annu Rev Biophys 51:409–429. https://doi.org/10.1146/annurev-biophys-090121-082906

Mittelbrunn M, Kroemer G (2021) Hallmarks of T cell aging. Nat Immunol 22:687–698. https://doi.org/10.1038/s41590-021-00927-z

Morán Luengo T, Mayer MP, Rüdiger SGD (2019) The Hsp70–Hsp90 chaperone cascade in protein folding. Trends Cell Biol 29:164–177. https://doi.org/10.1016/j.tcb.2018.10.004

Munasinghe M, Almotayri A, Thomas J et al (2021) Cocoa improves age-associated health and extends lifespan in C. elegans. NHA 6:73–86. https://doi.org/10.3233/NHA-200100

Naeini SH, Mavaddatiyan L, Kalkhoran ZR et al (2023) Alpha-ketoglutarate as a potent regulator for lifespan and healthspan: evidences and perspectives. Exp Gerontol 175:112154. https://doi.org/10.1016/j.exger.2023.112154

Oh S, Kwon D, Lee E (2011) Cytoprotective activity of elevated static pressure against oxidative stress in normal human fibroblasts. Mol Cell Toxicol 7:323–332. https://doi.org/10.1007/s13273-011-0038-5

Piper MDW, Partridge L (2018) Drosophila as a model for ageing. Biochim Biophys Acta (BBA) Mol Basis Dis 1864:2707–2717. https://doi.org/10.1016/j.bbadis.2017.09.016

Read A, Schröder M (2021) The unfolded protein response: an overview. Biology 10:384. https://doi.org/10.3390/biology10050384

Reinle K, Mogk A, Bukau B (2022) The diverse functions of small heat shock proteins in the proteostasis network. J Mol Biol 434:167157. https://doi.org/10.1016/j.jmb.2021.167157

Ruano D (2021) Proteostasis dysfunction in aged mammalian cells. The stressful role of inflammation. Front Mol Biosci 8:658742. https://doi.org/10.3389/fmolb.2021.658742

Sayed SMA, Siems K, Schmitz-Linneweber C et al (2021) Enhanced healthspan in caenorhabditis elegans treated with extracts from the traditional Chinese medicine plants Cuscuta chinensis Lam. and Eucommia ulmoides Oliv. Front Pharmacol 12:604435. https://doi.org/10.3389/fphar.2021.604435

Schmitt F, Eckert GP (2022) Caenorhabditis elegans as a model for the effects of phytochemicals on mitochondria and aging. Biomol Ther 12:1550. https://doi.org/10.3390/biom12111550

Schuck S (2020) Microautophagy—distinct molecular mechanisms handle cargoes of many sizes. J Cell Sci 133:jcs246322. https://doi.org/10.1242/jcs.246322

Sjödin S, Brinkmalm G, Öhrfelt A et al (2019) Endo-lysosomal proteins and ubiquitin CSF concentrations in Alzheimer's and Parkinson's disease. Alz Res Therapy 11:82. https://doi.org/10.1186/s13195-019-0533-9

Soltanmohammadi E, Farmaki E, Zhang Y et al (2021) Coordination in the unfolded protein response during aging in outbred deer mice. Exp Gerontol 144:111191. https://doi.org/10.1016/j.exger.2020.111191

Su Y, Wang T, Wu N et al (2019) Alpha-ketoglutarate extends Drosophila lifespan by inhibiting mTOR and activating AMPK. Aging 11:4183–4197. https://doi.org/10.18632/aging.102045

Sutphin GL, Backer G, Sheehan S et al (2017) *Caenorhabditis elegans* orthologs of human genes differentially expressed with age are enriched for determinants of longevity. Aging Cell 16:672–682. https://doi.org/10.1111/acel.12595

Swain U, Sindhu KK, Boda U et al (2012) Studies on the molecular correlates of genomic stability in rat brain cells following Amalakirasayana therapy. Mech Ageing Dev 133:112–117. https://doi.org/10.1016/j.mad.2011.10.006

Tae N, Hung TM, Kim O et al (2017) A cassaine diterpene alkaloid, 3β-acetyl-nor-erythrophlamide, suppresses VEGF-induced angiogenesis and tumor growth via inhibiting eNOS activation. Oncotarget 8:92346–92358. https://doi.org/10.18632/oncotarget.21307

Tang Y, Zhou Y, Fan S, Wen Q (2022) The multiple roles and therapeutic potential of HSP60 in cancer. Biochem Pharmacol 201:115096. https://doi.org/10.1016/j.bcp.2022.115096

Thomas C, Tampé R (2019) MHC I chaperone complexes shaping immunity. Curr Opin Immunol 58:9–15. https://doi.org/10.1016/j.coi.2019.01.001

Tittelmeier J, Nachman E, Nussbaum-Krammer C (2020) Molecular chaperones: a double-edged sword in neurodegenerative diseases. Front Aging Neurosci 12:581374. https://doi.org/10.3389/fnagi.2020.581374

Trindade de Paula M, Poetini Silva MR, Machado Araujo S et al (2016) High-fat diet induces oxidative stress and MPK2 and HSP83 gene expression in *Drosophila melanogaster*. Oxid Med Cell Longev 2016:1–12. https://doi.org/10.1155/2016/4018157

Wang L, Bergkvist L, Kumar R et al (2021a) Targeting chaperone/co-chaperone interactions with small molecules: a novel approach to tackle neurodegenerative diseases. Cells 10:2596. https://doi.org/10.3390/cells10102596

Wang X, Xie L, Zhu L (2021b) Clinicopathological significance of HSP70 expression in gastric cancer: a systematic review and meta-analysis. BMC Gastroenterol 21:437. https://doi.org/10.1186/s12876-021-01990-4

Watanabe Y, Taguchi K, Tanaka M (2020) Ubiquitin, autophagy and neurodegenerative diseases. Cells 9:2022. https://doi.org/10.3390/cells9092022

Witkowski JM, Bryl E, Fulop T (2021) Proteodynamics and aging of eukaryotic cells. Mech Ageing Dev 194:111430. https://doi.org/10.1016/j.mad.2021.111430

Wyżewski Z, Gregorczyk KP, Szczepanowska J, Szulc-Dąbrowska L (2018) Functional role of Hsp60 as a positive regulator of human viral infection progression. Acta Virol 62:33–40. https://doi.org/10.4149/av_2018_104

Xue N, Du T, Lai F et al (2022) Secreted HSP90α-LRP1 signaling promotes tumor metastasis and chemoresistance in pancreatic cancer. IJMS 23:5532. https://doi.org/10.3390/ijms23105532

Yanai S, Endo S (2021) Functional aging in male C57BL/6J mice across the life-span: a systematic behavioral analysis of motor, emotional, and memory function to define an aging phenotype. Front Aging Neurosci 13:697621. https://doi.org/10.3389/fnagi.2021.697621

Yang Z-Z, Yu Y-T, Lin H-R et al (2018) Lonicera japonica extends lifespan and healthspan in Caenorhabditis elegans. Free Radic Biol Med 129:310–322. https://doi.org/10.1016/j.freeradbiomed.2018.09.035

Yu S, Lee E, Tsogbadrakh B et al (2016) Prenatal hyperbaric normoxia treatment improves healthspan and regulates chitin metabolic genes in Drosophila melanogaster. Aging 8:2538–2550. https://doi.org/10.18632/aging.101084

Zhou J-B, Zheng Y-L, Zeng Y-X et al (2018) Marine derived xyloketal derivatives exhibit anti-stress and anti-ageing effects through HSF pathway in Caenorhabditis elegans. Eur J Med Chem 148:63–72. https://doi.org/10.1016/j.ejmech.2018.02.028

Zhu Y, Tchkonia T, Pirtskhalava T et al (2015) The Achilles' heel of senescent cells: from transcriptome to senolytic drugs. Aging Cell 14:644–658. https://doi.org/10.1111/acel.12344

Zia A, Farkhondeh T, Pourbagher-Shahri AM, Samarghandian S (2021) The role of curcumin in aging and senescence: molecular mechanisms. Biomed Pharmacother 134:111119. https://doi.org/10.1016/j.biopha.2020.111119

Chapter 4
NAD⁺ Boosting Strategies

Jared Rice, Sofie Lautrup, and Evandro F. Fang

Abstract Nicotinamide adenine dinucleotide (oxidized form, NAD^+) serves as a co-substrate and co-enzyme in cells to execute its key roles in cell signalling pathways and energetic metabolism, arbitrating cell survival and death. It was discovered in 1906 by Arthur Harden and William John Young in yeast extract which could accelerate alcohol fermentation. NAD acts as an electron acceptor and cofactor throughout the processes of glycolysis, Tricarboxylic Acid Cycle (TCA), β oxidation, and oxidative phosphorylation (OXPHOS). NAD has two forms: NAD^+ and NADH. NAD^+ is the oxidising coenzyme that is reduced when it picks up electrons. NAD^+ levels steadily decline with age, resulting in an increase in vulnerability to chronic illness and perturbed cellular metabolism. Boosting NAD^+ levels in various model organisms have resulted in improvements in healthspan and lifespan extension. These results have prompted a search for means by which NAD^+ levels in the body can be augmented by both internal and external means. The aim of this chapter is to provide an overview of NAD^+, appraise clinical evidence of its importance and success in potentially extending health- and lifespan, as well as to explore NAD^+ boosting strategies.

Keywords Ageing · Supplementation · Neurodegeneration · Caloric restriction · Oxidative stress · NAD^+ · Nicotinamide riboside · Nicotinamide mononucleotide

J. Rice · S. Lautrup (✉) · E. F. Fang (✉)
Department of Clinical Molecular Biology, University of Oslo and Akershus University Hospital, Lørenskog, Norway
e-mail: s.h.lautrup@medisin.uio.no; e.f.fang@medisin.uio.no

© The Author(s), under exclusive license to Springer Nature Switzerland AG 2024

V. I. Korolchuk, J. R. Harris (eds.), *Biochemistry and Cell Biology of Ageing: Part V, Anti-Ageing Interventions*, Subcellular Biochemistry 107,
https://doi.org/10.1007/978-3-031-66768-8_4

Introduction

In the past several years, there has been a growing fascination with the field of NAD^+ biology, which has revealed much in the way of its involvement with the functioning of living organisms. Additionally, the advantages of supplementing with NAD^+ precursors have become a popular intervention given the growing body of evidence suggesting that declines in NAD^+ levels play a role in age-associated vulnerabilities and disease (Fang et al. 2017; Imai and Guarente 2014; Lautrup et al. 2019). NAD^+ plays a crucial role in mitochondrial operation in multiple ways; as a co-enzyme for enzymes in the electron transport chain used in ATP production, a substrate for malate dehydrogenase (MDH), isocitrate dehydrogenase (IDH), and α-ketoglutarate dehydrogenase (α-KGDH), all essential parts of the Tricarboxylic Acid Cycle (TCA). Next, NAD^+ is used in redox reactions which create the proton gradient that results in ATP production. Moreover, NAD^+ is a co-substrate for mitochondria located proteins such as specific sirtuins, PARPs, and CD38. Interestingly, NAD^+ is consumed at increasing rates during the course of ageing by enzymes such as poly-ADP-ribose polymerases (PARPs), sirtuins, and cADP-ribose synthases (Bai and Canto 2012; Fang et al. 2014; Imai et al. 2000).

The extensive role of NAD^+ in this plethora of cellular processes indicates substantial promise for treating various pathophysiological conditions through NAD^+ precursor supplementation. Nevertheless, the ideal precursor and dosage remain not fully determined. Further difficulties with establishing baselines within populations come from the effects of human digestion and the microbiome on precursor metabolization (Gazzaniga et al. 2009). The recommended daily allowance for niacin is 16 mg daily for men and 14 mg daily for women, with 60 mg of tryptophan required to produce 1 mg of niacin (Mousa and Mousa 2023; Villines et al. 2012). However, a mounting body of evidence underscores the numerous advantageous and even therapeutic effects associated with significantly elevated NAD^+ synthesis rates via supraphysiological doses, which can be attained through supplementation with its intermediates versus diet.

NAD^+ Synthesis

NAD^+ is produced via three known major pathways in animals. The de novo synthesis of NAD^+ refers to its biosynthesis from simpler precursors within the cell. The first step in the *de novo* synthesis of NAD is the conversion of tryptophan to kynurenine by the enzyme tryptophan 2,3-dioxygenase after the amino acid has been transported into the cell via transporter SLC6A19. Kynurenine is converted into α-amino-β-carboxymuconate-ε-semialdehyde (ACMS), which is cyclized to form quinolinic acid (QA) that will be condensed by quinolinate phosphoribosyl transferase into nicotinic acid mononucleotide (NaMN). The next step is the conversion of NaMN to nicotinic acid adenine dinucleotide (NaAD) by nicotinamide mononucleotide adenylyl transferase (NMNAT). NaMN is converted to NaAD by one of the three isoforms of NMNAT. The particular isoform used depends on the location of NaMN conversion, with NMNAT1 located in the nucleus, NMNAT2 associated primarily with the cytoplasm and Golgi apparatus of central nervous system neurons (Cambronne and Kraus 2020) and NMNAT3 found within the mitochondria (Brazill et al. 2017). The final step in the *de novo* synthesis of NAD$^+$ is the condensation of NaAD using ATP to NAD$^+$ by NAD$^+$ synthase (NADSYN) (Katsyuba et al. 2018).

The Preiss-Handler pathway is a metabolic pathway that generates NAD$^+$ from nicotinic acid (NA). The pathway operates in the cytoplasm and is named after David Preiss and Joseph Handler who first described the pathway in 1958 (Preiss and Handler 1958). NA enters the cell via SLC5A8 and SLC22A3 transporters (Gopal et al. 2005). NA is converted to NaMN by nicotinic acid phosphoribosyl transferase (NAPRT), constituting the rate-limiting step in the Preiss-Handler pathway (Marletta et al. 2015). At this point the Preiss-Hanlder pathway intersects with the *de novo* pathway, using NMNATs to convert NaMN to NaAD and NADSYN for the final transition to NAD$^+$.

The salvage pathway synthesis of NAD$^+$ is the process by which cells recycle cellular nicotinamide mononucleotide (NMN) and intracellular nicotinamide (NAM) into NAD$^+$. The salvage pathway provides a way for cells to conserve and regenerate NAD$^+$, reducing the need for *de novo* synthesis from tryptophan. Production of cellular and tissue NAD$^+$ in healthy cells occurs most often via the salvage pathway (Chiarugi et al. 2012). Following import of nicotinamide riboside (NR) through transporters SLC29A1-4 or through members of the equilibrative nucleoside transporter (ENT) family (e.g. ENT1, ENT2 and ENT4) (Kropotov et al. 2021), NMN potentially through SLC12A8 (Grozio et al. 2019a, b), or through simple diffusion of NAM (Xie et al. 2020), NAM is converted to NMN by nicotinamide phosphoribosyltransferase (NAMPT), and NR is converted to NMN by nicotinamide riboside kinases (NRK). NMN is then converted to NAD$^+$ by the NMNATs (Rajman et al. 2018). Detailed description of these three classic NAD$^+$ synthetic pathways are available in recent review papers (Lautrup et al. 2019; Verdin 2015).

Recently, the "NRH salvage pathway" was discovered. Dihydronicotinamide riboside (NRH) is transported across cellular membranes, and is then intracellularly phosphorylated, forming reduced NMN, by a kinase, likely adenosine kinase (AK) (Yang et al. 2020). A metabolomic experiment wherein hepatocytes with AK inhibitor 5-IT treated with NRH showed diminished NAD$^+$ content compared to wild type cells, strengthening the involvement of AK (Giroud-Gerbetant et al. 2019). Reduced form of NMN is then converted into NADH via NMNATs and oxidised to NAD$^+$ (Yang et al. 2020).

Though no known mammalian transporter exists to directly transport NAD$^+$ through the phospholipid bilayer of the cell membrane, SLC25A51 has recently been identified as a potential transporter capable of moving NAD$^+$ through the mitochondrial membrane (Luongo et al. 2020). Moreover, experimental evidence has shown that loss of SLC25A51 decreases mitochondrial NAD$^+$ content and reduces mitochondrial respiration. Overexpression of SLC25A51 increases mitochondrial NAD$^+$ levels and restores NAD$^+$ uptake into yeast mitochondria that are lacking endogenous NAD$^+$ transporters (Luongo et al. 2020).

Functions of NAD^+

NAD^+ serves as a co-substrate for multiple protein families. PARPs respond to DNA damage signals and mark them by cleaving NAD^+ forming ADP-ribose and nicotinamide, and then linking ADP-ribose covalently to histones and other proteins to induce a DNA damage response (Imai and Guarente 2014). PARP has multiple isoforms that make use of NAD^+ as substrate (Navas and Carnero 2021). PARP1, a nuclear enzyme, is primarily involved in DNA repair and accounts for more than 90% of cellular PARP response in genotoxic situations (Shieh et al. 1998). PARP2, a cytoplasmic enzyme recognizes DNA strand breaks and facilitates recruitment of repair complexes, accounts for much of the remaining PARP response (Schreiber et al. 2002) with both PARP1 and PARP2 generating nicotinamide (NAM) as a byproduct.

Sirtuins (SIRTs) are the NAD^+ dependent deacetylases/deacylases that are conserved regulators of ageing and longevity and are essential to cellular metabolism (Dali-Youcef et al. 2007; Imai et al. 2000). The seven known mammalian SIRTs utilize NAD^+ to remove acyl or acetyl groups from lysines of their substrates including histones (Imai and Guarente 2014). Loss of sirtuin function has been associated with genomic instability and shorter lifespan (Bosch-Presegue and Vaquero 2014). SIRT1, SIRT6 and SIRT7 are localized to the nucleus where they regulate mechanisms of DNA repair; such as SIRT1 recruiting repair proteins like Ku70 and WRN to DNA double strand breaks (DSBs) (Das et al. 2019; Fang et al. 2016), SIRT6 associates with PARP1 to enhance DSB repair (Mao et al. 2011), and SIRT7-mediated deacetylation at DNA damage sites facilitates recruitment of 53BP1 to DSB's (Vazquez et al. 2017). Moreover, SIRT1 promotes the expression of genes involved in mitochondrial biogenesis by deacetylating and activating transcription factors such as PGC-1α and NRF1 which in turn stimulate the expression of genes involved in mitochondrial DNA replication, transcription, and translation, the assembly of the respiratory chain complexes (Fang et al. 2014; Gomes et al. 2013), and mitophagy (Tang 2016). In tissues consuming a large amount of ATP such as the brain, skeletal muscle, liver, kidney, and heart tissue, mitochondria are in higher abundance in order to maintain the proper balance between energy demand and supply (Fernandez-Vizarra et al. 2011; Kerr et al. 2017; Mattson et al. 2008). When cells become hypoxic, mitophagy is induced to decrease mitochondrial quantity, thereby adapting cellular metabolism to anaerobic conditions (Wu and Chen 2015). This catabolic process of mitophagy, which sequesters and degrades impaired mitochondria, ensures mitochondrial quality in multiple cell types (Dikic 2017; Pickles et al. 2018). NMN and NR supplementation in experimental models resulted in mitophagy induction, improving mitochondrial quality and function (Fang et al. 2019a, 2014). Moreover, NMN supplementation has been demonstrated to activate the mitochondrial unfolding protein response, increasing quality control of defective mitochondria, and ameliorating cellular dysfunction (Du et al. 2022). SIRT3, SIRT4 and SIRT5 are localised to the mitochondria where they also participate in cellular metabolism regulation (Schwer and Verdin 2008). For example, in the brain,

SIRT3 plays a pivotal role in adaptive responses of neurons to exercise, metabolic, and excitatory challenges (Cheng et al. 2016). The relationship between Sirtuins and cancer is also a subject of intensive studies as evidenced in a comprehensive review (Chalkiadaki and Guarente 2015). Of note, SIRT3 has been demonstrated to promote apoptosis by regulating mutant tumour suppressors in cancer cells (Li et al. 2017b; Tang et al. 2020) and function as a negative regulator of autophagy via AMPK inhibition, which leads to mTORC1 activation (Li et al. 2017b). Moreover, several studies indicate a positive correlation between increased SIRT1/SIRT7 expression and cancer (Barber et al. 2012; Chen et al. 2014; Herranz et al. 2013).

CD38/157 allows for the hydrolyzation of extracellular NAD^+ (De Flora et al. 2004). CD38 is a transmembrane bound glycoprotein that can function as both a receptor and an enzyme (Nooka et al. 2019). Of note, COD38 is not only present on cell surfaces but is also in various intracellular organelles, including the nucleus, highlighting its multi-faceted roles in cell signalling pathways and cellular functions (Adebanjo et al. 1999; Lee 2006). CD38 regulates cellular NAD^+ levels through its enzymatic activity. CD38 hydrolyses NAD^+ to generate cyclic ADP-ribose, which is a second messenger involved in the regulation of intracellular calcium levels (Adebanjo et al. 1999). This hydrolysis of NAD^+ results in a reduction of cellular NAD^+ levels (Bai and Canto 2012). In mouse studies, knock out of CD38 has been shown to alleviate mitochondrial damage and reverse diminishment of NAD^+ levels *in vivo* (Camacho-Pereira et al. 2016). Owing to its ecto-enzymatic activity, CD38 also regulates cellular NAD^+ levels by controlling the availability of NMN, consistent with emerging models which show part of the metabolism of nicotinamide nucleotides to occur in the extracellular space (Chini et al. 2020). Age-related upregulation of CD38 is suspected to be a major culprit in the decline of NAD^+ levels. Stimulation of macrophages by inflammatory factors by senescent cells in the liver and abdominal adipose tissue has been shown to drive expression of CD38 (Covarrubias et al. 2020). CD157 is a transmembrane bound paralog of CD38 that takes part in the NAD^+ salvage pathway, hydrolysing NAD^+ to NAM. As there are no known mammalian transporters of NAD^+ into the cell, the efforts of both CD157 and CD38 to transform extracellular NAD^+ into NAM allow for NAM to diffuse across the cell membrane to be used in the production of NAD^+ (Gasparrini et al. 2021). By stabilizing NAMPT, CD157 helps to increase NAD^+ production in the body (Bogan and Brenner 2008).

SARM1 is a recognized class of NADase that cleaves NAD^+ into NAM, ADPR, and cADPR via its TIR domain. It is primarily expressed throughout nervous system tissues (Essuman et al. 2017). SARM1 is activated by increases in the ratio of NMN to NAD^+ (Figley et al. 2021) and the rapid breakdown of NAD^+ induced by SARM1, which has intrinsic NAD^+ cleavage activity, is similar to that observed when PARP is activated in response to DNA damage (Kim et al. 2005). SIRTs, PARPs, CD38/CD157, and SARM1 compete with each other to consume cellular NAD^+ and as such, the hyperactivation of one enzyme can impair the activities of other NAD^+-dependent enzymes.

NAD$^+$ Precursors: Nicotinic acid (NA), Nicotinamide (NAM), Nicotinamide Mononucleotide (NMN), Nicotinamide Riboside (NR), and Dihydronicotinamide Riboside (NRH)

NA constitutes a functional component in the coenzymes NAD$^+$ and nicotinamide adenine dinucleotide phosphate (NADP$^+$), which helps facilitating oxidative processes within the body. NA was first identified in the early twentieth century when Conrad Elvehjem utilised nicotinamide to treat pellagra, a disease marked by dermatitis, diarrhoea and dementia, in canines (Elvehjem et al. 2002), leading to dietary supplementations which helped mostly eradication of this condition in industrialised countries. Noteworthy is that pellagrous dementia, which often result in hallucinations, paranoia, and aggressive behaviours, resembles symptoms of patients with Alzheimer's disease (AD) (Gasperi et al. 2019).

In 1955, Rudolf Altschul and his associates demonstrated that supplementation of 3 grams NA a day reduced serum cholesterol in humans, indicative of its potential as an anti-hyperlipidemic and more than just a treatment for pellagra. Experimental data have shown NA to reduce low density lipoprotein production, increase high-density lipoprotein levels, and increase apolipoprotein AI (APOA1) (Zeman et al. 2016). More recently, clinical data suggested NA was beneficial in the management of type-2 diabetes, hyperalgesia, obesity, and atherosclerosis (Godin et al. 2012; Heemskerk et al. 2014; Su et al. 2015).

NA supplementation in patients with mitochondrial myopathy elevated whole blood NAD$^+$ concentration 8.2-fold compared to baselines after 10 months of administration (Pirinen et al. 2020). Considering that NA resulted in elevated levels of muscle and blood NAM and ADP-ribose, it is probable that NA underwent metabolism into NAM through the salvage pathway and not the Preiss-Handler pathway, where it was utilised by enzymes that consume NAD$^+$ (Pirinen et al. 2020). Administration of NA in cachectic mice improved tissue NAD$^+$ levels, improved mitochondrial fitness and ameliorated wasting syndromes resulting from cancer and chemotherapy in mouse models (Beltra et al. 2023). These results indicate that NA serves as a potent enhancer of NAD$^+$ in mammals, regardless of their health status. A major unwanted side effect of NA is flushing via activation of a GPR109A receptor, but is elsewise tolerated well in humans provided the doses are not supraphysiological (Guyton and Bays 2007; Mularski et al. 2006).

NAM, another form of vitamin B3, is utilised by the salvage pathway to replenish intracellular NAD$^+$. It has been found to benefit patients who have recently suffered a stroke by increasing cortical NAD$^+$ levels and reducing PARP activity (Liu et al. 2009), leading to reduced blood vessel obstruction and improved brain function (Mokudai et al. 2000). However, the rate limiting nature of NAMPT which allows NAM to be processed into NMN, limits its candidacy for NAD$^+$ boosting (Wang et al. 2021). Moreover; high doses of NAM have been found to be toxic (Kanayama and Luo 2022), leading to kidney damage and cancer as abundance of NAM competes with PARP and SIRT activity (Hwang and Song 2020).

NMN is a nucleotide derived from ribose, NA, nicotinamide, and NR. In humans, NMN conversion into NAD^+ occurs through the salvage pathway (Bogan and Brenner 2008). NMN is primarily located in the nucleus, mitochondria, cytoplasm, and in the human body, it can be found in bodily fluids such as blood, urine, and organs like placenta tissue (Nadeeshani et al. 2022). It is naturally found in a variety of fruits and vegetables including avocado, broccoli, cabbage, cucumber, raw beef, shrimp, soybean, and tomato (Mills et al. 2016). NMN is absorbed from the gut into blood circulation within 2–3 minutes and transported into tissues within 10–30 minutes (Mills et al. 2016; Yoshino et al. 2011). The fast pharmacokinetics of NMN points to an effective transport across the cell membrane. Solute carrier family 12 member 8 (SLC12A8) encodes for a transmembrane protein that has been reported as a potential high-affinity transporter for NMN, though with controversy (Grozio et al. 2019a, b). Moreover, mRNA expression of the SLC12A8 gene is upregulated in response to NAD^+ decline, allowing cells to meet an urgent demand of NAD^+ biosynthesis (Grozio et al. 2019a). NMN has been shown to activate SIRT1, which in turn can activate the kinase PINK-1, which will then recruit Parkin to the damaged mitochondria, leading to their degradation through the lysosomal pathway (Fang 2019; Fang et al. 2016, 2014). Several studies have highlighted NMN supplementation improving mitochondrial content, function, morphology, and health via activation of SIRT1 in mice (Table 4.1).

NR, like the precursor NMN, has a high level of molecular availability given that it does not require conversion to enter the cell. Dietary sources of NR are limited (e.g. 0.23 mg NR in 8 fl oz. of milk) necessitating supplementation in order to produce geroprotective effects (Trammell et al. 2016). Oral supplementation with NR can increase NAD^+ levels in multiple tissues, along with increased sirtuin activity (Belenky et al. 2007), improved mitochondrial function (Canto et al. 2012), and improved regenerative potential of stem cells (Zhang et al. 2016). *In vitro* and *in vivo* experiments in embryonic kidney cells and mouse models, respectively, have shown NR to enhance the activity of SIRT1/SIRT3 and attenuated body weight increases from high fat diets (Canto et al. 2012). In two mouse models of mitochondrial myopathy, NR administered via food admix improved mitochondrial function, mass, induced OXPHOS-related gene expression and delayed disease symptoms (Cerutti et al. 2014; Khan et al. 2014). Although supplementation with NR up to 2000 mg/day has been well tolerated, future studies will have to address potential side-effects related to protracted administration of NR (Airhart et al. 2017).

NRH is the most recently discovered precursor for NAD^+ synthesis (Megarity et al. 2014). *In vivo* and *in vitro* experimentation demonstrating the efficacy of NRH as composed to NMN and NR showed NRH to increase NAD^+ concentrations in cell lines at least two-fold higher than other precursors and almost 400% higher compared to NR and NMN in mouse models (Yang et al. 2019). It is indicated that NRH uses different steps and enzymes to synthesize NAD^+ compared to other known precursors, suggesting a novel NRK1-independent pathway for NAD^+ synthesis (Giroud-Gerbetant et al. 2019). In this new route, reduced NMN (NMNH) is cleaved

Table 4.1 Known Benefits and Mechanisms of Action of NMN from (Fang et al. 2017) with updates

Tissue	Supplement dose	Benefits	Pathways affected	References
Brain	NMN 12 mM	Improved mitochondrial (mt) morphology & health, leading to neuron protection.	Increased NAD⁺, activation of SIRT1 & PGC-1α & degradation of BACE1. Increased LC3-II & altered fission/fusion balance.	Mouchiroud et al. (2013)
	NMN 5 mM in *C. elegans*	Restoration of mitophagy and mitochondrial homeostasis	In the Aβ-bearing worms, the NMN-based memory benefit was dependent on PINK1, Parkin, and DCT-1/NIX.	Fang et al. (2019b)
Eye	NMN, 100 or 300 mg/kg/d	Prevented age-associated decline of rod & cone cell function. Increased tear production.	Likely due to increased NAD⁺ level & sirtuin activity.	Camacho-Pereira et al. (2016)
Muscle	NMN 500 mg/kg/d NMN 300 mg/kg/d	Increased NAD⁺ & mt content, prevented mt myopathy	SIRT1 & SIRT3 activationImproved mt function & mitophagy mediated via SIRT1	Camacho-Pereira et al. (2016); Canto et al. (2012); Mills et al. (2016); Mouchiroud et al. (2013)
Fat	NMN 500 mg/kg/dNMN 300 mg/kg/d	Increased mt content & mt respiratory capacity	Increased NAD⁺ levels, SIRT1 & SIRT3 activation.	Canto et al. (2012); Gomes et al. (2013); Wang et al. (2022)
Liver	NMN 500 mg/kg/d NMN 300 mg/kg/d	Increased content & activity of mt	SIRT1 & SIRT3 activation	Camacho-Pereira et al. (2016); Canto et al. (2012); Gomes et al. (2013); Mills et al. (2016)
Healthspan and lifespan	5 mM NMN to treat *Drosophila*	Increased lifespan in both WT and premature ageing Werner syndrome flies	Mitophagy, stem cell, AMP, ULK1	Fang et al. (2019a)

by CD73 to NRH, which is then taken up by the same SLC29A transporters as NR (Zapata-Perez et al. 2021). More studies are needed to clarify the metabolism of NRH and its efficiency compared to other NAD^+ precursors.

NAD+ and Premature Ageing

NAD+ exhaustion plays a significant role in both normal ageing and accelerated ageing diseases. As NAD+ levels decline or NAD+-dependent enzymes are altered, measurable effects to each of the original hallmarks of ageing are observed (Table 4.2). As NAD+ levels decrease with age, aberrant protein aggregations accumulate as a result of increased oxidative stress from mitochondrial electron transport via NADPH activity (Ying 2008). Furthermore, decreased NAD+ levels can result in a decline in sirtuin activity, particularly SIRT1, SIRT3 and SIRT6, generating NAM as a byproduct with high concentrations inhibiting the ability of these proteins to assist in DNA repair (Lautrup et al. 2019).

Table 4.2 NAD+ impacts on the hallmarks of brain ageing (Lautrup et al. 2019; Schmauck-Medina et al. 2022)

Hallmarks of ageing	Pathways and conditions subject to alterations in NAD+	References
Altered intercellular communication	The effect of NAD+ augmentation on inflammation has been demonstrated in age-associated diseases including diabetes and AD by reducing pro-inflammatory cytokines and inflammasome component NLRP3	Fang et al. (2019b); Hou et al. (2018); Lee et al. (2015)
Altered mechanical properties	NAD+ metabolism affects SASP that suppresses inhibition of MAPK, contributing to alterations in fibroblasts seen in accelerated ageing	Tivey et al. (2013)
Cellular senescence	Senescence associated secretory proteins (SASP) secreted by senescent cells induce CD38-NADase activity in non-senescent cells (immune cells or endothelial cells) that ultimately lead to tissue NAD+ decline	Chini et al. (2020); Covarrubias et al. (2020)
Compromised autophagy/ mitophagy	Functional autophagy is required for NAD+ maintenance and precursor supplementation positively impacts age-related pathologies mediated by an upregulation in autophagy	Fang (2019); Fang et al. (2019b); Fang et al. (2014); Lee et al. (2008)
Deregulated nutrient signalling	NAD+ decrease affects AMPK/sirtuin/PGC1α pathways	Canto et al. (2009); Fulco et al. (2008); Price et al. (2012)
Dysregulation of RNA processing	Gene expression regulated via removal of intact NAD+ caps from RNA	Kiledjian (2018); Wolfram-Schauerte and Hofer (2023)
Epigenetic alterations	SIRT-1 dependent gene regulation through histone modification, DNA methylation, and the modulation of chromatin structure is affected by declining NAD+ levels during ageing	Zhang and Kraus (2010)

(continued)

Table 4.2 (continued)

Hallmarks of ageing	Pathways and conditions subject to alterations in NAD$^+$	References
Genomic instability	In accelerated ageing diseases (XPA, CS, A-T), mutation in DNA repair mechanisms leads to PARP1 hyperactivation and reduction in sirtuin activity. Age-dependent NAD+ decline decreases PARP1 activity by promoting the interaction with DBC1	Fang et al. (2014, 2016); Li et al. (2017a); Scheibye-Knudsen et al. (2014)
Inflammation	NAD$^+$ deficiency drives mitochondrial dysfunction resulting in inflammation	Doke et al. (2023); Minhas et al. (2019)
Loss of proteostasis	In AD models, mitophagy is enhanced by NAD$^+$ boosting strategies via a mechanism mediated by PINK1, PDR-1, and DCT-1. In mice and *C. elegans*, sirtuin-mediated UPR^{mt} response and FOXO signalling is activated by NAD$^+$ augmentation or PARP inhibition.	Fang (2019); Fang et al. (2016); Mouchiroud et al. (2013); Scheibye-Knudsen et al. (2014)
Microbiome disturbances	Microbial deamidation greatly affects metabolic route of precursors to NAD$^+$	Chellappa et al. (2022); Shats et al. (2020)
Mitochondrial dysfunction	Age-dependent CD38 increase affects the availability of NAD$^+$ to mitochondrial enzymes including SIRT3, leading to impairment of oxygen consumption. In mice and *C. elegans*, a decline in NAD$^+$ and accumulation of HIF-1α causes loss of OXPHOS subunits via impaired SIRT1-PGC-1α signalling	Camacho-Pereira et al. (2016); Gomes et al. (2013)
Stem cell exhaustion	NAD$^+$ supplementation rejuvenates aged gut, muscle, and hematopoietic stem cell pools by enhancing SIRT1 activity or mitophagy	Fang et al. (2019a); Igarashi et al. (2019); Rimmele et al. (2014); Vannini et al. (2019); Zhang et al. (2016); Zong et al. (2021)
Telomere attrition	Telomere shortening in livers of TERT KO mice leads to a p53-dependent repression of SIRT1-7 NAD$^+$ reduces hematopoietic impairment linked to short telomeres *in vivo*	Amano et al. (2019); Stock et al. (2023)

Decline in NAD$^+$ levels also adversely impact cellular energy metabolism, in particular the maintenance, repair, and biogenesis of mitochondria (Fang et al. 2014). Nuclear-mitochondrial communication, facilitated by PGC-1α, is also adversely impacted by a decline in free NAD$^+$ availability (Gomes et al. 2013). In both aged wild type mice, a reduction in NAD$^+$ levels play a role in the age-related decline of mitochondrial biogenesis due to impaired signalling between SIRT1-PGC-1α (Mouchiroud et al. 2013). As mechanisms of mitophagy are compromised, faulty mitochondria will accumulate in number, producing large volumes of reactive oxygen species which in turn damage cellular organelles leading to the systemic decline seen in premature ageing (Fang 2019). DNA repair deficient accelerated ageing diseases, such as xeroderma pigmentosum, ataxia telangiectasia, Cockayne syndrome and Werner syndrome (WS), exhibit NAD$^+$ depletion (Fang et al. 2019a; Scheibye-Knudsen et al. 2014). Characteristics of these disorders, except WS include multiple neurodegenerative pathologies such as choreoathetosis, oculomotor apraxia, cerebellar ataxia, and cerebellar atrophy (Fang et al. 2016). Moreover, the depletion of NAD$^+$ leads to a pseudohypoxic condition, disturbing PGC-1α/β--independent nuclear-mitochondrial communication, thereby contributing to a decline in mitochondrial function associated with ageing (Gomes et al. 2013). As hyperactive PARPs are inhibited, the bioavailable level of NAD$^+$ increases, hereby effectively abating disease pathologies (Fang et al. 2014).

NAD$^+$ and Neurodegeneration

Neurodegenerative diseases are distinguished by the progressive deterioration of selected populations of neurons over time. Diseases resulting in neurodegeneration can be classified according to principal clinical features, such as motor neuron disease or dementia, molecular incongruity, or physiological patterns of neurodegeneration. The most common type of dementia is AD which affects one in nine people (10.8%) over the age of 65 with a combined care cost of $345 billion projected to triple to $1 trillion by 2050 (Wong 2020).

AD is characterised by memory loss, aphasia, disorientation, and various other neurocognitive dysfunctions. Pathological hallmarks of AD include extracellular plaques of the protein Amyloid β (Aβ) and intracellular neurofibrillary tangles consisting of hyperphosphorylated Tau (Weller and Budson 2018). NAD$^+$ supplementation has been shown to reduce AD pathologies and DNA damage responses in preclinical studies. Moreover, in a DNA repair-deficient AD model, NR supplementation reduced Tau pathology, and neuroinflammation, and improved synaptic transmission and cognition (Hou et al. 2018). NR has also been shown effective against neuroinflammation and senescence in another AD mouse model designated APP/PS1 (Hou et al. 2021). In a separate study, 250 mg/kg applications of NR per day attenuated amyloid beta toxicity via PGC-1α mediated BACE1 degradation in the cortex and hippocampus of a Tg2576 AD mouse model (Gong et al. 2013). Neurons in AD afflicted brains accumulate dysfunctional or damaged mitochondria in part due to lack of mitophagy (Cummins et al. 2019). By supplementing with NAD$^+$ precursors, the balance between mitophagy and mitochondrial biogenesis could reduce AD pathologies via the NAD$^+$-SIRT1-PGC-1α pathway or the DAF-16/FOXO pathway (Fang et al. 2016). Taken together, these results show NAD$^+$ supplementation to be a safe and effective anti-AD strategy (Conze et al. 2016; Yoshino et al. 2018).

Parkinson's disease (PD) is a degenerative neurological condition that progressively worsens with age, leading to severe difficulties in controlling body movements. The cause of PD is not fully understood but a combination of genetic factors, such as *SNCA* mutation, and environmental factors, such as traumatic head injury, are thought to play a role (Kalia and Lang 2015). Mutations in mitochondrial proteins and the lysosomal enzyme β-glucocerebrosidase (GBA) gene, which exhibits altered NAD^+ metabolism (Kalia and Lang 2015), have also been implicated in the cause of PD (Zuo and Motherwell 2013). In fly models of PD with the GBA mutation, applications of NR prevented dopaminergic neuronal loss and ameliorated motor deficits (Schondorf et al. 2018). Motor impairments stem from the loss of dopaminergic neurons in the substantia nigra, where abnormal accumulation of α-synuclein fibrils occurs in both the cell body and neurites of these neurons (Kam et al. 2018). PD patient cells show hyperactivation of SARM1, with the TIR domain inducing axonal degeneration and depletion of axonal NAD^+ (Essuman et al. 2017), and α-synuclein competition with NAD^+ for binding with glyceraldehyde-3-phosphate dehydrogenase, which co-localizes with α-synuclein in Lewy bodies (Barinova et al. 2018). Given the decline in NAD^+ levels that is observed in several models of PD and the initial success of precursor supplementation in PD models, researchers want to know if replenishment of NAD^+ could be a therapeutic target for treatment of this disease.

Currently, two clinical trials are ongoing, using a proprietary formulation of crystalline form β-nicotinamide mononucleotide called MIB-626 for the treatment of patients with AD. The first trial will determine if MIB-626 is able to cross the blood brain barrier and if an oral 90-day regimen can elevate NAD^+ levels in the brain. Next, a trial for 180-day administration will be conducted to assess if the biomarkers of amyloid deposition, neuronal/axonal degeneration, synaptic function and neuroinflammation improve (NCT05040321). Additionally, a separate trial (NCT04430517) seeks to investigate the effects of NR on brain energy metabolism, oxidative stress, and cognitive function in those with mild AD and cognitive impairment. In regard to PD, two randomised and double-blinded clinical trials are currently ongoing to assess the efficacy of NAD^+ supplementation in patients with PD designated as NAD-PARK (NCT03816020), which has patients supplementing with oral NR for four weeks, and NO-PARK (NCT03568968), which seeks to evaluate if NR supplementation can delay the progression of PD. Initial data from the NAD-PARK study showed increased cerebral NAD^+ levels and altered cerebral metabolism in PD patients (Brakedal et al. 2022). Further trial results could allow for NAD^+ supplementation to be used clinically as an intervention against these debilitating diseases.

NAD^+ Boosting

The effects of NAD^+ precursor treatment on the healthspan and lifespan of *C. elegans* and yeast have been extensively studied, with both organisms exhibiting longevity extension after applications of NR, NMN or NAM (Belenky et al. 2007; Fang et al. 2016; Mitchell et al. 2018; Mouchiroud et al. 2013). Long term studies of precursor application in mice have mitigated age associated physical declines via improved insulin sensitivity, energy metabolism and improved lipid profiles (Mills et al. 2016). Precursor supplementation to boost NAD^+ is a burgeoning field of study but is not the only option for increasing NAD^+ levels (Fig. 4.1). This section explores biosynthesis modulators, enzyme inhibitors, lifestyle changes etc. that have also shown to be viable methods of increasing NAD^+ (Table 4.3).

Pharmacological approaches to boost NAD^+ levels include the inhibition of PARP1, CD38, modulation of SIRTs or inhibition of ACMS (Katsyuba et al. 2018) which results in increased *de novo* synthesis of NAD^+ from tryptophan. Quercetin, a naturally non-toxic flavonoid found in grapes, peaches and garlic, has been evaluated for its ability to increase SIRT/PGC-1α/NRF-1, wherein mitochondrial function was protected in Hepg2 cells as a result of an increase in intracellular NAD^+ via quercetin effects (Houghton et al. 2018). Another flavonoid, Luteolindin, has been shown to inhibit CD38 by mimicking the NMN part of the NAD^+ active site, preventing hydrolysis and increasing cellular NAD^+ levels (Kellenberger et al. 2011). PARP inhibitors have long been investigated in BRCA-mutated cancer cells, with attempts to selectively target and kill defective cellular repair mechanisms while sparing innocuous cells. Rucaparib treatments on ovarian, colorectal cell lines, and

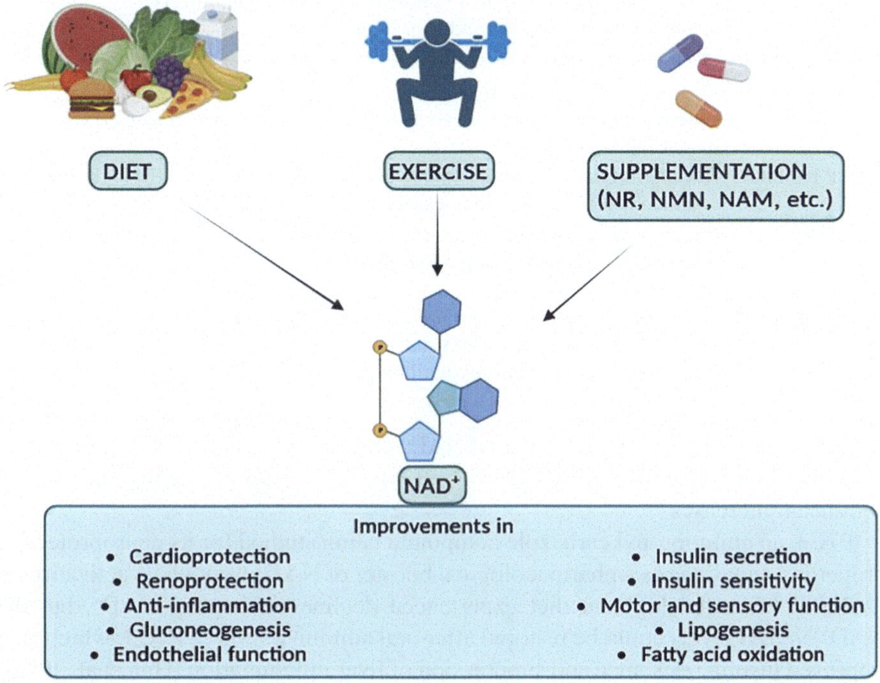

Fig. 4.1 Physiological improvements from NAD^+ boosting molecules. Raising NAD^+ levels through diet, exercise, supplementation with NAD^+ precursors or a combination of the three can improve noticeably on the health and survival of humans. These improvements include but are not limited to; cardioprotection, renoprotection, anti-inflammation, gluconeogenesis, endothelial function, motor and sensory function, lipogenesis, fatty acid oxidation and helps regulate pancreatic insulin secretion and insulin sensitivity in the muscle (created with BioRender.com)

Table 4.3 NAD$^+$ boosting strategies

NAD$^+$ boosting intervention	Methods and results of intervention on NAD$^+$ levels	References
NAD$^+$ precursors	The application of different NAD$^+$ precursors, including NR, NMN, NA, NAM, NRH etc.	Bonkowski and Sinclair (2016); Katsyuba et al. (2020); Lautrup et al. (2019); Reiten et al. (2021)
AMPK activator	Pharmacological activation of AMPK via AICAR enhanced SIRT1 activity by increasing cellular NAD$^+$ levels	Canto et al. (2009); Costford et al. (2010); Fulco et al. (2008)
Caloric restriction	NAD$^+$/NADH ratio in skeletal muscle increases significantly in response to SIRT1 activity during CR	Chen et al. (2008)
	NAD$^+$ levels increase in response to changes in AMPK activity as a result of caloric restriction	Canto et al. (2010)
Physical exertion	NAD$^+$ levels in exercised muscle increase in response to increase in NAMPT	Costford et al. (2010); de Guia et al. (2019); Lamb et al. (2020)
NAMPT activator	SBI-797812 IP injections increase NAD$^+$ levels	Gardell et al. (2019)
	Oral administration of P7C3 rejuvenated NAD$^+$ levels in compromised liver tissue	Hua et al. (2021)
CD38 Inhibitor	Quercetin administration modulates SIRT/PGC-1α, increasing NAD$^+$ levels	Houghton et al. (2018)
	Luteolindin prevents hydrolysis by CD38, increasing NAD$^+$ levels	Kellenberger et al. (2011)
PARP Inhibitor	Rucaparib/Olaparib administration causes significant NAD$^+$ repletion in cancer cells	Almeida et al. (2017); Bajrami et al. (2012)
ACMSD Inhibitor	Enhances *de novo* NAD$^+$ synthesis through inhibition of enzymes that participate in metabolism of tryptophan	Pellicciari et al. (2018)
Antioxidant	Pycnogenol supplementation pre-exercise yielded increases in NAD$^+$(H) and an increase in exertion capacity until fatigue	Mach et al. (2010)

liver tissue from mouse models yielded NAD$^+$ concentrations up to 78% higher than controls (Almeida et al. 2017), with all cell types and tissues showing statistically significant increases.

P7C3, an aminopropyl carbazole compound being studied for its neuroprotective properties, functions as a pharmacological booster of NAD$^+$ through its activation of NAMPT. Mice fed high fat diet experienced declines in hepatic NAD$^+$, but the NAD$^+$/NADH levels could be restored after oral administration of P7C3, which also improved insulin resistance and suppression of liver inflammation (Hua et al. 2021). What is found to be particularly advantageous about the use of P7C3 are its pharmacokinetics, bioavailability and its ability to cross the blood-brain barrier safely with doses several folds higher than the efficacious dose (Vazquez-Rosa et al. 2020).

A high-throughput screening using a chemical library of 57,004 compounds recently yielded a novel NAMPT activator called SBI-797812 (Gardell et al. 2019). The dose-dependent response of 0.4–10 mM in A549 lung carcinoma cells used routinely in the testing of NAMPT activators and intraperitoneal injections of 20 mg/kg body weight in mice both showed increases in NAD^+ activity up to five-fold (Gardell et al. 2019). Taken together, these studies suggest SBI-797812 to be a potent NAD^+ booster and underscore the importance of using high-throughput screenings and other advanced technologies to discover new geroprotectors.

Caloric restriction (CR) is a lifestyle adaption that grew in prominence after Clive McCay of Cornell University first conducted CR experiments on rats. He showed that rats given 30% less food than the control group, had a 30% increase in mean lifespan (Kemnitz 2011). In the time since then, CR has been shown to extend lifespan across many model organisms including worms, fruit flies, yeasts, and monkeys (Fontana et al. 2010). In contrast, links between metabolism alterations induced by consumption of high-fat diets and cellular NAD^+ quantity show marked decreases in tissue NAD^+ levels in addition to shortened lifespans in wild-type animal models (Bai et al. 2011; Canto et al. 2012; Kraus et al. 2014; Yang et al. 2014; Yoshino et al. 2011). CR has been shown in rat models to affect interactions between AMPK and NAMPT and alter NAD^+ precursor levels in the liver (Rodrigues et al. 2021), with CR shown to increase cellular and tissue NAD^+ levels (Qin et al. 2006). Rodents fasted for 48 hours show concomitant increases in both NAMPT and mitochondrial NAD^+ (Yang et al. 2007). CR has also been shown beneficial to rodent AD models. By reducing the caloric intake with 30% from four months of age to ten, the Tg2576 AD mice showed a significant elevation in SIRT1 protein content, a four-fold elevation in NAD^+ in the brain and significantly reduced amyloid β. Moreover, exercise and CR regulate modifiable alterations to gene expression via epigenetic mechanisms which can diminish the risk of developing AD and cognitive decline that can occur with ageing (Baumgart et al. 2015; Turner et al. 2015). Lastly, studies of Nigerian populations, which were shown to have the greatest quantitative levels of the APOE4 polymorphism, seldom showed AD cases which is hypothesized to be as the result of low cholesterol/calorie lifestyles (Sepehrnia et al. 1989). Taken together, this strongly suggests the efficacy of CR attenuation in AD-type neuropathological mice (Qin et al. 2006) and a potential lifestyle intervention for those who have a genetic predisposition to AD.

Physical exertion has long been implicated in the fluctuation of NAD^+ levels, with muscle contraction shown to decrease NADH (Godfraind-de Becker 1972). Circulatory occlusion experiments, which emulate isometric contractions, have found that post contraction NADH levels rise 100% with greater rises depending on the length of the contraction (Sahlin 1983). Further experiments show a 65% NADH reversion in one minute of recovery time (Henriksson et al. 1986). Examination of NAD^+ content between resting muscles and slow-twitch muscles, which are responsible for endurance based exercise performance, showed a positive linear relationship, indicative of different muscle types having different NAD^+ levels at rest and under exertion (Graham et al. 1978). In mouse models, slow-twitch muscle fibres recruited for swimming exercise, increased NAD^+ production in muscles (Canto et al. 2010). Interestingly, similar swimming exercises between young and aged rats showed a decrease in NAD^+ production (Koltai et al. 2010). However, nicotinamide supplementation allowed for extended swimming time in these rat models, indicative of NAD^+ functioning as a rate limiting factor in slow-twitch muscle exercise (Koltai et al. 2010). This suggests that supplementation with precursors may augment physical exertion capacity to allow for greater benefits from exercise.

Efforts to better understand mitochondrial function in real time *in vivo* monitoring via multiparametric fluorescent examination demonstrated that NAD^+ elevation happens primarily in the mitochondria (Mayevsky and Rogatsky 2007). Though oxidative phosphorylation drives NAD^+ production in muscle mitochondria during and post exertion (Sahlin et al. 1990), ageing muscle cells grow increasingly inefficient at NAD^+ replenishment necessitating consistency in both exercise and dietary supplementation of NAD^+ throughout life (Ji and Yeo 2022). Such data corroborate with the evidence of the importance of mitochondrial NAD+ in maintaining cell survival.

Challenges and Conclusions

Recently, intravenous administration of NAD^+ through private companies (NAD Treatment Center, NAD^+ Clinic etc.) have cropped up worldwide. These companies offer proprietary blends of NAD^+, several amino acids, vitamins and antioxidants priced at the time of this publication at £449 for a single infusion using NAD^+ Clinic and packages of up to $20000.00 for a series of treatments from NAD Treatment Center. Proprietary blends list the ingredients used within a product but the amounts are not declared, leaving it impossible to accurately calculate exposures and intakes of that ingredient or determine doses used in clinical trials from label information (Saldanha et al. 2023). Moreover, though many studies have evaluated the efficacy of NAD^+ precursors, only one examined directly infused NAD^+, and while changes in NAD^+ concentration within plasma and urine were assessed, no tissue biopsies were done to quantify changes in intracellular NAD^+ (Grant et al. 2019). This lack of transparency in the quantity of ingredients within these intravenous supplements and lack of clinical data suggest a word of caution to consumers who desire to augment their longevity through use of NAD^+ boosters.

NAD^+ boosting in most humans appears safe and beneficial. However, there exists concerns that elevating NAD^+ levels in humans with cancer could exacerbate their illness (Demarest et al. 2019). Many major NAD^+ producing metabolic enzymes (e.g. iNAMPT, eNAMPT, and/or NMNAT2) are overexpressed in a broad variety of cancers (Demarest et al. 2019), with an example of mouse models predisposed to pancreatic cancer developing more precancerous and cancerous growths after supplementation with NMN (Nacarelli et al. 2019). Further research shows inhibition of NAMPT *in vivo* has been demonstrated to decrease the growth and survival of pancreatic cancer cells (Chini et al. 2014). Moreover, synergistic cytotoxicity to tumour cells, such as utilising NAMPT inhibitor GMX1777 and radiation therapy, may prove an effective strategy against certain carcinomas (Kato et al. 2010). Increasing NAD^+ has been suggested to increase SASP production via suppression of AMPK and p53 thereby stimulating NF-κB via p38 MAPK to drive increases in cytokines (Mendelsohn and Larrick 2019), additional data in different systems, especially in animals, are needed. Currently, no studies exist that have shown NAD^+ supplementation to increase cancer risks in geriatric populations. These studies do more to highlight questions for scientists to address as time goes by and the field of NAD^+ augmentation research grows. Thorough exploration of NAD^+ precursors and boosters may potentially pave the way for its widespread use as a supplement and the development of innovative therapeutic approaches. These hold promise for addressing pathophysiological conditions that currently lack an effective treatment and potentially serve as a prophylactic against age-associated disease, reducing the exorbitant medical and societal costs associated with such.

Acknowledgement E.F.F. is supported by Cure Alzheimer's Fund (#282952), HELSE SØR-ØST (#2020001, #2021021, #2023093), the Research Council of Norway (#262175, #334361), Molecule AG/VITADAO (#282942), NordForsk Foundation (#119986), the National Natural Science Foundation of China (#81971327), Akershus University Hospital (#269901, #261973, #262960), the Civitan Norges Forskningsfond for Alzheimers sykdom (#281931), the Czech Republic-Norway KAPPA programme (with Martin Vyhnálek, #TO01000215), and the Rosa sløyfe/Norwegian Cancer Society & Norwegian Breast Cancer Society (#207819).

Conflict of Interests E.F.F. is the co-owner of Fang-S Consultation AS (Organization number 931 410 717); he has an MTA with LMITO Therapeutics Inc (South Korea), a CRADA arrangement with ChromaDex (USA), a commercialization agreement with Molecule AG/VITADAO; he is a consultant to Aladdin Healthcare Technologies (UK and Germany), the Vancouver Dementia Prevention Centre (Canada), Intellectual Labs (Norway), MindRank AI (China), and NYO3 (Norway and China).

References

Adebanjo OA, Anandatheerthavarada HK, Koval AP, Moonga BS, Biswas G, Sun L, Sodam BR, Bevis PJ, Huang CL, Epstein S, Lai FA, Avadhani NG, Zaidi M (1999) A new function for CD38/ADP-ribosyl cyclase in nuclear Ca2+ homeostasis. Nat Cell Biol 1(7):409–414. https://doi.org/10.1038/15640

Airhart SE, Shireman LM, Risler LJ, Anderson GD, Nagana Gowda GA, Raftery D, Tian R, Shen DD, O'Brien KD (2017) An open-label, non-randomized study of the pharmacokinetics of the nutritional supplement nicotinamide riboside (NR) and its effects on blood NAD+ levels in healthy volunteers. PLoS One 12(12):e0186459. https://doi.org/10.1371/journal.pone.0186459

Almeida GS, Bawn CM, Galler M, Wilson I, Thomas HD, Kyle S, Curtin NJ, Newell DR, Maxwell RJ (2017) PARP inhibitor rucaparib induces changes in NAD levels in cells and liver tissues as assessed by MRS. NMR Biomed 30(9). https://doi.org/10.1002/nbm.3736

Amano H, Chaudhury A, Rodriguez-Aguayo C, Lu L, Akhanov V, Catic A, Popov YV, Verdin E, Johnson H, Stossi F, Sinclair DA, Nakamaru-Ogiso E, Lopez-Berestein G, Chang JT, Neilson JR, Meeker A, Finegold M, Baur JA, Sahin E (2019) Telomere dysfunction induces Sirtuin repression that drives telomere-dependent disease. Cell Metab 29(6):1274–1290 e1279. https://doi.org/10.1016/j.cmet.2019.03.001

Bai P, Canto C (2012) The role of PARP-1 and PARP-2 enzymes in metabolic regulation and disease. Cell Metab 16(3):290–295. https://doi.org/10.1016/j.cmet.2012.06.016

Bai P, Canto C, Oudart H, Brunyanszki A, Cen Y, Thomas C, Yamamoto H, Huber A, Kiss B, Houtkooper RH, Schoonjans K, Schreiber V, Sauve AA, Menissier-de Murcia J, Auwerx J

(2011) PARP-1 inhibition increases mitochondrial metabolism through SIRT1 activation. Cell Metab 13(4):461–468. https://doi.org/10.1016/j.cmet.2011.03.004

Bajrami I, Kigozi A, Van Weverwijk A, Brough R, Frankum J, Lord CJ, Ashworth A (2012) Synthetic lethality of PARP and NAMPT inhibition in triple-negative breast cancer cells. EMBO Mol Med 4(10):1087–1096. https://doi.org/10.1002/emmm.201201250

Barber MF, Michishita-Kioi E, Xi Y, Tasselli L, Kioi M, Moqtaderi Z, Tennen RI, Paredes S, Young NL, Chen K, Struhl K, Garcia BA, Gozani O, Li W, Chua KF (2012) SIRT7 links H3K18 deacetylation to maintenance of oncogenic transformation. Nature 487(7405):114–118. https://doi.org/10.1038/nature11043

Barinova K, Khomyakova E, Semenyuk P, Schmalhausen E, Muronetz V (2018) Binding of alpha-synuclein to partially oxidized glyceraldehyde-3-phosphate dehydrogenase induces subsequent inactivation of the enzyme. Arch Biochem Biophys 642:10–22. https://doi.org/10.1016/j.abb.2018.02.002

Baumgart M, Snyder HM, Carrillo MC, Fazio S, Kim H, Johns H (2015) Summary of the evidence on modifiable risk factors for cognitive decline and dementia: a population-based perspective. Alzheimers Dement 11(6):718–726. https://doi.org/10.1016/j.jalz.2015.05.016

Belenky P, Racette FG, Bogan KL, McClure JM, Smith JS, Brenner C (2007) Nicotinamide riboside promotes Sir2 silencing and extends lifespan via Nrk and Urh1/Pnp1/Meu1 pathways to NAD+. Cell 129(3):473–484. https://doi.org/10.1016/j.cell.2007.03.024

Beltra M, Pollanen N, Fornelli C, Tonttila K, Hsu MY, Zampieri S, Moletta L, Corra S, Porporato PE, Kivela R, Viscomi C, Sandri M, Hulmi JJ, Sartori R, Pirinen E, Penna F (2023) NAD(+) repletion with niacin counteracts cancer cachexia. Nat Commun 14(1):1849. https://doi.org/10.1038/s41467-023-37595-6

Bogan KL, Brenner C (2008) Nicotinic acid, nicotinamide, and nicotinamide riboside: a molecular evaluation of NAD+ precursor vitamins in human nutrition. Annu Rev Nutr 28:115–130. https://doi.org/10.1146/annurev.nutr.28.061807.155443

Bonkowski MS, Sinclair DA (2016) Slowing ageing by design: the rise of NAD(+) and sirtuin-activating compounds. Nat Rev Mol Cell Biol 17(11):679–690. https://doi.org/10.1038/nrm.2016.93

Bosch-Presegue L, Vaquero A (2014) Sirtuins in stress response: guardians of the genome. Oncogene 33(29):3764–3775. https://doi.org/10.1038/onc.2013.344

Brakedal B, Dolle C, Riemer F, Ma Y, Nido GS, Skeie GO, Craven AR, Schwarzlmuller T, Brekke N, Diab J, Sverkeli L, Skjeie V, Varhaug K, Tysnes OB, Peng S, Haugarvoll K, Ziegler M, Gruner R, Eidelberg D, Tzoulis C (2022) The NADPARK study: a randomized phase I trial of nicotinamide riboside supplementation in Parkinson's disease. Cell Metab 34(3):396–407 e396. https://doi.org/10.1016/j.cmet.2022.02.001

Brazill JM, Li C, Zhu Y, Zhai RG (2017) NMNAT: It's an NAD(+) synthase… It's a chaperone… It's a neuroprotector. Curr Opin Genet Dev 44:156–162. https://doi.org/10.1016/j.gde.2017.03.014

Camacho-Pereira J, Tarrago MG, Chini CCS, Nin V, Escande C, Warner GM, Puranik AS, Schoon RA, Reid JM, Galina A, Chini EN (2016) CD38 dictates age-related NAD decline and mitochondrial dysfunction through an SIRT3-dependent mechanism. Cell Metab 23(6):1127–1139. https://doi.org/10.1016/j.cmet.2016.05.006

Cambronne XA, Kraus WL (2020) Location, location, location: compartmentalization of NAD(+) synthesis and functions in mammalian cells. Trends Biochem Sci 45(10):858–873. https://doi.org/10.1016/j.tibs.2020.05.010

Canto C, Gerhart-Hines Z, Feige JN, Lagouge M, Noriega L, Milne JC, Elliott PJ, Puigserver P, Auwerx J (2009) AMPK regulates energy expenditure by modulating NAD+ metabolism and SIRT1 activity. Nature 458(7241):1056–1060. https://doi.org/10.1038/nature07813

Canto C, Jiang LQ, Deshmukh AS, Mataki C, Coste A, Lagouge M, Zierath JR, Auwerx J (2010) Interdependence of AMPK and SIRT1 for metabolic adaptation to fasting and exercise in skeletal muscle. Cell Metab 11(3):213–219. https://doi.org/10.1016/j.cmet.2010.02.006

Canto C, Houtkooper RH, Pirinen E, Youn DY, Oosterveer MH, Cen Y, Fernandez-Marcos PJ, Yamamoto H, Andreux PA, Cettour-Rose P, Gademann K, Rinsch C, Schoonjans K, Sauve AA, Auwerx J (2012) The NAD(+) precursor nicotinamide riboside enhances oxidative metabolism and protects against high-fat diet-induced obesity. Cell Metab 15(6):838–847. https://doi.org/10.1016/j.cmet.2012.04.022

Cerutti R, Pirinen E, Lamperti C, Marchet S, Sauve AA, Li W, Leoni V, Schon EA, Dantzer F, Auwerx J, Viscomi C, Zeviani M (2014) NAD(+)-dependent activation of Sirt1 corrects the phenotype in a mouse model of mitochondrial disease. Cell Metab 19(6):1042–1049. https://doi.org/10.1016/j.cmet.2014.04.001

Chalkiadaki A, Guarente L (2015) The multifaceted functions of sirtuins in cancer. Nat Rev Cancer 15(10):608–624. https://doi.org/10.1038/nrc3985

Chellappa K, McReynolds MR, Lu W, Zeng X, Makarov M, Hayat F, Mukherjee S, Bhat YR, Lingala SR, Shima RT, Descamps HC, Cox T, Ji L, Jankowski C, Chu Q, Davidson SM, Thaiss CA, Migaud ME, Rabinowitz JD, Baur JA (2022) NAD precursors cycle between host tissues and the gut microbiome. Cell Metab 34(12):1947–1959 e1945. https://doi.org/10.1016/j.cmet.2022.11.004

Chen D, Bruno J, Easlon E, Lin SJ, Cheng HL, Alt FW, Guarente L (2008) Tissue-specific regulation of SIRT1 by calorie restriction. Genes Dev 22(13):1753–1757. https://doi.org/10.1101/gad.1650608

Chen X, Sun K, Jiao S, Cai N, Zhao X, Zou H, Xie Y, Wang Z, Zhong M, Wei L (2014) High levels of SIRT1 expression enhance tumorigenesis and associate with a poor prognosis of colorectal carcinoma patients. Sci Rep 4:7481. https://doi.org/10.1038/srep07481

Cheng A, Yang Y, Zhou Y, Maharana C, Lu D, Peng W, Liu Y, Wan R, Marosi K, Misiak M, Bohr VA, Mattson MP (2016) Mitochondrial SIRT3 mediates adaptive responses of neurons to exercise and metabolic and excitatory challenges. Cell Metab 23(1):128–142. https://doi.org/10.1016/j.cmet.2015.10.013

Chiarugi A, Dolle C, Felici R, Ziegler M (2012) The NAD metabolome--a key determinant of cancer cell biology. Nat Rev Cancer 12(11):741–752. https://doi.org/10.1038/nrc3340

Chini CC, Guerrico AM, Nin V, Camacho-Pereira J, Escande C, Barbosa MT, Chini EN (2014) Targeting of NAD metabolism in pancreatic cancer cells: potential novel therapy for pancreatic tumors. Clin Cancer Res 20(1):120–130. https://doi.org/10.1158/1078-0432.CCR-13-0150

Chini CCS, Peclat TR, Warner GM, Kashyap S, Espindola-Netto JM, de Oliveira GC, Gomez LS, Hogan KA, Tarrago MG, Puranik AS, Agorrody G, Thompson KL, Dang K, Clarke S, Childs BG, Kanamori KS, Witte MA, Vidal P, Kirkland AL et al (2020) CD38 ecto-enzyme in immune cells is induced during aging and regulates NAD(+) and NMN levels. Nat Metab 2(11):1284–1304. https://doi.org/10.1038/s42255-020-00298-z

Conze DB, Crespo-Barreto J, Kruger CL (2016) Safety assessment of nicotinamide riboside, a form of vitamin B(3). Hum Exp Toxicol 35(11):1149–1160. https://doi.org/10.1177/0960327115626254

Costford SR, Bajpeyi S, Pasarica M, Albarado DC, Thomas SC, Xie H, Church TS, Jubrias SA, Conley KE, Smith SR (2010) Skeletal muscle NAMPT is induced by exercise in humans. Am J Physiol Endocrinol Metab 298(1):E117–E126. https://doi.org/10.1152/ajpendo.00318.2009

Covarrubias AJ, Kale A, Perrone R, Lopez-Dominguez JA, Pisco AO, Kasler HG, Schmidt MS, Heckenbach I, Kwok R, Wiley CD, Wong HS, Gibbs E, Iyer SS, Basisty N, Wu Q, Kim IJ, Silva E, Vitangcol K, Shin KO et al (2020) Senescent cells promote tissue NAD(+) decline during ageing via the activation of CD38(+) macrophages. Nat Metab 2(11):1265–1283. https://doi.org/10.1038/s42255-020-00305-3

Cummins N, Tweedie A, Zuryn S, Bertran-Gonzalez J, Gotz J (2019) Disease-associated tau impairs mitophagy by inhibiting Parkin translocation to mitochondria. EMBO J 38(3). https://doi.org/10.15252/embj.201899360

Dali-Youcef N, Lagouge M, Froelich S, Koehl C, Schoonjans K, Auwerx J (2007) Sirtuins: the 'magnificent seven', function, metabolism and longevity. Ann Med 39(5):335–345. https://doi.org/10.1080/07853890701408194

Das D, Bristol ML, Smith NW, James CD, Wang X, Pichierri P, Morgan IM (2019) Werner helicase control of human papillomavirus 16 E1-E2 DNA replication is regulated by SIRT1 deacetylation. MBio 10(2). https://doi.org/10.1128/mBio.00263-19

De Flora A, Zocchi E, Guida L, Franco L, Bruzzone S (2004) Autocrine and paracrine calcium signaling by the CD38/NAD+/cyclic ADP-ribose system. Ann N Y Acad Sci 1028:176–191. https://doi.org/10.1196/annals.1322.021

de Guia RM, Agerholm M, Nielsen TS, Consitt LA, Sogaard D, Helge JW, Larsen S, Brandauer J, Houmard JA, Treebak JT (2019) Aerobic and resistance exercise training reverses age-dependent decline in NAD(+) salvage capacity in human skeletal muscle. Phys Rep 7(12):e14139. https://doi.org/10.14814/phy2.14139

Demarest TG, Truong GTD, Lovett J, Mohanty JG, Mattison JA, Mattson MP, Ferrucci L, Bohr VA, Moaddel R (2019) Assessment of NAD(+)metabolism in human cell cultures, erythrocytes, cerebrospinal fluid and primate skeletal muscle. Anal Biochem 572:1–8. https://doi.org/10.1016/j.ab.2019.02.019

Dikic I (2017) Proteasomal and Autophagic degradation systems. Annu Rev Biochem 86:193–224. https://doi.org/10.1146/annurev-biochem-061516-044908

Doke T, Mukherjee S, Mukhi D, Dhillon P, Abedini A, Davis JG, Chellappa K, Chen B, Baur JA, Susztak K (2023) NAD(+) precursor supplementation prevents mtRNA/RIG-I-dependent inflammation during kidney injury. Nat Metab 5(3):414–430. https://doi.org/10.1038/s42255-023-00761-7

Du SH, Shi J, Yu TY, Hu XX, He SM, Cao YY, Xie ZL, Liu SS, Li YT, Li N, Yu JB (2022) Nicotinamide mononucleotide ameliorates acute lung injury by inducing mitonuclear protein imbalance and activating the UPR(mt). Exp Biol Med (Maywood) 247(14):1264–1276. https://doi.org/10.1177/15353702221094235

Elvehjem CA, Madden RJ, Strong FM, Wolley DW (2002) The isolation and identification of the anti-black tongue factor. 1937. J Biol Chem 277(34):e22. https://www.ncbi.nlm.nih.gov/pubmed/12243127

Essuman K, Summers DW, Sasaki Y, Mao X, DiAntonio A, Milbrandt J (2017) The SARM1 toll/Interleukin-1 receptor domain possesses intrinsic NAD(+) cleavage activity that promotes pathological axonal degeneration. Neuron 93(6):1334–1343 e1335. https://doi.org/10.1016/j.neuron.2017.02.022

Fang EF (2019) Mitophagy and NAD(+) inhibit Alzheimer disease. Autophagy 15(6):1112–1114. https://doi.org/10.1080/15548627.2019.1596497

Fang EF, Scheibye-Knudsen M, Brace LE, Kassahun H, SenGupta T, Nilsen H, Mitchell JR, Croteau DL, Bohr VA (2014) Defective mitophagy in XPA via PARP-1 hyperactivation and NAD(+)/SIRT1 reduction. Cell 157(4):882–896. https://doi.org/10.1016/j.cell.2014.03.026

Fang EF, Kassahun H, Croteau DL, Scheibye-Knudsen M, Marosi K, Lu H, Shamanna RA, Kalyanasundaram S, Bollineni RC, Wilson MA, Iser WB, Wollman BN, Morevati M, Li J, Kerr JS, Lu Q, Waltz TB, Tian J, Sinclair DA et al (2016) NAD(+) replenishment improves lifespan and Healthspan in ataxia telangiectasia models via Mitophagy and DNA repair. Cell Metab 24(4):566–581. https://doi.org/10.1016/j.cmet.2016.09.004

Fang EF, Lautrup S, Hou Y, Demarest TG, Croteau DL, Mattson MP, Bohr VA (2017) NAD(+) in aging: molecular mechanisms and translational implications. Trends Mol Med 23(10):899–916. https://doi.org/10.1016/j.molmed.2017.08.001

Fang EF, Hou Y, Lautrup S, Jensen MB, Yang B, SenGupta T, Caponio D, Khezri R, Demarest TG, Aman Y, Figueroa D, Morevati M, Lee HJ, Kato H, Kassahun H, Lee JH, Filippelli D, Okur MN, Mangerich A et al (2019a) NAD(+) augmentation restores mitophagy and limits accelerated aging in Werner syndrome. Nat Commun 10(1):5284. https://doi.org/10.1038/s41467-019-13172-8

Fang EF, Hou Y, Palikaras K, Adriaanse BA, Kerr JS, Yang B, Lautrup S, Hasan-Olive MM, Caponio D, Dan X, Rocktaschel P, Croteau DL, Akbari M, Greig NH, Fladby T, Nilsen H, Cader MZ, Mattson MP, Tavernarakis N, Bohr VA (2019b) Mitophagy inhibits amyloid-beta and tau pathology and reverses cognitive deficits in models of Alzheimer's disease. Nat Neurosci 22(3):401–412. https://doi.org/10.1038/s41593-018-0332-9

Fernandez-Vizarra E, Enriquez JA, Perez-Martos A, Montoya J, Fernandez-Silva P (2011) Tissue-specific differences in mitochondrial activity and biogenesis. Mitochondrion 11(1):207–213. https://doi.org/10.1016/j.mito.2010.09.011

Figley MD, Gu W, Nanson JD, Shi Y, Sasaki Y, Cunnea K, Malde AK, Jia X, Luo Z, Saikot FK, Mosaiab T, Masic V, Holt S, Hartley-Tassell L, McGuinness HY, Manik MK, Bosanac T, Landsberg MJ, Kerry PS et al (2021) SARM1 is a metabolic sensor activated by an increased NMN/NAD(+) ratio to trigger axon degeneration. Neuron 109(7):1118–1136 e1111. https://doi.org/10.1016/j.neuron.2021.02.009

Fontana L, Partridge L, Longo VD (2010) Extending healthy life span--from yeast to humans. Science 328(5976):321–326. https://doi.org/10.1126/science.1172539

Fulco M, Cen Y, Zhao P, Hoffman EP, McBurney MW, Sauve AA, Sartorelli V (2008) Glucose restriction inhibits skeletal myoblast differentiation by activating SIRT1 through AMPK-mediated regulation of Nampt. Dev Cell 14(5):661–673. https://doi.org/10.1016/j.devcel.2008.02.004

Gardell SJ, Hopf M, Khan A, Dispagna M, Hampton Sessions E, Falter R, Kapoor N, Brooks J, Culver J, Petucci C, Ma CT, Cohen SE, Tanaka J, Burgos ES, Hirschi JS, Smith SR, Sergienko E, Pinkerton AB (2019) Boosting NAD(+) with a small molecule that activates NAMPT. Nat Commun 10(1):3241. https://doi.org/10.1038/s41467-019-11078-z

Gasparrini M, Sorci L, Raffaelli N (2021) Enzymology of extracellular NAD metabolism. Cell Mol Life Sci 78(7):3317–3331. https://doi.org/10.1007/s00018-020-03742-1

Gasperi V, Sibilano M, Savini I, Catani MV (2019) Niacin in the central nervous system: an update of biological aspects and clinical applications. Int J Mol Sci 20(4). https://doi.org/10.3390/ijms20040974

Gazzaniga F, Stebbins R, Chang SZ, McPeek MA, Brenner C (2009) Microbial NAD metabolism: lessons from comparative genomics. Microbiol Mol Biol Rev 73(3):529–541, Table of Contents. https://doi.org/10.1128/MMBR.00042-08

Giroud-Gerbetant J, Joffraud M, Giner MP, Cercillieux A, Bartova S, Makarov MV, Zapata-Perez R, Sanchez-Garcia JL, Houtkooper RH, Migaud ME, Moco S, Canto C (2019) A reduced form of nicotinamide riboside defines a new path for NAD(+) biosynthesis and acts as an orally bioavailable NAD(+) precursor. Mol Metab 30:192–202. https://doi.org/10.1016/j.molmet.2019.09.013

Godfraind-de Becker A (1972) Heat production and fluorescence changes of toad sartorius muscle during aerobic recovery after a short tetanus. J Physiol 223(3):719–734. https://doi.org/10.1113/jphysiol.1972.sp009871

Godin AM, Ferreira WC, Rocha LT, Ferreira RG, Paiva AL, Merlo LA, Nascimento EB Jr, Bastos LF, Coelho MM (2012) Nicotinic acid induces antinociceptive and anti-inflammatory effects in different experimental models. Pharmacol Biochem Behav 101(3):493–498. https://doi.org/10.1016/j.pbb.2012.02.012

Gomes AP, Price NL, Ling AJ, Moslehi JJ, Montgomery MK, Rajman L, White JP, Teodoro JS, Wrann CD, Hubbard BP, Mercken EM, Palmeira CM, de Cabo R, Rolo AP, Turner N, Bell EL, Sinclair DA (2013) Declining NAD(+) induces a pseudohypoxic state disrupting nuclear-mitochondrial communication during aging. Cell 155(7):1624–1638. https://doi.org/10.1016/j.cell.2013.11.037

Gong B, Pan Y, Vempati P, Zhao W, Knable L, Ho L, Wang J, Sastre M, Ono K, Sauve AA, Pasinetti GM (2013) Nicotinamide riboside restores cognition through an upregulation of proliferator-activated receptor-gamma coactivator 1alpha regulated beta-secretase 1 degradation and mitochondrial gene expression in Alzheimer's mouse models. Neurobiol Aging 34(6):1581–1588. https://doi.org/10.1016/j.neurobiolaging.2012.12.005

Gopal E, Fei YJ, Miyauchi S, Zhuang L, Prasad PD, Ganapathy V (2005) Sodium-coupled and electrogenic transport of B-complex vitamin nicotinic acid by slc5a8, a member of the Na/glucose co-transporter gene family. Biochem J 388(Pt 1):309–316. https://doi.org/10.1042/bj20041916

Graham T, Sjogaard G, Lollgen H, Saltin B (1978) NAD in muscle of man at rest and during exercise. Pflugers Arch 376(1):35–39. https://doi.org/10.1007/BF00585245

Grant R, Berg J, Mestayer R, Braidy N, Bennett J, Broom S, Watson J (2019) A pilot study investigating changes in the human plasma and urine NAD+ metabolome during a 6 hour intravenous infusion of NAD. Front Aging Neurosci 11:257. https://doi.org/10.3389/fnagi.2019.00257

Grozio A, Mills KF, Yoshino J, Bruzzone S, Sociali G, Tokizane K, Lei HC, Cunningham R, Sasaki Y, Migaud ME, Imai SI (2019a) Slc12a8 is a nicotinamide mononucleotide transporter. Nat Metab 1(1):47–57. https://doi.org/10.1038/s42255-018-0009-4

Grozio A, Mills K, Yoshino J, Bruzzone S, Sociali G, Tokizane K, Lei HC, Sasaki Y, Migaud M, Imai SI (2019b) Reply to: absence of evidence that Slc12a8 encodes a nicotinamide mononucleotide transporter. Nat Metab 1(7):662–665. https://doi.org/10.1038/s42255-019-0086-z

Guyton JR, Bays HE (2007) Safety considerations with niacin therapy. Am J Cardiol 99(6A):22C–31C. https://doi.org/10.1016/j.amjcard.2006.11.018

Heemskerk MM, Dharuri HK, van den Berg SA, Jonasdottir HS, Kloos DP, Giera M, van Dijk KW, van Harmelen V (2014) Prolonged niacin treatment leads to increased adipose tissue PUFA synthesis and anti-inflammatory lipid and oxylipin plasma profile. J Lipid Res 55(12):2532–2540. https://doi.org/10.1194/jlr.M051938

Henriksson J, Katz A, Sahlin K (1986) Redox state changes in human skeletal muscle after isometric contraction. J Physiol 380:441–451. https://doi.org/10.1113/jphysiol.1986.sp016296

Herranz D, Maraver A, Canamero M, Gomez-Lopez G, Inglada-Perez L, Robledo M, Castelblanco E, Matias-Guiu X, Serrano M (2013) SIRT1 promotes thyroid carcinogenesis driven by PTEN deficiency. Oncogene 32(34):4052–4056. https://doi.org/10.1038/onc.2012.407

Hou Y, Lautrup S, Cordonnier S, Wang Y, Croteau DL, Zavala E, Zhang Y, Moritoh K, O'Connell JF, Baptiste BA, Stevnsner TV, Mattson MP, Bohr VA (2018) NAD(+) supplementation normalizes key Alzheimer's features and DNA damage responses in a new AD mouse model with introduced DNA repair deficiency. Proc Natl Acad Sci USA 115(8):E1876–E1885. https://doi.org/10.1073/pnas.1718819115

Hou Y, Wei Y, Lautrup S, Yang B, Wang Y, Cordonnier S, Mattson MP, Croteau DL, Bohr VA (2021) NAD(+) supplementation reduces neuroinflammation and cell senescence in a transgenic mouse model of Alzheimer's disease via cGAS-STING. Proc Natl Acad Sci USA 118(37). https://doi.org/10.1073/pnas.2011226118

Houghton MJ, Kerimi A, Tumova S, Boyle JP, Williamson G (2018) Quercetin preserves redox status and stimulates mitochondrial function in metabolically-stressed HepG2 cells. Free Radic Biol Med 129:296–309. https://doi.org/10.1016/j.freeradbiomed.2018.09.037

Hua X, Sun DY, Zhang WJ, Fu JT, Tong J, Sun SJ, Zeng FY, Ouyang SX, Zhang GY, Wang SN, Li DJ, Miao CY, Wang P (2021) P7C3-A20 alleviates fatty liver by shaping gut microbiota and inducing FGF21/FGF1, via the AMP-activated protein kinase/CREB regulated transcription coactivator 2 pathway. Br J Pharmacol 178(10):2111–2130. https://doi.org/10.1111/bph.15008

Hwang ES, Song SB (2020) Possible adverse effects of high-dose nicotinamide: mechanisms and safety assessment. Biomol Ther 10(5). https://doi.org/10.3390/biom10050687

Igarashi M, Miura M, Williams E, Jaksch F, Kadowaki T, Yamauchi T, Guarente L (2019) NAD(+) supplementation rejuvenates aged gut adult stem cells. Aging Cell 18(3):e12935. https://doi.org/10.1111/acel.12935

Imai S, Guarente L (2014) NAD+ and sirtuins in aging and disease. Trends Cell Biol 24(8):464–471. https://doi.org/10.1016/j.tcb.2014.04.002

Imai S, Armstrong CM, Kaeberlein M, Guarente L (2000) Transcriptional silencing and longevity protein Sir2 is an NAD-dependent histone deacetylase. Nature 403(6771):795–800. https://doi.org/10.1038/35001622

Ji LL, Yeo D (2022) Maintenance of NAD+ homeostasis in skeletal muscle during aging and exercise. Cells 11(4). https://doi.org/10.3390/cells11040710

Kalia LV, Lang AE (2015) Parkinson's disease. Lancet 386(9996):896–912. https://doi.org/10.1016/S0140-6736(14)61393-3

Kam TI, Mao X, Park H, Chou SC, Karuppagounder SS, Umanah GE, Yun SP, Brahmachari S, Panicker N, Chen R, Andrabi SA, Qi C, Poirier GG, Pletnikova O, Troncoso JC, Bekris LM, Leverenz JB, Pantelyat A, Ko HS et al (2018) Poly(ADP-ribose) drives pathologic alpha-synuclein neurodegeneration in Parkinson's disease. Science 362(6414). https://doi.org/10.1126/science.aat8407

Kanayama M, Luo J (2022) CD38-induced apoptosis and mitochondrial damage is restored by nicotinamide in prostate cancer. Front Mol Biosci 9:890402. https://doi.org/10.3389/fmolb.2022.890402

Kato H, Ito E, Shi W, Alajez NM, Yue S, Lee C, Chan N, Bhogal N, Coackley CL, Vines D, Green D, Waldron J, Gullane P, Bristow R, Liu FF (2010) Efficacy of combining GMX1777 with radiation therapy for human head and neck carcinoma. Clin Cancer Res 16(3):898–911. https://doi.org/10.1158/1078-0432.CCR-09-1945

Katsyuba E, Mottis A, Zietak M, De Franco F, van der Velpen V, Gariani K, Ryu D, Cialabrini L, Matilainen O, Liscio P, Giacche N, Stokar-Regenscheit N, Legouis D, de Seigneux S, Ivanisevic J, Raffaelli N, Schoonjans K, Pellicciari R, Auwerx J (2018) De novo NAD(+) synthesis enhances mitochondrial function and improves health. Nature 563(7731):354–359. https://doi.org/10.1038/s41586-018-0645-6

Katsyuba E, Romani M, Hofer D, Auwerx J (2020) NAD(+) homeostasis in health and disease. Nat Metab 2(1):9–31. https://doi.org/10.1038/s42255-019-0161-5

Kellenberger E, Kuhn I, Schuber F, Muller-Steffner H (2011) Flavonoids as inhibitors of human CD38. Bioorg Med Chem Lett 21(13):3939–3942. https://doi.org/10.1016/j.bmcl.2011.05.022

Kemnitz JW (2011) Calorie restriction and aging in nonhuman primates. ILAR J 52(1):66–77. https://doi.org/10.1093/ilar.52.1.66

Kerr JS, Adriaanse BA, Greig NH, Mattson MP, Cader MZ, Bohr VA, Fang EF (2017) Mitophagy and Alzheimer's disease: cellular and molecular mechanisms. Trends Neurosci 40(3):151–166. https://doi.org/10.1016/j.tins.2017.01.002

Khan NA, Auranen M, Paetau I, Pirinen E, Euro L, Forsstrom S, Pasila L, Velagapudi V, Carroll CJ, Auwerx J, Suomalainen A (2014) Effective treatment of mitochondrial myopathy by nicotinamide riboside, a vitamin B3. EMBO Mol Med 6(6):721–731. https://doi.org/10.1002/emmm.201403943

Kiledjian M (2018) Eukaryotic RNA 5′-end NAD(+) capping and DeNADding. Trends Cell Biol 28(6):454–464. https://doi.org/10.1016/j.tcb.2018.02.005

Kim MY, Zhang T, Kraus WL (2005) Poly(ADP-ribosyl)ation by PARP-1: 'PAR-laying' NAD+ into a nuclear signal. Genes Dev 19(17):1951–1967. https://doi.org/10.1101/gad.1331805

Koltai E, Szabo Z, Atalay M, Boldogh I, Naito H, Goto S, Nyakas C, Radak Z (2010) Exercise alters SIRT1, SIRT6, NAD and NAMPT levels in skeletal muscle of aged rats. Mech Ageing Dev 131(1):21–28. https://doi.org/10.1016/j.mad.2009.11.002

Kraus D, Yang Q, Kong D, Banks AS, Zhang L, Rodgers JT, Pirinen E, Pulinilkunnil TC, Gong F, Wang YC, Cen Y, Sauve AA, Asara JM, Peroni OD, Monia BP, Bhanot S, Alhonen L, Puigserver P, Kahn BB (2014) Nicotinamide N-methyltransferase knockdown protects against diet-induced obesity. Nature 508(7495):258–262. https://doi.org/10.1038/nature13198

Kropotov A, Kulikova V, Nerinovski K, Yakimov A, Svetlova M, Solovjeva L, Sudnitsyna J, Migaud ME, Khodorkovskiy M, Ziegler M, Nikiforov A (2021) Equilibrative nucleoside transporters mediate the import of nicotinamide riboside and nicotinic acid riboside into human cells. Int J Mol Sci 22(3). https://doi.org/10.3390/ijms22031391

Lamb DA, Moore JH, Mesquita PHC, Smith MA, Vann CG, Osburn SC, Fox CD, Lopez HL, Ziegenfuss TN, Huggins KW, Goodlett MD, Fruge AD, Kavazis AN, Young KC, Roberts MD (2020) Resistance training increases muscle NAD(+) and NADH concentrations as well as NAMPT protein levels and global sirtuin activity in middle-aged, overweight, untrained individuals. Aging (Albany NY) 12(10):9447–9460. https://doi.org/10.18632/aging.103218

Lautrup S, Sinclair DA, Mattson MP, Fang EF (2019) NAD(+) in brain aging and neurodegenerative disorders. Cell Metab 30(4):630–655. https://doi.org/10.1016/j.cmet.2019.09.001

Lee HC (2006) Structure and enzymatic functions of human CD38. Mol Med 12(11–12):317–323. https://doi.org/10.2119/2006-00086.Lee

Lee IH, Cao L, Mostoslavsky R, Lombard DB, Liu J, Bruns NE, Tsokos M, Alt FW, Finkel T (2008) A role for the NAD-dependent deacetylase Sirt1 in the regulation of autophagy. Proc Natl Acad Sci USA 105(9):3374–3379. https://doi.org/10.1073/pnas.0712145105

Lee HJ, Hong YS, Jun W, Yang SJ (2015) Nicotinamide riboside ameliorates hepatic metaflammation by modulating NLRP3 Inflammasome in a rodent model of type 2 diabetes. J Med Food 18(11):1207–1213. https://doi.org/10.1089/jmf.2015.3439

Li J, Bonkowski MS, Moniot S, Zhang D, Hubbard BP, Ling AJ, Rajman LA, Qin B, Lou Z, Gorbunova V, Aravind L, Steegborn C, Sinclair DA (2017a) A conserved NAD(+) binding pocket that regulates protein-protein interactions during aging. Science 355(6331):1312–1317. https://doi.org/10.1126/science.aad8242

Li S, Dou X, Ning H, Song Q, Wei W, Zhang X, Shen C, Li J, Sun C, Song Z (2017b) Sirtuin 3 acts as a negative regulator of autophagy dictating hepatocyte susceptibility to lipotoxicity. Hepatology 66(3):936–952. https://doi.org/10.1002/hep.29229

Liu D, Gharavi R, Pitta M, Gleichmann M, Mattson MP (2009) Nicotinamide prevents NAD+ depletion and protects neurons against excitotoxicity and cerebral ischemia: NAD+ consumption by SIRT1 may endanger energetically compromised neurons. NeuroMolecular Med 11(1):28–42. https://doi.org/10.1007/s12017-009-8058-1

Luongo TS, Eller JM, Lu MJ, Niere M, Raith F, Perry C, Bornstein MR, Oliphint P, Wang L, McReynolds MR, Migaud ME, Rabinowitz JD, Johnson FB, Johnsson K, Ziegler M, Cambronne XA, Baur JA (2020) SLC25A51 is a mammalian mitochondrial NAD(+) transporter. Nature 588(7836):174–179. https://doi.org/10.1038/s41586-020-2741-7

Mach J, Midgley AW, Dank S, Grant RS, Bentley DJ (2010) The effect of antioxidant supplementation on fatigue during exercise: potential role for NAD+(H). Nutrients 2(3):319–329. https://doi.org/10.3390/nu2030319

Mao Z, Hine C, Tian X, Van Meter M, Au M, Vaidya A, Seluanov A, Gorbunova V (2011) SIRT6 promotes DNA repair under stress by activating PARP1. Science 332(6036):1443–1446. https://doi.org/10.1126/science.1202723

Marletta AS, Massarotti A, Orsomando G, Magni G, Rizzi M, Garavaglia S (2015) Crystal structure of human nicotinic acid phosphoribosyltransferase. FEBS Open Bio 5:419–428. https://doi.org/10.1016/j.fob.2015.05.002

Mattson MP, Gleichmann M, Cheng A (2008) Mitochondria in neuroplasticity and neurological disorders. Neuron 60(5):748–766. https://doi.org/10.1016/j.neuron.2008.10.010

Mayevsky A, Rogatsky GG (2007) Mitochondrial function in vivo evaluated by NADH fluorescence: from animal models to human studies. Am J Physiol Cell Physiol 292(2):C615–C640. https://doi.org/10.1152/ajpcell.00249.2006

Megarity CF, Gill JR, Caraher MC, Stratford IJ, Nolan KA, Timson DJ (2014) The two common polymorphic forms of human NRH-quinone oxidoreductase 2 (NQO2) have different biochemical properties. FEBS Lett 588(9):1666–1672. https://doi.org/10.1016/j.febslet.2014.02.063

Mendelsohn AR, Larrick JW (2019) Interacting NAD(+) and cell senescence pathways complicate antiaging therapies. Rejuvenation Res 22(3):261–266. https://doi.org/10.1089/rej.2019.2218

Mills KF, Yoshida S, Stein LR, Grozio A, Kubota S, Sasaki Y, Redpath P, Migaud ME, Apte RS, Uchida K, Yoshino J, Imai SI (2016) Long-term administration of nicotinamide mononucleotide mitigates age-associated physiological decline in mice. Cell Metab 24(6):795–806. https://doi.org/10.1016/j.cmet.2016.09.013

Minhas PS, Liu L, Moon PK, Joshi AU, Dove C, Mhatre S, Contrepois K, Wang Q, Lee BA, Coronado M, Bernstein D, Snyder MP, Migaud M, Majeti R, Mochly-Rosen D, Rabinowitz JD, Andreasson KI (2019) Macrophage de novo NAD(+) synthesis specifies immune function in aging and inflammation. Nat Immunol 20(1):50–63. https://doi.org/10.1038/s41590-018-0255-3

Mitchell SJ, Bernier M, Aon MA, Cortassa S, Kim EY, Fang EF, Palacios HH, Ali A, Navas-Enamorado I, Di Francesco A, Kaiser TA, Waltz TB, Zhang N, Ellis JL, Elliott PJ, Frederick DW, Bohr VA, Schmidt MS, Brenner C et al (2018) Nicotinamide improves aspects of Healthspan, but not lifespan, in mice. Cell Metab 27(3):667–676 e664. https://doi.org/10.1016/j.cmet.2018.02.001

Mokudai T, Ayoub IA, Sakakibara Y, Lee EJ, Ogilvy CS, Maynard KI (2000) Delayed treatment with nicotinamide (Vitamin B(3)) improves neurological outcome and reduces infarct volume

after transient focal cerebral ischemia in Wistar rats. Stroke 31(7):1679–1685. https://doi.org/10.1161/01.str.31.7.1679
Mouchiroud L, Houtkooper RH, Moullan N, Katsyuba E, Ryu D, Canto C, Mottis A, Jo YS, Viswanathan M, Schoonjans K, Guarente L, Auwerx J (2013) The NAD(+)/Sirtuin pathway modulates longevity through activation of mitochondrial UPR and FOXO signaling. Cell 154(2):430–441. https://doi.org/10.1016/j.cell.2013.06.016
Mousa TY, Mousa OY (2023) Nicotinic acid deficiency. StatPearls. https://www.ncbi.nlm.nih.gov/pubmed/32491681
Mularski RA, Grazer RE, Santoni L, Strother JS, Bizovi KE (2006) Treatment advice on the internet leads to a life-threatening adverse reaction: hypotension associated with niacin overdose. Clin Toxicol (Phila) 44(1):81–84. https://doi.org/10.1080/15563650500394928
Nacarelli T, Lau L, Fukumoto T, Zundell J, Fatkhutdinov N, Wu S, Aird KM, Iwasaki O, Kossenkov AV, Schultz D, Noma KI, Baur JA, Schug Z, Tang HY, Speicher DW, David G, Zhang R (2019) NAD(+) metabolism governs the proinflammatory senescence-associated secretome. Nat Cell Biol 21(3):397–407. https://doi.org/10.1038/s41556-019-0287-4
Nadeeshani H, Li J, Ying T, Zhang B, Lu J (2022) Nicotinamide mononucleotide (NMN) as an anti-aging health product - promises and safety concerns. J Adv Res 37:267–278. https://doi.org/10.1016/j.jare.2021.08.003
Navas LE, Carnero A (2021) NAD(+) metabolism, stemness, the immune response, and cancer. Signal Transduct Target Ther 6(1):2. https://doi.org/10.1038/s41392-020-00354-w
Nooka AK, Kaufman JL, Hofmeister CC, Joseph NS, Heffner TL, Gupta VA, Sullivan HC, Neish AS, Dhodapkar MV, Lonial S (2019) Daratumumab in multiple myeloma. Cancer 125(14):2364–2382. https://doi.org/10.1002/cncr.32065
Pellicciari R, Liscio P, Giacche N, De Franco F, Carotti A, Robertson J, Cialabrini L, Katsyuba E, Raffaelli N, Auwerx J (2018) Alpha-amino-beta-carboxymuconate-epsilon-semialdehyde decarboxylase (ACMSD) inhibitors as novel modulators of De novo nicotinamide adenine dinucleotide (NAD(+)) biosynthesis. J Med Chem 61(3):745–759. https://doi.org/10.1021/acs.jmedchem.7b01254
Pickles S, Vigie P, Youle RJ (2018) Mitophagy and quality control mechanisms in mitochondrial maintenance. Curr Biol 28(4):R170–R185. https://doi.org/10.1016/j.cub.2018.01.004
Pirinen E, Auranen M, Khan NA, Brilhante V, Urho N, Pessia A, Hakkarainen A, Kuula J, Heinonen U, Schmidt MS, Haimilahti K, Piirila P, Lundbom N, Taskinen MR, Brenner C, Velagapudi V, Pietilainen KH, Suomalainen A (2020) Niacin cures systemic NAD(+) deficiency and improves muscle performance in adult-onset mitochondrial myopathy. Cell Metab 31(6):1078–1090 e1075. https://doi.org/10.1016/j.cmet.2020.04.008
Preiss J, Handler P (1958) Biosynthesis of Diphosphopyridine nucleotide: II. Enzymatic aspects. J Biol Chem 233(2):493–500. https://doi.org/10.1016/S0021-9258(18)64790-8
Price NL, Gomes AP, Ling AJ, Duarte FV, Martin-Montalvo A, North BJ, Agarwal B, Ye L, Ramadori G, Teodoro JS, Hubbard BP, Varela AT, Davis JG, Varamini B, Hafner A, Moaddel R, Rolo AP, Coppari R, Palmeira CM et al (2012) SIRT1 is required for AMPK activation and the beneficial effects of resveratrol on mitochondrial function. Cell Metab 15(5):675–690. https://doi.org/10.1016/j.cmet.2012.04.003
Qin W, Yang T, Ho L, Zhao Z, Wang J, Chen L, Zhao W, Thiyagarajan M, MacGrogan D, Rodgers JT, Puigserver P, Sadoshima J, Deng H, Pedrini S, Gandy S, Sauve AA, Pasinetti GM (2006) Neuronal SIRT1 activation as a novel mechanism underlying the prevention of Alzheimer disease amyloid neuropathology by calorie restriction. J Biol Chem 281(31):21745–21754. https://doi.org/10.1074/jbc.M602909200
Rajman L, Chwalek K, Sinclair DA (2018) Therapeutic potential of NAD-boosting molecules: the in vivo evidence. Cell Metab 27(3):529–547. https://doi.org/10.1016/j.cmet.2018.02.011
Reiten OK, Wilvang MA, Mitchell SJ, Hu Z, Fang EF (2021) Preclinical and clinical evidence of NAD(+) precursors in health, disease, and ageing. Mech Ageing Dev 199:111567. https://doi.org/10.1016/j.mad.2021.111567

Rimmele P, Bigarella CL, Liang R, Izac B, Dieguez-Gonzalez R, Barbet G, Donovan M, Brugnara C, Blander JM, Sinclair DA, Ghaffari S (2014) Aging-like phenotype and defective lineage specification in SIRT1-deleted hematopoietic stem and progenitor cells. Stem Cell Reports 3(1):44–59. https://doi.org/10.1016/j.stemcr.2014.04.015

Rodrigues LGF, de Araujo LD, Roa SLR, Bueno AC, Uchoa ET, Antunes-Rodrigues J, Moreira AC, Elias LLK, de Castro M, Martins CS (2021) Restricted feeding modulates peripheral clocks and nutrient sensing pathways in rats. Arch Endocrinol Metab 65(5):549–561. https://doi.org/10.20945/2359-3997000000407

Sahlin K (1983) NADH and NADPH in human skeletal muscle at rest and during ischaemia. Clin Physiol 3(5):477–485. https://doi.org/10.1111/j.1475-097x.1983.tb00856.x

Sahlin K, Gorski J, Edstrom L (1990) Influence of ATP turnover and metabolite changes on IMP formation and glycolysis in rat skeletal muscle. Am J Phys 259(3 Pt 1):C409–C412. https://doi.org/10.1152/ajpcell.1990.259.3.C409

Saldanha LG, Dwyer JT, Hardy CJ, MacKay DJ (2023) Perspectives on the use of proprietary blends in dietary supplements. J Nutr 153(5):1305–1308. https://doi.org/10.1016/j.tjnut.2023.03.035

Scheibye-Knudsen M, Mitchell SJ, Fang EF, Iyama T, Ward T, Wang J, Dunn CA, Singh N, Veith S, Hasan-Olive MM, Mangerich A, Wilson MA, Mattson MP, Bergersen LH, Cogger VC, Warren A, Le Couteur DG, Moaddel R, Wilson DM 3rd et al (2014) A high-fat diet and NAD(+) activate Sirt1 to rescue premature aging in cockayne syndrome. Cell Metab 20(5):840–855. https://doi.org/10.1016/j.cmet.2014.10.005

Schmauck-Medina T, Moliere A, Lautrup S, Zhang J, Chlopicki S, Madsen HB, Cao S, Soendenbroe C, Mansell E, Vestergaard MB, Li Z, Shiloh Y, Opresko PL, Egly JM, Kirkwood T, Verdin E, Bohr VA, Cox LS, Stevnsner T et al (2022) New hallmarks of ageing: a 2022 Copenhagen ageing meeting summary. Aging (Albany NY) 14(16):6829–6839. https://doi.org/10.18632/aging.204248

Schondorf DC, Ivanyuk D, Baden P, Sanchez-Martinez A, De Cicco S, Yu C, Giunta I, Schwarz LK, Di Napoli G, Panagiotakopoulou V, Nestel S, Keatinge M, Pruszak J, Bandmann O, Heimrich B, Gasser T, Whitworth AJ, Deleidi M (2018) The NAD+ precursor nicotinamide riboside rescues mitochondrial defects and neuronal loss in iPSC and Fly models of Parkinson's disease. Cell Rep 23(10):2976–2988. https://doi.org/10.1016/j.celrep.2018.05.009

Schreiber V, Ame JC, Dolle P, Schultz I, Rinaldi B, Fraulob V, Menissier-de Murcia J, de Murcia G (2002) Poly(ADP-ribose) polymerase-2 (PARP-2) is required for efficient base excision DNA repair in association with PARP-1 and XRCC1. J Biol Chem 277(25):23028–23036. https://doi.org/10.1074/jbc.M202390200

Schwer B, Verdin E (2008) Conserved metabolic regulatory functions of sirtuins. Cell Metab 7(2):104–112. https://doi.org/10.1016/j.cmet.2007.11.006

Sepehrnia B, Kamboh MI, Adams-Campbell LL, Bunker CH, Nwankwo M, Majumder PP, Ferrell RE (1989) Genetic studies of human apolipoproteins. X. The effect of the apolipoprotein E polymorphism on quantitative levels of lipoproteins in Nigerian blacks. Am J Hum Genet 45(4):586–591. https://www.ncbi.nlm.nih.gov/pubmed/2491016

Shats I, Williams JG, Liu J, Makarov MV, Wu X, Lih FB, Deterding LJ, Lim C, Xu X, Randall TA, Lee E, Li W, Fan W, Li JL, Sokolsky M, Kabanov AV, Li L, Migaud ME, Locasale JW, Li X (2020) Bacteria boost mammalian host NAD metabolism by engaging the deamidated biosynthesis pathway. Cell Metab 31(3):564–579 e567. https://doi.org/10.1016/j.cmet.2020.02.001

Shieh WM, Ame JC, Wilson MV, Wang ZQ, Koh DW, Jacobson MK, Jacobson EL (1998) Poly(ADP-ribose) polymerase null mouse cells synthesize ADP-ribose polymers. J Biol Chem 273(46):30069–30072. https://doi.org/10.1074/jbc.273.46.30069

Stock AJ, Ayyar S, Kashyap A, Wang Y, Yanai H, Starost MF, Tanaka-Yano M, Bodogai M, Sun C, Wang Y, Gong Y, Puligilla C, Fang EF, Bohr VA, Liu Y, Beerman I (2023) Boosting NAD ameliorates hematopoietic impairment linked to short telomeres in vivo. Geroscience 45:2213–2228. https://doi.org/10.1007/s11357-023-00752-2

Su G, Sun G, Liu H, Shu L, Zhang J, Guo L, Huang C, Xu J (2015) Niacin suppresses progression of atherosclerosis by inhibiting vascular inflammation and apoptosis of vascular smooth muscle cells. Med Sci Monit 21:4081–4089. https://doi.org/10.12659/msm.895547

Tang BL (2016) Sirt1 and the mitochondria. Mol Cell 39(2):87–95. https://doi.org/10.14348/molcells.2016.2318

Tang X, Li Y, Liu L, Guo R, Zhang P, Zhang Y, Zhang Y, Zhao J, Su J, Sun L, Liu Y (2020) Sirtuin 3 induces apoptosis and necroptosis by regulating mutant p53 expression in small-cell lung cancer. Oncol Rep 43(2):591–600. https://doi.org/10.3892/or.2019.7439

Tivey HS, Brook AJ, Rokicki MJ, Kipling D, Davis T (2013) p38 (MAPK) stress signalling in replicative senescence in fibroblasts from progeroid and genomic instability syndromes. Biogerontology 14(1):47–62. https://doi.org/10.1007/s10522-012-9407-2

Trammell SA, Yu L, Redpath P, Migaud ME, Brenner C (2016) Nicotinamide riboside is a major NAD+ precursor vitamin in cow milk. J Nutr 146(5):957–963. https://doi.org/10.3945/jn.116.230078

Turner RS, Thomas RG, Craft S, van Dyck CH, Mintzer J, Reynolds BA, Brewer JB, Rissman RA, Raman R, Aisen PS, Cooperative A's D, S. (2015) A randomized, double-blind, placebo-controlled trial of resveratrol for Alzheimer disease. Neurology 85(16):1383–1391. https://doi.org/10.1212/WNL.0000000000002035

Vannini N, Campos V, Girotra M, Trachsel V, Rojas-Sutterlin S, Tratwal J, Ragusa S, Stefanidis E, Ryu D, Rainer PY, Nikitin G, Giger S, Li TY, Semilietof A, Oggier A, Yersin Y, Tauzin L, Pirinen E, Cheng WC et al (2019) The NAD-booster nicotinamide riboside potently stimulates hematopoiesis through increased mitochondrial clearance. Cell Stem Cell 24(3):405–418 e407. https://doi.org/10.1016/j.stem.2019.02.012

Vazquez BN, Thackray JK, Serrano L (2017) Sirtuins and DNA damage repair: SIRT7 comes to play. Nucleus 8(2):107–115. https://doi.org/10.1080/19491034.2016.1264552

Vazquez-Rosa E, Shin MK, Dhar M, Chaubey K, Cintron-Perez CJ, Tang X, Liao X, Miller E, Koh Y, Barker S, Franke K, Crosby DR, Schroeder R, Emery J, Yin TC, Fujioka H, Reynolds JD, Harper MM, Jain MK, Pieper AA (2020) P7C3-A20 treatment one year after TBI in mice repairs the blood-brain barrier, arrests chronic neurodegeneration, and restores cognition. Proc Natl Acad Sci USA 117(44):27667–27675. https://doi.org/10.1073/pnas.2010430117

Verdin E (2015) NAD(+) in aging, metabolism, and neurodegeneration. Science 350(6265):1208–1213. https://doi.org/10.1126/science.aac4854

Villines TC, Kim AS, Gore RS, Taylor AJ (2012) Niacin: the evidence, clinical use, and future directions. Curr Atheroscler Rep 14(1):49–59. https://doi.org/10.1007/s11883-011-0212-1

Wang X, He HJ, Xiong X, Zhou S, Wang WW, Feng L, Han R, Xie CL (2021) NAD(+) in Alzheimer's disease: molecular mechanisms and systematic therapeutic evidence obtained in vivo. Front Cell Dev Biol 9:668491. https://doi.org/10.3389/fcell.2021.668491

Wang X, Liu R, Wei C, Xu M, Li Y (2022) Exogenous nucleotides improved the oxidative stress and Sirt-1 protein level of brown adipose tissue on senescence-accelerated mouse Prone-8 (SAMP8) mice. Nutrients 14(14). https://doi.org/10.3390/nu14142796

Weller J, Budson A (2018) Current understanding of Alzheimer's disease diagnosis and treatment. F1000Res 7:1161. https://doi.org/10.12688/f1000research.14506.1

Wolfram-Schauerte M, Hofer K (2023) NAD-capped RNAs - a redox cofactor meets RNA. Trends Biochem Sci 48(2):142–155. https://doi.org/10.1016/j.tibs.2022.08.004

Wong W (2020) Economic burden of Alzheimer disease and managed care considerations. Am J Manag Care 26(8 Suppl):S177–S183. https://doi.org/10.37765/ajmc.2020.88482

Wu H, Chen Q (2015) Hypoxia activation of mitophagy and its role in disease pathogenesis. Antioxid Redox Signal 22(12):1032–1046. https://doi.org/10.1089/ars.2014.6204

Xie N, Zhang L, Gao W, Huang C, Huber PE, Zhou X, Li C, Shen G, Zou B (2020) NAD(+) metabolism: pathophysiologic mechanisms and therapeutic potential. Signal Transduct Target Ther 5(1):227. https://doi.org/10.1038/s41392-020-00311-7

Yang H, Yang T, Baur JA, Perez E, Matsui T, Carmona JJ, Lamming DW, Souza-Pinto NC, Bohr VA, Rosenzweig A, de Cabo R, Sauve AA, Sinclair DA (2007) Nutrient-sensitive mitochondrial NAD+ levels dictate cell survival. Cell 130(6):1095–1107. https://doi.org/10.1016/j.cell.2007.07.035

Yang SJ, Choi JM, Kim L, Park SE, Rhee EJ, Lee WY, Oh KW, Park SW, Park CY (2014) Nicotinamide improves glucose metabolism and affects the hepatic NAD-sirtuin pathway in a rodent model of obesity and type 2 diabetes. J Nutr Biochem 25(1):66–72. https://doi.org/10.1016/j.jnutbio.2013.09.004

Yang Y, Mohammed FS, Zhang N, Sauve AA (2019) Dihydronicotinamide riboside is a potent NAD(+) concentration enhancer in vitro and in vivo. J Biol Chem 294(23):9295–9307. https://doi.org/10.1074/jbc.RA118.005772

Yang Y, Zhang N, Zhang G, Sauve AA (2020) NRH salvage and conversion to NAD(+) requires NRH kinase activity by adenosine kinase. Nat Metab 2(4):364–379. https://doi.org/10.1038/s42255-020-0194-9

Ying W (2008) NAD+/NADH and NADP+/NADPH in cellular functions and cell death: regulation and biological consequences. Antioxid Redox Signal 10(2):179–206. https://doi.org/10.1089/ars.2007.1672

Yoshino J, Mills KF, Yoon MJ, Imai S (2011) Nicotinamide mononucleotide, a key NAD(+) intermediate, treats the pathophysiology of diet- and age-induced diabetes in mice. Cell Metab 14(4):528–536. https://doi.org/10.1016/j.cmet.2011.08.014

Yoshino J, Baur JA, Imai SI (2018) NAD(+) intermediates: the biology and therapeutic potential of NMN and NR. Cell Metab 27(3):513–528. https://doi.org/10.1016/j.cmet.2017.11.002

Zapata-Perez R, Wanders RJA, van Karnebeek CDM, Houtkooper RH (2021) NAD(+) homeostasis in human health and disease. EMBO Mol Med 13(7):e13943. https://doi.org/10.15252/emmm.202113943

Zeman M, Vecka M, Perlik F, Stankova B, Hromadka R, Tvrzicka E, Sirc J, Hrib J, Zak A (2016) Pleiotropic effects of niacin: current possibilities for its clinical use. Acta Pharma 66(4):449–469. https://doi.org/10.1515/acph-2016-0043

Zhang T, Kraus WL (2010) SIRT1-dependent regulation of chromatin and transcription: linking NAD(+) metabolism and signaling to the control of cellular functions. Biochim Biophys Acta 1804(8):1666–1675. https://doi.org/10.1016/j.bbapap.2009.10.022

Zhang H, Ryu D, Wu Y, Gariani K, Wang X, Luan P, D'Amico D, Ropelle ER, Lutolf MP, Aebersold R, Schoonjans K, Menzies KJ, Auwerx J (2016) NAD(+) repletion improves mitochondrial and stem cell function and enhances life span in mice. Science 352(6292):1436–1443. https://doi.org/10.1126/science.aaf2693

Zong L, Tanaka-Yano M, Park B, Yanai H, Turhan FT, Croteau DL, Tian J, Fang EF, Bohr VA, Beerman I (2021) NAD(+) augmentation with nicotinamide riboside improves lymphoid potential of Atm(−/−) and old mice HSCs. NPJ Aging Mech Dis 7(1):25. https://doi.org/10.1038/s41514-021-00078-3

Zuo L, Motherwell MS (2013) The impact of reactive oxygen species and genetic mitochondrial mutations in Parkinson's disease. Gene 532(1):18–23. https://doi.org/10.1016/j.gene.2013.07.085

Chapter 5
Unlocking the Potential of Senolytic Compounds: Advancements, Opportunities, and Challenges in Ageing-Related Research

Lilian Sales Gomez and Diana Jurk

Abstract Cellular senescence is recognised as a contributor to the ageing process and the development of multiple age-related conditions. Researchers have launched efforts to identify compounds capable to selectively kill senescent cells, known as senolytics, without affecting non senescent cells. As of now, over 40 compounds have demonstrated senolytic properties, offering promising prospects for reversing or ameliorating age-related conditions in preclinical studies.

This chapter presents the most recent developments in senolytic drug research, encompassing investigations spanning basic science, preclinical trials, and clinical studies. While many of these investigations have generated encouraging results in the realm of age-related interventions, this chapter also addresses potential challenges and pitfalls.

Keywords Cellular senescence · Ageing · Age-related diseases · Senolytic

Introduction

According to the World Health Organization (WHO), in 2020, the global population of people aged 60 years and older, exceeded the number of children under 5 years of age. Also, it is predicted that by 2050 the proportion of the world's population over 60 years will double. Even though global longevity has risen, there hasn't always been a corresponding improvement in overall health and well-being among older populations (WHO 2021). In fact, ageing is the major risk factor for the

L. S. Gomez · D. Jurk (✉)
Department of Physiology and Biomedical Engineering, Mayo Clinic, Rochester, MN, USA

Robert and Arlene Kogod Center on Aging, Mayo Clinic, Rochester, MN, USA
e-mail: Jurk.Diana@mayo.edu

© The Author(s), under exclusive license to Springer Nature Switzerland AG 2024

V. I. Korolchuk, J. R. Harris (eds.), *Biochemistry and Cell Biology of Ageing: Part V, Anti-Ageing Interventions*, Subcellular Biochemistry 107,
https://doi.org/10.1007/978-3-031-66768-8_5

development of the majority of age-related diseases (Marengoni et al. 2011; Barnett et al. 2012; López-Otín et al. 2013).

The geroscience hypothesis is a concept that suggests that understanding the biology of ageing, specifically the underlying molecular and cellular processes, can lead to significant advances in preventing or delaying a wide range of age-related diseases and conditions. It posits that targeting the fundamental mechanisms of ageing itself, rather than individual diseases, can result in more effective interventions to improve healthspan (Sierra 2016). Ageing is characterised by multiple molecular and cellular changes, including genomic instability, telomere attrition, epigenetic alterations, loss of proteostasis, dysfunctional macroautophagy, deregulated nutrient-sensing, mitochondrial dysfunction, stem cell exhaustion, altered intercellular communication, chronic inflammation, dysbiosis and cellular senescence (López-Otín et al. 2023). Within this array of alterations, this chapter will center its attention on cellular senescence, which is increasingly recognised as a promising focal point for interventions aimed at mitigating age-related comorbidities.

Cellular Senescence

A cell is considered senescent when it reaches a state of irreversible cell cycle arrest in response to different damaging stimuli (Muñoz-Espín and Serrano 2014). Senescent cells undergo profound transformations across various facets of their biology, encompassing alterations in morphology, metabolism, gene expression amongst many others (Muñoz-Espín and Serrano 2014). Senescent cells have been shown to play an important role as a safeguard to prevent proliferation of damaged cells and against tumorigenesis (van Deursen 2014). Also, these cells act as positive regulators of tissue remodelling and repair during development and adulthood (Storer et al. 2013; Muñoz-Espín and Serrano 2014, Gorgoulis et al. 2019). Senescent cells produce and secrete inflammatory factors, including chemokines which attract immune cells (Coppé et al. 2008). Through this mechanism, senescent cells are able to mediate their own immune clearance from tissues (Krizhanovsky et al. 2008; Kang et al. 2011).

Currently, there is a growing understanding that as individuals age, the immune system's diminished capacity to clear these cells from tissues leads to their accumulation (Ovadya et al. 2018; Gorgoulis et al. 2019). Senescent cells that manage to evade immune surveillance can play a role in promoting chronic inflammation through the SASP. This chronic SASP has the potential to not only foster chronic inflammation but also propagate senescence to neighbouring cells (Acosta et al. 2013). Additionally, it can have detrimental effects on the function of stem cells, disrupt the composition of the extracellular matrix, and even facilitate the initiation of tumorigenesis (Gorgoulis et al. 2019). For these reasons, it has been hypothesized that the accumulation of senescent cells can causally contribute to ageing and many age-related conditions.

To investigate this hypothesis, researchers have created mouse models that allow for the selective removal or elimination of senescent cells. For instance, researchers generated the INK-ATTAC or the p16 3MR mouse models, which are genetically engineered mouse models that can specifically eliminate p16 expressing senescent cells *in vivo* (Baker et al. 2011; Demaria et al. 2014). Using these models (as well as others not mentioned here), several groups investigated the role of elimination of p16-positive cells in various contexts, including ageing, age-related diseases, and tissue regeneration. These studies have established the proof-of-concept that targeting senescent cells could be a promising approach to ameliorate various age-related diseases. This concept has subsequently paved the way for the development of interventions that will be discussed in detail in the remainder of this chapter.

While we have gained substantial knowledge about cellular senescence since its initial discovery in 1961 (Hayflick and Moorhead 1961) through *in vitro* studies and have identified numerous markers for their detection, a significant hurdle in the field remains the reliable identification of senescent cells *in vivo*. Firstly, there is no single specific marker that unequivocally identifies a senescent cell. Consequently, the field has embraced the concept that a combination of markers must be assessed concurrently to accurately identify senescent cells. This multi-marker approach is essential for the reliable detection of senescent cells in various biological contexts (Gorgoulis et al. 2019). Moreover, cellular senescence is a heterogeneous phenotype, and its characteristics can vary depending on factors such as cell type, the tissue microenvironment, and the specific triggering stimuli. Consequently, the prevailing efforts in the field have been focused on systematically mapping senescent cells across diverse tissues during the ageing process, leveraging state-of-the-art multiomics technologies to comprehensively understand this complexity (Gurkar et al. 2023).

Despite our current limitations in fully understanding cellular senescence and its *in vivo* heterogeneity, substantial efforts are being made by both academia and industry to discover interventions capable of selectively targeting and eliminating these cells. These interventions, commonly referred to as senolytics, are at the forefront of research and development. In the subsequent sections of this chapter, we will delve into the reasoning behind the development of senolytic therapies and provide an overview of the current state of the field.

Senolytic Therapies

As previously discussed, the presence of senescent cells is strongly associated with the development of age-related dysfunction. Employing genetic approaches to eliminate these cells has been shown to enhance healthspan and delay the emergence of age-related traits in mice (Baker et al. 2011). Since genetic modification techniques are not practical or feasible in humans, pharmacological interventions were developed. This led to the emergence of a class of drugs commonly known as senolytic, which are capable of specifically targeting and eliminating senescent cells.

The conceptual foundation for the development of senolytic drugs originated from the discovery made by Eugenia Wang in 1995 (Wang 1995). Her work revealed that senescent cells, when cultured *in vitro*, exhibited resistance to apoptotic signals. This discovery provided the rationale for exploring compounds capable of overcoming this apoptosis resistance and inducing the selective death of senescent cells. James Kirkland's laboratory employed bioinformatic approaches to identify Senescence-associated anti-apoptotic pathways (SCAPs). Subsequently, they tested drugs with the potential to specifically target these pathways. This research effort led to the discovery that dasatinib, an inhibitor of multiple tyrosine kinases, was effective at eliminating senescent preadipocytes. Quercetin, a flavonoid found in various plants and foods known to target multiple SCAPs, displayed a higher degree of selectivity in killing senescent endothelial cells and mouse mesenchymal cells. *In vivo* experiments further demonstrated the effectiveness of a combination of dasatinib and quercetin (D + Q) in reducing the burden of senescent cells in three distinct mouse models characterised by high levels of senescence (Zhu et al. 2015).

Dasatinib + Quercetin

Since its discovery in 2015, D + Q has been the subject of extensive investigation as a senolytic therapy in preclinical models within the context of ageing and multiple age-related diseases. D + Q has been shown to improve functional outcomes in multiple organs and in the context of brain ageing, Alzheimer's disease, pulmonary fibrosis, musculoskeletal ageing, hepatic steatosis, diabetes amongst many others (Ogrodnik et al. 2017, 2021; Schafer et al. 2017; Zhang et al. 2019, 2022).

The promising outcomes observed in pre-clinical studies paved the way for the feasibility of testing D + Q in human trials, and in 2019, the first report of such a study was published. This study involved the recruitment of 14 patients diagnosed with Idiopathic Pulmonary Fibrosis (IPF), who received a combination of dasatinib (D) at a dose of 100 mg/day and quercetin (Q) at a dose of 1250 mg/day, administered 3 days a week over the course of 3 weeks. This study indicated that the treatment led to a clinically meaningful improvement in physical dysfunction among the patients. However, there were no observable changes in pulmonary function, clinical laboratory parameters, frailty index (FI-LAB), or self-reported health status. In addition, the effects of D + Q on circulating Senescence-Associated Secretory Phenotype (SASP) factors were inconclusive (Sciences 2016–2019; Justice et al. 2019). In another clinical trial, administering D only over an extended period to patients with systemic sclerosis resulted in a decreased SASP and other markers of senescence in skin biopsies (Martyanov et al. 2019). An additional study showed that treatment with the senolytic drug cocktail D + Q significantly decreased senescence-associated markers in human adipose tissue of individuals with diabetic kidney disease (Hickson 2016–2025; Hickson et al. 2019).

D + Q has also been shown to have beneficial effects in mice brains, in the context of age-related cognitive function and Alzheimer's disease. Therefore, a randomized, double-blind, placebo-controlled multicentre Phase II trial is currently underway to delve deeper into the safety, feasibility, and effectiveness of D + Q in influencing the advancement of Alzheimer's disease (Gonzales 2020–2023; Gonzales et al. 2022).

ABT-737 and ABT-263 (Navitoclax)

The development of BCL-2 family inhibitors initially emerged as a potential therapy for cancer treatment due to the central role of these proteins in regulating programmed cell death, and their overexpression in many cancer types. One such inhibitor, ABT-737, was discovered through nuclear magnetic resonance (NMR)-based screening, and it was found to be a BH3 mimetic capable of targeting anti-apoptotic proteins like Bcl-2, Bcl-XL, and Bcl-w, leading to cell death (Oltersdorf et al. 2005). Years later, this rationale was confirmed in senescent cells, where the increased presence of BCL-W and BCL-XL was found to underlie senescent cell resistance to cell death. ABT-737 treatment was shown to preferentially induce apoptosis in senescent cells, both *in vitro* and *in vivo*, effectively eliminating these cells from tissues (Yosef et al. 2016). However, ABT-737 faced limitations in terms of oral bioavailability and aqueous solubility, making it less suitable for clinical use. As a result, a derivative molecule with improved pharmaceutical properties, ABT-263 (Navitoclax), was developed. This second-generation BCL-2 family inhibitor is orally bioavailable and exhibits selective cytotoxicity to Bcl-2/Bcl-xL-dependent cells *in vitro*, as well as promoting tumour regression *in vivo* (Tse et al. 2008).

Navitoclax was tested in senescent human umbilical vein epithelial cells (HUVECs), IMR90 human lung fibroblasts, and murine embryonic fibroblasts (MEFs) cells and was shown to decrease the viability of these senescent cell types. On the other hand, navitoclax failed to induce cell death in senescent preadipocytes (Zhu et al. 2016). These results corroborate what was shown by siRNA targeting Bcl-xl, a member of the anti-apoptotic Bcl-2 family, which reduced viability of senescent HUVEC but had no effect on senescent human preadipocytes (Zhu et al. 2015).

To assess the potential beneficial effects of navitoclax in mammals, irradiated or naturally aged mice, were administered this drug. These mice exhibited a reduction in senescent cells, which was accompanied by improved functional outcomes in both irradiated and naturally aged mice (Chang et al. 2016).

Despite promising results, the use of ABT-737 and ABT-263 (Navitoclax) has raised safety concerns. Clinical trials and animal studies have indicated that thrombocytopenia, a condition characterised by low platelet levels, is a major side effect of these drugs (Zhang et al. 2007; Wilson et al. 2010; Schoenwaelder et al. 2011; Vogler et al. 2011). It was speculated that the non-specificity of both ABTs for all BCL-2 proteins family may be responsible for its side effects and a more specific inhibition of BCL-XL may maintain the intended effectiveness while preserving platelets and neutrophils thereby reducing toxicity (Leverson et al. 2015). Two more specific BCL-XL inhibitors were identified, the A-1331852 and A-1155463 (Wang et al. 2020), and both were shown to be senolytic in HUVECs and human lung fibroblasts (Zhu et al. 2017). However, it remains unclear whether these compounds have fewer side effects compared to ABT inhibitors. The balance between efficacy and safety in the development of senolytic drugs is a critical consideration in ongoing research and clinical trials.

HSP90 Inhibitors

Multiple studies have linked autophagy to the process of senescence. Autophagy encompasses various cellular processes aimed at promoting the degradation of intracellular components through lysosomal proteases (Korolchuk et al. 2017).

In one study, a library of autophagy regulators was screened, leading to the identification of HSP90 inhibitors as a novel class of drugs with senolytic effects (Fuhrmann-Stroissnigg et al. 2017). Heat shock protein 90 (HSP90) is a molecular chaperone expressed ubiquitously and plays a crucial role in stabilising and degrading proteins. As part of their mechanism, HSP90 inhibitors were found to downregulate the antiapoptotic phosphatidylinositol 3-kinase (PI3K)/AKT pathway. In this study, two compounds, geldanamycin and 17-AAG (also known as tanespimycin), were found to reduce the viability of senescent but not proliferative cells. Interestingly, geldanamycin was initially discovered in the bacterium *Streptomyces hygroscopicus* and was first used as an antibiotic. A synthetic derivative of geldanamycin, known as 17-DMAG, was also identified as an HSP90 inhibitor with improved bioavailability properties and was shown to act as a senolytic (Fuhrmann-Stroissnigg et al. 2017).

Ganetespib, a compound among the library of HSP90 inhibitors with demonstrated potential anticancer activity in preclinical tests (Youssef et al. 2023) was also found to exhibit senolytic activity (Fuhrmann-Stroissnigg et al. 2017).

Targeting p53

The mechanism by which senescent cells evade apoptosis is multifactorial, and understanding the pathways involved is crucial for the development of anti-ageing compounds. p53, also known as the tumour protein 53 or TP53, is a critical and well-studied protein that plays a central role in the regulation of multiple cellular functions including apoptosis. It is often referred to as the “guardian of the genome” because of its key role in maintaining genomic stability and preventing the development of cancer. Evidence suggests that targeting the p53 pathway indirectly may serve as a senolytic strategy.

Forkhead box protein O4 (FOXO4), which has been identified as a feature of senescence, is important for maintaining the viability of senescent cells (Baar et al. 2017). FOXO4 can promote senescence by interacting with p53, leading to the upregulation of the senescence regulator p21. Additionally, FOXO4 binds to p53 and prevents its translocation to the cytosol, thereby inhibiting apoptosis. In this way, p53 acts as a regulator of both apoptosis and senescence depending on its mode of regulation. Baar and colleagues designed a modified FOXO4 peptide known as D-(amino acid) retro inverso (FOXO4-DRI), which mimics the interaction surface of FOXO4 and p53, disrupting their interaction. This binding disruption allows p53 to translocate to the cytosol, where it can promote apoptosis in senescent cells. In aged animal models, the administration of FOXO4-DRI has been shown to alleviate frailty and loss of renal function (Baar et al. 2017). Another study demonstrated that FOXO4-DRI can alleviate the age-related onset of hypogonadism in aged male mice, likely by selectively inducing apoptosis in senescent Leydig cells as a result of disrupting the FOXO4/p53 interaction (Zhang et al. 2020).

The utility of FOXO4-DRI as a senolytic has also garnered attention in the field of bioengineering, particularly in the context of osteoarthritis (OA). OA is characterised by the degeneration of articular cartilage, and ageing and trauma are significant risk factors for its development (Wieland et al. 2005). One common procedure used to treat OA in humans is Autologous Chondrocyte Implantation (ACI). In this procedure, chondrocyte cells are expanded *in vitro* before being implanted into the patient. However, this expansion process can lead to the generation of senescent chondrocytes, which can negatively impact the quality and quantity of newly formed cartilage. A study tested FOXO4-DRI on chondrocytes isolated from healthy donors that underwent expansion. The results showed that treatment with FOXO4-DRI significantly reduced the level of senescence in these chondrocyte populations (Huang et al. 2021). This finding underscores the potential of senolytic compounds as a strategy to address diseases like OA and highlights the versatility of senolytics in various therapeutic applications.

In a mouse model of post-traumatic osteoarthritis (OA), local clearance of senescent cells using senolytic compound UBX0101 has been shown to mitigate the development of post-traumatic OA and promote a pro-regenerative environment (Jeon et al. 2017). UBX0101 also interferes with p53 activity, and its mechanism of action involves the inhibition of the MDM2/p53 interaction. MDM2 (Murine double minute 2) is a crucial negative regulator of p53 that facilitates the translocation of p53 from the nucleus to the cytoplasm for subsequent proteasomal degradation. To evaluate the safety, tolerability, and clinical effects of UBX0101, both single and repeat dose intra-articular administrations were assessed in patients suffering from moderate to severe painful knee osteoarthritis (OA). However, this phase 2 clinical trial failed to meet 12-week primary endpoint (Unity Biotechnology, Inc 2020).

MDM2 is regulated through ubiquitination, and when it becomes ubiquitinated, it is targeted for degradation by the ubiquitin-proteasome system. Ubiquitin-specific peptidase 7 (USP7) plays a critical role in deubiquitinating MDM2, which leads to the stabilisation of MDM2 and allows it to interact with p53. Small molecules such as P5091 or P22077 have been developed to inhibit the activity of USP7. These compounds have been shown to selectively induce apoptosis in senescent cells. They achieve this by reducing the expression of MDM2 and thereby restoring the activity of p53, which can trigger apoptosis in senescent cells (He et al. 2020a).

RG7112 (RO5045337), another small synthetic molecule that inhibits the MDM2/p53 interaction, has been tested as senolytic. It was reported that RG7112 selectively kills senescent intervertebral disc (IVD) cells through apoptosis (Cherif et al. 2020).

Fisetin

Quercetin, as previously mentioned, has been shown to induce apoptosis in senescent HUVECs, and its combination with Dasatinib extended the range of senescent cells targeted, with synergistic effects observed in specific cases (Zhu et al. 2015). Because quercetin is a naturally occurring flavonoid found in various fruits and vegetables, it raised the question of whether a related flavonoid, fisetin, might also possess senolytic properties. Studies have demonstrated that fisetin, like its analogue quercetin, has senolytic properties. Fisetin has been shown to induces cell death in senescent HUVECs by promoting apoptosis (Zhu et al. 2017).

In studies using aged mice, fisetin was found to reduce senescent cells in multiple tissues. This reduction in senescent cell burden led to the restoration of tissue homeostasis, a decrease in age-related pathology, and an extension of both median and maximum lifespan in the mice (Yousefzadeh et al. 2018). Importantly, fisetin also reduced senescence in a specific subset of cells within murine and human adipose tissue, highlighting its cell-type specificity. The mechanism underlying fisetin's senolytic effect appears to involve the phosphoinositide 3-kinase (PI3K) pathway. PI3K activation leads to the phosphorylation and activation of Akt, a signalling protein that prevents apoptosis in senescent cells. Fisetin is suggested to bind to PI3K, impairing Akt activation and ultimately leading to senescent cell death (Wong et al. 2023).

The senotherapeutic activity of fisetin observed in mice and human tissues, along with its potential health benefits, has paved the way for the investigation of fisetin in human clinical studies. Fisetin is currently undergoing phase 1 and 2 clinical trials to assess its safety and efficacy in treating mild to moderate osteoarthritis, an age-related cartilage degenerative disease (Institute 2020–2022). Additionally, fisetin is the subject of a phase 2 clinical trial aimed at evaluating its efficacy in alleviating frailty and inflammation in older adults (Kirkland 2018–2024).

The potential benefits of fisetin in reducing senescent cell burden and improving health outcomes in ageing have also expanded to research involving viral infections, including those induced by coronaviruses. In studies involving aged mice exposed to mouse hepatitis virus (MHV) or a β-coronavirus closely related to SARS-CoV-2, fisetin treatment was shown to have several positive effects including reducing senescent cell burden, inflammation and improving immune response (Camell et al. 2021). Importantly, treatment with fisetin decreased mortality in the aged mice exposed to the virus. Fisetin is currently undergoing a phase 2 clinical trial to determine if it can prevent an increase in disease progression and mitigate excessive inflammatory reactions in individuals infected with the SARS-CoV-2 virus (Kirkland 2020–2023). Similar studies are also in progress to assess the effects of fisetin in aged COVID-19 patients residing in nursing homes (Kirkland 2021–2022, 2023–2025). The investigation of fisetin in the context of viral infections, underscores its potential as a therapeutic agent with broader applications in the field of infectious diseases and ageing-related health conditions.

Additional Naturally Occurring Senolytic Compounds

Curcumin is a naturally existing compound derived from turmeric that has gained attention for its wide range of biological activities, including anti-inflammatory, anti-cancer, and potential therapeutic applications in age-related diseases (Bielak-Zmijewska et al. 2019). Additionally, curcumin and one of its metabolites, o-Vanillin, have been found to possess senolytic properties. Studies have demonstrated that curcumin and o-Vanillin have senolytic effects by clearing senescent intervertebral disc cells in cell cultures from isolated human discs cells. Moreover, these compounds have been shown to reduce the SASP, which is associated with chronic inflammation and conditions such as back pain (Cherif et al. 2019). Nonetheless, clinical application of curcumin faces challenges due to its limited water solubility, suboptimal oral absorption, and rapid liver metabolism in normal physiological conditions (Purpura et al. 2018). For this reason, some analogues of curcumin were developed with improved chemical properties. These analogues were tested for their potential senolytic activity. The EF-24 curcumin analogue showed to be the most potent senolytic agent among different curcumin analogues tested. EF-24 eliminates senescent cells by increasing the proteasome degradation of the Bcl-2 anti-apoptotic family proteins and consequently, inducing apoptosis (Li et al. 2019).

The same team of scientists that first showed clearance of senescent cells by ABT263 in aged mice (Chang et al. 2016) also explored the potential senolytic properties of **Piperlongumine** (PL). PL is a natural alkaloid/amide compound derived from various pepper species. Their research showed that PL can induce cell death in senescent human WI-38 fibroblasts through apoptosis without the need for the induction of reactive oxygen species (ROS) (Wang et al. 2016). However, another research group found that PL targets oxidation resistance 1 (OXR1), a crucial antioxidant protein. They demonstrated that OXR1 is upregulated in senescent human WI38 fibroblasts, and PL binds directly to OXR1, leading to its degradation. This process results in an increased production of ROS (Zhang et al. 2018). While the exact mechanism by which PL induces senescent cell death remains unclear, its senolytic effects, along with those of its analogues, have been well demonstrated (Liu et al. 2018).

GL-V9, a newly synthetic flavonoid derived from wogonin, has exhibited an anti-tumour effect in breast cancer cells (Guo et al. 2020). Studies have shown that

GL-V9 possesses high potency in eliminating senescent breast cancer cells by inducing apoptosis through a mechanism dependent on the generation of ROS. Furthermore, in the same study, researchers showed that GL-V9 has senolytic effects *in vivo* using a transgenic female mouse model that spontaneously develops breast cancer (Yang et al. 2021). Interestingly, it was also found that a nonlethal dose of GL-V9 promotes cellular senescence in malignant T-cells (Li et al. 2020). Further research is needed to fully understand the mechanisms behind senolysis and therapeutic potential of GL-V9 in various disease contexts.

In a recent screening of various plant extracts, it was discovered that ginger extract (derived from *Zingiber officinale Rosc.*) exhibited the ability to selectively induce cell death in senescent WI-38 human fibroblast cells, without affecting proliferating cells. Among the components found within ginger extract, **gingerenone A** was identified as a compound with the specific ability to eliminate senescent cells (Moaddel et al. 2022).

Cycloastragenol, a secondary metabolite extracted from the herb *Astragalus membranaceus*, has been studied and found to delay age-related physical dysfunction in mice. Its beneficial effects are attributed to its ability to induce apoptosis in senescent cells. The senolytic action of cycloastragenol has been associated with inhibition of the Bcl-2 protein and the PI3K/AKT/mTOR pathway which play crucial roles in regulating cell survival (Zhang et al. 2023).

Another compound that shows anti-ageing effects is the **Anthocyanin** extracted from the fruits of *Sambucus canadensis*. Anthocyanin reduced cell senescence burden potentially by inhibiting the activity of the PI3K/AKT/mTOR signalling pathway (Hu et al. 2023). Interestingly, cycloastragenol and Anthocyanin appear to act through a similar pathway target, despite having different chemical structures. Cycloastragenol is a triterpenoid saponin, while Anthocyanin is a flavonoid. Additionally, other studies have revealed an interesting aspect of cycloastragenol's mechanism of action, in particular, that it can function as a telomerase activator. This may also contribute to a reduction of cellular senescence by counteracting telomere shortening (Hu et al. 2023).

Cardiac Glycosides

Recent efforts to identify senolytic drugs have focused on medium to large throughput screenings, using libraries of compounds. In 2019, Triana-Martínez and colleagues employed a cell-based screening strategy to identify senolytic compounds using a chemical repositioning library. In their study, they successfully identified cardiac glycosides (CG) as a novel class of compounds with broad-spectrum senolytic activity. Through the application of a specific cut-off criteria, they found that out of the nine compounds exhibiting senolytic effects, the most promising one was Proscillaridin A. This compound is part of the GCs family and is derived from the foxglove plant, *Digitalis purpurea* (Triana-Martínez et al. 2019). Subsequent screening of CGs identified several compounds with senolytic effects, including ouabagenin, ouabain, bufalin, cinobufagin, peruvocide, digitoxin, digoxin, convallatoxins, k-stropanthin, and strophanthidin. Remarkably, they exhibited comparable senolytic activity across various senescent tumour and primary cell lines (Guerrero et al. 2019; Triana-Martínez et al. 2019).

The mechanism by which CGs have senolytic activity is likely due to their inhibition of the Na^+/K^+-ATPase, which leads to depolarisation of the plasma membrane causing a cationic imbalance and, consequently lowering the intracellular pH which sensitises senescent cells to cell-death (Guerrero et al. 2019; Triana-Martínez et al. 2019; Johmura et al. 2021). Another proposed mechanism of action for cardiac glycosides (CGs) involves the induction of apoptosis by increasing the expression of the pro-apoptotic BCL2-family protein NOXA within senescent cells (Guerrero et al. 2019).

Galactose Modified Prodrugs

Up to this point, we have summarised senolytic drugs that have shown efficacy in eliminating senescent cells and improving age-related conditions. However, it's essential to acknowledge that many of these senolytic drugs encounter significant challenges concerning specificity and broad-spectrum activity due to the highly heterogeneous nature of the senescent cell program. The complexity of senescence results in varying levels of susceptibility among different types of senescent cells to current senolytic drugs, as previously demonstrated (Childs et al. 2017; Zhu et al. 2017). This heterogeneity poses a considerable obstacle in the development of universal and highly targeted senolytic therapies. Researchers are actively working to address these challenges by finding novel strategies to selectively eliminate senescent cells. One of these strategies is the use of Prodrugs.

Prodrugs are defined as pharmaceutical compounds that are inactive in the intended pharmacological action and require conversion into bioactive agents through either metabolic or physicochemical changes (Wu 2009). Prodrugs can exist naturally or can be produced synthetically. For instance, modification of galactose has been frequently used to enhance the pharmacological properties and delivery of existing drugs. It can also be utilised to create prodrugs (Melisi et al. 2011).

As some senescent cells show high lysosomal β-galactosidase activity (Dimri et al. 1995), it was proposed that one could use this property of senescent cells to design more specific senolytic interventions. This strategy involves attaching a galactose moiety to various senolytic compounds. The galactose-modified prodrugs are designed to be preferentially processed through enzymatic cleavage by senescence-associated β-galactosidase (SA-β-gal), which is predominantly present in senescent cells (Dimri et al. 1995).

The first senolytic prodrug reported was gemcitabine modified with an acetyl galactose moiety and a non-toxic aromatic spacer to generate Synthesized Senescence-Specific Killing compound 1 (**SSK1**) (Cai et al. 2020). The Theragnostic Senolytic Prodrug (**TSPD**), another gemcitabine modified with a β-galactosyl group and the aromatic compound coumarin, was shown to attenuate kidney injury and improve its function through senolytic mechanisms in a mouse model of acute kidney injury (Song et al. 2022). The development of galactose-modified prodrugs has yielded promising results in the field of senolytic therapies, with several additional examples demonstrating enhanced selectivity for senescent cells:

Prodrug A, this prodrug is based on the modification of duocarmycin with galactose. In a previous study, the precursor of prodrug A, Duocarmycin SA, which lacked a galactose moiety, was found to kill both proliferating and senescent cells. However, prodrug A, which differs only in the addition of two galactose groups, showed a preference for eliminating senescent cells (Guerrero et al. 2020). Similarly, while Navitoclax has demonstrated high potency in killing senescent cells, it is associated with significant on-target hematological toxicity, including thrombocytopenia. To address this issue, a galacto-conjugated form of Navitoclax, named **Nav-Gal**, was developed. Nav-Gal functions as a prodrug with selective pro-apoptotic senolytic activity, exclusively released in senescent cells (González-Gualda et al. 2020). Another senolytic prodrug was created by modifying the clinically approved anti-cancer drug 5-fluorouracil with a β-D-galactosyl moiety. Known as **5FURGa**, this prodrug has demonstrated safety and effectiveness in improving various aspects of health in geriatric mice, including frailty, skeletal muscle function, muscle stem cell function, cognitive function, and survival (Doan et al. 2020).

These examples highlight the potential of galactose-modified prodrugs to preferentially target and eliminate senescent cells with high selectivity, both *in vitro* and *in vivo*. While galactose-modified prodrugs represent a promising approach to selectively target and eliminate senescent cells with high SA-β-galactosidase activity, it is important to acknowledge that this enzyme's activity may not be universally present in all types of senescent cells. Additionally, studies have shown that SA-β-galactosidase activity can also be detected in some non-senescent cells (Severino et al. 2000), which presents a challenge to achieving specificity. Given these considerations, it is crucial for researchers to explore alternative and more specific strategies for senolytic interventions. The heterogeneity of senescent cells, both in terms of their origins and the molecular pathways involved, necessitates a multifaceted approach to senolytic drug development. This might involve identifying novel biomarkers or molecular targets that are specific to particular types of senescent cells, as well as refining existing prodrug designs to enhance their selectivity.

PROTACS (Proteolysis Targeting Chimera Senolytics)

Emerging techniques that enable the design of senolytic drugs with enhanced selectivity and efficacy have garnered increased attention in the field of ageing research. One alternative approach that holds promise for the development of safer and more specific senolytics is based on the innovative concept of Proteolysis-Targeting Chimera (PROTAC) technology. PROTACs consists of the ligand specific domain to a target protein and an E3 ligase recognition domain which are fused via an optimized linker. The rationale behind PROTAC technology is to bring both the target protein and the E3 ligase complex in close proximity to each other. This proximity facilitates the ubiquitination of the target protein by the E3 ligase complex, marking it for degradation via the ubiquitin-proteasome system (Burslem and Crews 2020).

In the context of senolytic drug development, researchers have identified bromodomain and extra-terminal domain (BET) family protein inhibitors as having senolytic activity (Wakita et al. 2020). One such compound is ARV825, which is a small-molecule BET degrader designed as a heterobifunctional PROTAC. ARV825 functions by recruiting BET family proteins to the E3 ubiquitin ligase cereblon, leading to the degradation of BET family proteins (Lu et al. 2015). Wakita and colleagues demonstrated that ARV825 exhibited potent senolytic activity, surpassing other BET inhibitors in its ability to eliminate senescent cells. Furthermore, their research showed that treatment with ARV825 effectively eliminated chemotherapy-induced senescent cells and enhanced the efficacy of chemotherapy against xenograft tumours in mouse models (Wakita et al. 2020).

The application of PROTAC technology has also been extended to the well-known senolytic drug Navitoclax to address its side effect of thrombocytopenia. In this context, Navitoclax was transformed into PZ15227, a specific PROTAC designed to target Bcl-xl to the E3 ligase cereblon which is poorly expressed in platelets (He et al. 2020b). PZ15227 was shown to exhibit reduced toxicity to platelets while being more effective in eliminating senescent cells, particularly in aged mice, compared to Navitoclax (He et al. 2020b). The broader application of targeted protein degradation using PROTACs offers a promising tool in the field of senescence research. It enables researchers to target a variety of proteins involved in senescence pathways, potentially enhancing selectivity, and reducing toxicity to non-senescent cells.

Other Unclassified Senolytics

Histone deacetylases (HDACs) play a significant role in cancer biology, with abnormal expression associated with various malignancies. HDAC inhibitors have emerged as promising treatments for cancer, including multiple myelomas (Xu et al. 2007; Raedler 2016). Interestingly, senescence is characterized by decreased global histone acetylation (Li et al. 2013). Panobinostat, an FDA-approved HDAC inhibitor employed for various cancer types, has also been explored as a senolytic agent following chemotherapy (Samaraweera et al. 2017). Samaraweera and colleagues showed that **Panobinostat** effectively eliminated senescent cells that had accumulated in non-small cell lung cancer and neck squamous cell carcinoma lines after cisplatin or taxol chemotherapy treatment. This senolytic effect was associated with increased histone H3 acetylation and a reduction in BCL-XL expression (Samaraweera et al. 2017).

Additionally, **CUDC-907**, a heterobifunctional molecule targeting both HDACs and PI3K, a key player in the SCAPS signalling pathway, has been investigated in clinical trials due to its promising antitumor activity (Curis 2012–2020; Oki et al. 2017). Recent research demonstrated that CUDC-907 selectively induces senolysis at very low concentrations in a p53-dependent manner in various *in vitro* models of cellular senescence. Importantly, CUDC-907 has already been deemed safe for use in humans through clinical trials (Al-Mansour et al. 2023). However, there is no data available to establish its physiological benefits as a senolytic agent.

Mitochondria-targeted tamoxifen (**MitoTam**) has been shown to reduce senescence-associated markers across multiple tissues during ageing in mice. Mechanistically, it was proposed that senescent cells are susceptible to MitoTam-induced cell death due to very low expression of adenine nucleotide translocase-2 (Hubackova et al. 2019).

The screening of FDA-approved drugs for their potential senolytic properties is gaining increasing attention due to the advantage of these drugs having already demonstrated safety and efficacy for their intended uses. Notably, two antibiotics, **Azithromycin** and **Roxithromycin**, have been identified as novel clinically approved senolytic drugs through a streamlined screening strategy (Ozsvari et al. 2018). The observed senolytic effects of these drugs have been suggested to be linked to the autophagic and metabolic changes that occur within senescent cells. Interestingly, Erythromycin, another antibiotic screened, did not show any senolytic effect (Ozsvari et al. 2018).

The investigation into the potential therapeutic benefits of antibiotics as senolytic agents in the context of ageing-related interventions led to the testing of Roxithromycin in Idiopathic Pulmonary Fibrosis (IPF), a disease associated with cellular senescence (Schafer et al. 2017). In a mouse model of bleomycin-induced pulmonary fibrosis, this antibiotic attenuated lung injury, inflammation, and fibrosis. Importantly, this was accompanied by a decrease in the senescent cell phenotype within lung tissues (Zhang et al. 2021).

As mentioned earlier, ageing is a recognised risk factor for the onset of osteoarthritis, a condition marked by the deterioration of articular cartilage (Wieland et al. 2005) and is closely linked to the senescence of chondrocytes (Price et al. 2002). Fenofibrate, an agonist of peroxisome proliferator-activated receptor alpha (PPARα), has been identified as a senolytic agent through cell-based high throughput screening. **Fenofibrate** exhibits selectivity in eliminating senescent chondrocytes by increasing apoptosis (Nogueira-Recalde et al. 2019). In mouse models of osteoarthritis, PPARα is typically downregulated, but treatment with fenofibrate has been shown to protect against cartilage degradation, regulate autophagy, and reduce inflammation. This beneficial effect extends to osteoarthritis patients, where fenofibrate treatment has been associated with improved physical function and increased mobility (Nogueira-Recalde et al. 2019). Another notable senolytic example is **AT-406**, which has demonstrated efficacy in attenuating the progression of osteoarthritis. Senescent cells were found to upregulate antiapoptotic proteins such as c-IAP1, c-IAP2, and XIAP. AT-406, a small molecule inhibitor of antiapoptotic proteins, specifically induces apoptosis in senescent cells (Peilin et al. 2019).

Treatment with AT-406 in post-traumatic osteoarthritis rats resulted in a rescue of cartilage degeneration (Peilin et al. 2019).

In another high-throughput screening, R406, also known as **Tamatinib**, was identified as a senolytic compound. R406 demonstrated the ability to reduce the cell viability of replicative senescent diploid human dermal fibroblast cells. The study revealed that R406 induces cell death through the caspase-9-mediated intrinsic apoptotic pathway. Notably, unlike other well-known senolytics, the apoptotic pathway triggered by R406 does not involve the modulation of BCL-2 levels. Instead, it operates by inhibiting the phosphorylation of focal adhesion kinase (FAK) and p38 mitogen-activated protein kinase (MAPK), both of which play crucial roles in regulating cell survival (Cho et al. 2020). It should be noted however that R406 has not been tested *in vivo*.

Recently, the inhibition of glutaminolysis has emerged as a promising target for senolytic interventions. Studies have shown that senescent cells have a lower pH due to lysosomal membrane damage. This intracellular acidification in senescent cells is compensated for by an upregulation of glutaminase 1 (GLS1) expression. GLS1's activity induces ammonia production, helping senescent cells maintain their survival. However, treatment of senescent human fibroblasts with **BPTES**, a GLS1 inhibitor, results in senolysis. This suggests that GLS1 is essential for senescent cell survival and for neutralising intracellular acidosis. GLS1 is also highly expressed in senescent cells in mouse tissues. When BPTES was administered, it effectively eliminated senescent cells in aged mice, demonstrating its potential in ameliorating age-associated organ dysfunction (Johmura et al. 2021).

Machine learning has significantly facilitated the process of drug discovery and development by combining training data derived from biological screening or publicly available databases. In a recent study 2352 compounds were screened in silico by machine learning models on the basis of their chemical structure (Wong et al. 2023). Through this approach, the study identified novel senolytic compounds BRDK20733377, BRD-K56819078 and BRD-K44839765 which target the BCL2 protein family. This study also showed that one of the compounds, BRD-K56819078, significantly decreased mRNA expression of senescence-associated genes in the kidneys of aged mice.

Conclusions

As demonstrated throughout this chapter (Table 5.1), numerous studies, predominantly pre-clinical in nature, have provided compelling evidence of the potential of senolytics to enhance healthspan in various age-related disorders. The discovery that numerous FDA-approved drugs possess senolytic properties represents a significant breakthrough. It not only highlights the potential of these drugs for targeting senescent cells but also expedites their repurposing for clinical trials. However, many challenges remain in the field. It is evident that a one-size-fits-all senolytic approach is unlikely to effectively target every type of senescent cell. This challenge

Table 5.1 Summary of several senolytic drugs discovered to date and their mechanism of action

Class	Compound	Feature	Mechanism of action	Reference
BCL-2 family inhibitors	Dasatinib + Quercetin	Tyrosine kinase inhibitor + naturally occurring flavonoid	Inhibit anti-apoptotic pathway	Zhu et al. (2015)
	ABT-737	A BH3 mimetic inhibitor	Inhibit BCL-2, BCL-XL, BCL-W	Yosef et al. (2016)
	ABT-263 (Navitoclax)	A BH3 mimetic inhibitor	Inhibit BCL-2, BCL-XL, BCL-W	Zhu et al. (2016)
	A-1331852	Synthetic drug - *Cat. No.: HY-19741*	Inhibit BCL-XL	Zhu et al. (2017)
	A-1155463	Synthetic drug - *Cat. No.: HY-19725*	Inhibit BCL-XL	Zhu et al. (2017)
HSP90 inhibitors	Geldanamycin	Benzoquinone ansamycin antibiotic	Inhibit HSP90	Fuhrmann-Stroissnigg et al. (2017)
	17-AAG (tanespimycin)	Synthetic Geldanamycin analogue - *Cat. No.: HY-10211*	Inhibit HSP90	Fuhrmann-Stroissnigg et al. (2017)
	17-DMAG (alvespimycin)	Synthetic Geldanamycin analogue - *Cat. No.: HY-12024*	Disrupt HSP90-AKT interaction	Fuhrmann-Stroissnigg et al. (2017)
	Ganetespib	Synthetic drug - *Cat. No.: HY-15205*	Inhibit HSP90	Fuhrmann-Stroissnigg et al. (2017)
Targeting p53	FOXO4-DRI	Peptide antagonist	Disrupt FOXO4-p53 interaction	Baar et al. (2017), Zhang et al. (2020)
	UBX0101	Synthetic drug	Disrupt MDM2/p53 interaction	Jeon et al. (2017)
	RG7112 (RO5045337)	Synthetic drug - *Cat. No.: HY-10959*	Disrupt MDM2/p53 interaction	Cherif et al. (2020)
	P5091	Synthetic drug - *Cat. No.: HY-15667*	USP7 inhibitor	He et al. (2020a)
	P22077	Synthetic drug - *Cat. No.: HY-13865*	USP7 inhibitor	He et al. (2020a)

(continued)

Table 5.1 (continued)

Class	Compound	Feature	Mechanism of action	Reference
Natural products and analogues	Fisetin	Naturally occurring flavonoid in fruits and vegetables	BCL-2, PI3K/AKT, p53, NF-jB and more	Zhu et al. (2017), Yousefzadeh et al. (2018)
	Curcumin	Polyphenol found primarily in turmeric	Down-regulate Nrf2 and NF-kB pathways	Cherif et al. (2019)
	o-Vanillin	Curcumin metabolite	Down-regulate Nrf2 and NF-kB pathways	Cherif et al. (2019)
	EF-24	Curcumin analogue - *Cat. No. E8409*	BCL-2 family	Li et al. (2019)
	Piperlongumine and its analogue	Natural product isolated from Long pepper	Induce apoptosis, OXR1 degradation	Wang et al. (2016), Zhang et al. (2018)
	GL-V9	Synthetic flavonoid analogue of wogonin	alkalize lysosome and elevate ROS levels	Yang et al. (2021)
	Gingerenone A	Compound extracted from *Zingiber officinale Rosc.* (Ginger)	Induce Cleaved-caspase 3	Moaddel et al. (2022)
	Anthocyanin	Compound extracted from Sambucus canadensis (fruit)	Inhibition of PI3k/Akt/mTOR pathway	Hu et al. (2023)
	Cycloastragenol	Compound extracted from *Astragalus membrananceus* (herb)	Inhibition of BCL2 and PI3k/Akt/mTOR	Zhang et al. (2023)

(continued)

Table 5.1 (continued)

Class	Compound	Feature	Mechanism of action	Reference
Cardiac glycosides	Proscillaridin A	Derived from the foxglove plant (*Digitalis purpurea*)	Inhibition of Na+/K+ATPase	Triana-Martínez et al. (2019)
	Ouabain	Steroid hormone isolated naturally from *Strophanthus gratus*	up regulation of BCL-2 family protein NOXA, Inhibition of Na+/K+ATPase	Guerrero et al. (2019), Triana-Martínez et al. (2019)
	Ouabagenin	Steroidal lactone obtained by hydrolysis of ouabain	Inhibition of Na+/K+ATPase	Guerrero et al. (2019)
	Digoxin	Extracted from the leaves of *Digitalis lanata*	Inhibition of Na+/K+ATPase	Guerrero et al. (2019), Triana-Martínez et al. (2019)
	Bufalin	Cardiotonic steroid toxin originally isolated from Chinese toad venom	Inhibition of Na+/K+ATPase	Guerrero et al. (2019), Triana-Martínez et al. (2019)
	K-Stropanthin	Cardenolide found in species of the genus Strophanthus	Inhibition of Na+/K+ATPase	Guerrero et al. (2019)
	Strophanthidin	Cardenolides and a steroid aldehyde	Inhibition of Na+/K+ATPase	Guerrero et al. (2019)
Galactose modified drugs	SSK1	Galactose-modified Gemcitabine pro-drug	SA-b-galactosidase/blocking DNA synthesis	Cai et al. (2020)
	Pro-drug A (JHB75B)	Galactose-modified Duocarmycin pro-drug	SA-b-galactosidase/irreversible DNA alkylation	Guerrero et al. (2020)
	Nav-Gal	Galactose-modified Navitoclax pro-drug	SA-b-galactosidase/Inhibit BCL-2, BCL-XL, BCL-W	González-Gualda et al. (2020)
	5FURGa	Galactosyl-modified 5-fluorouridine (5-FU)	SA-b-galactosidase/inhibition of thymidylate synthase	Doan et al. (2020)
	TSPD	Galactose-modified Gemcitabine pro-drug	SA-b-galactosidase	Song et al. (2022)

(continued)

Table 5.1 (continued)

Class	Compound	Feature	Mechanism of action	Reference
PROTACs (Proteolysis targeting chimeras)	PZ15227	Pomalidomide E3 ligase cereblon (CRBN)-linked to ABT-263	recruiting an E3 ubiquitin ligase/ Degrade BCL-XL	He et al. (2020b)
	ARV825	Small molecule conjugating OTX015 with an E3 ligase cereblon (CRBN).	BET inhibitor (Degrade BRD4 - BET bamily)	Wakita et al. (2020)
Others unclassified senolytics	Fenofibrate	Derivatives of fibric acid	PPARa agonist	Nogueira-Recalde et al. (2019)
	Azithromycin	Antibiotic	Induce aerobic glycolysis and autophagy	Ozsvari et al. (2018)
	Roxithromycin	Antibiotic	Induce aerobic glycolysis and autophagy/Inhibit NADPH oxidase 4	Ozsvari et al. (2018), Zhang et al. (2021)
	Tamatinib (R406)	Active form of the prodrug fostamatinib - Immunotherapy	ATP-competitive (Type I) Syk inhibitor - Inhibit phosphorilation of FAK and p38MAPK	Cho et al. (2020)
	BPTES	Molecule inhibitor of Glutaminase	GLS1 inhibitor	Johmura et al. (2021)
	MitoTam	I odide, hydriodide, mitochondria-targeted analog of tamoxifen	Decrease Mitochondria membrane potential and inhibit OXPHOS	Hubackova et al. (2019)
	Panobinostat	Histone deacetylase inhibitor - Chemotherapy	Decreased expression of Bcl-xL and increased acetylated H3	Samaraweera et al. (2017)
	CUD-907	Heterobifunctional molecule inhibitor	Inhibition of HDACs and PI3K	Al-Mansour et al. (2023)
	AT-406	Small molecule inhibitor of the IAP genes	Inhibit c-IAP1, c-IAP2 and XIAP	Peilin et al. (2019)
Possible news senolytics	BRD-K56819078	Molecule screened	BCL-2 family inhibitors	Wong et al. (2023)
	BRD-K20733377	Molecule screened	BCL-2 family inhibitors	Wong et al. (2023)
	BRD-K44839765	Molecule screened	BCL-2 family inhibitors	Wong et al. (2023)

arises from the significant heterogeneity observed in the senescent cell phenotypes. Different senescent cell types may exhibit distinct characteristics and vulnerabilities, necessitating a diverse range of senolytic strategies to effectively address this heterogeneity. Moreover, it remains unclear whether eliminating all senescent cells is a desirable goal, considering their crucial roles in tumour suppression and tissue repair. In certain physiological contexts, the removal of senescent cells may have unintended negative consequences. Therefore, the timing of senolytic interventions becomes a critical factor to carefully consider. Balancing the removal of harmful senescent cells with the preservation of beneficial functions requires a nuanced approach to optimise the therapeutic outcomes.

References

Acosta JC, Banito A, Wuestefeld T, Georgilis A, Janich P, Morton JP, Athineos D, Kang TW, Lasitschka F, Andrulis M, Pascual G, Morris KJ, Khan S, Jin H, Dharmalingam G, Snijders AP, Carroll T, Capper D, Pritchard C, Inman GJ, Longerich T, Sansom OJ, Benitah SA, Zender L, Gil J (2013) A complex secretory program orchestrated by the inflammasome controls paracrine senescence. Nat Cell Biol 15(8):978–990

Al-Mansour F, Alraddadi A, He B, Saleh A, Poblocka M, Alzahrani W, Cowley S, Macip S (2023) Characterization of the HDAC/PI3K inhibitor CUDC-907 as a novel senolytic. Aging (Albany NY) 15(7):2373–2394

Baar MP, Brandt RMC, Putavet DA, Klein JDD, Derks KWJ, Bourgeois BRM, Stryeck S, Rijksen Y, van Willigenburg H, Feijtel DA, van der Pluijm I, Essers J, van Cappellen WA, van IWF, Houtsmuller AB, Pothof J, de Bruin RWF, Madl T, Hoeijmakers JHJ, Campisi J, de Keizer PLJ (2017) Targeted apoptosis of senescent cells restores tissue homeostasis in response to chemotoxicity and aging. Cell 169(1):132–147.e116

Baker DJ, Wijshake T, Tchkonia T, LeBrasseur NK, Childs BG, van de Sluis B, Kirkland JL, van Deursen JM (2011) Clearance of p16Ink4a-positive senescent cells delays ageing-associated disorders. Nature 479(7372):232–236

Barnett K, Mercer SW, Norbury M, Watt G, Wyke S, Guthrie B (2012) Epidemiology of multimorbidity and implications for health care, research, and medical education: a cross-sectional study. Lancet 380(9836):37–43

Bielak-Zmijewska A, Grabowska W, Ciolko A, Bojko A, Mosieniak G, Bijoch Ł, Sikora E (2019) The role of curcumin in the modulation of ageing. Int J Mol Sci 20(5)

Burslem GM, Crews CM (2020) Proteolysis-targeting chimeras as therapeutics and tools for biological discovery. Cell 181(1):102–114

Cai Y, Zhou H, Zhu Y, Sun Q, Ji Y, Xue A, Wang Y, Chen W, Yu X, Wang L, Chen H, Li C, Luo T, Deng H (2020) Elimination of senescent cells by β-galactosidase-targeted prodrug attenuates inflammation and restores physical function in aged mice. Cell Res 30(7):574–589

Camell CD, Yousefzadeh MJ, Zhu Y, Prata L, Huggins MA, Pierson M, Zhang L, O'Kelly RD, Pirtskhalava T, Xun P, Ejima K, Xue A, Tripathi U, Espindola-Netto JM, Giorgadze N, Atkinson EJ, Inman CL, Johnson KO, Cholensky SH, Carlson TW, LeBrasseur NK, Khosla S, O'Sullivan MG, Allison DB, Jameson SC, Meves A, Li M, Prakash YS, Chiarella SE, Hamilton SE, Tchkonia T, Niedernhofer LJ, Kirkland JL, Robbins PD (2021) Senolytics reduce coronavirus-related mortality in old mice. Science 373(6552)

Chang J, Wang Y, Shao L, Laberge R-M, Demaria M, Campisi J, Janakiraman K, Sharpless NE, Ding S, Feng W, Luo Y, Wang X, Aykin-Burns N, Krager K, Ponnappan U, Hauer-Jensen M, Meng A, Zhou D (2016) Clearance of senescent cells by ABT263 rejuvenates aged hematopoietic stem cells in mice. Nat Med 22(1):78–83

Cherif H, Bisson DG, Jarzem P, Weber M, Ouellet JA, Haglund L (2019) Curcumin and o-vanillin exhibit evidence of senolytic activity in human IVD cells in vitro. J Clin Med 8(4)

Cherif H, Bisson DG, Mannarino M, Rabau O, Ouellet JA, Haglund L (2020) Senotherapeutic drugs for human intervertebral disc degeneration and low back pain. Elife 9

Childs BG, Gluscevic M, Baker DJ, Laberge RM, Marquess D, Dananberg J, van Deursen JM (2017) Senescent cells: an emerging target for diseases of ageing. Nat Rev Drug Discov 16(10):718–735

Cho HJ, Yang EJ, Park JT, Kim JR, Kim EC, Jung KJ, Park SC, Lee YS (2020) Identification of SYK inhibitor, R406 as a novel senolytic agent. Aging (Albany NY) 12(9):8221–8240

Coppé JP, Patil CK, Rodier F, Sun Y, Muñoz DP, Goldstein J, Nelson PS, Desprez PY, Campisi J (2008) Senescence-associated secretory phenotypes reveal cell-nonautonomous functions of oncogenic RAS and the p53 tumor suppressor. PLoS Biol 6(12):2853–2868

Curis I (2012–2020) Study to assess the safety, tolerability and pharmacokinetics of fimepinostat (CUDC-907) in patients with lymphoma (NCT01742988)

Demaria M, Ohtani N, Youssef SA, Rodier F, Toussaint W, Mitchell JR, Laberge RM, Vijg J, Van Steeg H, Dollé ME, Hoeijmakers JH, de Bruin A, Hara E, Campisi J (2014) An essential role for senescent cells in optimal wound healing through secretion of PDGF-AA. Dev Cell 31(6):722–733

Dimri GP, Lee X, Basile G, Acosta M, Scott G, Roskelley C, Medrano EE, Linskens M, Rubelj I, Pereira-Smith O et al (1995) A biomarker that identifies senescent human cells in culture and in aging skin in vivo. Proc Natl Acad Sci U S A 92(20):9363–9367

Doan L, Paine P, Tran C, Parsons B, Hiller A, Joshua I, Collaco N, Vitari A, Morree AD, Ishak H, Jasper J, Gallop M, Luo J, Wyss-Coray T, Rando T, Quarta M (2020) Targeted senolytic prodrug is well tolerated and results in amelioration of frailty, muscle regeneration and cognitive functions in geriatric mice. Res Square

Fuhrmann-Stroissnigg H, Ling YY, Zhao J, McGowan SJ, Zhu Y, Brooks RW, Grassi D, Gregg SQ, Stripay JL, Dorronsoro A, Corbo L, Tang P, Bukata C, Ring N, Giacca M, Li X, Tchkonia T, Kirkland JL, Niedernhofer LJ, Robbins PD (2017) Identification of HSP90 inhibitors as a novel class of senolytics. Nat Commun 8(1):422

Gonzales M (2020–2023) Senolytic therapy to modulate progression of Alzheimer's disease (SToMP-AD) (NCT04063124)

Gonzales MM, Garbarino VR, Marques Zilli E, Petersen RC, Kirkland JL, Tchkonia T, Musi N, Seshadri S, Craft S, Orr ME (2022) Senolytic therapy to modulate the progression of Alzheimer's disease (SToMP-AD): a pilot clinical trial. J Prev Alzheimers Dis 9(1):22–29

González-Gualda E, Pàez-Ribes M, Lozano-Torres B, Macias D, Wilson JR 3rd, González-López C, Ou HL, Mirón-Barroso S, Zhang Z, Lérida-Viso A, Blandez JF, Bernardos A, Sancenón F, Rovira M, Fruk L, Martins CP, Serrano M, Doherty GJ, Martínez-Máñez R, Muñoz-Espín D (2020) Galacto-conjugation of Navitoclax as an efficient strategy to increase senolytic specificity and reduce platelet toxicity. Aging Cell 19(4):e13142

Gorgoulis V, Adams PD, Alimonti A, Bennett DC, Bischof O, Bishop C, Campisi J, Collado M, Evangelou K, Ferbeyre G, Gil J, Hara E, Krizhanovsky V, Jurk D, Maier AB, Narita M, Niedernhofer L, Passos JF, Robbins PD, Schmitt CA, Sedivy J, Vougas K, von Zglinicki T, Zhou D, Serrano M, Demaria M (2019) Cellular senescence: defining a path forward. Cell 179(4):813–827

Guerrero A, Guiho R, Herranz N, Uren A, Withers DJ, Martínez-Barbera JP, Tietze LF, Gil J (2020) Galactose-modified duocarmycin prodrugs as senolytics. Aging Cell 19(4):e13133

Guerrero A, Herranz N, Sun B, Wagner V, Gallage S, Guiho R, Wolter K, Pombo J, Irvine EE, Innes AJ, Birch J, Glegola J, Manshaei S, Heide D, Dharmalingam G, Harbig J, Olona A, Behmoaras J, Dauch D, Uren AG, Zender L, Vernia S, Martínez-Barbera JP, Heikenwalder M, Withers DJ, Gil J (2019) Cardiac glycosides are broad-spectrum senolytics. Nat Metab 1(11):1074–1088

Guo Y, Wei L, Zhou Y, Lu N, Tang X, Li Z, Wang X (2020) Flavonoid GL-V9 induces apoptosis and inhibits glycolysis of breast cancer via disrupting GSK-3β-modulated mitochondrial binding of HKII. Free Radic Biol Med 146:119–129

Gurkar AU, Gerencser AA, Mora AL, Nelson AC, Zhang AR, Lagnado AB, Enninful A, Benz C, Furman D, Beaulieu D, Jurk D, Thompson EL, Wu F, Rodriguez F, Barthel G, Chen H, Phatnani H, Heckenbach I, Chuang JH, Horrell J, Petrescu J, Alder JK, Lee JH, Niedernhofer LJ, Kumar M, Königshoff M, Bueno M, Sokka M, Scheibye-Knudsen M, Neretti N, Eickelberg O, Adams PD, Hu Q, Zhu Q, Porritt RA, Dong R, Peters S, Victorelli S, Pengo T, Khaliullin T, Suryadevara V, Fu X, Bar-Joseph Z, Ji Z, Passos JF (2023) Spatial mapping of cellular senescence: emerging challenges and opportunities. Nat Aging 3(7):776–790

Hayflick L, Moorhead PS (1961) The serial cultivation of human diploid cell strains. Exp Cell Res 25:585–621

He Y, Li W, Lv D, Zhang X, Zhang X, Ortiz YT, Budamagunta V, Campisi J, Zheng G, Zhou D (2020a) Inhibition of USP7 activity selectively eliminates senescent cells in part via restoration of p53 activity. Aging Cell 19(3):e13117

He Y, Zhang X, Chang J, Kim H-N, Zhang P, Wang Y, Khan S, Liu X, Zhang X, Lv D, Song L, Li W, Thummuri D, Yuan Y, Wiegand JS, Ortiz YT, Budamagunta V, Elisseeff JH, Campisi J, Almeida M, Zheng G, Zhou D (2020b) Using proteolysis-targeting chimera technology to reduce navitoclax platelet toxicity and improve its senolytic activity. Nat Commun 11(1):1996

Hickson LJ (2016–2025) Senescence in chronic kidney disease (NCT02848131)

Hickson LJ, Langhi Prata LGP, Bobart SA, Evans TK, Giorgadze N, Hashmi SK, Herrmann SM, Jensen MD, Jia Q, Jordan KL, Kellogg TA, Khosla S, Koerber DM, Lagnado AB, Lawson DK, LeBrasseur NK, Lerman LO, McDonald KM, McKenzie TJ, Passos JF, Pignolo RJ, Pirtskhalava T, Saadiq IM, Schaefer KK, Textor SC, Victorelli SG, Volkman TL, Xue A, Wentworth MA, Wissler Gerdes EO, Zhu Y, Tchkonia T, Kirkland JL (2019) Senolytics decrease senescent cells in humans: preliminary report from a clinical trial of Dasatinib plus Quercetin in individuals with diabetic kidney disease. EBioMedicine 47:446–456

Hu X, Yang Y, Tang S, Chen Q, Zhang M, Ma J, Qin J, Yu H (2023) Anti-aging effects of anthocyanin extracts of Sambucus canadensis caused by targeting mitochondrial-induced oxidative stress. Int J Mol Sci 24(2)

Huang Y, He Y, Makarcyzk MJ, Lin H (2021) Senolytic peptide FOXO4-DRI selectively removes senescent cells from in vitro expanded human chondrocytes. Front Bioeng Biotechnol 9:677576

Hubackova S, Davidova E, Rohlenova K, Stursa J, Werner L, Andera L, Dong L, Terp MG, Hodny Z, Ditzel HJ, Rohlena J, Neuzil J (2019) Selective elimination of senescent cells by mitochondrial targeting is regulated by ANT2. Cell Death Differ 26(2):276–290

Institute, S. P. R. (2020–2022) Senolytic drugs attenuate osteoarthritis-related articular cartilage degeneration: a clinical trial (NCT04210986)

Jeon OH, Kim C, Laberge RM, Demaria M, Rathod S, Vasserot AP, Chung JW, Kim DH, Poon Y, David N, Baker DJ, van Deursen JM, Campisi J, Elisseeff JH (2017) Local clearance of senescent cells attenuates the development of post-traumatic osteoarthritis and creates a pro-regenerative environment. Nat Med 23(6):775–781

Johmura Y, Yamanaka T, Omori S, Wang TW, Sugiura Y, Matsumoto M, Suzuki N, Kumamoto S, Yamaguchi K, Hatakeyama S, Takami T, Yamaguchi R, Shimizu E, Ikeda K, Okahashi N, Mikawa R, Suematsu M, Arita M, Sugimoto M, Nakayama KI, Furukawa Y, Imoto S, Nakanishi M (2021) Senolysis by glutaminolysis inhibition ameliorates various age-associated disorders. Science 371(6526):265–270

Justice JN, Nambiar AM, Tchkonia T, LeBrasseur NK, Pascual R, Hashmi SK, Prata L, Masternak MM, Kritchevsky SB, Musi N, Kirkland JL (2019) Senolytics in idiopathic pulmonary fibrosis: results from a first-in-human, open-label, pilot study. EBioMedicine 40:554–563

Kang TW, Yevsa T, Woller N, Hoenicke L, Wuestefeld T, Dauch D, Hohmeyer A, Gereke M, Rudalska R, Potapova A, Iken M, Vucur M, Weiss S, Heikenwalder M, Khan S, Gil J, Bruder D, Manns M, Schirmacher P, Tacke F, Ott M, Luedde T, Longerich T, Kubicka S, Zender L

(2011) Senescence surveillance of pre-malignant hepatocytes limits liver cancer development. Nature 479(7374):547–551

Kirkland J (2018–2024) Alleviation by Fisetin of frailty, inflammation, and related measures in older women (AFFIRM) (NCT03430037)

Kirkland J (2020–2023) COVID-FISETIN: pilot in SARS-CoV-2 of Fisetin to alleviate dysfunction and inflammation (NCT04476953)

Kirkland J (2021–2022) COVFIS-HOME: COVID-19 pilot study of Fisetin to alleviate dysfunction and decrease complications (COVFIS-HOME) (NCT04771611)

Kirkland J (2023–2025) COVID-FIS: pilot in COVID-19 (SARS-CoV-2) of Fisetin in older adults in nursing homes (COVID-FIS) (NCT04537299)

Korolchuk VI, Miwa S, Carroll B, von Zglinicki T (2017) Mitochondria in cell senescence: is Mitophagy the weakest link? EBioMedicine 21:7–13

Krizhanovsky V, Yon M, Dickins RA, Hearn S, Simon J, Miething C, Yee H, Zender L, Lowe SW (2008) Senescence of activated stellate cells limits liver fibrosis. Cell 134(4):657–667

Leverson JD, Phillips DC, Mitten MJ, Boghaert ER, Diaz D, Tahir SK, Belmont LD, Nimmer P, Xiao Y, Ma XM, Lowes KN, Kovar P, Chen J, Jin S, Smith M, Xue J, Zhang H, Oleksijew A, Magoc TJ, Vaidya KS, Albert DH, Tarrant JM, La N, Wang L, Tao ZF, Wendt MD, Sampath D, Rosenberg SH, Tse C, Huang DC, Fairbrother WJ, Elmore SW, Souers AJ (2015) Exploiting selective BCL-2 family inhibitors to dissect cell survival dependencies and define improved strategies for cancer therapy. Sci Transl Med 7(279):279ra240

Li H, Hu P, Wang Z, Wang H, Yu X, Wang X, Qing Y, Zhu M, Xu J, Li Z, Guo Q, Hui H (2020) Mitotic catastrophe and p53-dependent senescence induction in T-cell malignancies exposed to nonlethal dosage of GL-V9. Arch Toxicol 94(1):305–323

Li M, Durbin KR, Sweet SM, Tipton JD, Zheng Y, Kelleher NL (2013) Oncogene-induced cellular senescence elicits an anti-Warburg effect. Proteomics 13(17):2585–2596

Li W, He Y, Zhang R, Zheng G, Zhou D (2019) The curcumin analog EF24 is a novel senolytic agent. Aging (Albany NY) 11(2):771–782

Liu X, Wang Y, Zhang X, Gao Z, Zhang S, Shi P, Zhang X, Song L, Hendrickson H, Zhou D, Zheng G (2018) Senolytic activity of piperlongumine analogues: synthesis and biological evaluation. Bioorg Med Chem 26(14):3925–3938

López-Otín C, Blasco MA, Partridge L, Serrano M, Kroemer G (2013) The hallmarks of aging. Cell 153(6):1194–1217

López-Otín C, Blasco MA, Partridge L, Serrano M, Kroemer G (2023) Hallmarks of aging: an expanding universe. Cell 186(2):243–278

Lu J, Qian Y, Altieri M, Dong H, Wang J, Raina K, Hines J, Winkler JD, Crew AP, Coleman K, Crews CM (2015) Hijacking the E3 ubiquitin ligase Cereblon to efficiently target BRD4. Chem Biol 22(6):755–763

Marengoni A, Angleman S, Melis R, Mangialasche F, Karp A, Garmen A, Meinow B, Fratiglioni L (2011) Aging with multimorbidity: a systematic review of the literature. Ageing Res Rev 10(4):430–439

Martyanov V, Whitfield ML, Varga J (2019) Senescence signature in skin biopsies from systemic sclerosis patients treated with Senolytic therapy: potential predictor of clinical response? Arthritis Rheumatol 71(10):1766–1767

Melisi D, Curcio A, Luongo E, Morelli E, Rimoli MG (2011) D-galactose as a vector for prodrug design. Curr Top Med Chem 11(18):2288–2298

Moaddel R, Rossi M, Rodriguez S, Munk R, Khadeer M, Abdelmohsen K, Gorospe M, Ferrucci L (2022) Identification of gingerenone A as a novel senolytic compound. PLoS One 17(3):e0266135

Muñoz-Espín D, Serrano M (2014) Cellular senescence: from physiology to pathology. Nat Rev Mol Cell Biol 15(7):482–496

Nogueira-Recalde U, Lorenzo-Gómez I, Blanco FJ, Loza MI, Grassi D, Shirinsky V, Shirinsky I, Lotz M, Robbins PD, Domínguez E, Caramés B (2019) Fibrates as drugs with senolytic and autophagic activity for osteoarthritis therapy. EBioMedicine 45:588–605

Ogrodnik M, Evans SA, Fielder E, Victorelli S, Kruger P, Salmonowicz H, Weigand BM, Patel AD, Pirtskhalava T, Inman CL, Johnson KO, Dickinson SL, Rocha A, Schafer MJ, Zhu Y, Allison DB, von Zglinicki T, LeBrasseur NK, Tchkonia T, Neretti N, Passos JF, Kirkland JL, Jurk D (2021) Whole-body senescent cell clearance alleviates age-related brain inflammation and cognitive impairment in mice. Aging Cell 20(2):e13296

Ogrodnik M, Miwa S, Tchkonia T, Tiniakos D, Wilson CL, Lahat A, Day CP, Burt A, Palmer A, Anstee QM, Grellscheid SN, Hoeijmakers JHJ, Barnhoorn S, Mann DA, Bird TG, Vermeij WP, Kirkland JL, Passos JF, von Zglinicki T, Jurk D (2017) Cellular senescence drives age-dependent hepatic steatosis. Nat Commun 8(1):15691

Oki Y, Kelly KR, Flinn I, Patel MR, Gharavi R, Ma A, Parker J, Hafeez A, Tuck D, Younes A (2017) CUDC-907 in relapsed/refractory diffuse large B-cell lymphoma, including patients with MYC-alterations: results from an expanded phase I trial. Haematologica 102(11):1923–1930

Oltersdorf T, Elmore SW, Shoemaker AR, Armstrong RC, Augeri DJ, Belli BA, Bruncko M, Deckwerth TL, Dinges J, Hajduk PJ, Joseph MK, Kitada S, Korsmeyer SJ, Kunzer AR, Letai A, Li C, Mitten MJ, Nettesheim DG, Ng S, Nimmer PM, O'Connor JM, Oleksijew A, Petros AM, Reed JC, Shen W, Tahir SK, Thompson CB, Tomaselli KJ, Wang B, Wendt MD, Zhang H, Fesik SW, Rosenberg SH (2005) An inhibitor of Bcl-2 family proteins induces regression of solid tumours. Nature 435(7042):677–681

Ovadya Y, Landsberger T, Leins H, Vadai E, Gal H, Biran A, Yosef R, Sagiv A, Agrawal A, Shapira A, Windheim J, Tsoory M, Schirmbeck R, Amit I, Geiger H, Krizhanovsky V (2018) Impaired immune surveillance accelerates accumulation of senescent cells and aging. Nat Commun 9(1):5435

Ozsvari B, Nuttall JR, Sotgia F, Lisanti MP (2018) Azithromycin and Roxithromycin define a new family of "senolytic" drugs that target senescent human fibroblasts. Aging (Albany NY) 10(11):3294–3307

Peilin W, Songsong T, Chengyu Z, Zhi C, Chunhui M, Yinxian Y, Lei Z, Min M, Zongyi W, Mengkai Y, Jing X, Tao Z, Zhuoying W, Fei Y, Chengqing Y (2019) Directed elimination of senescent cells attenuates development of osteoarthritis by inhibition of c-IAP and XIAP. Biochim Biophys Acta Mol Basis Dis 1865(10):2618–2632

Price JS, Waters JG, Darrah C, Pennington C, Edwards DR, Donell ST, Clark IM (2002) The role of chondrocyte senescence in osteoarthritis. Aging Cell 1(1):57–65

Purpura M, Lowery RP, Wilson JM, Mannan H, Münch G, Razmovski-Naumovski V (2018) Analysis of different innovative formulations of curcumin for improved relative oral bioavailability in human subjects. Eur J Nutr 57(3):929–938

Raedler LA (2016) Farydak (Panobinostat): first HDAC inhibitor approved for patients with relapsed multiple myeloma. Am Health Drug Benefits 9(Spec Feature):84–87

Samaraweera L, Adomako A, Rodriguez-Gabin A, McDaid HM (2017) A novel indication for panobinostat as a senolytic drug in NSCLC and HNSCC. Sci Rep 7(1):1900

Schafer MJ, White TA, Iijima K, Haak AJ, Ligresti G, Atkinson EJ, Oberg AL, Birch J, Salmonowicz H, Zhu Y, Mazula DL, Brooks RW, Fuhrmann-Stroissnigg H, Pirtskhalava T, Prakash YS, Tchkonia T, Robbins PD, Aubry MC, Passos JF, Kirkland JL, Tschumperlin DJ, Kita H, LeBrasseur NK (2017) Cellular senescence mediates fibrotic pulmonary disease. Nat Commun 8(1):14532

Schoenwaelder SM, Jarman KE, Gardiner EE, Hua M, Qiao J, White MJ, Josefsson EC, Alwis I, Ono A, Willcox A, Andrews RK, Mason KD, Salem HH, Huang DC, Kile BT, Roberts AW, Jackson SP (2011) Bcl-xL-inhibitory BH3 mimetics can induce a transient thrombocytopathy that undermines the hemostatic function of platelets. Blood 118(6):1663–1674

Sciences, W. F. U. H. (2016–2019) Targeting pro-inflammatory cells in idiopathic pulmonary fibrosis: a human trial (IPF) (NCT02874989)

Severino J, Allen RG, Balin S, Balin A, Cristofalo VJ (2000) Is beta-galactosidase staining a marker of senescence in vitro and in vivo? Exp Cell Res 257(1):162–171

Sierra F (2016) The emergence of geroscience as an interdisciplinary approach to the enhancement of health span and life span. Cold Spring Harb Perspect Med 6(4):a025163

Song Y, Li X, Shi D, Sun T, Liu W, Li X, Qiao S, Chen X, Guo Y, Li J (2022) A senolysis-based theragnostic prodrug strategy towards chronic renal failure. Chem Sci 13(40):11738–11745

Storer M, Mas A, Robert-Moreno A, Pecoraro M, Ortells MC, Di Giacomo V, Yosef R, Pilpel N, Krizhanovsky V, Sharpe J, Keyes WM (2013) Senescence is a developmental mechanism that contributes to embryonic growth and patterning. Cell 155(5):1119–1130

Triana-Martínez F, Picallos-Rabina P, Da Silva-Álvarez S, Pietrocola F, Llanos S, Rodilla V, Soprano E, Pedrosa P, Ferreirós A, Barradas M, Hernández-González F, Lalinde M, Prats N, Bernadó C, González P, Gómez M, Ikonomopoulou MP, Fernández-Marcos PJ, García-Caballero T, Del Pino P, Arribas J, Vidal A, González-Barcia M, Serrano M, Loza MI, Domínguez E, Collado M (2019) Identification and characterization of cardiac glycosides as senolytic compounds. Nat Commun 10(1):4731

Tse C, Shoemaker AR, Adickes J, Anderson MG, Chen J, Jin S, Johnson EF, Marsh KC, Mitten MJ, Nimmer P, Roberts L, Tahir SK, Xiao Y, Yang X, Zhang H, Fesik S, Rosenberg SH, Elmore SW (2008) ABT-263: a potent and orally bioavailable Bcl-2 family inhibitor. Cancer Res 68(9):3421–3428

Unity Biotechnology, Inc (2020) Senolytic therapy to modulate progression of Alzheimer's disease (SToMP-AD) (NCT04229225)

van Deursen JM (2014) The role of senescent cells in ageing. Nature 509(7501):439–446

Vogler M, Hamali HA, Sun X-M, Bampton ETW, Dinsdale D, Snowden RT, Dyer MJS, Goodall AH, Cohen GM (2011) BCL2/BCL-XL inhibition induces apoptosis, disrupts cellular calcium homeostasis, and prevents platelet activation. Blood 117(26):7145–7154

Wakita M, Takahashi A, Sano O, Loo TM, Imai Y, Narukawa M, Iwata H, Matsudaira T, Kawamoto S, Ohtani N, Yoshimori T, Hara E (2020) A BET family protein degrader provokes senolysis by targeting NHEJ and autophagy in senescent cells. Nat Commun 11(1):1935

Wang E (1995) Senescent human fibroblasts resist programmed cell death, and failure to suppress bcl2 is involved. Cancer Res 55(11):2284–2292

Wang L, Doherty GA, Judd AS, Tao ZF, Hansen TM, Frey RR, Song X, Bruncko M, Kunzer AR, Wang X, Wendt MD, Flygare JA, Catron ND, Judge RA, Park CH, Shekhar S, Phillips DC, Nimmer P, Smith ML, Tahir SK, Xiao Y, Xue J, Zhang H, Le PN, Mitten MJ, Boghaert ER, Gao W, Kovar P, Choo EF, Diaz D, Fairbrother WJ, Elmore SW, Sampath D, Leverson JD, Souers AJ (2020) Discovery of A-1331852, a first-in-class, potent, and orally-bioavailable BCL-X(L) inhibitor. ACS Med Chem Lett 11(10):1829–1836

Wang Y, Chang J, Liu X, Zhang X, Zhang S, Zhang X, Zhou D, Zheng G (2016) Discovery of piperlongumine as a potential novel lead for the development of senolytic agents. Aging (Albany NY) 8(11):2915–2926

WHO (2021) World health statistics 2021: monitoring health for the SDGs, sustainable development goals, World Health. Organization 136

Wieland HA, Michaelis M, Kirschbaum BJ, Rudolphi KA (2005) Osteoarthritis—an untreatable disease? Nat Rev Drug Discov 4(4):331–344

Wilson WH, O'Connor OA, Czuczman MS, LaCasce AS, Gerecitano JF, Leonard JP, Tulpule A, Dunleavy K, Xiong H, Chiu YL, Cui Y, Busman T, Elmore SW, Rosenberg SH, Krivoshik AP, Enschede SH, Humerickhouse RA (2010) Navitoclax, a targeted high-affinity inhibitor of BCL-2, in lymphoid malignancies: a phase 1 dose-escalation study of safety, pharmacokinetics, pharmacodynamics, and antitumour activity. Lancet Oncol 11(12):1149–1159

Wong F, Omori S, Donghia NM, Zheng EJ, Collins JJ (2023) Discovering small-molecule senolytics with deep neural networks. Nat Aging 3(6):734–750

Wu KM (2009) A new classification of prodrugs: regulatory perspectives. Pharmaceuticals (Basel) 2(3):77–81

Xu WS, Parmigiani RB, Marks PA (2007) Histone deacetylase inhibitors: molecular mechanisms of action. Oncogene 26(37):5541–5552

Yang D, Tian X, Ye Y, Liang Y, Zhao J, Wu T, Lu N (2021) Identification of GL-V9 as a novel senolytic agent against senescent breast cancer cells. Life Sci 272:119196

Yosef R, Pilpel N, Tokarsky-Amiel R, Biran A, Ovadya Y, Cohen S, Vadai E, Dassa L, Shahar E, Condiotti R, Ben-Porath I, Krizhanovsky V (2016) Directed elimination of senescent cells by inhibition of BCL-W and BCL-XL. Nat Commun 7:11190

Yousefzadeh MJ, Zhu Y, McGowan SJ, Angelini L, Fuhrmann-Stroissnigg H, Xu M, Ling YY, Melos KI, Pirtskhalava T, Inman CL, McGuckian C, Wade EA, Kato JI, Grassi D, Wentworth M, Burd CE, Arriaga EA, Ladiges WL, Tchkonia T, Kirkland JL, Robbins PD, Niedernhofer LJ (2018) Fisetin is a senotherapeutic that extends health and lifespan. EBioMedicine 36:18–28

Youssef ME, Cavalu S, Hasan AM, Yahya G, Abd-Eldayem MA, Saber S (2023) Role of Ganetespib, an HSP90 inhibitor, in cancer therapy: from molecular mechanisms to clinical practice. Int J Mol Sci 24(5)

Zhang C, Xie Y, Chen H, Lv L, Yao J, Zhang M, Xia K, Feng X, Li Y, Liang X, Sun X, Deng C, Liu G (2020) FOXO4-DRI alleviates age-related testosterone secretion insufficiency by targeting senescent Leydig cells in aged mice. Aging (Albany NY) 12(2):1272–1284

Zhang H, Nimmer PM, Tahir SK, Chen J, Fryer RM, Hahn KR, Iciek LA, Morgan SJ, Nasarre MC, Nelson R, Preusser LC, Reinhart GA, Smith ML, Rosenberg SH, Elmore SW, Tse C (2007) Bcl-2 family proteins are essential for platelet survival. Cell Death Differ 14(5):943–951

Zhang P, Kishimoto Y, Grammatikakis I, Gottimukkala K, Cutler RG, Zhang S, Abdelmohsen K, Bohr VA, Misra Sen J, Gorospe M, Mattson MP (2019) Senolytic therapy alleviates Aβ-associated oligodendrocyte progenitor cell senescence and cognitive deficits in an Alzheimer's disease model. Nat Neurosci 22(5):719–728

Zhang X, Dong Y, Li WC, Tang BX, Li J, Zang Y (2021) Roxithromycin attenuates bleomycin-induced pulmonary fibrosis by targeting senescent cells. Acta Pharmacol Sin 42(12):2058–2068

Zhang X, Pearsall VM, Carver CM, Atkinson EJ, Clarkson BDS, Grund EM, Baez-Faria M, Pavelko KD, Kachergus JM, White TA, Johnson RK, Malo CS, Gonzalez-Suarez AM, Ayasoufi K, Johnson KO, Tritz ZP, Fain CE, Khadka RH, Ogrodnik M, Jurk D, Zhu Y, Tchkonia T, Revzin A, Kirkland JL, Johnson AJ, Howe CL, Thompson EA, LeBrasseur NK, Schafer MJ (2022) Rejuvenation of the aged brain immune cell landscape in mice through p16-positive senescent cell clearance. Nat Commun 13(1):5671

Zhang X, Zhang S, Liu X, Wang Y, Chang J, Zhang X, Mackintosh SG, Tackett AJ, He Y, Lv D, Laberge RM, Campisi J, Wang J, Zheng G, Zhou D (2018) Oxidation resistance 1 is a novel senolytic target. Aging Cell 17(4):e12780

Zhang Y, Gao D, Yuan Y, Zheng R, Sun M, Jia S, Liu J (2023) Cycloastragenol: a novel Senolytic agent that induces senescent cell apoptosis and restores physical function in TBI-aged mice. Int J Mol Sci 24(7)

Zhu Y, Doornebal EJ, Pirtskhalava T, Giorgadze N, Wentworth M, Fuhrmann-Stroissnigg H, Niedernhofer LJ, Robbins PD, Tchkonia T, Kirkland JL (2017) New agents that target senescent cells: the flavone, fisetin, and the BCL-X(L) inhibitors, A1331852 and A1155463. Aging (Albany NY) 9(3):955–963

Zhu Y, Tchkonia T, Fuhrmann-Stroissnigg H, Dai HM, Ling YY, Stout MB, Pirtskhalava T, Giorgadze N, Johnson KO, Giles CB, Wren JD, Niedernhofer LJ, Robbins PD, Kirkland JL (2016) Identification of a novel senolytic agent, navitoclax, targeting the Bcl-2 family of anti-apoptotic factors. Aging Cell 15(3):428–435

Zhu Y, Tchkonia T, Pirtskhalava T, Gower AC, Ding H, Giorgadze N, Palmer AK, Ikeno Y, Hubbard GB, Lenburg M, O'Hara SP, LaRusso NF, Miller JD, Roos CM, Verzosa GC, LeBrasseur NK, Wren JD, Farr JN, Khosla S, Stout MB, McGowan SJ, Fuhrmann-Stroissnigg H, Gurkar AU, Zhao J, Colangelo D, Dorronsoro A, Ling YY, Barghouthy AS, Navarro DC, Sano T, Robbins PD, Niedernhofer LJ, Kirkland JL (2015) The Achilles' heel of senescent cells: from transcriptome to senolytic drugs. Aging Cell 14(4):644–658

Chapter 6
Stem Cell Therapies and Ageing: Unlocking the Potential of Regenerative Medicine

Chen Rui, Mike K. S. Chan, and Thomas Skutella

Abstract A multifaceted biological process of ageing culminates in the gradual decline of tissue and organ functions, escalating vulnerability to age-related diseases. Stem cell therapies, standing at the frontier of regenerative medicine, hold the potential to mitigate the challenges induced by ageing. By harnessing the unique regenerative capabilities of stem cells, these therapies aim to renew and heal ageing or damaged cells and tissues, thereby bolstering their function. In this chapter, we explore the potential of stem cell-based interventions against age-related degeneration, emphasising their underlying mechanisms, challenges, and future possibilities. As elucidated by the Buck Institute for Research on Aging, ageing is characterised by an accrual of macromolecular damage, genomic instability, and loss of heterochromatin (Campisi et al. Nature 571:183–192, 2019). These aspects culminate in stem cell fatigue and a dwindling tissue regenerative capacity. However, with the advent of stem cell therapy and regenerative medicine, we now hold the tools to reverse these age-induced changes by rejuvenating stem cells, the keystones of tissue regeneration, and fostering their proliferation and differentiation.

Keywords Ageing · Age-related diseases · Genomic instability · Regeneration · Stem cells

---

C. Rui
Reproductive Medicine Center, The Affiliated Hospital of Qingdao University, Qingdao, China
e-mail: rui.chen@qdu.edu.cn

M. K. S. Chan
European Wellness International, Edenkoben, Germany

T. Skutella (✉)
Group for Regeneration and Reprogramming, Institute for Anatomy and Cell Biology, Medical Faculty, Heidelberg University, Heidelberg, Germany
e-mail: Thomas.Skutella@uni-heidelberg.de

© The Author(s), under exclusive license to Springer Nature Switzerland AG 2024

V. I. Korolchuk, J. R. Harris (eds.), *Biochemistry and Cell Biology of Ageing: Part V, Anti-Ageing Interventions*, Subcellular Biochemistry 107,
https://doi.org/10.1007/978-3-031-66768-8_6

Introduction

Globally, an ageing demographic poses a significant challenge to healthcare infrastructures, necessitating innovative methods to uplift the quality of life for the elderly. Stem cell therapies are seen as a beacon of hope for restoring tissues and organs debilitated by age-associated degeneration. This chapter outlines the foundational tenets of stem cell biology and their potential applications in ageing research.

Stem cells, celebrated for their prodigious capacity to self-renew and metamorphose into diverse cell types, are instrumental in upholding tissue equilibrium and facilitating tissue restoration and renewal. Yet, ageing casts a shadow on these cells, leading to a decrease in their numbers, functionality, and differentiation potential (Brunet et al. 2023). This decline is implicated in several age-induced challenges, encompassing heightened cancer risk, tissue decay, and a diminished tissue repair capacity. Grasping the nuances of how ageing interacts with stem cells and their ageing niche is imperative, shedding light on age-related diseases and unveiling potential stem cell-based therapeutic avenues.

Over the years, the spotlight on stem cell research has intensified, especially within the realm of ageing research. Reprogrammed adult stem cells, distinguished by their self-renewing and diverse differentiation capacities, have piqued interest not only for ageing studies but also for their prospects in regenerative medicine, tissue healing, gene therapy, and cell-based autologous therapies. Furthermore, adult tissue-specific stem cells, sourced from various organs, are pivotal to ageing research. Comprehending the architecture and functionality of stem cell microenvironments/niches across tissues can yield insights crucial for gene therapy, oncological treatments, and broader ageing research.

Stem Cells in the Context of Ageing

Of the more than 200 types of cells that make up the human body, most of them are terminally differentiated cells, which are highly differentiated so that they lose the ability to divide again and eventually age and die. Still, at the same time, the body also retains a portion of undifferentiated primitive cells, i.e., stem cells. A stem cell is an undifferentiated cell that has the potential to develop into different cell types in the body. Stem cells can undergo numerous cycles of cell division while maintaining their undifferentiated state. They possess two key properties: self-renewal, which is the ability to divide and produce more stem cells, and differentiation, which is the potential to mature into specialised cells with specific functions. Stem cells play a crucial role in maintaining tissue homeostasis by providing a continuous source of new cells to replace those lost due to normal physiological processes, injury, or disease. This balance between cell growth and decline is essential for the body's overall function and the integrity of organs and tissues. Here are reasons why stem cells are central to this balance. During Regeneration and Repair, stem cells are responsible for repairing damaged tissues. When an injury occurs, stem cells can divide and differentiate into the specific types of cells needed to replace the ones lost or damaged. This regenerative capability is vital for the recovery of tissues. Many tissues, such as the skin, blood, and the lining of the gut, have a high turnover rate, meaning their cells are constantly being replaced. Stem cells provide a reservoir of new cells to sustain this ongoing replacement, ensuring the tissues remain functional and healthy. Stem cells can be activated upon injury to proliferate and regenerate the affected tissue. This response is crucial to initiate the healing process. The body is a dynamic system where cells are regularly subjected to apoptosis (programmed cell death). Stem cells help maintain homeostasis by producing new cells to replace those that are lost, ensuring that the tissue remains in a steady state without atrophy or overgrowth. Because of their unique capabilities, stem cells are essential for the lifelong maintenance and healing of the body, making them a target of interest for therapeutic strategies to treat a wide array of diseases and injuries. The ability of stem cells to regenerate and produce a variety of cells within tissues gradually diminishes with ageing, and this imbalance leads to a decline in tissue function and overall health (Brunet et al. 2023). In terms of tissue function, stopping this trend improves the dysfunction caused by ageing (Campisi et al. 2019).

The functions and characteristics of stem cells make them a broad application prospect in clinical medical fields such as trauma repair, nerve regeneration, and anti-ageing. It has been proved that stem cells are effective in the treatment of cardiovascular diseases, metabolic diseases, Parkinson's syndrome, cirrhosis, and other diseases (Demurtas et al. 2021; Tran and Kahn 2010; Parmar et al. 2020; Yang et al. 2023). Stem cell anti-ageing is one of the 10 major scientific advances selected in 1999 by Science magazine, which triggered the revolution of "regenerative medicine." This revolution in regenerative medicine should not be underestimated.

As we can imagine, the anti-ageing effect of stem cells depends on the mobilisation of a sufficient number of ideal stem cells. The persistence of stem cells in the body makes them susceptible to the accumulation of ageing-related toxic metabolites, and stem cells in many tissues undergo profound changes with age, and eventually undergo cell death, senescence, or loss of regenerative function. These age-related changes in the stem cell function manifest in the ageing body's delayed response to tissue damage, dysregulation of proliferative activity, and a decline in cell renewal and tissue regeneration function. Stem cells in the ageing organism are subjected to stresses mainly from somatic gene mutations, epigenetic changes, and environmental factors (Sharpless and DePinho 2007). Additionally, ageing-related toxic metabolites such as reactive oxygen species (ROS), DNA or protein damage can affect the function or number of stem cells (López-Otín et al. 2023).

In contrast, injection of blood from young organisms into senescent organisms, calorie restriction, or the use of the life-prolonging drug rapamycin may also enhance stem cell function in senescent organisms to a certain extent, resulting in a more youthful cellular state in senescent organisms (Novak et al. 2021; Goodell and Rando 2015). The mechanism may involve reprogramming of the epigenome, reduction of inflammatory mediators, improvement of mitochondrial function, and re-balance of stem cell proliferation (Goodell and Rando 2015). However, the repair of aged epigenome and damaged DNA requires the identification of significant regulatory genes. Furthermore, the process of cellular autophagy, another hallmark of ageing, must be selectively controlled. Due to the complexity of the mechanisms and means of realisation, there is still a long way to go to achieve the anti-ageing effect through the intervention of endogenous stem cells. The other direction of stem cell anti-ageing, specifically the use of exogenous adult stem cells to correct the problem of body ageing, is more widely researched, and applications are more advanced.

Current research focuses on several types of anti-ageing stem cells. Many studies have found that stem cells that exist throughout the life process of the organism have antioxidant, anti-wrinkle, and wound healing functions (Wang et al. 2020; Chen et al. 2018). The most direct evidence is that dermal multipotent stem cells (DMSCs) prevent skin ageing by increasing the level of type I collagen, increasing skin thickness (Wang et al. 2019), and can work together with dermal stem cells (DSCs) to increase the number of skin precursor cells and epidermal regeneration (Zhong et al. 2011). The exogenous adult stem cells that can be used for anti-ageing mainly include adipose stem cells, bone marrow mesenchymal stem cells, and mesenchymal stem cells derived from gestational tissues. Highly undifferentiated embryonic stem cells also have a small number of applications, among which pluripotent stem cells have a wide range of sources and do not involve ethical issues, and have become the most researched and applied stem cell types in the international arena at present.

Importantly, the anti-ageing effect of stem cells as a transplanted cell source depends on mobilizing a sufficient number of ideal stem cells and their integration into the host or target tissue. Transplanted stem cells integrate into the host tissue through a process known as engraftment. These include the following stages:

(a) Migration: Post-transplantation, stem cells must migrate towards the damaged or target health tissues. This function is typically facilitated by certain chemical signals (cytokines and chemokines) released by the damaged tissue and helps direct the stem cells to that location.
(b) Extravasation and Occupation: blood stem cells must cross the blood vessel barrier to move into the tissue. This process is known as extravasation. After extravasation, the cells situate themselves within the tissue.
(c) Proliferation and Differentiation: Once at their destination, the transplanted stem cells may multiply and proliferate. Then, under the influence of specific growth and differentiation factors within this microenvironment, they will potentially differentiate into the necessary cell type to replace the damaged or lost cells in the target tissue.
(d) Functional Integration: Finally, the newly differentiated cells must integrate into the host tissue and adopt the functional characteristics of the surrounding cells. They start to work alongside the native cells to restore normal tissue functions. It is noteworthy to mention that not all stem cells have equal potential for migration, extravasation and engraftment, and their ability to integrate into new

tissue can depend on the type of stem cell, the specific target tissue, and the patient's health condition.

Classification of Stem Cells that Can Be Used for Anti-ageing Applications

1. Adipose stem cells: Adipose tissue is a renewable source of stem cells, most of the adipose stem cells originate from the fibrous tissue and blood vessel wall in the adipose tissue, and a few are free in the adipose interstitium. Adipose stem cells can differentiate into a variety of tissues such as bone, adipose, cardiac muscle, nerve, etc., while releasing important growth factors and cytokines to promote wound healing, modulate inflammatory response, reduce scar formation and anti-ageing (Wang et al. 2023).

In anti-ageing applications, adipose stem cells can not only be used as fillers, but also as permanent regenerative agents with a high differentiation capacity, resulting in significant anti-ageing effects and long-lasting efficacy. Animal experiments on the effects of allogeneic transplantation on free radicals in ageing model rats revealed that rats injected intravenously with adipose-derived stem cells had elevated levels of superoxide dismutase (SOD), enhanced antioxidant capacity, and slowed down the ageing process (Yang et al. 2015). The transplantation of adipose MSCs was effective in improving the symptoms of patients with facial muscular dystrophy and facial seborrhea, and the effect of its differentiation from hyaluronic acid was obvious. The transplantation of adipose MSCs is effective in improving symptoms in patients with facial muscle atrophy and facial lipoatrophy, and its use in combination with hyaluronic acid can also achieve the repair of photodamaged skin (Chen et al. 2020). The convenience of autologous transplantation is an irreplaceable advantage of the anti-ageing application of adipose stem cells, and there is almost no safety problem in the process of using them.

2. Bone marrow mesenchymal stem cells: including hematopoietic stem cells, mesenchymal stromal cells, endothelial cells, etc., is a heterogeneous group of pluripotent mesenchymal cells, which can be used for the repair and regeneration of tissue damage caused by ionizing radiation and other tissue injuries. In the regeneration of tissue injuries, it can be differentiated into functional parenchymal cells, and create a microenvironment for the growth of other cells (Gong et al. 2016). Bone marrow MSCs have been used as a powerful tool for the regeneration of diseased or injured tissues for the potential differentiation of chondrocytes, osteoblasts, adipocytes, neuronal cells, etc.
3. Gestational tissue-derived MSCs possess greater proliferative capacity than adult-derived MSCs and higher stem cell yield than bone marrow. These cells can regulate several biological processes such as angiogenesis, morphogenesis,

tissue regeneration, and cell survival. As an important realisation of cell-based tissue regeneration therapies, the effects of transplantation of such exogenous MSCs on reducing scarring and promoting tissue regeneration have been demonstrated in model animals or clinical trials (Zhang and Lai 2020; Liu et al. 2021). It is worth noting that stem cells from umbilical cord blood/cord have the advantages of no tumour contamination, high proliferative capacity, easy collection, low immunoreactivity, and no detrimental effect on the donor, which can be an ideal source of anti-ageing stem cells.

4. Embryonic stem cells: Embryos are the main source of pluripotent stem cells, which can be induced to differentiate into blood, endothelial, cardiac muscle, bone, nerve, and other cells. Animal experiments have shown that the implantation of human embryonic stem cells into damaged tissues of ageing mice can produce soluble proteins that activate anti-ageing signals and exert anti-ageing effects (Dorronsoro et al. 2021). However, the clinical application of embryonic stem cells is limited due to ethical issues.
5. Other: cells such as human induced pluripotent stem cells (iPSCs), have also been shown to treat specific skin diseases or rejuvenate the skin (Mandai et al. 2017). iPSCs have been explored for their potential in anti-ageing research. Due to their unique ability to self-renew indefinitely and differentiate into virtually any type of cell in the human body, iPSCs represent a promising way to replace cells or tissues damaged due to ageing. Schwartz, S. D. et al. reported 18-month results using iPSC-derived retinal pigment epithelium for macular degeneration. This study proved that cells derived from iPSCs could be safely transplanted into humans (Schwartz et al. 2015). The Cyranoski study reported an experiment in Japan where researchers used iPSC-derived dopaminergic progenitors to treat Parkinson's Disease (Cyranoski 2019). Research led by Liu et al. demonstrated how individual iPSC lines with longer telomeres increased the capacity to regenerate tissues in a mouse with premature ageing symptoms. This relates directly to ageing and potential anti-ageing implications (Liu et al. 2012). These studies are helping to advance our understanding of and the therapeutic potential of iPSCs for age-related diseases and anti-ageing strategies. However, we should remember that transforming the iPSC-based therapies into routine clinical reality is complex and ethical, and safety concerns need to be addressed carefully.

Several preclinical studies—and even some early-stage clinical trials—have used iPSCs to treat conditions typically associated with ageing, such as macular degeneration, Parkinson's disease, and heart disease. For instance, in 2014, Japanese scientists transplanted retinal cells derived from iPSCs into a patient with age-related macular degeneration (Mandai et al. 2017). Studies of iPSCs are also helping to deepen our understanding of the ageing process itself, including how cells become senescent and why certain age-related diseases occur. Moreover, iPSC technology could be used to generate cells for *in vitro* testing of anti-ageing compounds. However, using iPSCs in anti-ageing therapies is still early, with significant safety and ethical obstacles to overcome. For example, one primary concern is that the reprogramming needed to create iPSCs could potentially increase the risk of cancer.

So, while iPSCs are an exciting area of anti-ageing research, much more work is needed.

Mechanism of Stem Cell Anti-ageing

Stem cells contribute to anti-ageing by offering multidirectional differentiation capabilities within the tissue microenvironment. They can transform into various tissue cells, effectively replacing those that have aged or died. Additionally, stem cells possess a potent secretory function, releasing growth factors and cytokines that enhance the body's defences against free radicals. These secretions support angiogenesis and foster the proliferation and differentiation of cells. They also have anti-inflammatory properties and can influence chemotaxis, ultimately regulating cellular adhesion and migration processes. Through these mechanisms, stem cells stimulate the regeneration and repair of tissue cells, accelerating wound healing, tissue remodelling, and effectively combating the ageing process.

1. Adipose stem cells: They can secrete a variety of neuroprotective factors, such as the tissue inhibitor of metalloproteinase-1 (TIMP-1) and the secreted protein acidic and rich in cysteine (SPARC), which inhibits phototropic cell death after photoreceptor damage and retinal dysfunction through its effect on AKT Ser473 phosphorylation, exerts neuroprotective effects in retinal injury and effectively corrects degenerative retinopathy (Sugitani et al. 2013). Other cytokines secreted by adipose stem cells, such as vascular endothelial growth factor, epidermal growth factor, fibroblast growth factor and cytokines for anti-fibrosis, mainly play a role in promoting vascular regeneration and vascular stabilisation, elevating the function of skin fibroblasts, immune regulation, and neutralising ROS, among other mechanisms. Adipose-derived stem cells can also be differentiated into a variety of skin cells, etc. (Kim et al. 2011), which is currently one the most promising and practical new means of anti-ageing.
2. Bone marrow mesenchymal stem cells (MSCs): the absence of growth factors can induce cell autophagy and senescence, and downregulate cell stemness through the inhibition of AKT, also known as the PI3K-Akt signalling pathway and the extracellular regulated protein kinases (ERK) signalling pathways. Bone marrow MSCs can secrete a wide range of growth factors and cytokines and have even been compared to growth factor factories. These growth factors and cytokines can play an autocrine role by regulating the proliferation and differentiation of MSCs. For example, fibroblast growth factor 2 (FGF-2) and FGF-4 can rapidly induce the activation of AKT, followed by the activation of ERK, and the cellular proliferation is greatly enhanced, while hepatocyte growth factor (HGF) maintains cell differentiation potential (Hou et al. 2013; Wang et al. 2022), thus promoting healing of damaged tissues and organs. MSCs can also secrete cytokines essential for hematopoietic function such as Interleukin 6 (IL-6), IL-12, IL-14, leukaemia inhibitory factor, granulocyte colony-stimulating factor, as

well as neurotrophic factors such as glial-derived neurotrophic factor, nerve growth factor, etc. (Kim et al. 2005), which can play an anti-ageing role from multiple perspectives.

3. Gestational tissue-derived MSCs: MSCs isolated from placental and umbilical cord tissues can be cultured *in vitro*, and the secretion of a variety of angiogenesis- and wound healing-related factors could be detected in their supernatants (Heo et al. 2016), including IL-6, IL-8, transforming growth factor-β1 (TGF-β1) and monocyte chemoattractant protein 1 (MCP1) (Arici et al. 1995), vascular endothelial growth factor (VEGF), granulocyte-macrophage colony-stimulating factor (GM-CSF), etc. These factors provide the molecular basis for the maintenance of the *in vivo* microenvironment, as well as for clinical applications in trauma and ulcer therapy.

In summary, stem cell loss and malfunction and senescence, lead to skin ageing. Stem cell-based therapies, or supplemented with appropriate medications, can control the ageing process by deregulating signalling cascades including telomerase shortening, estrogen loss, and excess ROS production. Self-renewal, multidirectional differentiation, and secretion of paracrine factors are the mechanisms by which various stem cells exert their anti-ageing effects and promote skin rejuvenation.

There are some unfavourable aspects of the use of stem cells, such as ethical issues in the use of embryonic stem cells and standards for large-scale stem cell preparation. Considering these challenges, leveraging regenerative cells obtained from an individual's own adipose and subcutaneous tissues is anticipated to emerge as a conventional approach for skin regeneration. To circumvent the diminishment of stem cell potency due to ageing, it is prudent to initiate early preservation of autologous stem cells. Nevertheless, there is a pressing necessity to regulate the industry's growth, enhance management practices, and guarantee controllable quality along with reliable evaluation standards. This is especially true concerning the safety assessments and the verification of the cells' stemness.

The utilisation of exogenous adult stem cells for anti-ageing applications presents promising avenues, particularly when considering the integration of cytokine replacement therapy (Fig. 6.1). This approach aims to lower the technical barriers associated with stem cell therapies by focusing on replacing or supplementing key cytokines—proteins that act as signalling molecules in cellular communication. Cytokines play a crucial role in the regulation of inflammation, immunity, and the regeneration processes of the body. In the context of ageing, the production and activity of beneficial cytokines may decline, leading to reduced cellular repair and increased susceptibility to age-related pathologies. By administering the appropriate cytokines derived from stem cells, it may be possible to simulate or enhance the anti-ageing effects that stem cells provide, such as promoting tissue repair, reducing inflammation, and potentially improving the function of ageing cells.

Furthermore, exploring anti-ageing from the perspective of stem cell mobilization opens new research pathways. Mobilization refers to the process of stimulating stem cells to leave their niches and enter the bloodstream, increasing their

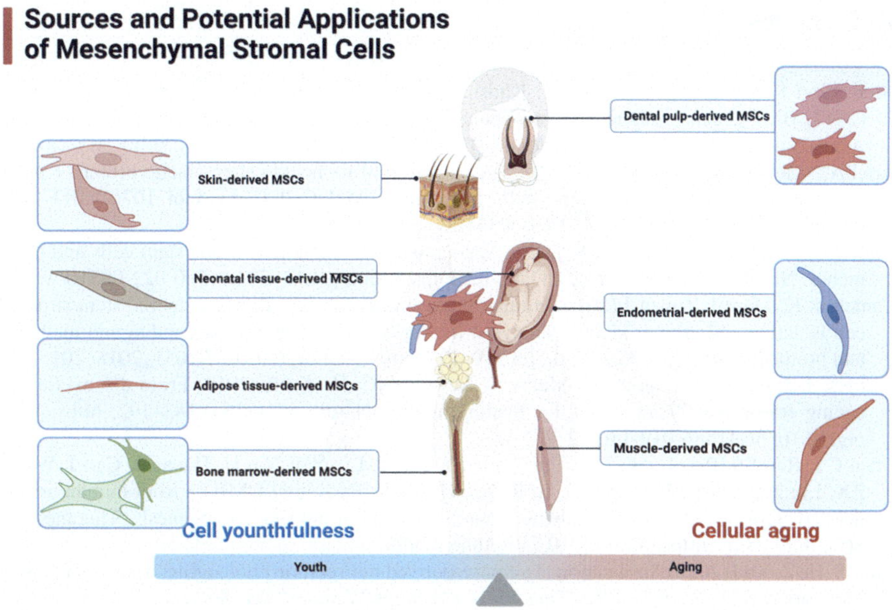

Fig. 6.1 Sources and potential applications of mesenchymal stromal cells

availability to repair and regenerate tissues. Identifying novel targets for intervention that can effectively promote stem cell mobilization could lead to the development of innovative pharmaceuticals, offering less invasive and more accessible anti-ageing treatments. As scientists unveil the mechanisms by which stem cells are recruited to sites of damage and how their regenerative capabilities can be harnessed or enhanced, these findings could culminate in ground-breaking drugs. Additionally, this line of research holds the potential to unravel the complex biological processes that underlie the ageing mechanism, possibly discovering key interventions that could delay or even reverse aspects of ageing, contributing to longevity and improved quality of life.

Shown above are schematic diagrams of the structural morphology of mesenchymal stem cells from different tissues, derived from dental pulp (Ullah et al. 2016), endometrium (Ai et al. 2017), muscle (Čamernik et al. 2019), skin (Jeremias Tda et al. 2014), placenta (Talwadekar et al. 2015), adipose tissue (Wu et al. 2016) and bone marrow. The quality and quantity of these cells decline with age, so supplementation of these cells at the appropriate age has the opportunity to prolong the balance between the quantity and quality of these cells in order to intervene in the ageing of specific tissues.

References

Ai J, Azizi E, Shamsian A, Eslami A, Khoshzaban A, Ebrahimi-Barough S, Ai A, Alizadeh A (2017) BMP-2 can promote the osteogenic differentiation of human endometrial stem cells. Asian Biomed 8(1):21–29. https://doi.org/10.5372/1905-7415.0704.258

Arici A, MacDonald PC, Casey ML (1995) Regulation of monocyte chemotactic protein-1 gene expression in human endometrial cells in cultures. Mol Cell Endocrinol 107(2):189–197. https://doi.org/10.1016/0303-7207(94)03442-v

Brunet A, Goodell MA, Rando TA (2023) Ageing and rejuvenation of tissue stem cells and their niches. Nat Rev Mol Cell Biol 24(1):45–62. https://doi.org/10.1038/s41580-022-00510-w

Čamernik K, Marc J, Zupan J (2019) Human skeletal muscle-derived mesenchymal stem/stromal cell isolation and growth kinetics analysis. In: Turksen K (ed) Stem cells and aging: methods and protocols. Springer, New York, pp 119–129. https://doi.org/10.1007/7651_2018_201

Campisi J, Kapahi P, Lithgow GJ, Melov S, Newman JC, Verdin E (2019) From discoveries in ageing research to therapeutics for healthy ageing. Nature 571(7764):183–192. https://doi.org/10.1038/s41586-019-1365-2

Chen CY, Rao SS, Ren L, Hu XK, Tan YJ, Hu Y, Luo J, Liu YW, Yin H, Huang J, Cao J, Wang ZX, Liu ZZ, Liu HM, Tang SY, Xu R, Xie H (2018) Exosomal DMBT1 from human urine-derived stem cells facilitates diabetic wound repair by promoting angiogenesis. Theranostics 8(6):1607–1623. https://doi.org/10.7150/thno.22958

Chen S, He Z, Xu J (2020) Application of adipose-derived stem cells in photoaging: basic science and literature review. Stem Cell Res Ther 11(1):491. https://doi.org/10.1186/s13287-020-01994-z

Cyranoski D (2019) Japan's approval of stem-cell treatment for spinal-cord injury concerns scientists. Nature 565(7741):544–545. https://doi.org/10.1038/d41586-019-00178-x

Demurtas J, Fanelli GN, Romano SL, Solari M, Yang L, Soysal P, López Sánchez GF, Grabovac I, Smith L, Zorzi A, Luchini C, Veronese N (2021) Stem cells for treatment of cardiovascular diseases: an umbrella review of randomized controlled trials. Ageing Res Rev 67:101257. https://doi.org/10.1016/j.arr.2021.101257

Dorronsoro A, Santiago FE, Grassi D, Zhang T, Lai RC, McGowan SJ, Angelini L, Lavasani M, Corbo L, Lu A, Brooks RW, Garcia-Contreras M, Stolz DB, Amelio A, Boregowda SV, Fallahi M, Reich A, Ricordi C, Phinney DG, Huard J, Lim SK, Niedernhofer LJ, Robbins PD (2021) Mesenchymal stem cell-derived extracellular vesicles reduce senescence and extend health span in mouse models of aging. Aging Cell 20(4):e13337. https://doi.org/10.1111/acel.13337

Gong W, Guo M, Han Z, Wang Y, Yang P, Xu C, Wang Q, Du L, Li Q, Zhao H, Fan F, Liu Q (2016) Mesenchymal stem cells stimulate intestinal stem cells to repair radiation-induced intestinal injury. Cell Death Dis 7(9):e2387. https://doi.org/10.1038/cddis.2016.276

Goodell MA, Rando TA (2015) Stem cells and healthy aging. Science 350(6265):1199–1204. https://doi.org/10.1126/science.aab3388

Heo JS, Choi Y, Kim HS, Kim HO (2016) Comparison of molecular profiles of human mesenchymal stem cells derived from bone marrow, umbilical cord blood, placenta and adipose tissue. Int J Mol Med 37(1):115–125. https://doi.org/10.3892/ijmm.2015.2413

Hou C, Shen L, Huang Q, Mi J, Wu Y, Yang M, Zeng W, Li L, Chen W, Zhu C (2013) The effect of heme oxygenase-1 complexed with collagen on MSC performance in the treatment of diabetic ischemic ulcer. Biomaterials 34(1):112–120. https://doi.org/10.1016/j.biomaterials.2012.09.022

Jeremias Tda S, Machado RG, Visoni SB, Pereima MJ, Leonardi DF, Trentin AG (2014) Dermal substitutes support the growth of human skin-derived mesenchymal stromal cells: potential tool for skin regeneration. PLoS One 9(2):e89542. https://doi.org/10.1371/journal.pone.0089542

Kim DH, Yoo KH, Choi KS, Choi J, Choi SY, Yang SE, Yang YS, Im HJ, Kim KH, Jung HL, Sung KW, Koo HH (2005) Gene expression profile of cytokine and growth factor during differentiation of bone marrow-derived mesenchymal stem cell. Cytokine 31(2):119–126. https://doi.org/10.1016/j.cyto.2005.04.004

Kim JH, Jung M, Kim HS, Kim YM, Choi EH (2011) Adipose-derived stem cells as a new therapeutic modality for ageing skin. Exp Dermatol 20(5):383–387. https://doi.org/10.1111/j.1600-0625.2010.01221.x

Liu GH, Qu J, Suzuki K, Nivet E, Li M, Montserrat N, Yi F, Xu X, Ruiz S, Zhang W, Wagner U, Kim A, Ren B, Li Y, Goebl A, Kim J, Soligalla RD, Dubova I, Thompson J, Yates J 3rd, Esteban CR, Sancho-Martinez I, Izpisua Belmonte JC (2012) Progressive degeneration of human neural stem cells caused by pathogenic LRRK2. Nature 491(7425):603–607. https://doi.org/10.1038/nature11557

Liu H, Jiang C, La B, Cao M, Ning S, Zhou J, Yan Z, Li C, Cui Y, Ma X, Wang M, Chen L, Yu Y, Chen F, Zhang Y, Wu H, Liu J, Qin L (2021) Human amnion-derived mesenchymal stem cells improved the reproductive function of age-related diminished ovarian reserve in mice through Ampk/FoxO3a signaling pathway. Stem Cell Res Ther 12(1):317. https://doi.org/10.1186/s13287-021-02382-x

López-Otín C, Blasco MA, Partridge L, Serrano M, Kroemer G (2023) Hallmarks of aging: an expanding universe. Cell 186(2):243–278. https://doi.org/10.1016/j.cell.2022.11.001

Mandai M, Watanabe A, Kurimoto Y, Hirami Y, Morinaga C, Daimon T, Fujihara M, Akimaru H, Sakai N, Shibata Y, Terada M, Nomiya Y, Tanishima S, Nakamura M, Kamao H, Sugita S, Onishi A, Ito T, Fujita K, Kawamata S, Go MJ, Shinohara C, Hata KI, Sawada M, Yamamoto M, Ohta S, Ohara Y, Yoshida K, Kuwahara J, Kitano Y, Amano N, Umekage M, Kitaoka F, Tanaka A, Okada C, Takasu N, Ogawa S, Yamanaka S, Takahashi M (2017) Autologous induced stem-cell-derived retinal cells for macular degeneration. N Engl J Med 376(11):1038–1046. https://doi.org/10.1056/NEJMoa1608368

Novak JSS, Baksh SC, Fuchs E (2021) Dietary interventions as regulators of stem cell behavior in homeostasis and disease. Genes Dev 35(3–4):199–211. https://doi.org/10.1101/gad.346973.120

Parmar M, Grealish S, Henchcliffe C (2020) The future of stem cell therapies for Parkinson disease. Nat Rev Neurosci 21(2):103–115. https://doi.org/10.1038/s41583-019-0257-7

Schwartz SD, Regillo CD, Lam BL, Eliott D, Rosenfeld PJ, Gregori NZ, Hubschman JP, Davis JL, Heilwell G, Spirn M, Maguire J, Gay R, Bateman J, Ostrick RM, Morris D, Vincent M, Anglade E, Del Priore LV, Lanza R (2015) Human embryonic stem cell-derived retinal pigment epithelium in patients with age-related macular degeneration and Stargardt's macular dystrophy: follow-up of two open-label phase 1/2 studies. Lancet 385(9967):509–516. https://doi.org/10.1016/s0140-6736(14)61376-3

Sharpless NE, DePinho RA (2007) How stem cells age and why this makes us grow old. Nat Rev Mol Cell Biol 8(9):703–713. https://doi.org/10.1038/nrm2241

Sugitani S, Tsuruma K, Ohno Y, Kuse Y, Yamauchi M, Egashira Y, Yoshimura S, Shimazawa M, Iwama T, Hara H (2013) The potential neuroprotective effect of human adipose stem cells conditioned medium against light-induced retinal damage. Exp Eye Res 116:254–264. https://doi.org/10.1016/j.exer.2013.09.013

Talwadekar MD, Kale VP, Limaye LS (2015) Placenta-derived mesenchymal stem cells possess better immunoregulatory properties compared to their cord-derived counterparts–a paired sample study. Sci Rep 5(1):15784. https://doi.org/10.1038/srep15784

Tran TT, Kahn CR (2010) Transplantation of adipose tissue and stem cells: role in metabolism and disease. Nat Rev Endocrinol 6(4):195–213. https://doi.org/10.1038/nrendo.2010.20

Ullah I, Subbarao RB, Kim E-J, Bharti D, Jang S-J, Park J-S, Shivakumar SB, Lee S-L, Kang D, Byun J-H, Park B-W, Rho G-J (2016) In vitro comparative analysis of human dental stem cells from a single donor and its neuronal differentiation potential evaluated by electrophysiology. Life Sci 154:39–51. https://doi.org/10.1016/j.lfs.2016.04.026

Wang X, Jiang B, Sun H, Zheng D, Zhang Z, Yan L, Li E, Wu Y, Xu RH (2019) Noninvasive application of mesenchymal stem cell spheres derived from hESC accelerates wound healing in a CXCL12-CXCR4 axis-dependent manner. Theranostics 9(21):6112–6128. https://doi.org/10.7150/thno.32982

Wang T, Jian Z, Baskys A, Yang J, Li J, Guo H, Hei Y, Xian P, He Z, Li Z, Li N, Long Q (2020) MSC-derived exosomes protect against oxidative stress-induced skin injury via adaptive regulation of the NRF2 defense system. Biomaterials 257:120264. https://doi.org/10.1016/j.biomaterials.2020.120264

Wang P, Cui Y, Wang J, Liu D, Tian Y, Liu K, Wang X, Liu L, He Y, Pei Y, Li L, Sun L, Zhu Z, Chang D, Jia J, You H (2022) Mesenchymal stem cells protect against acetaminophen hepatotoxicity by secreting regenerative cytokine hepatocyte growth factor. Stem Cell Res Ther 13(1):94. https://doi.org/10.1186/s13287-022-02754-x

Wang T, Li Y, Zhu Y, Liu Z, Huang L, Zhao H, Zhou Z, Wu Q (2023) Anti-aging mechanism of different age donor-matched adipose-derived stem cells. Stem Cell Res Ther 14(1):192. https://doi.org/10.1186/s13287-023-03415-3

Wu P-H, Chung H-Y, Wang J-H, Shih J-C, Kuo MY-P, Chang P-C, Huang Y-D, Wang P-C, Chang C-C (2016) Amniotic membrane and adipose-derived stem cell co-culture system enhances bone regeneration in a rat periodontal defect model. J Formos Med Assoc 115(3):186–194. https://doi.org/10.1016/j.jfma.2015.02.002

Yang C, Du YK, Wang J, Luan P, Yang QL, Huang WH, Yuan L (2015) Transplanted adipose-derived stem cells ameliorate testicular dysfunction in A D-galactose-induced aging rat model. J Cell Physiol 230(10):2403–2414. https://doi.org/10.1002/jcp.24970

Yang X, Li Q, Liu W, Zong C, Wei L, Shi Y, Han Z (2023) Mesenchymal stromal cells in hepatic fibrosis/cirrhosis: from pathogenesis to treatment. Cell Mol Immunol 20(6):583–599. https://doi.org/10.1038/s41423-023-00983-5

Zhang Q, Lai D (2020) Application of human amniotic epithelial cells in regenerative medicine: a systematic review. Stem Cell Res Ther 11(1):439. https://doi.org/10.1186/s13287-020-01951-w

Zhong J, Hu N, Xiong X, Lei Q, Li L (2011) A novel promising therapy for skin aging: dermal multipotent stem cells against photoaged skin by activation of TGF-β/Smad and p38 MAPK signaling pathway. Med Hypotheses 76(3):343–346. https://doi.org/10.1016/j.mehy.2010.10.035

Chapter 7
Diet-Modifiable Redox Alterations in Ageing and Cancer

Christopher Hine, Anand Kumar Patel, and András K. Ponti

Abstract With ageing comes some of life's best and worst moments. Those lucky enough to live out into the seventh, eighth, and nineth decades and perhaps beyond have more opportunities to experience the wonders and joys of the world. As the world's population shifts towards more and more of these individuals, this is something to be celebrated. However, it is not without negative consequences. Advanced age also ushers in health decline and the burden of non-communicable diseases such as cancer, heart disease, stroke, and organ function decay. Thus, alleviating or at least dampening the severity of ageing as a whole, as well as these individual age-related disorders will enable the improvement in lifespan and healthspan. In the following chapter, we delve into hypothesised causes of ageing and experimental interventions that can be taken to slow their progression. We also highlight cellular and subcellular mechanisms of ageing with a focus on protein thiol oxidation and posttranslational modifications that impact cellular homeostasis and the advent and progression of ageing-related cancers. By having a better understanding of the mechanisms of ageing, we can hopefully develop effective, safe, and efficient therapeutic modalities that can be used prophylactically and/or concurrent to the onset of ageing.

Keywords Ageing · Antioxidant systems · Diet · Reactive oxygen species · Redox

C. Hine (✉) · A. K. Ponti
Department of Cardiovascular and Metabolic Sciences, Cleveland Clinic Lerner Research Institute, Cleveland, OH, USA

Department of Molecular Medicine, Cleveland Clinic Lerner College of Medicine, Cleveland, OH, USA
e-mail: hinec@ccf.org; PONTIA2@ccf.org

A. K. Patel
Department of Cardiovascular and Metabolic Sciences, Cleveland Clinic Lerner Research Institute, Cleveland, OH, USA

Cardiovascular Genetics Lab, Department of Biotechnology, Bhupat and Jyoti Mehta School of Biosciences, Indian Institute of Technology Madras, Chennai, India

© The Author(s), under exclusive license to Springer Nature Switzerland AG 2024

V. I. Korolchuk, J. R. Harris (eds.), *Biochemistry and Cell Biology of Ageing: Part V, Anti-Ageing Interventions*, Subcellular Biochemistry 107,
https://doi.org/10.1007/978-3-031-66768-8_7

Introduction

As the global population surpassed 8-billion individuals in November of 2022 despite an ever declining birth rate (Llarena and Hine 2021; Oluwole et al. 2023), it becomes paramount to analyse the demographic shifts in aged populations that have resulted in this momentous event which will ultimately influence current and future public health strategies and medical innovations. Those over the age of 65 are some of the most vulnerable to non-communicable as well as communicable diseases including cardiovascular disease, neurodegeneration, cancer, inflammation, and sepsis (Niccoli and Partridge 2012). Thus, the ever-increasing proportion and absolute number of people in this age category calls for effective geroscience approaches to slow the ageing processes and provide therapeutic avenues to treat these ageing-related diseases. Given there are numerous underlying causes and hallmarks for ageing (Lopez-Otin et al. 2023a), this may be the reason clinical anti-ageing therapies for humans have seemingly been out of reach and currently unavailable. However, exploratory research from the lab on model systems and in human epidemiological studies point to some promising central targets and interventions, mainly those centred on controlling pathological oxidative damage to tissues, cells, and biological macromolecules by targeting nutrient sensing and metabolic hormonal signalling (Hine et al. 2018; Luo et al. 2017; Xiao et al. 2020). Thus, examining and safely enhancing the cellular mechanisms that maintain redox homeostasis during ageing through reducing the production of reactive oxygen species or conversely augmenting anti-oxidant processes may deliver hope for addressing ageing as the major risk factor in the most deadly and debilitating diseases. In the following chapter, we provide an overview on the biology of ageing with an emphasis on oxidative damage, followed by examples of dietary and nutritional interventions that counteract the ageing process, and then we examine the diet-redox-ageing axis as it pertains to protein damage and cancer.

Oxidative Stress and Ageing

Harman's Theory of Ageing: Role of Free Radicals

Free radicals and their biological roles have been known to biologists dating back to at least the late eighteenth century. Famously in 1956, it was Denham Harman who postulated a theory of ageing based on free radicals (Harman 1956). However, it does not mean that there was no speculation on the link between ageing and metabolism before Harman, as in 1928 Pearl presented his rate of living hypothesis (Lints 1989). According to Pearl, the lifespan of an individual is inversely related to the metabolic rate. While Pearl did not explain any precise mechanism on what causes metabolic rates to modify ageing, it can be hypothesised that disruptions in both metabolic and redox homeostasis may be closely linked and act as causative factors in advancing the ageing process.

According to Harman's free radical theory of ageing, a living organism accumulates irreparable damage to biological macromolecules such as DNA, RNA, proteins, lipids, and carbohydrates by reactive oxygen and nitrogen species and free radicals generated during both normal and pathological metabolic processes. While the general premise of the free radical theory of ageing still has merit, it should be noted it has been amended, expanded upon, and even challenged as ageing research has become more sophisticated in the past 20 years (Gladyshev 2014). When Harman introduced his free radical idea of ageing, there was little experimental evidence to support his theory. However, a group of evolutionally conserved enzymes under the common term of Superoxide Dismutases (SODs), was later discovered to scavenge the anions enzymatically (McCord and Fridovich 1969). It was a ground-breaking discovery, an essential milestone of ageing research, and perhaps the first solid experimental evidence to support Harman's free radical theory. But how? The discovery of such enzymes inside cells indicated that there has to be a mechanism through which anions are continuously generated; otherwise, any unnecessary protein would theoretically disappear due to evolutionary pressure. Later, reports from several groups confirmed the presence of superoxide radicals in cells generated inherently, and the existence of mechanisms for the checks and balances of these radicals at the intracellular level, as explained in the next section.

Oxygen is a Blessing in Disguise for Living Organisms

The idea that oxygen is dangerous for living organisms sounds counterintuitive initially; however, when we delve deeper into cellular metabolism, we would enlighten ourselves about the extent of damage oxygen can cause to biomolecules. Therefore, oxygen is a blessing in disguise for living organisms as it is simultaneously both vital and dangerous. To understand this contradiction, we have to understand the forms of oxygen and their interaction at the subcellular level. As eukaryotic cells are compartmentalised, the flow and concentration of oxygen in these compartments is not equal and thus must be regulated. Oxygen exists in various forms, such as the diatomic form O_2 that we breathe, O_3, which is ozone primarily found in the ozone layer, and in radical forms such as $O_2^{\cdot-}$ and the superoxide and peroxide ions and radicals. The radical form is the most reactive, and while important as a signalling molecule, is a threat to biomolecules when enhanced generation and lack of removal are left unchecked (Keyer and Imlay 1996).

These reactive or radical forms of oxygen are grouped as reactive oxygen species (ROS). The ROS are an integral part of the cellular system and are generated during metabolic processes and neutralized by antioxidants or cellular enzymes utilising reducing equivalents and/or glutathione. The majority of ROS under normal conditions are generated in the cells' mitochondria while transferring electrons through the various protein complexes involved in ATP generation. This series of the flow of electrons mediated by multiple proteins to generate ATP is called the Electron Transport Chain (ETC).

Under physiological concentrations, ROS takes part in a plethora of signalling pathways and biological processes. For instance, it is involved in the differentiation of cardiomyocytes, embryonic stem cells, and smooth muscle cells (Sinenko et al. 2021). These are so important for maintaining cellular homeostasis that if the delicate balance between ROS and antioxidants, which are scavengers of ROS, is shifted towards antioxidants, the cell undergoes reductive stress (Xiao and Loscalzo 2020).

This infers that organisms that can either reduce the production of radical species or suppress its effect are expected to achieve a higher maximum life span. It can further be extrapolated that maintenance of endogenous antioxidant production and/or utilisation into older ages must slow down the ageing process as they suppress the cellular overloading of ROS. In the following Tables 7.1 and 7.2, we present cellular sources of ROS, as well as cellular sources of antioxidant defences.

ROS Creates and/or Strengthens the Foundation of the Pillars of Ageing

Seven factors are assumed to be critical drivers of ageing in multicellular and/or mammalian organisms. These factors are stress, metabolism, macromolecular damage, proteostasis, epigenetic modifications, inflammation, and stem cell regeneration. Altogether, these highly interdependent stressors are called pillars of ageing as

Table 7.1 Cellular Sources of Reactive Oxygen Species (ROS)

Source	Radical species	Location	References
Complex I	Superoxide	Mitochondria	Brand (2010)
Complex II	Superoxide	Mitochondria	Brand (2010)
Complex III	Superoxide	Mitochondria	Brand (2010)
Glycerol-3-phosphate dehydrogenase	Superoxide	Mitochondria	Brand (2010)
Pyruvate dehydrogenase	Superoxide	Mitochondria	Brand (2010)
2-oxoglutarate dehydrogenase	Superoxide	Mitochondria	Brand (2010)
Cytochrome P450 (CYP)	Superoxide	Mitochondria	Yasui et al. (2005)
NADH-cytochrome b5 reductase	Superoxide	Mitochondria	Whatley et al. (1998)
Dihydroorotate dehydrogenase	Superoxide	Mitochondria	Hey-Mogensen et al. (2014)
Xanthine oxidoreductase	Superoxide and peroxynitrile	Peroxisomes	Lee et al. (2014b)
Urate oxidase	Superoxide and peroxynitrile	Peroxisomes	Sautin and Johnson (2008)
Mitochondrial aconitase	Hydroxyl	Mitochondria	Vasquez-Vivar et al. (2000)

Table 7.2 Cellular Sources of Antioxidant Defence

Source	Location	References
Enzymatic Sources		
Superoxide dismutase	Mitochondria and Peroxisomes	Sharifi-Rad et al. (2020)
Catalase	Peroxisomes and Cytoplasm	Sharifi-Rad et al. (2020)
Glutathione peroxidase	Mitochondria and Peroxisomes	Sharifi-Rad et al. (2020)
Peroxiredoxins	Peroxisomes	Sharifi-Rad et al. (2020)
Non-enzymatic		
Glutathione	Produced in cytosol and translocated in mitochondria	Sharifi-Rad et al. (2020)
Thioredoxins	Cytosol	Sharifi-Rad et al. (2020)
Hydrogen Sulphide	Cytosol and mitochondria	Olas (2017)

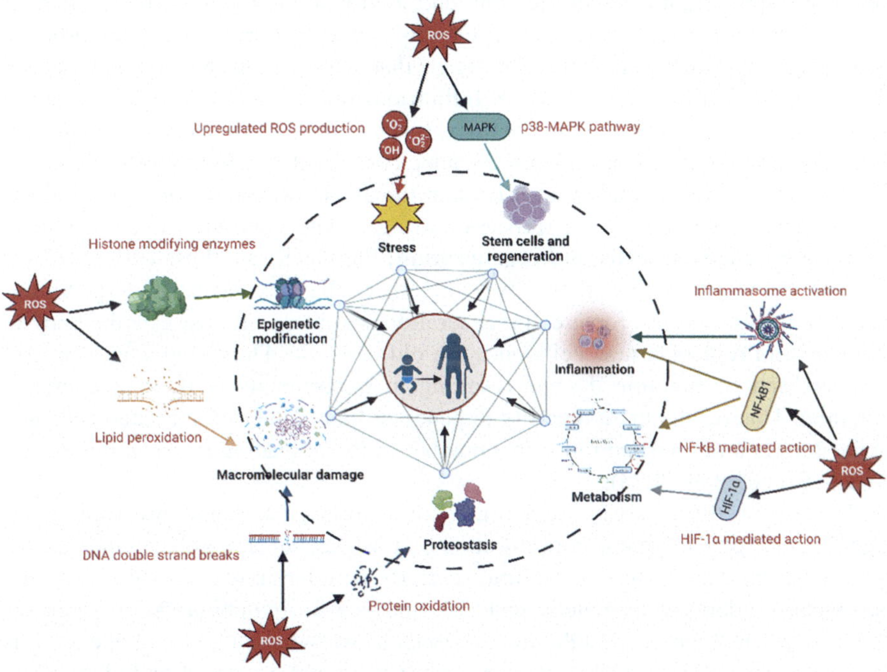

Fig. 7.1 Reactive Oxygen Species (ROS) and their relation to the seven pillars of ageing

described Kennedy and colleagues (Kennedy et al. 2014). How do these factors affect ageing, and how are they interconnected? To understand this, let us briefly described these factors below and depict them in Fig. 7.1.

i. Stress: It can be defined as any stimulus or event, external or internal that leads to a damaging situation unless resistance and resilience are in place. These types of cellular stress include sudden changes in temperature (heat shock), pH or osmolarity, ROS, nutrient deprivation or overload, and xenobiotic build-up. **ii. Metabolism:** It includes all the anabolic and catabolic processes in a cell that are necessary for maintaining cellular homeostasis for breaking down or producing macromolecules for growth, energy, protection, and signalling. **iii. Macromolecular damage:** Macromolecules are the biomolecules necessary for a cell's structural and functional integrity, such as DNA, RNA, lipids, and proteins. Damage to these molecules includes oxidative, misfolding (in the case of proteins), and glycation. **iv. Proteostasis:** It is the homeostasis of the proteome in a cell. In other words, it is a balance between producing and maintaining necessary and functional proteins and eliminating unwanted proteins that could lead to further misfolding, aggregation, and/or amyloid formation. **v. Epigenetic modification:** Epigenetic modifications are genetic sequence-independent but mitotically or meiotically heritable changes with chemical modifications on DNA and associated histone proteins and include but are not limited to DNA methylation, chromatin remodelling, and histone acetylation, phosphorylation, ubiquitination, and methylation. Together, these changes can act transiently to regulate gene expression. However, during the ageing process, there are changes in the epigenetic landscape that drive cellular survival and renewal (Pan et al. 2023; Yang et al. 2023). **vi. Inflammation:** It is a physiological response against any injury or infection in the body. It is mediated by the host's immune system, and when utilized and resolved under controlled conditions offers defence from foreign as well as endogenous cellular threats. However, uncontrolled inflammation exacerbates the ageing process and age-related chronic diseases (Chung et al. 2019). **vii. Stem cells and regeneration:** Stem cells are those cells that have the potential to differentiate themselves into other cell types and can undergo unlimited cell division cycles. Stem cell regeneration is strategically used by the body to generate and replenish new cells once the old cells die. During ageing, stem cell numbers and or function decline resulting in impairments in their regenerative potential. This results in an inability to generate functional differentiated cells and in the correct amounts and ratios in a number of tissue leading to organ failure and frailty (Brunet et al. 2023).

As mentioned previously, ROS is not just a stressor but is also involved in cell signalling at physiological concentrations. However, an unregulated increase in ROS levels beyond physiological limits perturbs cellular homeostasis through lipid peroxidation, damage to genetic material, and protein oxidation. As macromolecules are a direct target of ROS, and due to their overwhelming role in ageing, it is intuitive to speculate that ROS could be an activator and a connecting link between the seven pillars of ageing. ROS is an essential mediator of Nuclear Factor Kappa B (NF-kB) mediated inflammation (Marui et al. 1993), which can be activated by any dietary factors (Baker et al. 2011; Xu et al. 2003), or H_2O_2 (Canty Jr. et al. 1999). The activation of NF-kB is followed by its translocation to the nucleus and results in metabolic reprogramming. NLRP3, a causative factor for sterile inflammation (Cassel and Sutterwala 2010) is also activated by mitochondrial ROS (Heid et al.

2013), whereas cytoplasmic ROS is an activator of AMP-mediated protein kinases (AMPK) (Wang et al. 2012), which regulates a plethora of metabolic pathways such as glycolysis and lipid metabolism and physiological functions like cell growth and autophagy. Hypoxia-inducible factor (HIF-1α), an oxygen-sensitive protein under the influence of mitochondrial ROS, switches the cellular metabolism to anaerobic glycolysis to inhibit the Krebs cycle with an upregulation of glucose transporters (Robey et al. 2005), (Solaini et al. 2010).

Due to their reactivity and susceptibility of DNA towards oxidation, ROS can introduce single and double-strand breaks in the DNA and could be a significant cause of genotoxic-driven diseases like cancer (Sharma et al. 2016). The cell membrane is oxidized and loses its integrity and functionality when ROS reacts with the unsaturated lipid bilayer through lipid peroxidation. Beyond damaging the genetic material, ROS also epigenetically regulates the genome by modifying histone acetylases, deacetylases, and methyl transferases. The histone modifications act as switch for the accessibility of transcriptional machinery (Kietzmann et al. 2017). Methylation of the genome has been linked with the transcriptional repression of genes (Anastasiadi et al. 2018). Atherosclerosis is a common cardiovascular disease caused by lipid peroxidation (Gianazza et al. 2021). Protein oxidation leads to its aggregation (Vasconcellos et al. 2016), and the aggregation of proteins oxidized by ROS is also a common cause of age-related diseases such as Alzheimer's disease (Irvine et al. 2008).

Removal of the damaged tissue and its regeneration is vital for keeping an organ healthy and functional. Stem cells play a crucial role in tissue regeneration due to their pluripotency and capacity to divide an infinite number of times. Pathological levels of ROS can inhibit the quiescent state of hematopoietic stem cells probably by upregulating MAPK pathway, thus leading to a significant reduction in the stem cell population (Ito et al. 2006). The ROS affects cellular homeostasis with a multi-targeted approach and simultaneously regulates a plethora of cellular events. This makes it a key therapeutic target, a goal that has been achieved by antioxidant interventions. However, overwhelming antioxidant levels can suppress ROS beyond the physiological limit eliciting reductive stress, which is caused due to highly reduced environment (Peris et al. 2019). As detailed further in the below sections, Dietary restriction (DR) interventions without malnutrition, rather than pharmacological treatments or genetic manipulations, are well-studied approaches in the lab and an upcoming avenue in the clinic to take advantage as well as counter increasing ROS contents during ageing with minimal pathological outcomes.

Dietary Interventions for Ageing Studies in Model Organisms

Over a century of research on the interactions between ageing, diet, and ageing-related disorders in a variety of organisms, animal models, and human trials have undoubtedly found enormous health benefits associated with controlled food intake, aka dietary restriction (DR) and such that it is considered a gold standard for maintaining a healthy and lengthy life (Fontana and Partridge 2015). Is there any infallible dietary regimen optimised for maximum health benefits? It is somewhat difficult to answer this question at present as most of the DR studies have been conducted in rodents or invertebrates and only a few studies have used non-human primates or humans ourselves. Humans and other primates are evolutionarily closer and share common feeding and ageing patterns, unlike rodents, who are nocturnal in behaviour and do not naturally face many of the same age-driven diseases such as cardiovascular and cerebrovascular issues and diabetes as do humans. Nonetheless, the results of these studies cannot be cancelled as the similar molecular basis of the health benefits points towards an evolutionarily conserved mechanism across different taxa. Here in this section, we briefly discussed different types of DR, their role in intracellular redox balance, and examples of their use in model organism studies.

Calorie Restriction (CR)

CR is generally defined as the reduction of daily overall caloric consumption without malnutrition. Various CR regimens in rodents starting from 10% up to 60% have been implicated in lifespan extension in many studies, with effectiveness somewhat dependent on strain, age at commencing the diet, and sex (Mitchell et al. 2016; Weindruch et al. 1986). CR is also a dietary stress that exists in nature, such as famine, during which the animals have to go through short episodes of food shortage (Kirk 2001). The earliest record citing a link between CR and health benefits in vertebrates can be traced back to the 1930s, specifically from McCay's work (McCay et al. 1989). Following this, many groups replicated the study and reported similar observations (McDonald and Ramsey 2010). As per the observation in the animal models, the health benefits associated with CR include but are not limited to metabolic reprogramming (Kirchner et al. 2012) and increased fat-burning (Bruss et al. 2010), reduced lipid accumulation and redistribution patterns (Kim et al. 2009), (Jiang et al. 2005), but also improves the glucose tolerance and increases the insulin sensitivity in the mice, monkeys, and humans.

However, the outcome significantly depends on the age of the organisms. After extrapolating the results of rodents onto humans, it can be roughly assumed that the earlier the CR is introduced in life, the better the outcome in terms of longevity. For instance, it is estimated if started at the age of 25 years, a 20% CR can increase human life by five years, whereas the same intensity of CR, if started at 55 years, will only increase the lifespan by two months (Redman and Ravussin 2011). The NIA has performed the most comprehensive human study elucidating the role of CR via a comprehensive assessment of the long-term effect of reduced intake of energy (CALERIE) is a two-phase study. In the first phase, non-obese individuals from the age group of 25–45 were subjected to a targeted 25% CR diet (Rickman et al. 2011). The second phase of the study included individuals of 21–45 years of age subjected to 25% CR for 2 years (Kraus et al. 2019). While the 25% CR benchmark was not reached by most individuals, even having a 10–15% CR corroborated many the anti-ageing effects found in the rodent models.

Another study involved an artificial biosphere, including all biomes of the earth to mimic the earth-like conditions. The idea was to determine the effect of feeding habits on human lifespan in the absence of unnatural stressors. The human subjects of age group 27–42 years residing here were fed a 30% CR diet for two years and were isolated from the outside world (Walford et al. 1992). The outcomes of the studies were similar to CALERIE and demonstrated that even though in isolation, the subjects did not show any adverse effects on mental and physical health due to the diets. One parameter in this study worth noting is that the active thyroid hormone (TH) triiodothyronine (T3) levels were reduced in the CR subjects (Walford et al. 2002). The increase in TH has been directly associated with the downregulation in expression and or activity of the hydrogen sulphide (H_2S) and cysteine generating enzyme cystathionine γ-lyase (CTH, CSE, or CGL, aka cystathionase), thus potentially attenuating redox homeostasis and serving as a hallmark of ageing (Barja 2013; Hine et al. 2017; Malaeb et al. 2022; Tyshkovskiy et al. 2019; Xiao et al. 2020). As will be described in greater detail in Section "Diet Impacts Oxidative Stress in Ageing", H_2S regulates the redox state of protein cysteine residues and other polysulfide species. In the context of diet, a long-term 40% CR diet in mice decreased the levels of carbonyl damage of proteins and increased sulfhydryl/persulfidated protein content of several tissues (Bithi et al. 2021; Dubey et al. 1996; Forster et al. 2000).

Intermittent Fasting (IF)

While CR is responsible for health benefits, the role of fasting in all CR-implemented diets cannot be neglected, as animals fed the CR diet tend to consume food faster than the controls (Mitchell et al. 2019). This leads to a long gap between each feeding for CR-fed animals, whereas the control animals consume food for a longer duration and, thus, a lesser gap between the two feeds. Intermittent fasting (IF) has been widely implicated in health studies, possibly due to less adverse effects and incidence of malnutrition and maximum health benefits. In this fasting method, it is suggested to have a significant gap between two meals. This technique was found to be helpful in gaining health benefits without reducing significant body weight as the overall calorie consumption is equal to the controls.

The idea of an association between intermittent fasting and health benefits originated with Carlson (Carlson and Hoelzel 1946) when it was reported a 5:2 fasting regime providing 25% and 15% lifespan extension in male and female rats compared to the control fed littermates. Intermittent fasting also improves cardiovascular health by reducing the damage and remodelling caused to cardiomyocytes due to myocardial ischemia (Ahmet et al. 2005). IF also protects neurons by reducing the load of reactive oxygen species, increasing insulin sensitivity, and modulating blood glucose levels post-feeding (Anson et al. 2003). It also improves the adaptive stress response, bioenergetics, and the removal and repair of damaged molecules and organelles (Mattson et al. 2014). This regimen is categorised into different types based on the fasting periods. The first practice includes fasting every day for 16 or 14 hours and eating time restricted to 8 or 10 hours, respectively. The former method is called 16:8, and the latter is known as the 14:10 fasting method. Another multi-day type of IF include weekly fasting, also known as the 5:2 method. This technique consists of two days fasting period and five non-fasting days. The fasting should not be done on consecutive days, which means that there should be at least one non-fasting day in between. Alternate day fasting is another type in which fasting is suggested every alternate day, and has been shown to provide numerous metabolic and cognitive improvements in aged mice (Henderson et al. 2021, 2023).

Protein Restriction (PR)

Proteins are the building blocks of living organisms and an integral part of the diet. Consumption of higher protein than recommended can lead to several health issues over time (Delimaris 2013). The suggested daily intake of protein for a healthy young individual is 0.8 grams per kg body weight. While protein deficiency causes severe disorders in the body, a regulated cut down of its intake leads to several health benefits, including improved longevity. For instance, female mice fed an iso-caloric diet but with proteins such that it contributed only 18% of total calories consumed had a higher chance of mammary tumour development and progression compared to the mice fed protein content equal to 4%–7% of total calorie intake (Delimaris 2013). A 40% decrease in protein intake with 8.5% CR was found to reduce mitochondrial ROS production, mitochondrial complex 1, oxidative DNA damage, and degree of membrane unsaturation (Ayala et al. 2007; Sanz et al. 2004). However, chronic PR should be avoided as the detrimental effects of long-term PR may outweigh the short-term benefits (Savitikadi et al. 2023). Irrespective of the source, whether natural or radiation-induced, the protein restricted diet provided significant protection against protein oxidation in rats. The 40% of total dietary restriction (including protein restriction) led to better resistance against oxidative damage than 25% or 40% only CR (no protein restriction) (Youngman et al. 1992). Patients with end-stage renal disease suffer from low blood thiol and high carbonyl contents in their blood. The protein restriction increases the sulfhydryl content in these patients with a simultaneous reduction in the pathogenic load of oxidized proteins (Fanti et al. 2015). Intermittent protein fasting helped in the proliferation or differentiation of the pancreatic β-cells and thus reduced the fasting glucose levels and improved glucose homeostasis in Zucker diabetic fatty rats (Li et al. 2023). Supporting these findings, a study involving different combinations of carbohydrates, protein, and fat found the ratio of the combination rather than restriction to be a decisive factor. A high carbohydrate, moderate fat, and low protein diet was found to be beneficial to health and extended lifespan (Solon-Biet et al. 2014), suggesting that dietary proteins reduction without malnutrition may be a key to slow the ageing process.

Methionine and Essential Amino Acids Restriction (MetR or SAAR)

PR is an effective means of DR, but because amino acids are the building blocks of proteins, it is important to ask if removing some of these amino acids can deliver the same effect as that of whole protein restriction? Interestingly, a plethora of studies in recent years have identified methionine and branched-chain amino acids as the causative amino acids driving benefits of PR. The health benefits of methionine restriction in rodents were first investigated by Orentreich et al. in 1993 (Orentreich et al. 1993). The reduction of methionine from 0.87 to 0.17% and removal of cysteine without any CR increased the lifespan of male Fischer rats by 30% (Orentreich et al. 1993). Methioninase is a bacterial enzyme that degrades methionine to generate ammonia, α-ketobutyrate, and methanthiol (Sharma et al. 2014). Methionine degradation can limit methionine availability to the cells even with a methionine-rich diet. Genetically introducing the methioninase coding gene in fruit flies leads to better phenotypes in aged flies than compared methionine-restricted flies (Parkhitko et al. 2021). The benefits associated with the transgenic model were also independent of the amino acid levels, which does not hold true for wild-type methionine-restricted flies. More recently, intermittent MetR, with a feeding pattern of 0.12% methionine (restricted methionine) for three days and *ad libitum* control feed (0.86% methionine) for four days of the week, was found to confer metabolic and health benefits with a 45% increase in the lifespan in mice, which was comparable to what was obtained in continuous methionine restriction in mice (Plummer and Johnson 2022).

As methionine is an essential amino acid (EAA), it is logical to speculate if restriction of other EAAs can also be geroprotective. In support of this, studies have found that the potential of protein restriction, even without any calorie deficit, increases when branched-chain amino acids alone are restricted from the diet (Richardson et al. 2021; Solon-Biet et al. 2014). Another finding that reconciled with this narrowed the benefits of protein restriction down to threonine and tryptophan deprivation being necessary for a systemic metabolic reprogramming (Yap et al. 2020). In summary, these studies indicate that dietary proteins, in general, and EAAs, in particular, are superior gatekeepers of dietary restriction-based healthy ageing.

Dietary Interventions in Yeast

Saccharomyces cerevisiae, or budding yeast, is a unicellular eukaryote that has been used in genetic studies for nearly 100 years. It rapidly doubles, is genetically malleable, and despite being a unicellular organism it has similar metabolic processes to humans and is thus a great model to understand the fundamental mechanisms of dietary and nutrient restrictions. In yeast, CR can be achieved by restricting the glucose content to 0.5%, which is usually 2% in the standard growth medium. The 1.5% decrease in the glucose content can increase the cell's lifespan by 25% (Lin et al. 2002). The increase in the lifespan and other associated benefits was accompanied by slowing down metabolism specifically in the glycolytic cycle and increasing cellular respiration (Lin et al. 2002) or gluconeogenesis (Lin et al. 2001). Respiration rate could be an important decisive factor that mechanistically links DR and chronological lifespan in yeast (Smith Jr. et al. 2007), which is reflected by an increasing NAD+/NADH ratio in dietary-restricted yeasts (Anderson et al. 2003), (Xiao and Loscalzo 2020). Glucose restriction also decreases the bioavailability of methionine by perturbing the expression of genes involved in methionine synthesis/uptake (Zou et al. 2020), which creates a methionine restriction-like condition, a geroprotective approach repeatedly proven across various models (Lee et al. 2016). Reducing the methionine and ATP reservoir in yeast and creating a CR-like condition by stimulating S-adenosyl-L-methionine (SAM) synthesis increases the lifespan in a Snf1 dependent manner (Ogawa et al. 2016).

In contrast, the methionine restricted diets and growth media are mostly believed to act via Target of Rapamycin (TOR) and whose inhibition was enough to achieve the CR benefits with a regular diet (Mohammad and Titorenko 2021). Cell-to-cell communication provides survival benefits against abiotic stress in unicellular organisms. A similar observation was also made in yeast grown in glucose-restricted media as the CR benefits were lost in the mother cell isolated from the neighbouring cells but still grown on the glucose-restricted media (Mei and Brenner 2015). This alludes to the importance of extracellular factors in mediating CR benefits.

Dietary Restriction in Flies

Drosophila melanogaster, also known as the fruit fly or vinegar fly, are highly preferred invertebrate models for ageing research due to their relatively short lifespan, large number of offspring, genetic malleability, and multi-tissue and stem cell systems making it convenient to study any biological phenomena. The dietary restriction-mediated lifespan extension has been a well-known phenomenon in fruit flies since the early twentieth century, even before Harman proposed his theory on ageing. There is substantial evidence that the benefits of DR in flies are often a result of a trade-off between fecundity and longevity (Mair et al. 2005), (Krittika and

Yadav 2020), (Burger et al. 2007). This means that the flies will tend to survive more during a restricted food supply with the trade-off being less eggs being produced.

Interestingly, detailed insight into the possible mechanisms does indicate the involvement of metabolic factors and diet composition that drive ageing. The benefits of DR in flies may not necessarily be associated with a reduction in overall calories, but instead via the reduction of protein intake, as the lifespan benefits are greater for flies fed isocaloric food composed of yeast than sucrose (Grandison et al. 2009; Mair et al. 2005). The DR-associated lifespan extension and reduced fecundity rate of the flies also vanish upon supplementation of either essential amino acids or methionine, suggesting a pivotal role of these amino acids for the survival and reproductive health of flies (Grandison et al. 2009). The genetic model of methionine restriction in flies unequivocally proves the longevity benefits of methionine restriction (Parkhitko et al. 2021).

Beyond this, the protein:carbohydrate ratio has been recently proposed to be the determinant of DR benefits. A lower protein:carbohydrate is essential for longevity, whereas a higher protein:carbohydrate will result in a shorter lifespan and supplementation of cholesterol (an essential entity for reproduction), irrespective of the caloric count of the food, rescues the reproductive output of flies (Zanco et al. 2021). The mechanism for this does not advocate the involvement of TOR signalling that has been observed in other studies; instead, it supports an indirect nutrient trade-off theory and macronutrient balance. This mechanism may be relevant to *Drosophila* but obscure to other organisms due to the incapability of flies to synthesise cholesterol in their body (Seegmiller et al. 2002), therefore making it a limiting factor for growth and development. However, it does not question the relevance of universally relevant mechanisms such as TOR and insulin-like growth factor (IGF) as flies devoid of the *Drosophila* Insulin-like growth factor-like peptide (DILP2) gene have extended lifespan (Behrendt et al. 2011). Further, methionine restriction also increases the lifespan in flies facilitated by the TOR pathway (Lee et al. 2014a).

Dietary Restriction in Worms

Caenorhabditis elegans (*C.elegans*) is a nematode worm and contains homologs of two-thirds of all human disease genes, thus making it one of the most preferred non-vertebrate models whose potential has been used extensively for studying the genetic basis of longevity (Zhang et al. 2020). Supporting this, many genetic models of longevity have been developed in *C.elegans* (Lakowski and Hekimi 1998) as mentioned below. DR was found to increase the lifespan of the worms by 45%, simultaneously providing oxidative stress resistance but surprisingly unaffecting the rate of ageing as determined by the changes in the rate of pharynx pumping and locomotion. Interestingly, the basal rate of these parameters was higher in the DR worms than *ad libitum* diet-fed worms (Lee et al. 2006). Genetically modifying the worms to consume less food by mutating *eat* genes reduces overall caloric consumption due to the defective pharynx. *eat-2* mutants live longer than other *eat* mutants and non-mutants (Lakowski and Hekimi 1998), which is mediated by Protein skinhead-1 (SKN-1)-Neuropeptide-Like Protein-7 (NLP-7) (SKN-1-NLP-7) axis and independent of Insulin-like signalling (*age-1* mutants) or metabolic slowdown (*clk-1* mutants) (Park et al. 2010). Furthermore, as observed in the flies, methionine restriction also increases the lifespan in worms by perturbed AMPK signalling triggered by accumulating intracellular SAM caused due to supplementary S-adenosyl-L-homocysteine (Ogawa et al. 2022).

Dietary Restriction in Rodents

In 1935, McCay and colleagues developed a rodent 2-step retarded growth model using a CR approach (McCay et al. 1989). As a first step, the rats were fed *ad libitum* or CR diets after weaning. The animals in the CR group were then fed *ad libitum* diet after 766 days, which was the second step of the model. The mean lifespan (MLS) and maximum lifespan (MALS) varied by more than 300 days between males of each group. However, differences in MLS and MALS between in females between the groups were marginal, thus making it a first sex-dependent CR response milestone study. The metabolic rate in these mice was significantly higher than *ad libitum* diet-fed mice, a partially contradictory finding as the metabolic activity was mostly unchanged in long-term CR and may not be a sole determinant of CR-based longevity (McCarter et al. 1985).

CR is perhaps the oldest and most commonly known dietary intervention used in rodents. A number of studies indicate that the reduction in calorie intake in rodents is associated with a reduced propensity of these animals to age-related diseases or disorders, a common cause of death. For instance, the hepatic insulin sensitivity increases with a reduction in the visceral fat mass in the Sprague-Dawley (SD) rats subjected to moderate CR up to 18 months of age (Barzilai et al. 1998). In rats chronically exposed to carcinogens, the 40% reduction in calories reduced the incidence of mammary and colon tumour-formation to one-fourth and half, respectively (Klurfeld et al. 1987). Chronic CR decreased incidences of spontaneous hepatoma, whereas graded caloric intake abrogated chemical induction of skin tumours in male mice (Tannenbaum and Silverstone 1949). CR also reduced the prevalence of age-related vascular diseases in rodent models by possibly improving endothelial function, inhibiting vascular remodelling, perivascular adipose tissue dysfunction, and regulating adipokine production (Garcia-Prieto and Fernandez-Alfonso 2016).

While the effect of CR on the metabolic rate of the organism is arguable, the transcriptomic and metabolic reprogramming is a common point of agreement in various studies (Tyshkovskiy et al. 2019), (Wahl and LaRocca 2021), (Kirchner et al. 2012), (Martin-Montalvo and de Cabo 2013), (Kuhla et al. 2014), (Solon-Biet et al. 2015). In addition, the protein persulfidome also partially contributes to lifespan extension in mice (Bithi et al. 2021). Short-term dietary interventions in mice with either a low protein-high carbohydrate (LPHC) *ad libitum* or CR diet improved metabolic health but with opposing metabolic outcomes (Solon-Biet et al. 2015).

Adult life-long CR reprogrammed the hepatic fat metabolism with reduced lipogenesis by augmented lipolysis and ketogenesis in female rats (Kuhla et al. 2014). This was accompanied by a 24-fold increase in the fibroblast growth factor-21 (FGF-21) expression. FGF-21 is a candidate gene for longevity and healthy ageing that may act via the insulin signalling pathway (Mendelsohn and Larrick 2012). The reduction of protein intake without CR promoted healthy ageing and longevity in male mice but did not elicit the geroprotective response in mice genetically devoid of FGF-21, thus underlining the importance of this gene in dietary restriction benefits in vertebrates (Hill et al. 2022). In contrast, insulin and other factors such as IGF-1, T4, and glucose mediate the health benefits of methionine restriction (Miller et al. 2005). Methionine restriction, like other dietary interventions, also alters the transcriptome and reprograms the metabolism to maximize the lifespan of Hutchinson-Gilford progeria syndrome (HGPS) mice (Barcena et al. 2018).

Dietary Restriction in Non-human Primates

Due to the evolutionary proximity and genetic similarity of non-human primates, including rhesus monkeys who have nearly 93% similarity with the human genome and demonstrate similar ageing patterns, monkeys make a relatable model for ageing and dietary studies. However, only a handful of reports examine the translatability of the geroscience approaches used in flies, worms, and rodents in non-human primate studies. Amongst the most influential studies examining CR in non-human primates include those by the University of Wisconsin and the National Institute of Aging (NIA). The dietary regimen followed for the monkeys of the Wisconsin study was a young onset CR with a gradual reduction of 10% calories each month for three months to finally achieve a 30% reduction in the calories in the third month, which was continued for the rest of the life (Colman et al. 2009). The NIA study included young and old onset-30% CR diet for ten years (Mattison et al. 2012). The Wisconsin study reported that monkeys on CR diets benefitted with improved longevity. Additionally, other incidences of age-related diseases such as diabetes (100%), neoplasia (50%), and cardiovascular diseases (50%) were reduced in the CR group concomitant with improved metabolism, cognition, and reduced body fat. In contrast, the NIA study found slightly less improvements with CR over the control fed group.

These differences in study outcomes were surprising at first and offered insight into how the set-up of dietary restriction studies can impact the lifespan and healthspan result. They were eventually resolved in follow-up studies investigating the timing of the feeding (Mattison et al. 2017; Mitchell et al. 2019). Additional studies advocating the health benefits of CR in these animals include improvements in insulin sensitivity and improved glucose clearance under 30% CR in adult monkeys (Kemnitz et al. 1994), which is consistent with the observation with CR in rodent models.

Diet Impacts Oxidative Stress in Ageing

As ageing is associated with cellular damage to biomolecules and DR can slow down the ageing process and age-associated disorders, it is important to ask if the dietary intervention can reduce the load of oxidative stress. Intriguingly, this hypothesis was investigated by several research groups in the past. Initially, it was believed that the benefits of CR could be associated with, at least in part, decreased body temperature (Duffy et al. 1989), (Duffy et al. 1990), (Rikke et al. 2003), (Lane et al. 1996), which also means a slowed metabolic rate and ROS production (Tirza et al. 2020), (Weidemann et al. 1994), (Jorjani and Ozturk 1999), (Moore et al. 1997). In contrast, an increase in ROS generation, even with a slowed metabolic rate and a drop in the temperature is also reported (Ali et al. 2010). In some cases, a reduced antioxidant level rather than increased ROS production was associated with ageing. Walsh et al. (2014) have comprehensively reviewed the effects of DR on the antioxidant system and ROS generation. According to them, the notion of DR being protective against ROS-mediated injury faded as the existing literature had contradictory as well as complementary findings. Nevertheless, the reports that found ROS to be a crucial player in age-associated studies have witnessed a reduction in the levels of these factors and the associated damages with dietary interventions. For instance, 59 obese individuals subjected to a very low-calorie diet for an initial six weeks, followed by a low-calorie diet for the next 18 weeks, showed a significant improvement in antioxidants levels and a significant reduction in ROS production (Lopez-Domenech et al. 2019).

Metformin treatment, an anti-glycaemic drug that causes mitohormesis (a mechanism of dealing with stress of higher intensity by predisposing the cells to low-intensity stress) and can be thought of as a DR-mimetic. Metformin enhanced the lifespan of *C.elegans* by modulating methionine metabolism by mimicking methionine restriction (Cabreiro et al. 2013). Methionine is an essential amino acid, and methionine restriction concurrent with cysteine restriction has been found to deliver similar benefits to protein-restricted diets to extend lifespan (Barcena et al. 2018; Richie et al. 1994; Sun et al. 2009). Surprisingly, metformin also required peroxiredoxin-2 (PRDX-2) (an enzyme that neutralises H_2O_2) as the lifespan expansion benefits disappear in its absence (De Haes et al. 2014).

Alterations in sulphur amino acid metabolism via the transsulfuration pathway can be considered a common phenomenon of anti-ageing studies (Hine et al. 2018; Tyshkovskiy et al. 2019). Hydrogen sulphide (H_2S), a bioactive gaseous signalling molecule and product of sulphur amino acid metabolism, is a widely accepted agent for protection against ischemia-induced oxidative injuries (Hine et al. 2015; Sun et al. 2017). H_2S is enzymatically produced by transsulfuration pathway enzymes, cystathionine gamma-lyase (CTH, aka CGL or CSE), cystathionine beta-synthase (CBS), and 3-Mercaptopyruvate sulphurtransferase (MPST) and non-enzymatically by iron in the blood (Filipovic et al. 2018; Yang et al. 2019). CR and methionine restriction were found to enhance plasma H_2S levels (Duan et al. 2022; Hine et al. 2015). In mice with a CR diet, enhanced H_2S production was associated with life-prolonging effects as the animals lacking Cystathionine gamma-lyase (CTH) did not fully benefit from CR for protein thiol protection (Bithi et al. 2021), resistance to ischemic stress (Hine et al. 2015), adaptation to fasting (Hine et al. 2017), and stimulation of angiogenesis in aged skeletal muscle (Longchamp et al. 2018). Additionally, the redox sensitive nuclear factor erythroid 2-related factor-2 (Nrf-2) was found to be activated by H_2S and facilitate its translocation into the nucleus (explained in detail in the later section). Nrf2 binds to the antioxidant response elements (ARE) of the promoter of the antioxidant genes and activates the gene expression during oxidative stress (Hine and Mitchell 2012). The takeaway message from this section is that geroprotective approaches, such as dietary interventions, execute their biological effects by mechanistically manipulating ROS levels. This upholds the idea that ROS could be a connecting link of all the pillars of ageing. Next, we will delve deeper to unearth fundamental chemistry behind the biological action of ROS and ageing.

Implication of Oxidation States of Cysteine Residues in Ageing and Dietary Interventions

Cysteine is a sulphur-containing amino acid, and this residue's tendency to actively alter between different oxidation states (Garrido Ruiz et al. 2022) makes it one of the crucial determinants of a protein's structure and function. Interestingly, changes in the oxidative state of cysteine residues are also one of the important molecular changes associated with ageing and other oxidative events (Zivanovic et al. 2019). The global quantitative mapping of cysteine's oxidation in old and aged mice found nearly 34,000 cysteine residues on nearly 9400 proteins that are sensitive to redox changes (Xiao et al. 2020). The redox sensitivity of these cysteines was affected by the surrounding amino acid, and the cysteine capable of forming a disulphide bond was prone to higher oxidation.

A comprehensive analysis of the proteins containing highly modifiable cysteine residues revealed highly coordinated tissue-dependent and age-specific redox networks. However, the overall oxidation of the cysteine residues in flies did not change with age, but a 24-hour fasting or paraquat treatment, but not H_2O_2 treatment, enhanced cysteine oxidation. This is surprising as for both paraquat and H_2O_2; the pathophysiological effects are mediated by ROS (Menger et al. 2015). However, protein persulfidation increased in response to H_2O_2, which was a result of a nucleophilic attack of the thiol group on sulfenylated cysteine (reversible oxidation) but not sulfinylated or sulfonylated cysteine that are irreversible in nature (Zivanovic et al. 2019). The formation of a persulfidated cysteine also provides a mechanistic basis for thioredoxins driven pathway that safeguards the cysteine residues from hyper-oxidation and proteins from degradation and/or aggregation.

Due to the somewhat reversible nature of oxidised cysteine residues, different modifications can act as switch and activate or inactivate proteins in response to changes in the intracellular environment as displayed in Fig. 7.2. Protein persulfidation decreases with ageing in rodents and human cells and is upregulated in dietary-restricted mice (Bithi et al. 2021; Zivanovic et al. 2020). Altogether, these observations suggest that the redox state of cysteine residues in proteins could be a crucial determinant of the protective effects of therapeutic or dietary interventions. Let us further look into the details, along with a few examples from existing literature that could provide insights into the role of persulfidation at molecular levels.

Maintenance of proteostasis is an important process during oxidative stress and ageing. The hyperoxidized proteins are shunted towards degradation pathways during oxidative injury to maintain the healthy protein pool. Three dynamic modifications of cysteine residues are persulfidation, S-nitrosylation, and S-glutathionylation. Persulfidation reduces the oxidation of proteins through a mechanism that involves a series of modifications in the partially oxidised cysteine residues (residues carrying the -SOH group, also known as sulfenylation). The addition of the –SOH group makes it a favourable site for a thiol attack which adds an –SH group, thus forming –SSH, also known as persulfidation. The presence of the -SSH group on cysteine enables thioredoxins to restore the protein back to its native form (Filipovic et al. 2018; Zivanovic et al. 2019). However, if the oxidation of cysteine is not interrupted by H_2S or glutathione, the cysteine may be fully oxidised and may lead to protein degradation. The other two modifications of the cysteine are caused due to the direct attack of radical nitrogen species or oxidised glutathione.

However, unlike persulfidation, both of these modifications do not need an oxidised intermediate. S-nitrosylation has been studied in the context of cardiovascular disease (Chung et al. 2013) and holds great relevance due to the important role of

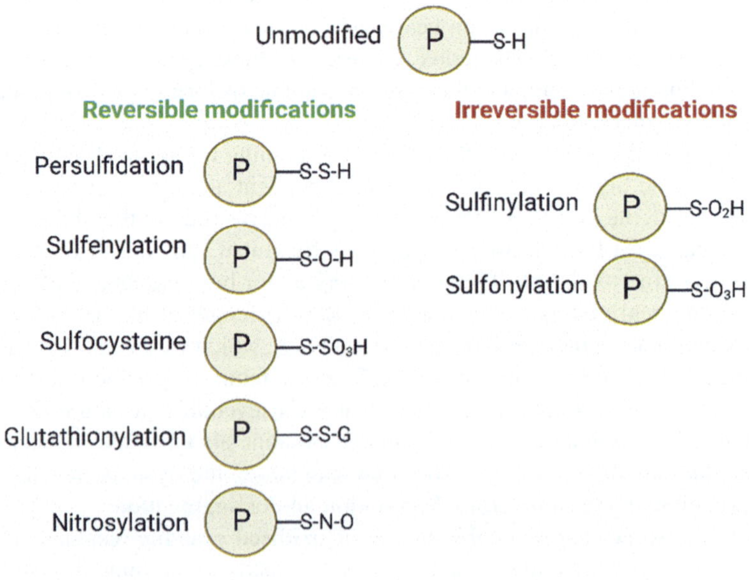

Fig. 7.2 Oxidative states and modifications of protein cysteine residues and thiols

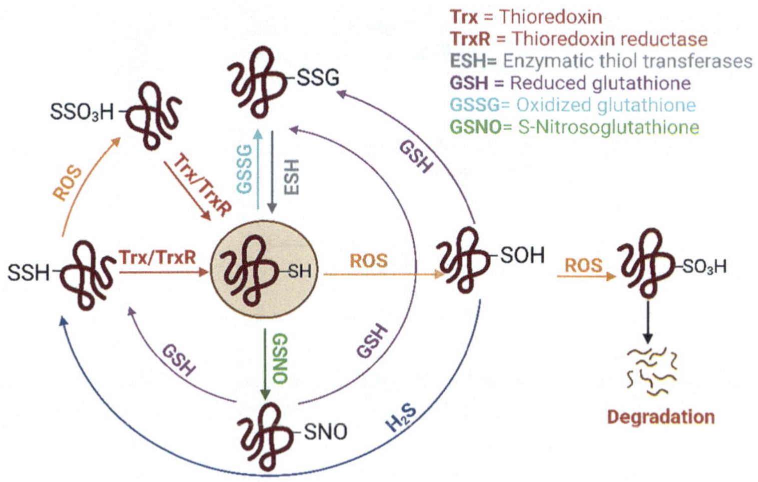

Fig. 7.3 Fates of proteins with different redox modifications of cysteine residues

nitric oxide (NO) produced by endothelial nitric oxide synthase (eNOS) in the cardiovascular system (Loscalzo and Welch 1995). Similarly, the oxidised form of glutathione is responsible for S-glutathionylation of the cysteine residues and has been implicated in pathologies like diabetes mellitus, liver injury, brain injury and cardiovascular disorders (Xiong et al. 2011). The reversal of diverse oxidative modifications in proteins are mediated by their convergence to persulfidated state, thus emphasising the role of persulfidation in redox balance of cysteine residues. The cyclic nature of protein thiol oxidation and modifications is depicted in Fig. 7.3.

H_2S Manifests Its Signalling via Persulfidation

H_2S is a well-known scavenger of ROS, and its physiological effects are mediated by regulating ROS homeostasis at the cellular level. However, by virtue of its chemical nature, it also reacts with the partially oxidised cysteine residues of proteins. In recent years, persulfidation has been recognised as a vital mechanism of its action and a dynamic post-translational modification (Filipovic et al. 2018). Also known as S-sulfhydration, persulfidation is a post-translational modification of the partially oxidised cysteine residues by adding a $-S_{(n)}H$ group to the existing sulphur atom (Finkel 2012). Like other post-translational modifications, persulfidation also affects the structure, function, and stability of the proteins. With age, it is reported in human cells and rodent tissues that protein persulfidation decreases, while thiol oxidation increases (Malaeb et al. 2022; Zivanovic et al. 2020). This shift in protein thiol status is depicted below in Fig. 7.4.

The interaction of the transcription factor Nrf-2 and Kelch-like ECH-associated protein 1 (Keap-1) in the cytoplasm is an excellent example of a redox sensitive

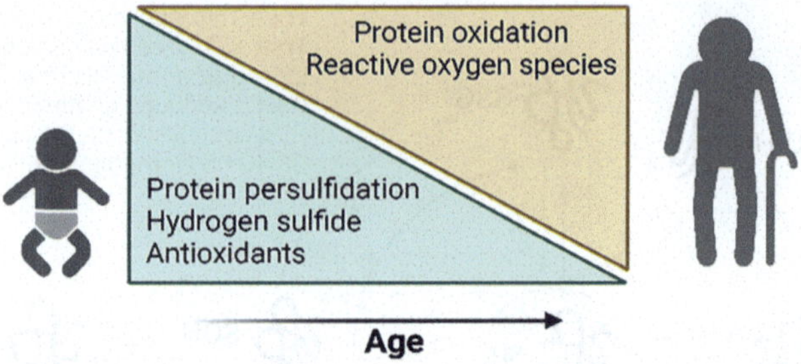

Fig. 7.4 Shifts in protein persulfidation and protein cysteine thiol oxidation with age

switch mechanism that ultimately stimulates a phase II cellular antioxidant and detoxification process (Hine and Mitchell 2012). Nrf-2 is a well-known ARE-binding protein that is activated upon oxidative overload, but under physiological conditions, the translocation of Nrf-2 to the nucleus is inhibited by its binding to Keap1 (Nguyen et al. 2009). When the levels of ROS rise, Keap1 is persulfidated at Cys151 residue; the Nrf-2 uncouples itself from Keap1, which facilitates its translocation to the nucleus and activation of antioxidant genes expression (Yang et al. 2013). Another example of cysteine oxidation state acting as a switch is the modification of CTH itself. Persulfidation of CTH upregulates the enzyme activity of CTH, whereas S-nitrosylation reduces it (Bithi et al. 2021; Fan et al. 2019). An example for how H_2S interacts with the NRF2/Keap-1 system and ultimately restores redox homeostasis is provided in Fig. 7.5.

Persulfidation of Proteins May Be Pivotal for the Protective Actions of DR

Oxidative load increases with ageing (Gorni and Finco 2020). Naturally, CTH expression decreases with age, as does circulating sulphide levels and the persulfidome (Bithi et al. 2021; Kabil et al. 2011; Malaeb et al. 2022; Zivanovic et al. 2019). In addition to this, the reduced plasma H_2S also correlates with pathophysiological conditions such as diabetes, hypertension, and obesity, which are age-related diseases and simultaneously render an individual more prone to oxidative injuries. While few studies have explored a link between ageing and H_2S, its vitality in oxidative injury models has been widely implicated (Hine et al. 2015; Magierowska et al. 2022; Ng et al. 2017; Wang et al. 2015). Therapeutic interventions by external donors of H_2S have been found to effectively protect the major organs, such as the liver, kidney, heart, and brain, against oxidative injuries and reduce the oxidative load caused due to ischemic conditions. The gut microbiota is an essential contributor to plasma H_2S and other reactive sulphur species (RSS). Antibiotics treatment reduces gut microbiota populations, which reduces the levels of RSS in the blood,

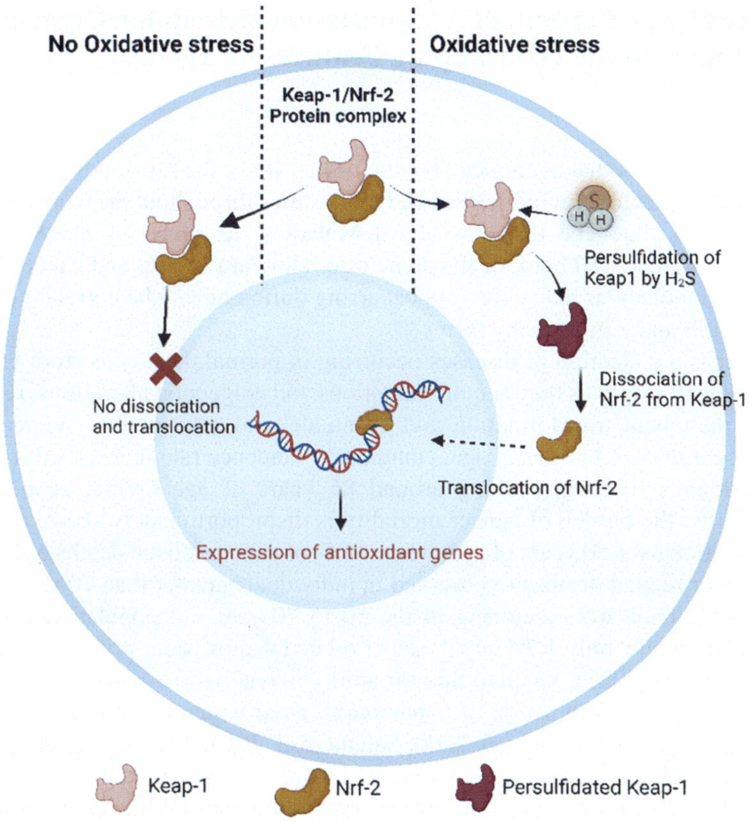

Fig. 7.5 Persulfidation regulates the translocation of Nrf-2

which makes the liver more prone to inflammation-triggered oxidative damage (Uchiyama et al. 2022).

Although a plethora of studies exploring the therapeutic effects of H_2S against oxidative injuries, as mentioned above, do provide molecular mechanisms that are centred around an individual protein or signalling axis, only a few groups of researchers have been able to provide a more comprehensive picture at the levels of transcriptome and persulfidome. In this direction, our group recently reported that rodents missing the CTH gene (CTH Knockout (CTH KO)) had decreased protein persulfidation in a number tissues including the liver, kidney, brain, and skeletal muscle under CR compared to wildtype mice under CR (Bithi et al. 2021). Interestingly, protein persulfidation of enzymes involved in energy metabolism and glycolytic flux also protects the pancreatic β cells from integrated stress response by metabolic reprogramming (Gao et al. 2015). Interestingly, transcriptional activation of CTH (Tyshkovskiy et al. 2019) and metabolic reprogramming (Anderson and Weindruch 2007) is a hallmark of anti-ageing dietary interventions.

Biological Age-dependent Accumulation of Cellular Damage Contributes to the Overlapping Features of Ageing and Cancer

While it is obvious that ageing can be one of the, if not the most, powerful driving forces and risk factors for mortalities and morbidities throughout the world, it is also said cancer is the "The Emperor of All Maladies" as stated by Dr. Siddhartha Mukherjee in his 2010 book of the same title. How are ageing and cancer linked, and can the molecular redox changes occurring during ageing be correlative and/or causative of cancer in its many forms?

Cancer is a collection of diseases occurring in normal, host cells from virtually every tissue of origin as they acquire mutations and epigenetic alterations, resulting in their malignant transformation and eventual tumour formation. While cancer strikes regardless of biological age, cumulative incidence rate increases throughout an individual's lifetime, peaking around 85 years of age (White et al. 2014). Additionally, the burden of cancer mortality is disproportionately skewed towards the population over 50 years of age, where 90% of cancer related deaths occur. Half of all cancer related deaths are observed in individuals greater than 70 years of age and an additional 40% occurring in the 50 to 70 year old population (Editorial 2022). Meanwhile, only 1.7% of all cancer related deaths occur prior to 40 years of age (Siegel et al. 2023). Much of this variability in receiving a cancer diagnosis can be attributed to a combination of behavioural, environmental, and genetic factors (Fearon 1997; Lichtenstein et al. 2000; Saletta et al. 2015; Tran et al. 2022; Weeden et al. 2023). However, ageing remains one of the most significant risk factors for many of the common forms of cancer observed in humans (White et al. 2014). The relationship between ageing and increased cancer incidence as well as mortality can be clearly observed in the case of the debilitating brain cancer glioblastoma where the median age of diagnosis is between 68 and 70 years, a trend which is closely mirrored by mortality rates (Kim et al. 2021).

Upon closer examination of the overlapping features observed with advanced biological age and that of cancer, it is clear that there are several ageing-related factors that could be contributing to this phenomenon. Many of these ageing-related factors contributing to occurrence and progression of cancer have matching or nearly equivalent features between the 12 hallmarks of ageing, and the 14 hallmarks or enabling characteristics of cancer (Hanahan 2022; Lopez-Otin et al. 2023a). López-Otín et al. provided a systematic analysis on much of what is currently known in the overlapping fields of ageing and cancer, which they described as the "Meta-hallmarks of ageing and cancer" (Lopez-Otin et al. 2023b). These "meta-hallmarks" of cancer and ageing contribute to the malignant transformation of normal cells and eventual tumour development through accumulation of cellular damage. The observed cellular damage is the result of endogenous metabolic free radical production and exogenous influences ranging from UV exposure to dietary factors, all of which lead to increased oxidative stress, a mediator of DNA damage, oxidative modification of proteins, and lipid peroxidation.

Age-induced Accumulation of Oxidative Stress Triggers Cancer Associated DNA Damage

As living organisms age, the genomic material faces many challenges which lead to activation of DNA damage responses, and an increased mutational burden associated with both ageing and cancer incidence (Huang et al. 2022; Martincorena et al. 2018; Podolskiy et al. 2016). These age related increases in genomic instability and chromosomal abnormalities correlate with evidence of observed increases in reactive oxygen species (ROS) generation, oxidative damage, mitochondrial dysfunction, and oxidative stress (Capel et al. 2005; Gorni and Finco 2020; Gouspillou et al. 2014; Herrero and Barja 2001; Navarro and Boveris 2004). The ageing-related DNA damage often leads to mutations in key tumour suppressor and oncogenes involved in preventing the replication of damaged cells enabling the uncontrolled growth of cancerous cells experiencing a gain of function. Cancer associated mutational signatures, characterised by single nucleotide variations (SNVs) and insertion-deletion mutations (INDELs) can be induced with exposure to high levels of ROS (Pilati et al. 2017; Rose Li et al. 2020; Viel et al. 2017). Additionally, accumulation of ROS has been shown to induce epigenetic alterations related to the direct and indirect modification of DNA bases or histones (Kietzmann et al. 2017; Ziech et al. 2011). ROS-mediated hypermethylation of many gene regulatory elements in the presence of a more global hypomethylation state are associated with many cancer mutational signatures, leading to alterations in expression of tumour suppressor and oncogenes (Lu and Lu 2022; Madugundu et al. 2014; Pastukh et al. 2015; Seddon et al. 2021). While increase in ROS contributes to the accumulation of DNA damage associated with ageing and a decrease in cell fitness, it often has the opposite effect, driving oncogenesis and subsequent tumour formation.

Oxidation of Proteins Promotes Cancer Progression

Redox changes are also able to post-translationally modify cysteine residues of proteins in reversible and irreversible fashions. These protein modifications occur in the form of oxidation which controls their activity similar to what is observed with phosphorylation. Cysteine modifications are widely observed throughout the eukaryotic proteome, and they likely have far reaching effects (Weerapana et al. 2010). ROS-dependent protein modifications can further influence genomic stability through targeting of histones, ultimately leading to alterations in chromatin structure, a function essential for maintaining genome stability (García-Giménez et al. 2013; Khan et al. 2016). It has also been observed that ROS can alter the activity of DNA methyltransferases, Tet family proteins, histone methyltransferases, histone demethylases, histone acetyltransferases, and histone deacetylases (Aerbajinai et al. 2007; Ah Kang et al. 2016; Coulter et al. 2013; Gào et al. 2019; Hickok et al.

2013; Kloypan et al. 2015; Mahalingaiah et al. 2016; O'Hagan et al. 2011; Yoshino et al. 2011).

The effect on such epigenetic regulatory mechanisms can promote a more migratory and invasive phenotype through the epithelial-mesenchymal transition to promote cancer metastasis (Kamiya et al. 2016). Oxidation of proteins via ROS signalling can also promote cell growth and proliferation, while preventing cell death via pathways involving PI3K/AKT/mTOR, MAPK/ERK and AP-1 (Klaunig et al. 2010; Lee et al. 2002; Reczek and Chandel 2017; Salmeen et al. 2003; Seth and Rudolph 2006; Weinberg et al. 2010). It has also been demonstrated that continuous exposure to free radical species generates chronic inflammation capable of inducing carcinogenesis and promoting early tumour initiation while providing a tumour supportive microenvironment at later stages. Cancer cells have been found to secrete hydrogen peroxide into the surrounding environment to induce oxidative stress in cancer-associated fibroblasts (CAFs), causing them to shift their metabolic processes while promoting myofibroblast differentiation (Martinez-Outschoorn et al. 2011; Toullec et al. 2010). Methylmalonic acid, an age associated metabolite increases generation of ROS to activate CAFs in the tumour microenvironment so that they release IL-6, which functions to activate JAK/STAT3 and TGFβ signalling pathways in tumour cell, promoting a more aggressive behaviour (Gomes et al. 2020; Li et al. 2022). ROS in the tumour microenvironment is also capable of inducing HIF1α activation, leading to upregulation of VEGF responsible for mediating angiogenesis (Arbiser et al. 2002; Gerald et al. 2004). Production of ROS in the tumour microenvironment by both the malignant cells as well as the CAFs also leads to increased oxidative stress that promotes the conversion of nearby monocytes into myeloid-derived suppressor cells (MDSCs), which leads to suppression of cytotoxic T cells and generation of a immunosuppressive environment (Cheung and Vousden 2022; Hayes et al. 2020; Xing et al. 2022).

Much of the evidence points to a system by which accumulation of ROS in the tumour microenvironment functions to support growth, survival, infiltration, and metastasis. While accumulation of ROS with age contributes to the development of cancer through mechanisms such as genomic instability, epigenetic remodelling, metabolic reprogramming, and immune suppression as previously mentioned, too high of levels can trigger cell death and cell cycle arrest in cancer cells.

Therefore, cancer cells have adapted antioxidant defences such as the activation of the transcription factor Nrf2, which occurs through modification of cysteine residues on its regulatory protein, Keap1 (Itoh et al. 1999). Additional regulatory mechanisms are utilised by cancer cells to tightly maintain ROS homeostasis including NADPH, glutathione synthesis and utilisation genes (GSH), peroxiredoxins (PRXs), glutathione peroxidases (GPXs), superoxide dismutases (SODs), and catalase (CAT) (Reczek and Chandel 2017). In their absence or inactivity, accumulation of age-related ROS would trigger cancer cell death. The evidence therefore suggests a contextually driven balancing act between the pro-tumorigenic and tumour suppressive functions of ROS.

Altered Redox Homeostasis, a Means to Decouple Cancer from Biological Ageing

The need for redox homeostasis in cancer suggests that the association between ageing and cancer is modifiable to an extent through means of attenuating oxidative stress. Additional evidence supporting this notion is the probability of developing an invasive cancer, where approximately 40.9% of men and 39.1% of women will have such a diagnosis, suggesting the behavioural and environmental modifications can alter risk and severity of a cancer diagnosis (Siegel et al. 2023). As accumulation of ROS and subsequent cellular damage occurs with increasing biological age, and incidence of cancer as well as prognosis get worse with age, it points to the potential for using geroscience based approaches for the prevention and treatment of neoplasms.

One such approach is the use of nutritional interventions. These nutritional interventions can take the form of fasting regimens with the goal of reducing caloric supply through either caloric restriction, intermittent fasting, or time restricted feeding. Some dietary modifications are a more global nutritional intervention-based approach, such as the Mediterranean, vegetarian, vegan, and palaeolithic diets, where the goal is to alter eating patterns. Macronutrient intervention including low-carb, ketogenic, low-protein, and low-fat in addition to micronutrient depletion nutritional interventions such as methionine restriction are alternative approaches to inducing alterations in nutrient sensing pathways. The final major form of nutritional intervention to prevent or treat cancer is micronutrient supplementation where the goal is to add in vitamins, spermidine, oligo-elements, polyphenols, or omega-3 fatty acids, in addition to other supplements. The use of nutritional intervention in preventing and treating cancer are covered in extensive detail by Montégut et al., where they reviewed preclinical and clinical models testing the efficacy of nutrition-based anticancer treatments (Montégut et al. 2022). The focus here will be on the usage of fasting-based regimens including caloric restriction to prevent the occurrence or slow the progression of cancer through a reduction of oxidative stress.

Nutritional Intervention in Preventing Cancer

The anti-ageing effects of caloric restriction are largely due to a reduction in the generation of ROS and inhibition of chronic inflammation. Therefore, it is plausible that a reduction in age-associated ROS resulting from caloric restriction would be protective against development of cancer. The first scientific report of the effects of caloric restriction on preventing cancer was given in 1909 by Moreschi (1909). This report was followed by the findings of Peyton Rous in 1914, where it was observed that tumours implanted in mice on a "low diet" grew at a slower rate and with decreased frequency (Rous 1914). Since the effects of caloric restriction on preventing tumour implantation in mice, many additional studies performed in various rodent models have yielded similar results (Tannenbaum 1940, 1996). Such models have included neuroblastoma xenograft models in CD-1 nude mice, where caloric restriction reduced tumour growth and prolonged survival (Morscher et al. 2015). Caloric restriction has also been found to delay the onset of spontaneous tumorigenesis in p53 deficient mice (Berrigan et al. 2002; Hursting et al. 1997, 1994).

Reduction in overall calorie intake has also been found to reduce the occurrence of spontaneous mammary tumours in mice during development (Dirx et al. 2003; Engelman et al. 1994; Fernandes et al. 1995). Meanwhile, the anti-tumour effects of caloric restriction in rhesus monkeys has provided mixed results with one 20-yearlong study showing a 50% reduction in cancer incidence and another performed at the NIA showing no such improvement (Colman et al. 2009; Mattison et al. 2012). When looking in human, caloric restriction as a means of preventing or treating cancer are typically investigated through population studies. Comparisons between individuals consuming fewer calories in Okinawa, Japan versus residents of mainland Japan showed that those with lower caloric intake had reduced risk of cancer (Kagawa 1978). Clinical trials such as the CALERIE study, investigating the benefits of caloric restriction in healthy, non-obese individuals showed similar metabolic changes as were observed in preclinical models (Heilbronn et al. 2006; Redman et al. 2009). These results indicate that similar anti-tumour benefits may be possible through caloric restriction in humans. Interestingly, these results are accompanied by a reduction in generation of ROS (Redman et al. 2018). Achieving caloric restriction in patients undergoing chemotherapy, radiation, and surgery, or even healthy individuals becomes problematic to maintain. Possible alternatives include time restricted fasting regimens.

An alternative to caloric restriction, fasting provides many of the benefits while being easier to adhere to. These nutritional interventions can be performed short-term at more frequent intervals, or long-term but less frequently. Using fasting or fasting-mimicking diets, it was found that fasting along with vitamin C display potent anticancer effects against *KRAS* mutated cancer cells (Di Tano et al. 2020). Time restricted caloric intake in rats has been found to reduce DNA replication and increase apoptosis, resulting in fewer pre-neoplastic liver foci. Meanwhile, returning to *ad libitum* feeding schedules resulted in normalisation to controls (Grasl-Kraupp et al. 1994).

Fasting has also been found to increase the sensitivity of cancer cells to therapy. These effects have been observed in models of glioma where sensitivity to temozolomide and radiotherapy were increased in response to fasting (Safdie et al. 2012). Fasting induced drug sensitivity has been observed in various other models including melanoma, breast cancer, neuroblastoma, and pancreatic cancer (D'Aronzo et al. 2015; Lee et al. 2012, 2010; Salvadori et al. 2021). Additional work has found that fasting is also capable of increasing the effectiveness of immunotherapy against breast cancer (Cortellino et al. 2022). Fasting and fasting-mimicking diets have also been found to reduce the risk of experiencing unintended side effects of standard therapy (Blaževitš et al. 2023; Cortellino et al. 2022; Di Tano and Longo 2022). These results suggest that fasting regimens may be suitable for prevention or treatment of cancer in combination with standard forms of therapy. Furthermore, in human trials involving cancer patients fasting has been found to be well tolerated, but additional analysis is needed to determine efficacy in extending progression free survival (Fay-Watt et al. 2023; Marinac et al. 2016; Vernieri et al. 2022). Therefore, fasting may be a more suitable alternative to caloric restriction which is difficult to maintain. However, recent findings indicate that caloric restriction may be a more effective approach than cycles of fasting (Pomatto-Watson et al. 2021).

The multitude of studies investigating the utility of nutritional intervention in preventing and treating cancer point to tightly regulated molecular pathways that overlap with their effectiveness in slowing the onset of ageing related disease. Dietary intervention, whether it be caloric restriction or fasting induce metabolic changes that decrease production of growth factors and hormones (Longo and Fontana 2010). Dietary restriction also exerts its effects by suppressing production of pro-inflammatory factors, while simultaneously increasing production of anti-inflammatory factors (Cangemi et al. 2016). Altered caloric intake also delay ageing-related declines in immune responses while promoting anti-tumoral immunity (Buono et al. 2020; Messaoudi et al. 2008).

It is also clear that there is an increase in anti-oxidant production which decreases ROS mediated accumulation of cellular damage (Luo et al. 2017). The molecular mechanisms involve metabolic reprogramming, which forces cancer cells to increase oxidative phosphorylation, resulting in increased production of ROS which disrupts redox homeostasis. This alteration in redox homeostasis triggers DNA damage and cell death. Additionally, restricting food intake disrupts nutrient sensing pathways involving IGF-1/PI3K/AKT, mTOR, sirtuins, NF-kB, and AMPK (Vidoni et al. 2021). Other mechanisms by which nutritional intervention may be impacting tumour development and growth include the production H_2S. In glioblastoma H_2S has been found to act as a tumour suppressor, and mice fed on a high-fat diet were found to have decreased levels resulting in increased presence of cancer stem cells and a greater disease progression (Silver et al. 2021a, b). However, the function of H_2S varies depending upon its context. While H_2S functions as a tumour suppressor in glioblastoma (Silver et al. 2021a, b), the activity of CTH has also been shown to promote glioblastoma invasion while also limiting overall tumour burden (Garcia et al. 2023). Additionally, dependent on the H_2S concentration it has been found to increase tumour growth in hepatocellular carcinoma (Silver et al. 2021b; Wu et al. 2017). Thus, H_2S and sulphides pose a somewhat complicated paradox in the realm of cancer growth, progression, and metastasis. Extensive future studies are needed to elucidate the mechanistic and physiological roles of H_2S in cancer and how geroscience intervention may be used to modulate its levels to prevent or treat cancer.

Conclusions

While the search for the literal and figurative fountains of youth have been unattainable, it is not without hope we will someday behold effective interventions and medical technologies to prolong healthy life and ward off the onset and severity of ageing-associated diseases. Doing so will require better understandings of the main topics we have covered in this body of work, mainly dietary restriction and its underlying mechanisms of action, control and maintenance of redox homeostasis during ageing, and the spectrum of cellular and subcellular events that occur to drive the ageing process. Developing ways to harness and enhance endogenous signalling metabolites and molecules, such as H_2S and NAD+ that are stimulated by dietary restriction, pose as two very promising targets that need more resources dedicated to their study and applications.

Acknowledgments The authors wish to thank the National Institutes of Health (NIH) for providing the following grant funding that supported this work: R01HL148352 and R01NS127374. Figures were created with BioRender.

References

Aerbajinai W, Zhu J, Gao Z, Chin K, Rodgers GP (2007) Thalidomide induces γ-globin gene expression through increased reactive oxygen species–mediated p38 MAPK signaling and histone H4 acetylation in adult erythropoiesis. Blood 110:2864–2871. https://doi.org/10.1182/blood-2007-01-065201

Ah Kang K, Jing Piao M, Seong Ryu Y, Kyoung Kang H, Young Chang W, Sam Keum Y, Won Hyun J (2016) Interaction of DNA demethylase and histone methyltransferase upregulates Nrf2 in 5-fluorouracil-resistant colon cancer cells. Oncotarget 7(26):40594–40620

Ahmet I, Wan R, Mattson MP, Lakatta EG, Talan M (2005) Cardioprotection by intermittent fasting in rats. Circulation 112:3115–3121. https://doi.org/10.1161/CIRCULATIONAHA.105.563817

Ali SS, Marcondes MC, Bajova H, Dugan LL, Conti B (2010) Metabolic depression and increased reactive oxygen species production by isolated mitochondria at moderately lower temperatures. J Biol Chem 285:32522–32528. https://doi.org/10.1074/jbc.M110.155432

Anastasiadi D, Esteve-Codina A, Piferrer F (2018) Consistent inverse correlation between DNA methylation of the first intron and gene expression across tissues and species. Epigenetics Chromatin 11:37. https://doi.org/10.1186/s13072-018-0205-1

Anderson RM, Weindruch R (2007) Metabolic reprogramming in dietary restriction. Interdiscip Top Gerontol 35:18–38. https://doi.org/10.1159/000096554

Anderson RM, Bitterman KJ, Wood JG, Medvedik O, Sinclair DA (2003) Nicotinamide and PNC1 govern lifespan extension by calorie restriction in Saccharomyces cerevisiae. Nature 423:181–185. https://doi.org/10.1038/nature01578

Anson RM, Guo Z, de Cabo R, Iyun T, Rios M, Hagepanos A, Ingram DK, Lane MA, Mattson MP (2003) Intermittent fasting dissociates beneficial effects of dietary restriction on glucose metabolism and neuronal resistance to injury from calorie intake. Proc Natl Acad Sci USA 100:6216–6220. https://doi.org/10.1073/pnas.1035720100

Arbiser JL, Petros J, Klafter R, Govindajaran B, McLaughlin ER, Brown LF, Cohen C, Moses M, Kilroy S, Arnold RS, Lambeth JD (2002) Reactive oxygen generated by Nox1 triggers the angiogenic switch. Proc Natl Acad Sci 99:715–720. https://doi.org/10.1073/pnas.022630199

Ayala V, Naudi A, Sanz A, Caro P, Portero-Otin M, Barja G, Pamplona R (2007) Dietary protein restriction decreases oxidative protein damage, peroxidizability index, and mitochondrial complex I content in rat liver. J Gerontol A Biol Sci Med Sci 62:352–360. https://doi.org/10.1093/gerona/62.4.352

Baker RG, Hayden MS, Ghosh S (2011) NF-kappaB, inflammation, and metabolic disease. Cell Metab 13:11–22. https://doi.org/10.1016/j.cmet.2010.12.008

Barcena C, Quiros PM, Durand S, Mayoral P, Rodriguez F, Caravia XM, Marino G, Garabaya C, Fernandez-Garcia MT, Kroemer G, Freije JMP, Lopez-Otin C (2018) Methionine restriction extends lifespan in progeroid mice and alters lipid and bile acid metabolism. Cell Rep 24:2392–2403. https://doi.org/10.1016/j.celrep.2018.07.089

Barja G (2013) Updating the mitochondrial free radical theory of aging: an integrated view, key aspects, and confounding concepts. Antioxid Redox Signal 19:1420–1445. https://doi.org/10.1089/ars.2012.5148

Barzilai N, Banerjee S, Hawkins M, Chen W, Rossetti L (1998) Caloric restriction reverses hepatic insulin resistance in aging rats by decreasing visceral fat. J Clin Invest 101:1353–1361. https://doi.org/10.1172/JCI485

Behrendt U, Schumann P, Hamada M, Suzuki KI, Sproer C, Ulrich A (2011) Reclassification of Leifsonia ginsengi (Qiu et al. 2007) as Herbiconiux ginsengi gen. nov., comb. nov. and description of Herbiconiux solani sp. nov., an actinobacterium associated with the phyllosphere of Solanum tuberosum L. Int J Syst Evol Microbiol 61:1039–1047. https://doi.org/10.1099/ijs.0.021352-0

Berrigan D, Perkins SN, Haines DC, Hursting SD (2002) Adult-onset calorie restriction and fasting delay spontaneous tumorigenesis in p53-deficient mice. Carcinogenesis 23:817–822. https://doi.org/10.1093/carcin/23.5.817

Bithi N, Link C, Henderson YO, Kim S, Yang J, Li L, Wang R, Willard B, Hine C (2021) Dietary restriction transforms the mammalian protein persulfidome in a tissue-specific and cystathionine gamma-lyase-dependent manner. Nat Commun 12:1745. https://doi.org/10.1038/s41467-021-22001-w

Blaževitš O, Di Tano M, Longo VD (2023) Fasting and fasting mimicking diets in cancer prevention and therapy. Trends in Cancer 9:212–222. https://doi.org/10.1016/j.trecan.2022.12.006

Brand MD (2010) The sites and topology of mitochondrial superoxide production. Exp Gerontol 45:466–472. https://doi.org/10.1016/j.exger.2010.01.003

Brunet A, Goodell MA, Rando TA (2023) Ageing and rejuvenation of tissue stem cells and their niches. Nat Rev Mol Cell Biol 24:45–62. https://doi.org/10.1038/s41580-022-00510-w

Bruss MD, Khambatta CF, Ruby MA, Aggarwal I, Hellerstein MK (2010) Calorie restriction increases fatty acid synthesis and whole body fat oxidation rates. Am J Physiol Endocrinol Metab 298:E108–E116. https://doi.org/10.1152/ajpendo.00524.2009

Buono R, Tucci J, Cutri R, Guidi N, Mangul S, Pellegrini M, Fruman DA, Mittelman SD, Longo V (2020) Fasting mimicking diet promotes immunotherapy-associated changes and inhibits autophagy to promote cancer free survival in acute lymphoblastic leukemia models. J Immunol 204:240.210. https://doi.org/10.4049/jimmunol.204.Supp.240.10

Burger JM, Hwangbo DS, Corby-Harris V, Promislow DE (2007) The functional costs and benefits of dietary restriction in Drosophila. Aging Cell 6:63–71. https://doi.org/10.1111/j.1474-9726.2006.00261.x

Cabreiro F, Au C, Leung KY, Vergara-Irigaray N, Cochemé HM, Noori T, Weinkove D, Schuster E, Greene ND, Gems D (2013) Metformin retards aging in C. elegans by altering microbial folate and methionine metabolism. Cell 153:228–239. https://doi.org/10.1016/j.cell.2013.02.035

Cangemi A, Fanale D, Rinaldi G, Bazan V, Galvano A, Perez A, Barraco N, Massihnia D, Castiglia M, Vieni S, Bronte G, Mirisola M, Russo A (2016) Dietary restriction: could it be considered as speed bump on tumor progression road? Tumour Biol 37:7109–7118. https://doi.org/10.1007/s13277-016-5044-8

Canty TG Jr, Boyle EM Jr, Farr A, Morgan EN, Verrier ED, Pohlman TH (1999) Oxidative stress induces NF-kappaB nuclear translocation without degradation of IkappaBalpha. Circulation 100:II361–II364. https://doi.org/10.1161/01.cir.100.suppl_2.ii-361

Capel F, Rimbert V, Lioger D, Diot A, Rousset P, Mirand PP, Boirie Y, Morio B, Mosoni L (2005) Due to reverse electron transfer, mitochondrial H2O2 release increases with age in human vastus lateralis muscle although oxidative capacity is preserved. Mech Ageing Dev 126:505–511. https://doi.org/10.1016/j.mad.2004.11.001

Carlson AJ, Hoelzel F (1946) Apparent prolongation of the life span of rats by intermittent fasting. J Nutr 31:363–375. https://doi.org/10.1093/jn/31.3.363

Cassel SL, Sutterwala FS (2010) Sterile inflammatory responses mediated by the NLRP3 inflammasome. Eur J Immunol 40:607–611. https://doi.org/10.1002/eji.200940207

Cheung EC, Vousden KH (2022) The role of ROS in tumour development and progression. Nat Rev Cancer 22:280–297. https://doi.org/10.1038/s41568-021-00435-0

Chung HS, Wang SB, Venkatraman V, Murray CI, Van Eyk JE (2013) Cysteine oxidative post-translational modifications: emerging regulation in the cardiovascular system. Circ Res 112:382–392. https://doi.org/10.1161/CIRCRESAHA.112.268680

Chung HY, Kim DH, Lee EK, Chung KW, Chung S, Lee B, Seo AY, Chung JH, Jung YS, Im E, Lee J, Kim ND, Choi YJ, Im DS, Yu BP (2019) Redefining chronic inflammation in aging and age-related diseases: proposal of the Senoinflammation concept. Aging Dis 10:367–382. https://doi.org/10.14336/AD.2018.0324

Colman RJ, Anderson RM, Johnson SC, Kastman EK, Kosmatka KJ, Beasley TM, Allison DB, Cruzen C, Simmons HA, Kemnitz JW, Weindruch R (2009) Caloric restriction delays disease onset and mortality in rhesus monkeys. Science 325:201–204. https://doi.org/10.1126/science.1173635

Cortellino S, Raveane A, Chiodoni C, Delfanti G, Pisati F, Spagnolo V, Visco E, Fragale G, Ferrante F, Magni S, Iannelli F, Zanardi F, Casorati G, Bertolini F, Dellabona P, Colombo MP, Tripodo C, Longo VD (2022) Fasting renders immunotherapy effective against low-immunogenic breast cancer while reducing side effects. Cell Rep 40:111256. https://doi.org/10.1016/j.celrep.2022.111256

Coulter JB, O'Driscoll CM, Bressler JP (2013) Hydroquinone increases 5-Hydroxymethylcytosine formation through ten eleven translocation 1 (TET1) 5-Methylcytosine dioxygenase. J Biol Chem 288:28792–28800. https://doi.org/10.1074/jbc.M113.491365

D'Aronzo M, Vinciguerra M, Mazza T, Panebianco C, Saracino C, Pereira SP, Graziano P, Pazienza V (2015) Fasting cycles potentiate the efficacy of gemcitabine treatment in in vitro and in vivo pancreatic cancer models. Oncotarget 6:18545–18557. https://doi.org/10.18632/oncotarget.4186

De Haes W, Frooninckx L, Van Assche R, Smolders A, Depuydt G, Billen J, Braeckman BP, Schoofs L, Temmerman L (2014) Metformin promotes lifespan through mitohormesis via the peroxiredoxin PRDX-2. Proc Natl Acad Sci USA 111:E2501–E2509. https://doi.org/10.1073/pnas.1321776111

Delimaris I (2013) Adverse effects associated with protein intake above the recommended dietary allowance for adults. ISRN Nutrition 2013:126929. https://doi.org/10.5402/2013/126929

Di Tano M, Longo VD (2022) Chapter three - fasting and cancer: from yeast to mammals. In: Buqué A, Galluzzi L (eds) International review of cell and molecular biology. Academic Press, pp 81–106

Di Tano M, Raucci F, Vernieri C, Caffa I, Buono R, Fanti M, Brandhorst S, Curigliano G, Nencioni A, de Braud F, Longo VD (2020) Synergistic effect of fasting-mimicking diet and vitamin C against KRAS mutated cancers. Nat Commun 11:2332. https://doi.org/10.1038/s41467-020-16243-3

Dirx MJM, Zeegers MPA, Dagnelie PC, van den Bogaard T, van den Brandt PA (2003) Energy restriction and the risk of spontaneous mammary tumors in mice: a meta-analysis. Int J Cancer 106:766–770. https://doi.org/10.1002/ijc.11277

Duan J, Xiang L, Yang Z, Chen L, Gu J, Lu K, Ma D, Zhao H, Yi B, Zhao H, Ning J (2022) Methionine restriction prevents lipopolysaccharide-induced acute lung injury via modulating CSE/H(2)S pathway. Nutrients 14. https://doi.org/10.3390/nu14020322

Dubey A, Forster MJ, Lal H, Sohal RS (1996) Effect of age and caloric intake on protein oxidation in different brain regions and on behavioral functions of the mouse. Arch Biochem Biophys 333:189–197. https://doi.org/10.1006/abbi.1996.0380

Duffy PH, Feuers RJ, Leakey JA, Nakamura K, Turturro A, Hart RW (1989) Effect of chronic caloric restriction on physiological variables related to energy metabolism in the male Fischer 344 rat. Mech Ageing Dev 48:117–133. https://doi.org/10.1016/0047-6374(89)90044-4

Duffy PH, Feuers RJ, Hart RW (1990) Effect of chronic caloric restriction on the circadian regulation of physiological and behavioral variables in old male B6C3F1 mice. Chronobiol Int 7:291–303. https://doi.org/10.1080/07420529009064635

Editorial NA (2022) The importance of aging in cancer research. Nat Aging 2:365–366. https://doi.org/10.1038/s43587-022-00231-x

Engelman RW, Day NK, Good RA (1994) Calorie intake during mammary development influences cancer risk: lasting inhibition of C3H/HeOu mammary tumorigenesis by peripubertal calorie restriction1. Cancer Res 54:5724–5730

Fan J, Zheng F, Li S, Cui C, Jiang S, Zhang J, Cai J, Cui Q, Yang J, Tang X, Xu G, Geng B (2019) Hydrogen sulfide lowers hyperhomocysteinemia dependent on cystathionine gamma lyase S-sulfhydration in ApoE-knockout atherosclerotic mice. Br J Pharmacol 176:3180–3192. https://doi.org/10.1111/bph.14719

Fanti P, Giustarini D, Rossi R, Cunningham SE, Folli F, Khazim K, Cornell J, Matteucci E, Bansal S (2015) Dietary intake of proteins and calories is inversely associated with the oxidation state of plasma thiols in end-stage renal disease patients. J Ren Nutr 25:494–503. https://doi.org/10.1053/j.jrn.2015.06.003

Fay-Watt V, O'Connor S, Roshan D, Romeo AC, Longo VD, Sullivan FJ (2023) The impact of a fasting mimicking diet on the metabolic health of a prospective cohort of patients with prostate cancer: a pilot implementation study. Prostate Cancer Prostatic Dis 26:317–322. https://doi.org/10.1038/s41391-022-00528-3

Fearon ER (1997) Human cancer syndromes: clues to the origin and nature of cancer. Science 278:1043–1050. https://doi.org/10.1126/science.278.5340.1043

Fernandes G, Chandrasekar B, Troyer DA, Venkatraman JT, Good RA (1995) Dietary lipids and calorie restriction affect mammary tumor incidence and gene expression in mouse mammary tumor virus/v-Ha-ras transgenic mice. Proc Natl Acad Sci 92:6494–6498. https://doi.org/10.1073/pnas.92.14.6494

Filipovic MR, Zivanovic J, Alvarez B, Banerjee R (2018) Chemical biology of H2S signaling through persulfidation. Chem Rev 118:1253–1337. https://doi.org/10.1021/acs.chemrev.7b00205

Finkel T (2012) From sulfenylation to sulfhydration: what a thiolate needs to tolerate. Sci Signal 5:pe10. https://doi.org/10.1126/scisignal.2002943

Fontana L, Partridge L (2015) Promoting health and longevity through diet: from model organisms to humans. Cell 161:106–118. https://doi.org/10.1016/j.cell.2015.02.020

Forster MJ, Sohal BH, Sohal RS (2000) Reversible effects of long-term caloric restriction on protein oxidative damage. J Gerontol A Biol Sci Med Sci 55:B522–B529. https://doi.org/10.1093/gerona/55.11.b522

Gao XH, Krokowski D, Guan BJ, Bederman I, Majumder M, Parisien M, Diatchenko L, Kabil O, Willard B, Banerjee R, Wang B, Bebek G, Evans CR, Fox PL, Gerson SL, Hoppel CL, Liu M, Arvan P, Hatzoglou M (2015) Quantitative H2S-mediated protein sulfhydration reveals metabolic reprogramming during the integrated stress response. elife 4:e10067. https://doi.org/10.7554/eLife.10067

Gào X, Zhang Y, Burwinkel B, Xuan Y, Holleczek B, Brenner H, Schöttker B (2019) The associations of DNA methylation alterations in oxidative stress-related genes with cancer incidence and mortality outcomes: a population-based cohort study. Clin Epigenetics 11:14. https://doi.org/10.1186/s13148-018-0604-y

Garcia JH, Akins EA, Jain S, Wolf KJ, Zhang J, Choudhary N, Lad M, Shukla P, Rios J, Seo K, Gill SA, Carson WH, Carrete LR, Zheng AC, Raleigh DR, Kumar S, Aghi MK (2023) Multi-omic screening of invasive GBM cells reveals targetable transsulfuration pathway alterations. J Clin Invest 134. https://doi.org/10.1172/JCI170397

García-Giménez JL, Olaso G, Hake SB, Bönisch C, Wiedemann SM, Markovic J, Dasí F, Gimeno A, Pérez-Quilis C, Palacios Ò, Capdevila M, Viña J, Pallardó FV (2013) Histone H3 Glutathionylation in proliferating mammalian cells destabilizes nucleosomal structure. Antioxid Redox Signal 19:1305–1320. https://doi.org/10.1089/ars.2012.5021

Garcia-Prieto CF, Fernandez-Alfonso MS (2016) Caloric restriction as a strategy to improve vascular dysfunction in metabolic disorders. Nutrients 8. https://doi.org/10.3390/nu8060370

Garrido Ruiz D, Sandoval-Perez A, Rangarajan AV, Gunderson EL, Jacobson MP (2022) Cysteine oxidation in proteins: structure, biophysics, and simulation. Biochemistry 61:2165–2176. https://doi.org/10.1021/acs.biochem.2c00349

Gerald D, Berra E, Frapart YM, Chan DA, Giaccia AJ, Mansuy D, Pouysségur J, Yaniv M, Mechta-Grigoriou F (2004) JunD reduces tumor angiogenesis by protecting cells from oxidative stress. Cell 118:781–794. https://doi.org/10.1016/j.cell.2004.08.025

Gianazza E, Brioschi M, Martinez Fernandez A, Casalnuovo F, Altomare A, Aldini G, Banfi C (2021) Lipid peroxidation in atherosclerotic cardiovascular diseases. Antioxid Redox Signal 34:49–98. https://doi.org/10.1089/ars.2019.7955

Gladyshev VN (2014) The free radical theory of aging is dead. Long live the damage theory! Antioxid Redox Signal 20:727–731. https://doi.org/10.1089/ars.2013.5228

Gomes AP, Ilter D, Low V, Endress JE, Fernandez-Garcia J, Rosenzweig A, Schild T, Broekaert D, Ahmed A, Planque M, Elia I, Han J, Kinzig C, Mullarky E, Mutvei AP, Asara J, de Cabo R, Cantley LC, Dephoure N, Fendt SM, Blenis J (2020) Age-induced accumulation of methylmalonic acid promotes tumour progression. Nature 585:283–287. https://doi.org/10.1038/s41586-020-2630-0

Gorni D, Finco A (2020) Oxidative stress in elderly population: a prevention screening study. Aging Medicine 3:205–213. https://doi.org/10.1002/agm2.12121

Gouspillou G, Bourdel-Marchasson I, Rouland R, Calmettes G, Biran M, Deschodt-Arsac V, Miraux S, Thiaudiere E, Pasdois P, Detaille D, Franconi J-M, Babot M, Trézéguet V, Arsac L, Diolez P (2014) Mitochondrial energetics is impaired in vivo in aged skeletal muscle. Aging Cell 13:39–48. https://doi.org/10.1111/acel.12147

Grandison RC, Piper MD, Partridge L (2009) Amino-acid imbalance explains extension of lifespan by dietary restriction in Drosophila. Nature 462:1061–1064. https://doi.org/10.1038/nature08619

Grasl-Kraupp B, Bursch W, Ruttkay-Nedecky B, Wagner A, Lauer B, Schulte-Hermann R (1994) Food restriction eliminates preneoplastic cells through apoptosis and antagonizes carcinogenesis in rat liver. Proc Natl Acad Sci 91:9995–9999. https://doi.org/10.1073/pnas.91.21.9995

Hanahan D (2022) Hallmarks of cancer: new dimensions. Cancer Discov 12:31–46. https://doi.org/10.1158/2159-8290.Cd-21-1059

Harman D (1956) Aging: a theory based on free radical and radiation chemistry. J Gerontol 11:298–300

Hayes JD, Dinkova-Kostova AT, Tew KD (2020) Oxidative stress in cancer. Cancer Cell 38:167–197. https://doi.org/10.1016/j.ccell.2020.06.001

Heid ME, Keyel PA, Kamga C, Shiva S, Watkins SC, Salter RD (2013) Mitochondrial reactive oxygen species induces NLRP3-dependent lysosomal damage and inflammasome activation. J Immunol 191:5230–5238. https://doi.org/10.4049/jimmunol.1301490

Heilbronn LK, de Jonge L, Frisard MI, DeLany JP, Larson-Meyer DE, Rood J, Nguyen T, Martin CK, Volaufova J, Most MM, Greenway FL, Smith SR, Deutsch WA, Williamson DA, Ravussin E, Pennington CT (2006) Effect of 6-month calorie restriction on biomarkers of longevity, metabolic adaptation, and oxidative stress in overweight individuals: a randomized controlled trial. JAMA 295:1539–1548. https://doi.org/10.1001/jama.295.13.1539

Henderson YO, Bithi N, Link C, Yang J, Schugar R, Llarena N, Brown JM, Hine C (2021) Late-life intermittent fasting decreases aging-related frailty and increases renal hydrogen sulfide production in a sexually dimorphic manner. Geroscience 43:1527–1554. https://doi.org/10.1007/s11357-021-00330-4

Henderson YO, Bithi N, Yang J, Link C, Zhang A, Baron B, Maina E, Hine C (2023) A long-term obesogenic high-fat diet in mice partially dampens the anti-frailty benefits of late-life intermittent fasting. Geroscience 45:1247–1262. https://doi.org/10.1007/s11357-022-00678-1

Herrero A, Barja G (2001) Effect of aging on mitochondrial and nuclear DNA oxidative damage in the heart and brain throughout the life-span of the rat. J Am Aging Assoc 24:45–50. https://doi.org/10.1007/s11357-001-0006-4

Hey-Mogensen M, Goncalves RL, Orr AL, Brand MD (2014) Production of superoxide/H2O2 by dihydroorotate dehydrogenase in rat skeletal muscle mitochondria. Free Radic Biol Med 72:149–155. https://doi.org/10.1016/j.freeradbiomed.2014.04.007

Hickok JR, Vasudevan D, Antholine WE, Thomas DD (2013) Nitric oxide modifies global histone methylation by inhibiting Jumonji C Domain-containing demethylases. J Biol Chem 288:16004–16015. https://doi.org/10.1074/jbc.M112.432294

Hill CM, Albarado DC, Coco LG, Spann RA, Khan MS, Qualls-Creekmore E, Burk DH, Burke SJ, Collier JJ, Yu S, McDougal DH, Berthoud HR, Munzberg H, Bartke A, Morrison CD (2022) FGF21 is required for protein restriction to extend lifespan and improve metabolic health in male mice. Nat Commun 13:1897. https://doi.org/10.1038/s41467-022-29499-8

Hine CM, Mitchell JR (2012) NRF2 and the phase II response in acute stress resistance induced by dietary restriction. J Clin Exp Pathol S4. https://doi.org/10.4172/2161-0681.S4-004

Hine C, Harputlugil E, Zhang Y, Ruckenstuhl C, Lee BC, Brace L, Longchamp A, Treviño-Villarreal JH, Mejia P, Ozaki CK, Wang R, Gladyshev VN, Madeo F, Mair WB, Mitchell JR (2015) Endogenous hydrogen sulfide production is essential for dietary restriction benefits. Cell 160:132–144. https://doi.org/10.1016/j.cell.2014.11.048

Hine C, Kim HJ, Zhu Y, Harputlugil E, Longchamp A, Matos MS, Ramadoss P, Bauerle K, Brace L, Asara JM, Ozaki CK, Cheng SY, Singha S, Ahn KH, Kimmelman A, Fisher FM, Pissios P, Withers DJ, Selman C, Wang R, Yen K, Longo VD, Cohen P, Bartke A, Kopchick JJ, Miller R, Hollenberg AN, Mitchell JR (2017) Hypothalamic-pituitary axis regulates hydrogen sulfide production. Cell Metab 25:1320–1333.e5. https://doi.org/10.1016/j.cmet.2017.05.003

Hine C, Zhu Y, Hollenberg AN, Mitchell JR (2018) Dietary and endocrine regulation of endogenous hydrogen sulfide production: implications for longevity. Antioxid Redox Signal 28:1483–1502. https://doi.org/10.1089/ars.2017.7434

Huang Z, Sun S, Lee M, Maslov AY, Shi M, Waldman S, Marsh A, Siddiqui T, Dong X, Peter Y, Sadoughi A, Shah C, Ye K, Spivack SD, Vijg J (2022) Single-cell analysis of somatic mutations in human bronchial epithelial cells in relation to aging and smoking. Nat Genet 54:492–498. https://doi.org/10.1038/s41588-022-01035-w

Hursting SD, Perkins SN, Phang JM (1994) Calorie restriction delays spontaneous tumorigenesis in p53-knockout transgenic mice. Proc Natl Acad Sci 91:7036–7040. https://doi.org/10.1073/pnas.91.15.7036

Hursting SD, Perkins SN, Brown CC, Haines DC, Phang JM (1997) Calorie restriction induces a p53-independent delay of spontaneous carcinogenesis in p53-deficient and wild-type Mice1. Cancer Res 57:2843–2846

Irvine GB, El-Agnaf OM, Shankar GM, Walsh DM (2008) Protein aggregation in the brain: the molecular basis for Alzheimer's and Parkinson's diseases. Mol Med 14:451–464. https://doi.org/10.2119/2007-00100.Irvine

Ito K, Hirao A, Arai F, Takubo K, Matsuoka S, Miyamoto K, Ohmura M, Naka K, Hosokawa K, Ikeda Y, Suda T (2006) Reactive oxygen species act through p38 MAPK to limit the lifespan of hematopoietic stem cells. Nat Med 12:446–451. https://doi.org/10.1038/nm1388

Itoh K, Wakabayashi N, Katoh Y, Ishii T, Igarashi K, Engel JD, Yamamoto M (1999) Keap1 represses nuclear activation of antioxidant responsive elements by Nrf2 through binding to the amino-terminal Neh2 domain. Genes Dev 13:76–86. https://doi.org/10.1101/gad.13.1.76

Jiang T, Liebman SE, Lucia MS, Phillips CL, Levi M (2005) Calorie restriction modulates renal expression of sterol regulatory element binding proteins, lipid accumulation, and age-related renal disease. J Am Soc Nephrol 16:2385–2394. https://doi.org/10.1681/ASN.2004080701

Jorjani P, Ozturk SS (1999) Effects of cell density and temperature on oxygen consumption rate for different mammalian cell lines. Biotechnol Bioeng 64:349–356. https://doi.org/10.1002/(sici)1097-0290(19990805)64:3<349::aid-bit11>3.0.co;2-v

Kabil H, Kabil O, Banerjee R, Harshman LG, Pletcher SD (2011) Increased transsulfuration mediates longevity and dietary restriction in Drosophila. Proc Natl Acad Sci USA 108:16831–16836. https://doi.org/10.1073/pnas.1102008108

Kagawa Y (1978) Impact of westernization on the nutrition of Japanese: changes in physique, cancer, longevity and centenarians. Prev Med 7:205–217. https://doi.org/10.1016/0091-7435(78)90246-3

Kamiya T, Goto A, Kurokawa E, Hara H, Adachi T (2016) Cross talk mechanism among EMT, ROS, and histone acetylation in Phorbol Ester-treated human breast cancer MCF-7 cells. Oxidative Med Cell Longev 2016:1284372. https://doi.org/10.1155/2016/1284372

Kemnitz JW, Roecker EB, Weindruch R, Elson DF, Baum ST, Bergman RN (1994) Dietary restriction increases insulin sensitivity and lowers blood glucose in rhesus monkeys. Am J Phys 266:E540–E547. https://doi.org/10.1152/ajpendo.1994.266.4.E540

Kennedy BK, Berger SL, Brunet A, Campisi J, Cuervo AM, Epel ES, Franceschi C, Lithgow GJ, Morimoto RI, Pessin JE, Rando TA, Richardson A, Schadt EE, Wyss-Coray T, Sierra F (2014) Geroscience: linking aging to chronic disease. Cell 159:709–713. https://doi.org/10.1016/j.cell.2014.10.039

Keyer K, Imlay JA (1996) Superoxide accelerates DNA damage by elevating free-iron levels. Proc Natl Acad Sci USA 93:13635–13640. https://doi.org/10.1073/pnas.93.24.13635

Khan MA, Alam K, Dixit K, Rizvi MMA (2016) Role of peroxynitrite induced structural changes on H2B histone by physicochemical method. Int J Biol Macromol 82:31–38. https://doi.org/10.1016/j.ijbiomac.2015.10.085

Kietzmann T, Petry A, Shvetsova A, Gerhold JM, Gorlach A (2017) The epigenetic landscape related to reactive oxygen species formation in the cardiovascular system. Br J Pharmacol 174:1533–1554. https://doi.org/10.1111/bph.13792

Kim JY, Kim DH, Choi J, Park JK, Jeong KS, Leeuwenburgh C, Yu BP, Chung HY (2009) Changes in lipid distribution during aging and its modulation by calorie restriction. Age 31:127–142. https://doi.org/10.1007/s11357-009-9089-0

Kim M, Ladomersky E, Mozny A, Kocherginsky M, O'Shea K, Reinstein ZZ, Zhai L, Bell A, Lauing KL, Bollu L, Rabin E, Dixit K, Kumthekar P, Platanias LC, Hou L, Zheng Y, Wu J, Zhang B, Hrachova M, Merrill SA, Mrugala MM, Prabhu VC, Horbinski C, James CD, Yamini B, Ostrom QT, Johnson MO, Reardon DA, Lukas RV, Wainwright DA (2021) Glioblastoma as an age-related neurological disorder in adults. Neuro-Oncol Adv 3. https://doi.org/10.1093/noajnl/vdab125

Kirchner H, Hofmann SM, Fischer-Rosinsky A, Hembree J, Abplanalp W, Ottaway N, Donelan E, Krishna R, Woods SC, Muller TD, Spranger J, Perez-Tilve D, Pfluger PT, Tschop MH, Habegger KM (2012) Caloric restriction chronically impairs metabolic programming in mice. Diabetes 61:2734–2742. https://doi.org/10.2337/db11-1621

Kirk KL (2001) Dietary restriction and aging: comparative tests of evolutionary hypotheses. J Gerontol A Biol Sci Med Sci 56:B123–B129. https://doi.org/10.1093/gerona/56.3.b123

Klaunig JE, Kamendulis LM, Hocevar BA (2010) Oxidative stress and oxidative damage in carcinogenesis. Toxicol Pathol 38:96–109. https://doi.org/10.1177/0192623309356453

Kloypan C, Srisa-art M, Mutirangura A, Boonla C (2015) LINE-1 hypomethylation induced by reactive oxygen species is mediated via depletion of S-adenosylmethionine. Cell Biochem Funct 33:375–384. https://doi.org/10.1002/cbf.3124

Klurfeld DM, Weber MM, Kritchevsky D (1987) Inhibition of chemically induced mammary and colon tumor promotion by caloric restriction in rats fed increased dietary fat. Cancer Res 47:2759–2762

Kraus WE, Bhapkar M, Huffman KM, Pieper CF, Krupa Das S, Redman LM, Villareal DT, Rochon J, Roberts SB, Ravussin E, Holloszy JO, Fontana L, Investigators C (2019) 2 years of calorie restriction and cardiometabolic risk (CALERIE): exploratory outcomes of a multicentre, phase 2, randomised controlled trial. Lancet Diabetes Endocrinol 7:673–683. https://doi.org/10.1016/S2213-8587(19)30151-2

Krittika S, Yadav P (2020) Dietary protein restriction deciphers new relationships between lifespan, fecundity and activity levels in fruit flies Drosophila melanogaster. Sci Rep 10:10019. https://doi.org/10.1038/s41598-020-66372-4

Kuhla A, Hahn S, Butschkau A, Lange S, Wree A, Vollmar B (2014) Lifelong caloric restriction reprograms hepatic fat metabolism in mice. J Gerontol A Biol Sci Med Sci 69:915–922. https://doi.org/10.1093/gerona/glt160

Lakowski B, Hekimi S (1998) The genetics of caloric restriction in Caenorhabditis elegans. Proc Natl Acad Sci USA 95:13091–13096

Lane MA, Baer DJ, Rumpler WV, Weindruch R, Ingram DK, Tilmont EM, Cutler RG, Roth GS (1996) Calorie restriction lowers body temperature in rhesus monkeys, consistent with a postulated anti-aging mechanism in rodents. Proc Natl Acad Sci USA 93:4159–4164. https://doi.org/10.1073/pnas.93.9.4159

Lee S-R, Yang K-S, Kwon J, Lee C, Jeong W, Rhee SG (2002) Reversible inactivation of the tumor suppressor PTEN by H_2O_2*. J Biol Chem 277:20336–20342. https://doi.org/10.1074/jbc.M111899200

Lee GD, Wilson MA, Zhu M, Wolkow CA, de Cabo R, Ingram DK, Zou S (2006) Dietary deprivation extends lifespan in Caenorhabditis elegans. Aging Cell 5:515–524. https://doi.org/10.1111/j.1474-9726.2006.00241.x

Lee C, Safdie FM, Raffaghello L, Wei M, Madia F, Parrella E, Hwang D, Cohen P, Bianchi G, Longo VD (2010) Reduced levels of IGF-I mediate differential protection of normal and cancer cells in response to fasting and improve chemotherapeutic index. Cancer Res 70:1564–1572. https://doi.org/10.1158/0008-5472.Can-09-3228

Lee C, Raffaghello L, Brandhorst S, Safdie FM, Bianchi G, Martin-Montalvo A, Pistoia V, Wei M, Hwang S, Merlino A, Emionite L, de Cabo R, Longo VD (2012) Fasting cycles retard growth of tumors and sensitize a range of cancer cell types to chemotherapy. Sci Transl Med 4:124ra127. https://doi.org/10.1126/scitranslmed.3003293

Lee BC, Kaya A, Ma S, Kim G, Gerashchenko MV, Yim SH, Hu Z, Harshman LG, Gladyshev VN (2014a) Methionine restriction extends lifespan of Drosophila melanogaster under conditions of low amino-acid status. Nat Commun 5:3592. https://doi.org/10.1038/ncomms4592

Lee MC, Velayutham M, Komatsu T, Hille R, Zweier JL (2014b) Measurement and characterization of superoxide generation from xanthine dehydrogenase: a redox-regulated pathway of radical generation in ischemic tissues. Biochemistry 53:6615–6623. https://doi.org/10.1021/bi500582r

Lee BC, Kaya A, Gladyshev VN (2016) Methionine restriction and life-span control. Ann N Y Acad Sci 1363:116–124. https://doi.org/10.1111/nyas.12973

Li Z, Low V, Luga V, Sun J, Earlie E, Parang B, Shobana Ganesh K, Cho S, Endress J, Schild T, Hu M, Lyden D, Jin W, Guo C, Dephoure N, Cantley LC, Laughney AM, Blenis J (2022) Tumor-produced and aging-associated oncometabolite methylmalonic acid promotes cancer-associated fibroblast activation to drive metastatic progression. Nat Commun 13:6239. https://doi.org/10.1038/s41467-022-33862-0

Li C, Zhu J, Wei S, Ye X, Yang L, Wang Z, Chen Y (2023) Intermittent protein restriction improves glucose homeostasis in Zucker diabetic fatty rats and single-cell sequencing reveals distinct changes in beta cells. J Nutr Biochem 114:109275. https://doi.org/10.1016/j.jnutbio.2023.109275

Lichtenstein P, Holm NV, Verkasalo PK, Iliadou A, Kaprio J, Koskenvuo M, Pukkala E, Skytthe A, Hemminki K (2000) Environmental and heritable factors in the causation of cancer — analyses of cohorts of twins from Sweden, Denmark, and Finland. N Engl J Med 343:78–85. https://doi.org/10.1056/nejm200007133430201

Lin SS, Manchester JK, Gordon JI (2001) Enhanced gluconeogenesis and increased energy storage as hallmarks of aging in Saccharomyces cerevisiae. J Biol Chem 276:36000–36007. https://doi.org/10.1074/jbc.M103509200

Lin SJ, Kaeberlein M, Andalis AA, Sturtz LA, Defossez PA, Culotta VC, Fink GR, Guarente L (2002) Calorie restriction extends Saccharomyces cerevisiae lifespan by increasing respiration. Nature 418:344–348. https://doi.org/10.1038/nature00829

Lints FA (1989) The rate of living theory revisited. Gerontology 35:36–57. https://doi.org/10.1159/000212998

Llarena N, Hine C (2021) Reproductive longevity and aging: Geroscience approaches to maintain Long-term ovarian fitness. J Gerontol A Biol Sci Med Sci 76:1551–1560. https://doi.org/10.1093/gerona/glaa204

Longchamp A, Mirabella T, Arduini A, MacArthur MR, Das A, Trevino-Villarreal JH, Hine C, Ben-Sahra I, Knudsen NH, Brace LE, Reynolds J, Mejia P, Tao M, Sharma G, Wang R, Corpataux JM, Haefliger JA, Ahn KH, Lee CH, Manning BD, Sinclair DA, Chen CS, Ozaki CK, Mitchell JR (2018) Amino acid restriction triggers angiogenesis via GCN2/ATF4 regulation of VEGF and H2S production. Cell 173:117–129.e14. https://doi.org/10.1016/j.cell.2018.03.001

Longo VD, Fontana L (2010) Calorie restriction and cancer prevention: metabolic and molecular mechanisms. Trends Pharmacol Sci 31:89–98. https://doi.org/10.1016/j.tips.2009.11.004

Lopez-Domenech S, Martinez-Herrera M, Abad-Jimenez Z, Morillas C, Escribano-Lopez I, Diaz-Morales N, Banuls C, Victor VM, Rocha M (2019) Dietary weight loss intervention improves subclinical atherosclerosis and oxidative stress markers in leukocytes of obese humans. Int J Obes 43:2200–2209. https://doi.org/10.1038/s41366-018-0309-5

Lopez-Otin C, Blasco MA, Partridge L, Serrano M, Kroemer G (2023a) Hallmarks of aging: an expanding universe. Cell 186:243–278. https://doi.org/10.1016/j.cell.2022.11.001

Lopez-Otin C, Pietrocola F, Roiz-Valle D, Galluzzi L, Kroemer G (2023b) Meta-hallmarks of aging and cancer. Cell Metab 35:12–35. https://doi.org/10.1016/j.cmet.2022.11.001

Loscalzo J, Welch G (1995) Nitric oxide and its role in the cardiovascular system. Prog Cardiovasc Dis 38:87–104. https://doi.org/10.1016/s0033-0620(05)80001-5

Lu MJ, Lu Y (2022) 5-Hydroxymethylcytosine (5hmC) at or near cancer mutation hot spots as potential targets for early cancer detection. BMC Res Notes 15:143. https://doi.org/10.1186/s13104-022-06028-w

Luo H, Chiang H-H, Louw M, Susanto A, Chen D (2017) Nutrient sensing and the oxidative stress response. Trends Endocrinol Metab 28:449–460. https://doi.org/10.1016/j.tem.2017.02.008

Madugundu GS, Cadet J, Wagner JR (2014) Hydroxyl-radical-induced oxidation of 5-methylcytosine in isolated and cellular DNA. Nucleic Acids Res 42:7450–7460. https://doi.org/10.1093/nar/gku334

Magierowska K, Korbut E, Wojcik-Grzybek D, Bakalarz D, Sliwowski Z, Cieszkowski J, Szetela M, Torregrossa R, Whiteman M, Magierowski M (2022) Mitochondria-targeted hydrogen sulfide donors versus acute oxidative gastric mucosal injury. J Control Release 348:321–334. https://doi.org/10.1016/j.jconrel.2022.05.051

Mahalingaiah PKS, Ponnusamy L, Singh KP (2016) Oxidative stress-induced epigenetic changes associated with malignant transformation of human kidney epithelial cells. Oncotarget 8(7):11127

Mair W, Piper MD, Partridge L (2005) Calories do not explain extension of life span by dietary restriction in Drosophila. PLoS Biol 3:e223. https://doi.org/10.1371/journal.pbio.0030223

Malaeb H, Choucair I, Wang Z, Li XS, Li L, Boyd WC, Hine C, Tang WHW, Gogonea V, Hazen SL (2022) Stable isotope dilution mass spectrometry quantification of hydrogen sulfide and thiols in biological matrices. Redox Biol 55:102401. https://doi.org/10.1016/j.redox.2022.102401

Marinac CR, Nelson SH, Breen CI, Hartman SJ, Natarajan L, Pierce JP, Flatt SW, Sears DD, Patterson RE (2016) Prolonged nightly fasting and breast cancer prognosis. JAMA Oncol 2:1049–1055. https://doi.org/10.1001/jamaoncol.2016.0164

Martincorena I, Fowler JC, Wabik A, Lawson ARJ, Abascal F, Hall MWJ, Cagan A, Murai K, Mahbubani K, Stratton MR, Fitzgerald RC, Handford PA, Campbell PJ, Saeb-Parsy K,

Jones PH (2018) Somatic mutant clones colonize the human esophagus with age. Science 362:911–917. https://doi.org/10.1126/science.aau3879
Martinez-Outschoorn UE, Lin Z, Trimmer C, Flomenberg N, Wang C, Pavlides S, Pestell RG, Howell A, Sotgia F, Lisanti MP (2011) Cancer cells metabolically "fertilize" the tumor microenvironment with hydrogen peroxide, driving the Warburg effect. Cell Cycle 10:2504–2520. https://doi.org/10.4161/cc.10.15.16585
Martin-Montalvo A, de Cabo R (2013) Mitochondrial metabolic reprogramming induced by calorie restriction. Antioxid Redox Signal 19:310–320. https://doi.org/10.1089/ars.2012.4866
Marui N, Offermann MK, Swerlick R, Kunsch C, Rosen CA, Ahmad M, Alexander RW, Medford RM (1993) Vascular cell adhesion molecule-1 (VCAM-1) gene transcription and expression are regulated through an antioxidant-sensitive mechanism in human vascular endothelial cells. J Clin Invest 92:1866–1874. https://doi.org/10.1172/JCI116778
Mattison JA, Roth GS, Beasley TM, Tilmont EM, Handy AM, Herbert RL, Longo DL, Allison DB, Young JE, Bryant M, Barnard D, Ward WF, Qi W, Ingram DK, de Cabo R (2012) Impact of caloric restriction on health and survival in rhesus monkeys from the NIA study. Nature 489:318–321. https://doi.org/10.1038/nature11432
Mattison JA, Colman RJ, Beasley TM, Allison DB, Kemnitz JW, Roth GS, Ingram DK, Weindruch R, de Cabo R, Anderson RM (2017) Caloric restriction improves health and survival of rhesus monkeys. Nat Commun 8:14063. https://doi.org/10.1038/ncomms14063
Mattson MP, Allison DB, Fontana L, Harvie M, Longo VD, Malaisse WJ, Mosley M, Notterpek L, Ravussin E, Scheer FA, Seyfried TN, Varady KA, Panda S (2014) Meal frequency and timing in health and disease. Proc Natl Acad Sci USA 111:16647–16653. https://doi.org/10.1073/pnas.1413965111
McCarter R, Masoro EJ, Yu BP (1985) Does food restriction retard aging by reducing the metabolic rate? Am J Phys 248:E488–E490. https://doi.org/10.1152/ajpendo.1985.248.4.E488
McCay CM, Crowell MF, Maynard LA (1989) The effect of retarded growth upon the length of life span and upon the ultimate body size. 1935. Nutrition 5:155–171. discussion 172
McCord JM, Fridovich I (1969) Superoxide dismutase. An enzymic function for erythrocuprein (hemocuprein). J Biol Chem 244:6049–6055
McDonald RB, Ramsey JJ (2010) Honoring Clive McCay and 75 years of calorie restriction research. J Nutr 140:1205–1210. https://doi.org/10.3945/jn.110.122804
Mei SC, Brenner C (2015) Calorie restriction-mediated replicative lifespan extension in yeast is non-cell autonomous. PLoS Biol 13:e1002048. https://doi.org/10.1371/journal.pbio.1002048
Mendelsohn AR, Larrick JW (2012) Fibroblast growth factor-21 is a promising dietary restriction mimetic. Rejuvenation Res 15:624–628. https://doi.org/10.1089/rej.2012.1392
Menger KE, James AM, Cocheme HM, Harbour ME, Chouchani ET, Ding S, Fearnley IM, Partridge L, Murphy MP (2015) Fasting, but not aging, dramatically alters the redox status of cysteine residues on proteins in Drosophila melanogaster. Cell Rep 13:1285. https://doi.org/10.1016/j.celrep.2015.10.048
Messaoudi I, Fischer M, Warner J, Park B, Mattison J, Ingram DK, Totonchy T, Mori M, Nikolich-Žugich J (2008) Optimal window of caloric restriction onset limits its beneficial impact on T-cell senescence in primates. Aging Cell 7:908–919. https://doi.org/10.1111/j.1474-9726.2008.00440.x
Miller RA, Buehner G, Chang Y, Harper JM, Sigler R, Smith-Wheelock M (2005) Methionine-deficient diet extends mouse lifespan, slows immune and lens aging, alters glucose, T4, IGF-I and insulin levels, and increases hepatocyte MIF levels and stress resistance. Aging Cell 4:119–125. https://doi.org/10.1111/j.1474-9726.2005.00152.x
Mitchell SJ, Madrigal-Matute J, Scheibye-Knudsen M, Fang E, Aon M, Gonzalez-Reyes JA, Cortassa S, Kaushik S, Gonzalez-Freire M, Patel B et al (2016) Effects of sex, strain, and energy intake on hallmarks of aging in mice. Cell Metab 23:1093–1112. https://doi.org/10.1016/j.cmet.2016.05.027
Mitchell SJ, Bernier M, Mattison JA, Aon MA, Kaiser TA, Anson RM, Ikeno Y, Anderson RM, Ingram DK, de Cabo R (2019) Daily fasting improves health and survival in male

mice independent of diet composition and calories. Cell Metab 29:221–228.e3. https://doi.org/10.1016/j.cmet.2018.08.011

Mohammad K, Titorenko VI (2021) Caloric restriction creates a metabolic pattern of chronological aging delay that in budding yeast differs from the metabolic design established by two other geroprotectors. Oncotarget 12:608–625. https://doi.org/10.18632/oncotarget.27926

Montégut L, de Cabo R, Zitvogel L, Kroemer G (2022) Science-driven nutritional interventions for the prevention and treatment of cancer. Cancer Discov 12:2258–2279. https://doi.org/10.1158/2159-8290.CD-22-0504

Moore A, Mercer J, Dutina G, Donahue CJ, Bauer KD, Mather JP, Etcheverry T, Ryll T (1997) Effects of temperature shift on cell cycle, apoptosis and nucleotide pools in CHO cell batch cultues. Cytotechnology 23:47–54. https://doi.org/10.1023/A:1007919921991

Moreschi C (1909) Beziehungen zwischen Ernahrung und Tumorwachstum. Ztschr f Inmunitits-forsch u exper Therap 2(651):675

Morscher RJ, Aminzadeh-Gohari S, Feichtinger RG, Mayr JA, Lang R, Neureiter D, Sperl W, Kofler B (2015) Inhibition of neuroblastoma tumor growth by ketogenic diet and/or calorie restriction in a CD1-nu mouse model. PLoS One 10:e0129802. https://doi.org/10.1371/journal.pone.0129802

Navarro A, Boveris A (2004) Rat brain and liver mitochondria develop oxidative stress and lose enzymatic activities on aging. Am J Phys Regul Integr Comp Phys 287:R1244–R1249. https://doi.org/10.1152/ajpregu.00226.2004

Ng HH, Yildiz GS, Ku JM, Miller AA, Woodman OL, Hart JL (2017) Chronic NaHS treatment decreases oxidative stress and improves endothelial function in diabetic mice. Diab Vasc Dis Res 14:246–253. https://doi.org/10.1177/1479164117692766

Nguyen T, Nioi P, Pickett CB (2009) The Nrf2-antioxidant response element signaling pathway and its activation by oxidative stress. J Biol Chem 284:13291–13295. https://doi.org/10.1074/jbc.R900010200

Niccoli T, Partridge L (2012) Ageing as a risk factor for disease. Curr Biol 22:R741–R752. https://doi.org/10.1016/j.cub.2012.07.024

O'Hagan HM, Wang W, Sen S, Destefano Shields C, Lee SS, Zhang YW, Clements EG, Cai Y, Van Neste L, Easwaran H, Casero RA, Sears CL, Baylin SB (2011) Oxidative damage targets complexes containing DNA methyltransferases, SIRT1, and polycomb members to promoter CpG Islands. Cancer Cell 20:606–619. https://doi.org/10.1016/j.ccr.2011.09.012

Ogawa T, Tsubakiyama R, Kanai M, Koyama T, Fujii T, Iefuji H, Soga T, Kume K, Miyakawa T, Hirata D, Mizunuma M (2016) Stimulating S-adenosyl-l-methionine synthesis extends lifespan via activation of AMPK. Proc Natl Acad Sci USA 113:11913–11918. https://doi.org/10.1073/pnas.1604047113

Ogawa T, Masumura K, Kohara Y, Kanai M, Soga T, Ohya Y, Blackwell TK, Mizunuma M (2022) S-adenosyl-L-homocysteine extends lifespan through methionine restriction effects. Aging Cell 21:e13604. https://doi.org/10.1111/acel.13604

Olas B (2017) Hydrogen sulfide as a "double-faced" compound: one with pro- and antioxidant effect. Adv Clin Chem 78:187–196. https://doi.org/10.1016/bs.acc.2016.07.005

Oluwole O, Ibidapo O, Arowosola T, Raji F, Zandonadi RP, Alasqah I, Lho LH, Han H, Raposo A (2023) Sustainable transformation agenda for enhanced global food and nutrition security: a narrative review. Front Nutr 10:1226538. https://doi.org/10.3389/fnut.2023.1226538

Orentreich N, Matias JR, DeFelice A, Zimmerman JA (1993) Low methionine ingestion by rats extends life span. J Nutr 123:269–274

Pan C, Zhou F, Zhang L (2023) The loss of epigenetic information: not only consequences but a cause of mammalian aging. Signal Transduct Target Ther 8:140. https://doi.org/10.1038/s41392-023-01412-9

Park SK, Link CD, Johnson TE (2010) Life-span extension by dietary restriction is mediated by NLP-7 signaling and coelomocyte endocytosis in *C. elegans*. FASEB J 24:383–392. https://doi.org/10.1096/fj.09-142984

Parkhitko AA, Wang L, Filine E, Jouandin P, Leshchiner D, Binari R, Asara JM, Rabinowitz JD, Perrimon N (2021) A genetic model of methionine restriction extends Drosophila health- and lifespan. Proc Natl Acad Sci USA 118. https://doi.org/10.1073/pnas.2110387118

Pastukh V, Roberts JT, Clark DW, Bardwell GC, Patel M, Al-Mehdi A-B, Borchert GM, Gillespie MN (2015) An oxidative DNA "damage" and repair mechanism localized in the VEGF promoter is important for hypoxia-induced VEGF mRNA expression. Am J Phys Lung Cell Mol Phys 309:L1367–L1375. https://doi.org/10.1152/ajplung.00236.2015

Peris E, Micallef P, Paul A, Palsdottir V, Enejder A, Bauza-Thorbrugge M, Olofsson CS, Wernstedt Asterholm I (2019) Antioxidant treatment induces reductive stress associated with mitochondrial dysfunction in adipocytes. J Biol Chem 294:2340–2352. https://doi.org/10.1074/jbc.RA118.004253

Pilati C, Shinde J, Alexandrov LB, Assié G, André T, Hélias-Rodzewicz Z, Ducoudray R, Le Corre D, Zucman-Rossi J, Emile J-F, Bertherat J, Letouzé E, Laurent-Puig P (2017) Mutational signature analysis identifies MUTYH deficiency in colorectal cancers and adrenocortical carcinomas. J Pathol 242:10–15. https://doi.org/10.1002/path.4880

Plummer JD, Johnson JE (2022) Intermittent methionine restriction reduces IGF-1 levels and produces similar healthspan benefits to continuous methionine restriction. Aging Cell 21:e13629. https://doi.org/10.1111/acel.13629

Podolskiy DI, Lobanov AV, Kryukov GV, Gladyshev VN (2016) Analysis of cancer genomes reveals basic features of human aging and its role in cancer development. Nat Commun 7:12157. https://doi.org/10.1038/ncomms12157

Pomatto-Watson LCD, Bodogai M, Bosompra O, Kato J, Wong S, Carpenter M, Duregon E, Chowdhury D, Krishna P, Ng S, Ragonnaud E, Salgado R, Gonzalez Ericsson P, Diaz-Ruiz A, Bernier M, Price NL, Biragyn A, Longo VD, de Cabo R (2021) Daily caloric restriction limits tumor growth more effectively than caloric cycling regardless of dietary composition. Nat Commun 12:6201. https://doi.org/10.1038/s41467-021-26431-4

Reczek CR, Chandel NS (2017) The two faces of reactive oxygen species in cancer. Annu Rev Cancer Biol 1:79–98. https://doi.org/10.1146/annurev-cancerbio-041916-065808

Redman LM, Ravussin E (2011) Caloric restriction in humans: impact on physiological, psychological, and behavioral outcomes. Antioxid Redox Signal 14:275–287. https://doi.org/10.1089/ars.2010.3253

Redman LM, Heilbronn LK, Martin CK, de Jonge L, Williamson DA, Delany JP, Ravussin E, for the Pennington, C.t. (2009) Metabolic and behavioral compensations in response to caloric restriction: implications for the maintenance of weight loss. PLoS One 4:e4377. https://doi.org/10.1371/journal.pone.0004377

Redman LM, Smith SR, Burton JH, Martin CK, Il'yasova D, Ravussin E (2018) Metabolic slowing and reduced oxidative damage with sustained caloric restriction support the rate of living and oxidative damage theories of aging. Cell Metab 27:805–815.e804. https://doi.org/10.1016/j.cmet.2018.02.019

Richardson NE, Konon EN, Schuster HS, Mitchell AT, Boyle C, Rodgers AC, Finke M, Haider LR, Yu D, Flores V, Pak HH, Ahmad S, Ahmed S, Radcliff A, Wu J, Williams EM, Abdi L, Sherman DS, Hacker T, Lamming DW (2021) Lifelong restriction of dietary branched-chain amino acids has sex-specific benefits for frailty and lifespan in mice. Nat Aging 1:73–86. https://doi.org/10.1038/s43587-020-00006-2

Richie JP, Leutzinger Y, Parthasarathy S, Malloy V, Orentreich N, Zimmerman JA (1994) Methionine restriction increases blood glutathione and longevity in F344 rats. FASEB J 8:1302–1307

Rickman AD, Williamson DA, Martin CK, Gilhooly CH, Stein RI, Bales CW, Roberts S, Das SK (2011) The CALERIE study: design and methods of an innovative 25% caloric restriction intervention. Contemp Clin Trials 32:874–881. https://doi.org/10.1016/j.cct.2011.07.002

Rikke BA, Yerg JE 3rd, Battaglia ME, Nagy TR, Allison DB, Johnson TE (2003) Strain variation in the response of body temperature to dietary restriction. Mech Ageing Dev 124:663–678. https://doi.org/10.1016/s0047-6374(03)00003-4

Robey IF, Lien AD, Welsh SJ, Baggett BK, Gillies RJ (2005) Hypoxia-inducible factor-1alpha and the glycolytic phenotype in tumors. Neoplasia 7:324–330. https://doi.org/10.1593/neo.04430

Rose Li Y, Halliwill KD, Adams CJ, Iyer V, Riva L, Mamunur R, Jen K-Y, del Rosario R, Fredlund E, Hirst G, Alexandrov LB, Adams D, Balmain A (2020) Mutational signatures in tumours induced by high and low energy radiation in Trp53 deficient mice. Nat Commun 11:394. https://doi.org/10.1038/s41467-019-14261-4

Rous P (1914) The influence of diet on transplanted and spontaneous mouse tumors. J Exp Med 20:433–451. https://doi.org/10.1084/jem.20.5.433

Safdie F, Brandhorst S, Wei M, Wang W, Lee C, Hwang S, Conti PS, Chen TC, Longo VD (2012) Fasting enhances the response of glioma to chemo- and radiotherapy. PLoS One 7:e44603. https://doi.org/10.1371/journal.pone.0044603

Saletta F, Dalla Pozza L, Byrne JA (2015) Genetic causes of cancer predisposition in children and adolescents. Transl Pediatr 4:67–75. https://doi.org/10.3978/j.issn.2224-4336.2015.04.08

Salmeen A, Andersen JN, Myers MP, Meng T-C, Hinks JA, Tonks NK, Barford D (2003) Redox regulation of protein tyrosine phosphatase 1B involves a sulphenyl-amide intermediate. Nature 423:769–773. https://doi.org/10.1038/nature01680

Salvadori G, Zanardi F, Iannelli F, Lobefaro R, Vernieri C, Longo VD (2021) Fasting-mimicking diet blocks triple-negative breast cancer and cancer stem cell escape. Cell Metab 33:2247–2259. e2246. https://doi.org/10.1016/j.cmet.2021.10.008

Sanz A, Caro P, Barja G (2004) Protein restriction without strong caloric restriction decreases mitochondrial oxygen radical production and oxidative DNA damage in rat liver. J Bioenerg Biomembr 36:545–552. https://doi.org/10.1007/s10863-004-9001-7

Sautin YY, Johnson RJ (2008) Uric acid: the oxidant-antioxidant paradox. Nucleosides Nucleotides Nucleic Acids 27:608–619. https://doi.org/10.1080/15257770802138558

Savitikadi P, Gogulothu R, Ismail A, Reddy GB, Reddy VS (2023) Combined prenatal to postnatal protein restriction augments protein quality control processes and proteolysis in the muscle of rat offspring. J Nutr Biochem 114:109273. https://doi.org/10.1016/j.jnutbio.2023.109273

Seddon AR, Liau Y, Pace PE, Miller AL, Das AB, Kennedy MA, Hampton MB, Stevens AJ (2021) Genome-wide impact of hydrogen peroxide on maintenance DNA methylation in replicating cells. Epigenetics Chromatin 14:17. https://doi.org/10.1186/s13072-021-00388-6

Seegmiller AC, Dobrosotskaya I, Goldstein JL, Ho YK, Brown MS, Rawson RB (2002) The SREBP pathway in Drosophila: regulation by palmitate, not sterols. Dev Cell 2:229–238. https://doi.org/10.1016/s1534-5807(01)00119-8

Seth D, Rudolph J (2006) Redox regulation of MAP kinase phosphatase 3. Biochemistry 45:8476–8487. https://doi.org/10.1021/bi060157p

Sharifi-Rad M, Anil Kumar NV, Zucca P, Varoni EM, Dini L, Panzarini E, Rajkovic J, Tsouh Fokou PV, Azzini E, Peluso I, Prakash Mishra A, Nigam M, El Rayess Y, Beyrouthy ME, Polito L, Iriti M, Martins N, Martorell M, Docea AO, Setzer WN, Calina D, Cho WC, Sharifi-Rad J (2020) Lifestyle, oxidative stress, and antioxidants: back and forth in the pathophysiology of chronic diseases. Front Physiol 11:694. https://doi.org/10.3389/fphys.2020.00694

Sharma B, Singh S, Kanwar SS (2014) L-methionase: a therapeutic enzyme to treat malignancies. Biomed Res Int 2014:506287. https://doi.org/10.1155/2014/506287

Sharma V, Collins LB, Chen TH, Herr N, Takeda S, Sun W, Swenberg JA, Nakamura J (2016) Oxidative stress at low levels can induce clustered DNA lesions leading to NHEJ mediated mutations. Oncotarget 7:25377–25390. https://doi.org/10.18632/oncotarget.8298

Siegel RL, Miller KD, Wagle NS, Jemal A (2023) Cancer statistics, 2023. CA Cancer J Clin 73:17–48. https://doi.org/10.3322/caac.21763

Silver DJ, Lathia JD, Hine C (2021a) Hydrogen sulfide operates as a glioblastoma suppressor and is lost under high fat diet. Mol Cell Oncol 8:1973312. https://doi.org/10.1080/23723556.2021.1973312

Silver DJ, Roversi GA, Bithi N, Wang SZ, Troike KM, Neumann CK, Ahuja GK, Reizes O, Brown JM, Hine C, Lathia JD (2021b) Severe consequences of a high-lipid diet include hydrogen

sulfide dysfunction and enhanced aggression in glioblastoma. J Clin Invest 131. https://doi.org/10.1172/jci138276
Sinenko SA, Starkova TY, Kuzmin AA, Tomilin AN (2021) Physiological signaling functions of reactive oxygen species in stem cells: from flies to man. Front Cell Dev Biol 9:714370. https://doi.org/10.3389/fcell.2021.714370
Smith DL Jr, McClure JM, Matecic M, Smith JS (2007) Calorie restriction extends the chronological lifespan of Saccharomyces cerevisiae independently of the Sirtuins. Aging Cell 6:649–662. https://doi.org/10.1111/j.1474-9726.2007.00326.x
Solaini G, Baracca A, Lenaz G, Sgarbi G (2010) Hypoxia and mitochondrial oxidative metabolism. Biochim Biophys Acta 1797:1171–1177. https://doi.org/10.1016/j.bbabio.2010.02.011
Solon-Biet SM, McMahon AC, Ballard JW, Ruohonen K, Wu LE, Cogger VC, Warren A, Huang X, Pichaud N, Melvin RG, Gokarn R, Khalil M, Turner N, Cooney GJ, Sinclair DA, Raubenheimer D, Le Couteur DG, Simpson SJ (2014) The ratio of macronutrients, not caloric intake, dictates cardiometabolic health, aging, and longevity in ad libitum-fed mice. Cell Metab 19:418–430. https://doi.org/10.1016/j.cmet.2014.02.009
Solon-Biet SM, Mitchell SJ, Coogan SC, Cogger VC, Gokarn R, McMahon AC, Raubenheimer D, de Cabo R, Simpson SJ, Le Couteur DG (2015) Dietary protein to carbohydrate ratio and caloric restriction: comparing metabolic outcomes in mice. Cell Rep 11:1529–1534. https://doi.org/10.1016/j.celrep.2015.05.007
Sun L, Sadighi Akha AA, Miller RA, Harper JM (2009) Life-span extension in mice by preweaning food restriction and by methionine restriction in middle age. J Gerontol A Biol Sci Med Sci 64:711–722. https://doi.org/10.1093/gerona/glp051
Sun X, Wang W, Dai J, Jin S, Huang J, Guo C, Wang C, Pang L, Wang Y (2017) A long-term and slow-releasing hydrogen sulfide donor protects against myocardial ischemia/reperfusion injury. Sci Rep 7:3541. https://doi.org/10.1038/s41598-017-03941-0
Tannenbaum A (1940) The initiation and growth of tumors: introduction. I. Effects of underfeeding1. Am J Cancer 38:335–350. https://doi.org/10.1158/ajc.1940.335
Tannenbaum A (1996) The dependence of tumor formation on the composition of the calorie-restricted diet as well as on the degree of restriction. Nutrition 12:653–655. https://doi.org/10.1016/0899-9007(96)90484-0
Tannenbaum A, Silverstone H (1949) The influence of the degree of caloric restriction on the formation of skin tumors and hepatomas in mice. Cancer Res 9:724–727
Tirza G, Solodeev I, Sela M, Greenberg I, Pasmanik-Chor M, Gur E, Shani N (2020) Reduced culture temperature attenuates oxidative stress and inflammatory response facilitating expansion and differentiation of adipose-derived stem cells. Stem Cell Res Ther 11:35. https://doi.org/10.1186/s13287-019-1542-0
Toullec A, Gerald D, Despouy G, Bourachot B, Cardon M, Lefort S, Richardson M, Rigaill G, Parrini M-C, Lucchesi C, Bellanger D, Stern M-H, Dubois T, Sastre-Garau X, Delattre O, Vincent-Salomon A, Mechta-Grigoriou F (2010) Oxidative stress promotes myofibroblast differentiation and tumour spreading. EMBO Mol Med 2:211–230. https://doi.org/10.1002/emmm.201000073
Tran KB, Lang JJ, Compton K, Xu R, Acheson AR, Henrikson HJ, Kocarnik JM, Penberthy L, Aali A, Abbas Q et al (2022) The global burden of cancer attributable to risk factors, 2010–19: a systematic analysis for the global burden of disease study 2019. Lancet 400:563–591. https://doi.org/10.1016/s0140-6736(22)01438-6
Tyshkovskiy A, Bozaykut P, Borodinova AA, Gerashchenko MV, Ables GP, Garratt M, Khaitovich P, Clish CB, Miller RA, Gladyshev VN (2019) Identification and application of gene expression signatures associated with lifespan extension. Cell Metab 30:573–593.e8. https://doi.org/10.1016/j.cmet.2019.06.018
Uchiyama J, Akiyama M, Hase K, Kumagai Y, Kim YG (2022) Gut microbiota reinforce host antioxidant capacity via the generation of reactive sulfur species. Cell Rep 38:110479. https://doi.org/10.1016/j.celrep.2022.110479

Vasconcellos LR, Dutra FF, Siqueira MS, Paula-Neto HA, Dahan J, Kiarely E, Carneiro LA, Bozza MT, Travassos LH (2016) Protein aggregation as a cellular response to oxidative stress induced by heme and iron. Proc Natl Acad Sci USA 113:E7474–E7482. https://doi.org/10.1073/pnas.1608928113

Vasquez-Vivar J, Kalyanaraman B, Kennedy MC (2000) Mitochondrial aconitase is a source of hydroxyl radical. An electron spin resonance investigation. J Biol Chem 275:14064–14069. https://doi.org/10.1074/jbc.275.19.14064

Vernieri C, Fucà G, Ligorio F, Huber V, Vingiani A, Iannelli F, Raimondi A, Rinchai D, Frigè G, Belfiore A, Lalli L, Chiodoni C, Cancila V, Zanardi F, Ajazi A, Cortellino S, Vallacchi V, Squarcina P, Cova A, Pesce S, Frati P, Mall R, Corsetto PA, Rizzo AM, Ferraris C, Folli S, Garassino MC, Capri G, Bianchi G, Colombo MP, Minucci S, Foiani M, Longo VD, Apolone G, Torri V, Pruneri G, Bedognetti D, Rivoltini L, de Braud F (2022) Fasting-mimicking diet is safe and reshapes metabolism and antitumor immunity in patients with cancer. Cancer Discov 12:90–107. https://doi.org/10.1158/2159-8290.Cd-21-0030

Vidoni C, Ferraresi A, Esposito A, Maheshwari C, Dhanasekaran DN, Mollace V, Isidoro C (2021) Calorie restriction for cancer prevention and therapy: mechanisms, expectations, and efficacy. J Cancer Prev 26:224–236. https://doi.org/10.15430/jcp.2021.26.4.224

Viel A, Bruselles A, Meccia E, Fornasarig M, Quaia M, Canzonieri V, Policicchio E, Urso ED, Agostini M, Genuardi M, Lucci-Cordisco E, Venesio T, Martayan A, Diodoro MG, Sanchez-Mete L, Stigliano V, Mazzei F, Grasso F, Giuliani A, Baiocchi M, Maestro R, Giannini G, Tartaglia M, Alexandrov LB, Bignami M (2017) A specific mutational signature associated with DNA 8-Oxoguanine persistence in MUTYH-defective colorectal cancer. EBioMedicine 20:39–49. https://doi.org/10.1016/j.ebiom.2017.04.022

Wahl D, LaRocca TJ (2021) Transcriptomic effects of Healthspan-promoting dietary interventions: current evidence and future directions. Front Nutr 8:712129. https://doi.org/10.3389/fnut.2021.712129

Walford RL, Harris SB, Gunion MW (1992) The calorically restricted low-fat nutrient-dense diet in biosphere 2 significantly lowers blood glucose, total leukocyte count, cholesterol, and blood pressure in humans. Proc Natl Acad Sci USA 89:11533–11537. https://doi.org/10.1073/pnas.89.23.11533

Walford RL, Mock D, Verdery R, MacCallum T (2002) Calorie restriction in biosphere 2: alterations in physiologic, hematologic, hormonal, and biochemical parameters in humans restricted for a 2-year period. J Gerontol A Biol Sci Med Sci 57:B211–B224. https://doi.org/10.1093/gerona/57.6.b211

Walsh ME, Shi Y, Van Remmen H (2014) The effects of dietary restriction on oxidative stress in rodents. Free Radic Biol Med 66:88–99. https://doi.org/10.1016/j.freeradbiomed.2013.05.037

Wang S, Song P, Zou MH (2012) AMP-activated protein kinase, stress responses and cardiovascular diseases. Clin Sci 122:555–573. https://doi.org/10.1042/CS20110625

Wang Y, Shi S, Dong S, Wu J, Song M, Zhong X, Liu Y (2015) Sodium hydrosulfide attenuates hyperhomocysteinemia rat myocardial injury through cardiac mitochondrial protection. Mol Cell Biochem 399:189–200. https://doi.org/10.1007/s11010-014-2245-6

Weeden CE, Hill W, Lim EL, Grönroos E, Swanton C (2023) Impact of risk factors on early cancer evolution. Cell 186:1541–1563. https://doi.org/10.1016/j.cell.2023.03.013

Weerapana E, Wang C, Simon GM, Richter F, Khare S, Dillon MBD, Bachovchin DA, Mowen K, Baker D, Cravatt BF (2010) Quantitative reactivity profiling predicts functional cysteines in proteomes. Nature 468:790–795. https://doi.org/10.1038/nature09472

Weidemann R, Ludwig A, Kretzmer G (1994) Low temperature cultivation--a step towards process optimisation. Cytotechnology 15:111–116. https://doi.org/10.1007/BF00762385

Weinberg F, Hamanaka R, Wheaton WW, Weinberg S, Joseph J, Lopez M, Kalyanaraman B, Mutlu GM, Budinger GRS, Chandel NS (2010) Mitochondrial metabolism and ROS generation are essential for Kras-mediated tumorigenicity. Proc Natl Acad Sci 107:8788–8793. https://doi.org/10.1073/pnas.1003428107

Weindruch R, Walford RL, Fligiel S, Guthrie D (1986) The retardation of aging in mice by dietary restriction: longevity, cancer, immunity and lifetime energy intake. J Nutr 116:641–654

Whatley SA, Curti D, Das Gupta F, Ferrier IN, Jones S, Taylor C, Marchbanks RM (1998) Superoxide, neuroleptics and the ubiquinone and cytochrome b5 reductases in brain and lymphocytes from normals and schizophrenic patients. Mol Psychiatry 3:227–237. https://doi.org/10.1038/sj.mp.4000375

White MC, Holman DM, Boehm JE, Peipins LA, Grossman M, Henley SJ (2014) Age and cancer risk: a potentially modifiable relationship. Am J Prev Med 46:S7–S15. https://doi.org/10.1016/j.amepre.2013.10.029

Wu D, Li M, Tian W, Wang S, Cui L, Li H, Wang H, Ji A, Li Y (2017) Hydrogen sulfide acts as a double-edged sword in human hepatocellular carcinoma cells through EGFR/ERK/MMP-2 and PTEN/AKT signaling pathways. Sci Rep 7:5134. https://doi.org/10.1038/s41598-017-05457-z

Xiao W, Loscalzo J (2020) Metabolic responses to reductive stress. Antioxid Redox Signal 32:1330–1347. https://doi.org/10.1089/ars.2019.7803

Xiao H, Jedrychowski MP, Schweppe DK, Huttlin EL, Yu Q, Heppner DE, Li J, Long J, Mills EL, Szpyt J, He Z, Du G, Garrity R, Reddy A, Vaites LP, Paulo JA, Zhang T, Gray NS, Gygi SP, Chouchani ET (2020) A quantitative tissue-specific landscape of protein redox regulation during aging. Cell 180:968–983. https://doi.org/10.1016/j.cell.2020.02.012

Xing F, Hu Q, Qin Y, Xu J, Zhang B, Yu X, Wang W (2022) The relationship of redox with hallmarks of cancer: the importance of homeostasis and context. Front Oncol 12:862743

Xiong Y, Uys JD, Tew KD, Townsend DM (2011) S-glutathionylation: from molecular mechanisms to health outcomes. Antioxid Redox Signal 15:233–270. https://doi.org/10.1089/ars.2010.3540

Xu H, Barnes GT, Yang Q, Tan G, Yang D, Chou CJ, Sole J, Nichols A, Ross JS, Tartaglia LA, Chen H (2003) Chronic inflammation in fat plays a crucial role in the development of obesity-related insulin resistance. J Clin Invest 112:1821–1830. https://doi.org/10.1172/JCI19451

Yang G, Zhao K, Ju Y, Mani S, Cao Q, Puukila S, Khaper N, Wu L, Wang R (2013) Hydrogen sulfide protects against cellular senescence via S-sulfhydration of Keap1 and activation of Nrf2. Antioxid Redox Signal 18:1906–1919. https://doi.org/10.1089/ars.2012.4645

Yang J, Minkler P, Grove D, Wang R, Willard B, Dweik R, Hine C (2019) Non-enzymatic hydrogen sulfide production from cysteine in blood is catalyzed by iron and vitamin B6. Commun Biol 2:194. https://doi.org/10.1038/s42003-019-0431-5

Yang JH, Hayano M, Griffin PT, Amorim JA, Bonkowski MS, Apostolides JK, Salfati EL, Blanchette M, Munding EM, Bhakta M et al (2023) Loss of epigenetic information as a cause of mammalian aging. Cell 186:305–326. https://doi.org/10.1016/j.cell.2022.12.027

Yap YW, Rusu PM, Chan AY, Fam BC, Jungmann A, Solon-Biet SM, Barlow CK, Creek DJ, Huang C, Schittenhelm RB, Morgan B, Schmoll D, Kiens B, Piper MDW, Heikenwalder M, Simpson SJ, Broer S, Andrikopoulos S, Muller OJ, Rose AJ (2020) Restriction of essential amino acids dictates the systemic metabolic response to dietary protein dilution. Nat Commun 11:2894. https://doi.org/10.1038/s41467-020-16568-z

Yasui H, Hayashi S, Sakurai H (2005) Possible involvement of singlet oxygen species as multiple oxidants in p450 catalytic reactions. Drug Metab Pharmacokinet 20:1–13. https://doi.org/10.2133/dmpk.20.1

Yoshino J, Mills KF, Yoon MJ, Imai S (2011) Nicotinamide mononucleotide, a key NAD(+) intermediate, treats the pathophysiology of diet- and age-induced diabetes in mice. Cell Metab 14:528–536. https://doi.org/10.1016/j.cmet.2011.08.014

Youngman LD, Park JY, Ames BN (1992) Protein oxidation associated with aging is reduced by dietary restriction of protein or calories. Proc Natl Acad Sci USA 89:9112–9116. https://doi.org/10.1073/pnas.89.19.9112

Zanco B, Mirth CK, Sgro CM, Piper MD (2021) A dietary sterol trade-off determines lifespan responses to dietary restriction in Drosophila melanogaster females. elife 10. https://doi.org/10.7554/eLife.62335

Zhang S, Li F, Zhou T, Wang G, Li Z (2020) Caenorhabditis elegans as a useful model for studying aging mutations. Front Endocrinol 11:554994. https://doi.org/10.3389/fendo.2020.554994
Ziech D, Franco R, Pappa A, Panayiotidis MI (2011) Reactive oxygen species (ROS)—induced genetic and epigenetic alterations in human carcinogenesis. Mutat Res 711:167–173. https://doi.org/10.1016/j.mrfmmm.2011.02.015
Zivanovic J, Kouroussis E, Kohl JB, Adhikari B, Bursac B, Schott-Roux S, Petrovic D, Miljkovic JL, Thomas-Lopez D, Jung Y, Miler M, Mitchell S, Milosevic V, Gomes JE, Benhar M, Gonzalez-Zorn B, Ivanovic-Burmazovic I, Torregrossa R, Mitchell JR, Whiteman M, Schwarz G, Snyder SH, Paul BD, Carroll KS, Filipovic MR (2019) Selective Persulfide detection reveals evolutionarily conserved antiaging effects of S-Sulfhydration. Cell Metab 30:1152–1170. https://doi.org/10.1016/j.cmet.2019.10.007
Zivanovic J, Kouroussis E, Kohl JB, Adhikari B, Bursac B, Schott-Roux S, Petrovic D, Miljkovic JL, Thomas-Lopez D, Jung Y, Miler M, Mitchell S, Milosevic V, Gomes JE, Benhar M, Gonzalez-Zorn B, Ivanovic-Burmazovic I, Torregrossa R, Mitchell JR, Whiteman M, Schwarz G, Snyder SH, Paul BD, Carroll KS, Filipovic MR (2020) Selective Persulfide detection reveals evolutionarily conserved antiaging effects of S-Sulfhydration. Cell Metab 31:207. https://doi.org/10.1016/j.cmet.2019.12.001
Zou K, Rouskin S, Dervishi K, McCormick MA, Sasikumar A, Deng C, Chen Z, Kaeberlein M, Brem RB, Polymenis M, Kennedy BK, Weissman JS, Zheng J, Ouyang Q, Li H (2020) Life span extension by glucose restriction is abrogated by methionine supplementation: cross-talk between glucose and methionine and implication of methionine as a key regulator of life span. Sci Adv 6:eaba1306. https://doi.org/10.1126/sciadv.aba1306

Chapter 8
The Role of Calorie Restriction in Modifying the Ageing Process through the Regulation of SIRT1 Expression

Monia Kittana, Vasso Apostolopoulos, and Lily Stojanovska

Abstract Calorie restriction (CR), as a dietary approach of reducing caloric intake while maintaining nutritional adequacy, has gained significant attention due to its potential role in promoting longevity and enhancing health. Central to the beneficial effects of CR is SIRT1. SIRT1 belongs to a family of NAD+ dependent deacetylases and plays an important role in regulating various cellular processes, including histone deacetylation, oxidative stress response, and mitochondrial biogenesis. This chapter reviews the evidence regarding the effect of CR on SIRT1 expression and mitochondrial biogenesis. Both pre-clinical and human studies have consistently demonstrated that CR promotes an increase in SIRT1 expression and activity in different tissues. This is also associated with other favourable health outcomes, such as delayed neurodegeneration and improved cognitive function. Moderate CR (25% restriction) has shown an impact on promoting mitochondrial biogenesis, reflected in markers such as mitochondrial DNA and transcription factors. However, this is reviewed in light of some methodological limitations, as data varied in response to different CR regimens. Herein, we highlight the potential of CR in up-regulating SIRT1 and promoting mitochondrial biogenesis, which can have significant implications for developing strategies to manage and promote healthy ageing.

Keywords Calorie restriction · Aging · Ageing · Sirtuins · SIRT1 · Mitochondrial biogenesis · Healthy ageing

---

M. Kittana
Department of Nutrition and Health, College of Medicine and Health Sciences, United Arab Emirates University, Al Ain, United Arab Emirates
e-mail: 202090016@uaeu.ac.ae

V. Apostolopoulos (✉) · L. Stojanovska
Institute for Health and Sport, Victoria University, Melbourne, VIC, Australia
e-mail: vasso.apostolopoulos@vu.edu.au; lily.stojanovska@vu.edu.au

© The Author(s), under exclusive license to Springer Nature Switzerland AG 2024

V. I. Korolchuk, J. R. Harris (eds.), *Biochemistry and Cell Biology of Ageing: Part V, Anti-Ageing Interventions*, Subcellular Biochemistry 107,
https://doi.org/10.1007/978-3-031-66768-8_8

Introduction

Calorie restriction (CR) refers to the dietary method of reducing food intake, while maintaining adequate nutrition. Various beneficial effects of CR have been shown in mammals, specifically with regards to increasing the lifespan, linked with the function of a group proteins referred to as Sirtuins (Cohen et al. 2009). Sirtuins are a group of proteins that mainly function as NAD+-dependent deacetylases (Davenport et al. 2014), as they catalyse the reaction of removing an acetyl group (CH_3CO), and transferring it to ADP-ribose, forming O-acetyl-ADP-ribose (Dang 2014); a molecule which plays a role in several functions, such as heterochromatin formation and silencing, and modulating the cellular redox state (Dang 2014), as it has been linked with metabolic shifts reducing endogenous reactive oxygen species (ROS), preventing oxidative damage (Tong and Denu 2010).

There are seven sirtuins in mammals; SIRT1 though SIRT7 (Dang 2014; Wątroba et al. 2017). Sirtuins are either mitochondrial proteins (SIRT3, 4, and 5), or nuclear enzymes (SIRT1, 6, and 7), or both (SIRT2) (Wątroba et al. 2017). Although they differ in their profiles of action, they all share the same catalytic reaction and their use of NAD+ as a co-substrate (Wątroba et al. 2017). Of these sirtuins, SIRT1 is the most extensively studied (Dang 2014; Wątroba et al. 2017). SIRT1 proteins have been suggested to play an important role in promoting health and longevity. Through the studies of epigenetics, calorie restriction (CR) had been found to regulate the expression of SIRT1, leading to the consequent effect of delaying ageing in various species, such as yeast, worms, mice, and, more recently, humans. SIRT1's main function is that it deacetylases histones (proteins that associate with the DNA and help it to condense to chromatin), and this histone deacetylation affects many epigenetic and non-epigenetic targets (Nevoral et al. 2019), which increased the interest in SIRT1 in its role in influencing cell viability by regulating functions related to the cell cycle, and oxidative stress response (Nevoral et al. 2019). One of the main pathways related to ageing is the effect SIRT1 produces in regulating mitochondrial biogenesis. Mitochondria play an important role in regulating ageing due to its production of ROS as a by-product of the electron transport chain (ETC) (Guarente 2008). ROS damage macromolecules (lipids, protein, and DNA) progressively over time, these include mitochondrial constituents as well due to their proximity to the ROS (Guarente 2008). This eventually leads to a decline in cell function (Guarente 2008). With advanced aged, there is a decrease in the rate of mitochondrial biogenesis (Chistiakov et al. 2014), however the process of mitochondrial biogenesis is critical for maintaining energy production and reducing endogenous oxidative process, thus promoting healthy ageing (López-Lluch et al. 2008).

Through this process, SIRT1 plays an important role in DNA damage prevention through regulating oxidative stress, and it also promotes DNA damage repair (Wątroba et al. 2017). Thus, it gained attention due its role in promoting longevity (Singh et al. 2018). Mitochondrial biogenesis, a keyword regarding the role of SIRT1, is a process that maintains the number and size of mitochondria, including its genome (Bhatti et al. 2017). SIRT1 deacetylases and activates Peroxisome proliferator-activated receptor coactivator-1α (PGC-1α), a regulator of mitochondrial biogenesis, in an energy deprived state. PGC-1α interacts with various transcription factors, which regulate the transcription of mitochondrial enzymes and the synthesis of mitochondrial DNA (Bhatti et al. 2017). Thus, sirtuins may now serve as targets in the management of diseases that are regulated by the cellular redox state (Singh et al. 2018). Many studies attributed several metabolic conditions (insulin resistance, obesity, stroke, and heart disease) to a change in the structure and number of mitochondria, partly due to ROS accumulation (Bhatti et al. 2017). Several factors are increasingly being identified as sirtuins-activators, such as resveratrol, which has been investigated for its positive effect on cell viability due to the activating of SIRT1, mimicking the effect of the energy-deprived state (Nevoral et al. 2019). Many other factors such as some polyphenols, and other artificial, higher-potency stimulators have also been investigated for the same purpose (Villalba and Alcaín 2012). Interestingly, sirtuins, including SIRT1, are also activated by moderate under-nutrition, which is achieved through a moderate calorie deficit, which stimulates mitochondrial biogenesis. Increased biogenesis can reduce the effect of mitochondrial dysfunction (Kauppila et al. 2017). The aim of this article is to critically review the evidence regarding this association.

The Effects of Calorie Restriction

CR stands as one of the promising approaches in longevity and ageing research, as it modulates various molecular and cellular pathways that contribute to delaying the onset of ageing-related diseases and extending a healthy lifespan. One of the pathways, the up-regulation of SIRT1 expression and levels, has been documented in studies based on animal models, showing significant increases in SIRT1 expression, which was also associated with other positive health outcomes such as improved cognitive ability (Geng et al. 2011) and neurological function (Ran et al. 2015). Translating CR's potential for human longevity is challenging considering its implementation strategy and sustainability. Nonetheless, findings from recent human studies (from 2005 to 2018) suggest that moderate CR provides beneficial effects in regulating SIRT1 levels and promoting mitochondrial biogenesis. The studies applied either a randomized controlled-trial or a quasi-experiment study design. The number of participants was relatively small, and therefore the results are interpreted carefully, as it ranged from only 4 to 36 participants in the CR groups. Different methods of CR were used, including moderate approaches such as 25% CR from baseline, 500–1000 kcal reduction from baseline intake, and alternate day fasting, in addition to more restrictive approaches including limiting caloric intake to 1000 kcal/day and fasting for 6 days. The following section will review the evidence regarding SIRT1 expression and levels, and mitochondrial biogenesis changes in pre-clinical studies and human studies. Figure 8.1 provides a schematic representation of the interplay between calorie restriction and SIRT1 in the process of ageing.

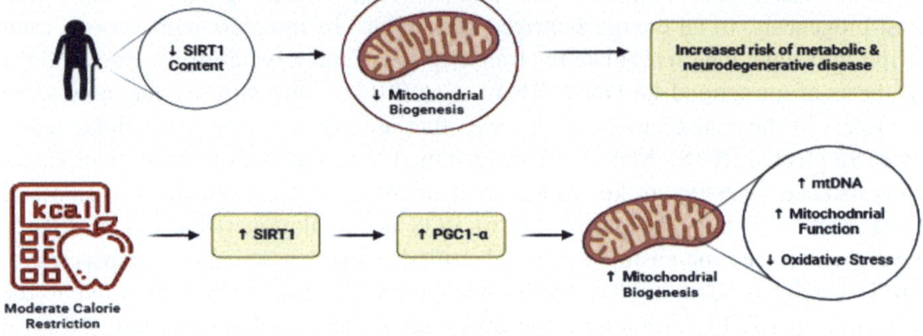

Fig. 8.1 Schematic representation of the interplay between calorie restriction, SIRT1, and mitochondrial biogenesis

Evidence from Pre-Clinical Studies on Calorie Restriction

Evidence from earlier studies establishes CR as a strategy of extending the lifespan of virtually all organisms. CR's induction of SIRT1 is postulated to be an evolutionary response which increases the long-term function and survival of critical cell types (Cohen et al. 2004). With ageing, SIRT1 levels are significantly reduced, which may be associated with the physiological changes that accompanies ageing. For example, SIRT1 protein content in the hippocampus was shown to decrease by 41% in aged Wistar male rats, which was a significant reduction as compared to young rats (Quintas et al. 2012). Similar protein expression declines and SIRT1 activity down-regulation was observed in the adipose tissue of aged mice (Xu et al. 2014). Due to SIRT1's inhibitory action on Nuclear factor kappa-B (NF-κB) signalling, its decreased expression is thought to contribute to the activated inflammatory responses in the ageing brain (Quintas et al. 2012). On the other hand, noticeable consequences of increased SIRT1 expression following CR included a delayed onset of neurodegeneration (Gräff et al. 2013), improved neurological function score (Ran et al. 2015), and a slowed decline in cognitive ability (Geng et al. 2011).

Animal studies have shown CR to activate SIRT1 and lead to an increase in SIRT1 expression and SIRT1 activity in different tissues (Cohen et al. 2004; Kume et al. 2010; Chen et al. 2010; Gräff et al. 2013; Yu et al. 2014; Xu et al. 2014; Ma et al. 2015; Li et al. 2017; Ding et al. 2017). For example, SIRT1 protein content in the hippocampus was significantly increased by 223% following a moderate CR regimen of 80% of *ad libitum*-fed mates (Quintas et al. 2012). In diabetic rats, which initially show a much lower expression rates of SIRT1, CR was found to significantly increase both mRNA and protein expressions of SIRT1 in islet beta cells and decreased the apoptosis ratio of islet beta cells, leading to notable increases in insulin sensitivity (Deng et al. 2010). Pre-clinical studies were conducted to elucidate the mechanism of SIRT1 as a mediator for CR's physiological response. It was observed that in SIRT1-null mice, the energy generation system was defective, indicated by the presence of hypermetabolism and lethargy (Boily et al. 2008). When challenged with a 40% CR, SIRT1-null mice were unable to adapt and reduced their metabolic rate, as compared to their normal counterparts, which maintained their metabolic rate, and increased their physical activity (Boily et al. 2008). The up-regulation of physical activity was previously demonstrated as a well-established behavioural consequence of CR, which may be affected by SIRT1 activity in the brain in mediating the changes observed in physical activity and behaviour (Cohen et al. 2009).

Evidence from Human Studies on Calorie Restriction

SIRT1 mRNA Expression Levels

Human studies of CR show an overall trend of increased expression of SIRT1 mRNA in various body tissues. mRNA expression was reported in 6 studies, 4 of which reported statistically significant increases (Heilbronn et al. 2005; Civitarese et al. 2007; Pedersen et al. 2008; Rappou et al. 2016). Longer-term studies (≥6 months) showed significant increases in mRNA expressions, and both of these studies applied moderate CR approaches (Civitarese et al. 2007; Rappou et al. 2016). However, the results in short-term studies (6–30 days) were controversial. Rappou et al. (2016) reported that mRNA levels were higher at baseline in the reference group, who were lean compared with the obese volunteers. Interestingly, after follow-up period of 12 months, some participants regained their weight and others maintained their loss. mRNA levels were higher in the weight loss maintainers, leading to the conclusion that mRNA inversely follows the trend of BMI (Rappou et al. 2016). BMI is significantly correlated with CR, and therefore CR and weight loss should be maintained for preserving SIRT1 expression, as it is not maintained after reverting back to the baseline diet.

Regarding SIRT1 levels, three studies reported SIRT1 levels, all of which showed a significant increase. Despite the reported insignificant changes in SIRT1 mRNA expression in Mansur et al. (2017) and Kitada et al. (2013), both studies showed a significant increase in SIRT1 protein levels. It was previously reported that SIRT1 increases may be due to protein stabilisation, and not due to higher genetic expression. This hypothesis was also presented by Kitada et al., as they showed that phosphorylated SIRT1 was higher after CR as compared to baseline levels. Post-transcriptional phosphorylation is essential for the regulation of SIRT1 enzymatic activity (Sasaki et al. 2008), which consequently leads to the desired effects in regulating cellular functions such as oxidative stress. Mansur et al. (2017) only reported SIRT1 as serum levels, but the authors provided the same justification for the unmatched SIRT1 mRNA expression and serum levels results (Mansur et al. 2017). These studies did have limitations in their design. Kitada et al. only recruited 4 volunteers, and therefore the effects should be interpreted carefully. Mansur et al. followed a stricter approach of only 1000 kcal/day. A recent animal-model study concluded that more positive results were shown with moderate approaches (25% CR) compared with more strict approaches (45% CR) (Yu et al. 2014). Moreover, moderate approaches may also be easier to adhere to on the longer-term, especially in today's modern society, where adhering to a very restricted CR regimen may prove to be challenging.

Mitochondrial Biogenesis

The effect of CR on mitochondrial biogenesis, mediated by the increased expression of SIRT1, was indirectly measured in 3 studies by reporting data for mitochondrial DNA (mtDNA) which signifies higher mitochondrial mass indicating a larger number of mitochondria. Mitochondrial Transcription Factor A (TFAM), which signifies higher transcription of mtDNA, PGC1-α, which suggests an indication of mitochondrial biogenesis, and mitochondrial enzymes such as citrate synthase, beta-hydroxyl-CoA dehydrogenase, and cytochrome-C oxidase also serve as markers for mitochondrial mass. PGC1-α and mtDNA were reported in three studies, two of which showed a significant increase following CR (Civitarese et al. 2007; Kitada et al. 2013). Surprisingly, only one study reported a decrease in mtDNA (Heilbronn et al. 2005). However, the type of CR used in this study was alternate day fasting, and while it did restrict caloric intake during fasting, on feasting days the energy intake was not controlled, and subjects were instructed to eat *ad libitum*. Moreover, subjects that were recruited were non-obese at baseline (BMI: 20–30 kg/m^2), and mtDNA decrease was proportional to the weight loss achieved. Therefore, authors suggested that CR may reduce mtDNA in non-obese subjects. In contrast, the studies that did report an increase in mtDNA both used a 25% CR from baseline. One study explored the changes in TFAM, which was significantly increased (Civitarese et al. 2007). This is consistent with their findings of increased mtDNA. Increased mitochondrial mass in response to CR is essential, as lower mass is associated with increased workload, which may lead to a higher production of ROS and promoting ageing. Civitarese et al. showed in their study that DNA damage was actually reduced after CR, however, they did not specifically define how DNA damage is evaluated (Civitarese et al. 2007).

Regarding mitochondrial enzymes, it is expected that enzyme activity increases with increased mitochondrial mass. However, two studies assessed the changes in citrate synthase, beta-hydroxyl-CoA dehydrogenase, and cytochrome-C oxidase, all of which were statistically insignificant [20,21]. Civitarese et al. concluded mitochondrial mass may be better assessed using mtDNA and other non-enzymatic markers. Previously, it was also reported that weight loss does not affect oxidative enzyme activity (Kempen et al. 1998).

Conclusion

The ageing process and controlling it is one of the main interests of the field of epigenetics. With a growing ageing population, a need to investigate factors that influence ageing and longevity control has been more pressing. One promising factor is calorie restriction, and its promotion of an energy-deprived state in mammals. There has been compelling evidence from animal studies, but relatively limited evidence from human studies. Through this review, human studies of CR show an overall trend of increased SIRT1 expression and levels, which through upregulating mitochondrial biogenesis and boosting mitochondrial activity, may progressively reduce ROS damage and preserve cellular function. The evidence is inconclusive due to the inconsistencies in the methods of delivering CR interventions, however, moderate approaches are promising in achieving results in the longer-term. Longer-term longitudinal studies are required to review the effect of calorie restriction in upregulating SIRT1 expression in human subjects, in order to promote the development of preventive therapeutic interventions to slow down ageing in the future.

References

Bhatti JS, Bhatti GK, Reddy PH (2017) Mitochondrial dysfunction and oxidative stress in metabolic disorders - a step towards mitochondria based therapeutic strategies. Biochim Biophys Acta Mol basis Dis 1863:1066–1077. https://doi.org/10.1016/j.bbadis.2016.11.010

Boily G, Seifert EL, Bevilacqua L et al (2008) SirT1 regulates energy metabolism and response to caloric restriction in mice. PLoS One 3:e1759. https://doi.org/10.1371/journal.pone.0001759

Chen Y-R, Fang S-R, Fu Y-C et al (2010) Calorie restriction on insulin resistance and expression of SIRT1 and SIRT4 in rats. This paper is one of a selection of papers published in this special issue entitled "Second international symposium on recent advances in basic, clinical, and social medicine" and has undergone the Journal's usual peer review process. Biochem Cell Biol 88:715–722. https://doi.org/10.1139/O10-010

Chistiakov DA, Sobenin IA, Revin VV et al (2014) Mitochondrial aging and age-related dysfunction of mitochondria. Biomed Res Int 2014:1–7. https://doi.org/10.1155/2014/238463

Civitarese AE, Carling S, Heilbronn LK et al (2007) Calorie restriction increases muscle mitochondrial biogenesis in healthy humans. PLoS Med 4:e76. https://doi.org/10.1371/journal.pmed.0040076

Cohen HY, Miller C, Bitterman KJ et al (2004) Calorie restriction promotes mammalian cell survival by inducing the SIRT1 deacetylase. Science 305:390–392. https://doi.org/10.1126/science.1099196

Cohen DE, Supinski AM, Bonkowski MS et al (2009) Neuronal SIRT1 regulates endocrine and behavioral responses to calorie restriction. Genes Dev 23:2812–2817. https://doi.org/10.1101/gad.1839209

Dang W (2014) The controversial world of sirtuins. Drug Discov Today Technol 12:e9–e17. https://doi.org/10.1016/j.ddtec.2012.08.003

Davenport AM, Huber FM, Hoelz A (2014) Structural and functional analysis of human SIRT1. J Mol Biol 426:526–541. https://doi.org/10.1016/j.jmb.2013.10.009

Deng X, Cheng J, Zhang Y et al (2010) Effects of caloric restriction on SIRT1 expression and apoptosis of islet beta cells in type 2 diabetic rats. Acta Diabetol 47:177–185. https://doi.org/10.1007/s00592-009-0159-7

Ding S, Jiang J, Zhang G et al (2017) Resveratrol and caloric restriction prevent hepatic steatosis by regulating SIRT1-autophagy pathway and alleviating endoplasmic reticulum stress in high-fat diet-fed rats. PLoS One 12:e0183541. https://doi.org/10.1371/journal.pone.0183541

Geng Y-Q, Li T-T, Liu X-Y et al (2011) SIRT1 and SIRT5 activity expression and behavioral responses to calorie restriction. J Cell Biochem 112:3755–3761. https://doi.org/10.1002/jcb.23315

Gräff J, Kahn M, Samiei A et al (2013) A dietary regimen of caloric restriction or pharmacological activation of SIRT1 to delay the onset of neurodegeneration. J Neurosci 33:8951–8960. https://doi.org/10.1523/JNEUROSCI.5657-12.2013

Guarente L (2008) Mitochondria—a nexus for aging, calorie restriction, and sirtuins? Cell 132:171–176. https://doi.org/10.1016/j.cell.2008.01.007

Heilbronn LK, Civitarese AE, Bogacka I et al (2005) Glucose tolerance and skeletal muscle gene expression in response to alternate day fasting. Obes Res 13:574–581. https://doi.org/10.1038/oby.2005.61

Kauppila TES, Kauppila JHK, Larsson N-G (2017) Mammalian mitochondria and aging: an update. Cell Metab 25:57–71. https://doi.org/10.1016/j.cmet.2016.09.017

Kempen KP, Saris WH, Kuipers H et al (1998) Skeletal muscle metabolic characteristics before and after energy restriction in human obesity: fibre type, enzymatic beta-oxidative capacity and fatty acid-binding protein content. Eur J Clin Investig 28:1030–1037. https://doi.org/10.1046/j.1365-2362.1998.00383.x

Kitada M, Kume S, Takeda-Watanabe A et al (2013) Calorie restriction in overweight males ameliorates obesity-related metabolic alterations and cellular adaptations through anti-aging effects, possibly including AMPK and SIRT1 activation. Biochim Biophys Acta 1830:4820–4827. https://doi.org/10.1016/j.bbagen.2013.06.014

Kume S, Uzu T, Horiike K et al (2010) Calorie restriction enhances cell adaptation to hypoxia through Sirt1-dependent mitochondrial autophagy in mouse aged kidney. J Clin Invest 120:1043–1055. https://doi.org/10.1172/JCI41376

Li J, Zhang C-X, Liu Y-M et al (2017) A comparative study of anti-aging properties and mechanism: resveratrol and caloric restriction. Oncotarget 8:65717–65729. https://doi.org/10.18632/oncotarget.20084

López-Lluch G, Irusta PM, Navas P, de Cabo R (2008) Mitochondrial biogenesis and healthy aging. Exp Gerontol 43:813–819. https://doi.org/10.1016/j.exger.2008.06.014

Ma L, Dong W, Wang R et al (2015) Effect of caloric restriction on the SIRT1/mTOR signaling pathways in senile mice. Brain Res Bull 116:67–72. https://doi.org/10.1016/j.brainresbull.2015.06.004

Mansur AP, Roggerio A, Goes MFS et al (2017) Serum concentrations and gene expression of sirtuin 1 in healthy and slightly overweight subjects after caloric restriction or resveratrol supplementation: a randomized trial. Int J Cardiol 227:788–794. https://doi.org/10.1016/j.ijcard.2016.10.058

Nevoral J, Landsmann L, Stiavnicka M et al (2019) Epigenetic and non-epigenetic mode of SIRT1 action during oocyte meiosis progression. J Anim Sci Biotechnol 10:67. https://doi.org/10.1186/s40104-019-0372-3

Pedersen SB, Ølholm J, Paulsen SK et al (2008) Low Sirt1 expression, which is upregulated by fasting, in human adipose tissue from obese women. Int J Obes 32:1250–1255. https://doi.org/10.1038/ijo.2008.78

Quintas A, de Solís AJ, Díez-Guerra FJ et al (2012) Age-associated decrease of SIRT1 expression in rat hippocampus: prevention by late onset caloric restriction. Exp Gerontol 47:198–201. https://doi.org/10.1016/j.exger.2011.11.010

Ran M, Li Z, Yang L et al (2015) Calorie restriction attenuates cerebral ischemic injury via increasing SIRT1 synthesis in the rat. Brain Res 1610:61–68. https://doi.org/10.1016/j.brainres.2015.03.043

Rappou E, Jukarainen S, Rinnankoski-Tuikka R et al (2016) Weight loss is associated with increased NAD(+)/SIRT1 expression but reduced PARP activity in white adipose tissue. J Clin Endocrinol Metab 101:1263–1273. https://doi.org/10.1210/jc.2015-3054

Sasaki T, Maier B, Koclega KD et al (2008) Phosphorylation regulates SIRT1 function. PLoS One 3:e4020. https://doi.org/10.1371/journal.pone.0004020

Singh CK, Chhabra G, Ndiaye MA et al (2018) The role of sirtuins in antioxidant and redox signaling. Antioxid Redox Signal 28:643–661. https://doi.org/10.1089/ars.2017.7290

Tong L, Denu JM (2010) Function and metabolism of sirtuin metabolite O-acetyl-ADP-ribose. Biochim Biophys Acta 1804:1617–1625. https://doi.org/10.1016/j.bbapap.2010.02.007

Villalba JM, Alcaín FJ (2012) Sirtuin activators and inhibitors. Biofactors 38:349–359. https://doi.org/10.1002/biof.1032

Wątroba M, Dudek I, Skoda M et al (2017) Sirtuins, epigenetics and longevity. Ageing Res Rev 40:11–19. https://doi.org/10.1016/j.arr.2017.08.001

Xu C, Cai Y, Fan P et al (2014) Calorie restriction prevents metabolic aging caused by abnormal SIRT1 function in adipose tissues. Diabetes 64:1576–1590. https://doi.org/10.2337/db14-1180

Yu W, Zhou H-F, Lin R-B et al (2014) Short-term calorie restriction activates SIRT1-4 and -7 in cardiomyocytes in vivo and in vitro. Mol Med Rep 9:1218–1224. https://doi.org/10.3892/mmr.2014.1944

Chapter 9
Resveratrol and Its Analogues: Anti-ageing Effects and Underlying Mechanisms

Dan-Dan Zhou, Jin Cheng, Jiahui Li, Si-Xia Wu, Ruo-Gu Xiong, Si-Yu Huang, Peter Chi-Keung Cheung, and Hua-Bin Li

Abstract Ageing is a natural process accompanied by functional and structural decline of diverse tissues and organs, which could cause susceptibility to various diseases and death. The anti-ageing interventions have aroused huge research interest with the rapid rise of ageing population in the world. Resveratrol, a polyphenolic stilbene, could be naturally isolated from various plants, such as grapes, blueberries, and peanuts. Many studies indicated that resveratrol possessed a broad spectrum of bioactivities, especially anti-ageing activity. A lot of attention has also been focused on resveratrol analogues because they have a similar structure to resveratrol, which may confer them a potent anti-ageing effect. The anti-ageing mechanisms of resveratrol and its analogues are complex and multifactorial, involving suppressing oxidative stress, ameliorating inflammation, activating SIRT1 pathway, reducing DNA damage, etc. In this chapter, the anti-ageing effects of resveratrol and its analogues are summarised with special attention paid to the underlying mechanisms. Further

D.-D. Zhou
Food & Nutritional Sciences Program, School of Life Sciences, Chinese University of Hong Kong, Hong Kong, China

School of Public Health, Sun Yat-Sen University, Guangzhou, China
e-mail: 1155203812@link.cuhk.edu.hk

J. Cheng · S.-X. Wu · R.-G. Xiong · S.-Y. Huang · H.-B. Li (✉)
School of Public Health, Sun Yat-Sen University, Guangzhou, China
e-mail: chengj225@mail2.sysu.edu.cn; wusx6@mail2.sysu.edu.cn; xiongrg@mail2.sysu.edu.cn; huangsy9@mail2.sysu.edu.cn; lihuabin@mail.sysu.edu.cn

J. Li
School of Chinese Medicine, Li Ka Shing Faculty of Medicine, The University of Hong Kong, Hong Kong, China
e-mail: lijiahui@hku.hk

P. C.-K. Cheung
Food & Nutritional Sciences Program, School of Life Sciences, Chinese University of Hong Kong, Hong Kong, China
e-mail: petercheung@cuhk.edu.hk

© The Author(s), under exclusive license to Springer Nature Switzerland AG 2024

V. I. Korolchuk, J. R. Harris (eds.), *Biochemistry and Cell Biology of Ageing: Part V, Anti-Ageing Interventions*, Subcellular Biochemistry 107,
https://doi.org/10.1007/978-3-031-66768-8_9

understanding of these small molecules could provide the necessary scientific basis for their development into anti-ageing agents.

Keywords Resveratrol · Pterostilbene · Piceatannol · Ageing · Longevity · Anti-ageing

Introduction

According to the data from the World Population Prospects 2022, the current global population is estimated to be around 8 billion with the population aged 65 and above taking up a 10% share of the whole population, which would soar to 16% in 2050 (United Nations 2023). Addressing the issue of the ever-increasing ageing population has become incredibly urgent for the world in recent decades. In this context, finding the ways to delay human ageing, improve the quality of life of the elderly, and reduce the occurrence of age-related diseases in the ageing population has aroused tremendous interest of researchers. Ageing is a natural process characterised by functional impairments and greater vulnerability to death, which could be caused by the accumulation of oxidative injury to macromolecules (DNA, lipids, proteins) by reactive oxygen species (ROS)/reactive nitrogen species (RNS) (Zhou et al. 2021). In addition, a new update of the hallmarks of ageing published in *Cell* has proposed twelve ageing hallmarks, including genomic instability, epigenetic alterations, deregulated nutrient-sensing, telomere attrition, loss of proteostasis, disabled macroautophagy, mitochondrial dysfunction, altered intercellular communication, cellular senescence, stem cell exhaustion, chronic inflammation, and dysbiosis (Lopez-Otin et al. 2023). Indeed, ageing is caused by a vast array of factors, which are interconnected and interdependent with each other. Each of the factors should be regarded as a point of entry for future research on the ageing process, and also for the exploitation of new anti-ageing agents. Noticeably, bioactive compounds from natural products with great antioxidant and anti-inflammatory activities could be developed into effective anti-ageing agents.

Resveratrol is one of the most studied compounds in stilbene family, which are a group of polyphenol substances with the same 1,2-stilbene nucleus (Wu et al. 2022b; Zhou et al. 2021). Resveratrol is a naturally synthesized phytoalexin found in many plants (such as grapes, berries, peanuts, and soy), which could protect plants from various environmental stressors and pathogenic attacks, such as UV irradiation and fungal infection (Guo et al. 2021; Zhou et al. 2022). Besides, many studies showed that resveratrol could exhibit a variety of pharmacological effects and health benefits, involving anti-oxidation, anti-inflammation, cardio-protection, neuroprotection, hepatoprotection, anti-diabetes, anti-obesity, anti-cancer, and anti-ageing (Guo et al. 2021; Zhou et al. 2021). The anti-ageing mechanisms underlying the promising anti-ageing activity of resveratrol are complex and multifactorial, including ameliorating oxidative stress, activating sirtuin1 (SIRT1) protein, reducing DNA damage and so on (Hosoda et al. 2023; Zhou et al. 2021). Besides,

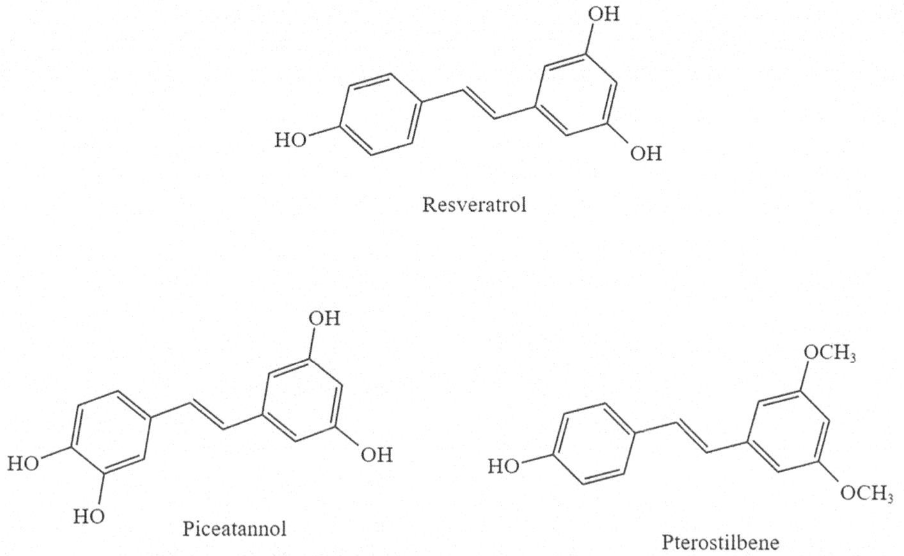

Fig. 9.1 Chemical structures of resveratrol (trans-3,4′,5-trihydroxystilbene) and its representative analogues piceatannol (trans-3,3′,4′,5-tetrahydroxystilbene) as well as pterostilbene (trans-3,5-dimethoxystilbene)

growing research interest also focused on the anti-ageing effects of resveratrol analogues as they have similar structure to resveratrol, which may confer important biological activities on them (Lephart and Andrus 2017; Naini et al. 2019). Here, the structures of resveratrol and some of its representative analogues are presented in Fig. 9.1.

In this chapter, the anti-ageing effects of resveratrol and its analogues are summarised based on the recent results with an emphasis on the mechanisms of action. It is anticipated that this chapter could provide the necessary scientific basis for the application of resveratrol and its analogues in anti-ageing interventions.

Anti-ageing Effects of Resveratrol and Its Analogues

Resveratrol and its analogues have shown a variety of beneficial effects, among which anti-ageing effects have aroused huge interest of the scientific community (Table 9.1). A study found that resveratrol significantly prolonged the lifespan of both female and male silkworms with mean lifespan and maximum lifespan elongating by 1.52 days (3.18%) and 1.20 days (2.31%) in females and 1.57 days (3.24%) and 2.13 days (3.89%) in males, respectively. This increased lifespan was found to be associated with the antioxidant activity of resveratrol via the SIRT7-forkhead box protein O (FOXO)-glutathione S-transferase (GST) signalling

Table 9.1 Anti-ageing effects and mechanisms of resveratrol and its analogues

Name	Study type	Model & Treatment	Effects and Mechanisms	Ref.
Amelioration of Oxidative Stress				
Resveratrol	*In vitro*	Aged oocytes: 2 μmol/L for 24 h or 48 h	Delayed postovulatory ageing through modulating oxidative stress (increased the gene expression of CAT, GPX and SOD1)	Abbasi et al. (2021)
Resveratrol	*In vivo*	Wistar male rats: 10 mg/kg, cannula administration for 18 months	Decreased ROS and LPO; Improved deficits of locomotor activity, short-term and long-term recognition memory in aged rats	Juarez et al. (2023)
Resveratrol	*In vivo*	6-month-old Sprague-Dawley rats: 1.25 mg/day for 5 months	Improved age-related cognitive performance via increasing cerebral blood flow and downregulating NO and ROS production in the brain	Garrigue et al. (2021)
Resveratrol	*In vivo, in vitro*	D-galactose-induced ageing mice TCMK-1 cells: 22, 44, and 88 μM	Increased the cell viability and repaired the kidney damage and brain damage of ageing mice via decreasing MDA and enhancing SOD and CAT activities	Chu et al. (2021)
Structurally modified resveratrol analogues (butyrate, isobutyrate, palmitoate, acetate, and diacetate)	*In vitro*	Epiderm full thickness skin cultures: 1%	Butyrate could most significantly inhibit the levels of ageing biomarker S100 calcium-binding protein A8 and increase the levels of anti-ageing biomarker TGFB1; Acetate and diacetate had the most consistent stimulatory influence on CAT, SOD, MT1H, and MT2H	Lephart and Andrus (2017)
Pterostilbene	*In vivo*	Male and Female *Drosophila Melanogaster*: 50, 100 and 200 μM	Increased the average lifespan of both genders with a higher effect in females; Inhibited oxidative stress and up-regulated the antioxidant enzymes Ho e Trxr-1only in male flies	Beghelli et al. (2022)

(continued)

Table 9.1 (continued)

Name	Study type	Model & Treatment	Effects and Mechanisms	Ref.
Pterostilbene	*In vivo*	Wild-type flies (w1118 outcrossed with OregonR): 0, 50, 100, 200, 250, 300, 400, and 500 μM	Reversed the negative effects of AgNPs on longevity and age-related function by decreasing ROS and activating antioxidant pathways	Chen et al. (2021)
Piceatannol	*In vivo*	D-galactose induced ageing Kunming mice: 20 mg/kg, gavage for 8 weeks	Alleviated behavioural and neurological deficits in ageing mice; Enhanced SOD and CAT activity in the hippocampal and cortical tissues, as well as reduced MDA	Zhang et al. (2018)
Piceatannol	*In vivo*	*Caenorhabditis elegans*: 50 and 100 μM	Delayed the age-related decline of pumping rate and locomotive activity, and protected the worms from oxidative stress via DAF-16	Shen et al. (2017)
Depression of Inflammation				
Resveratrol	*In vitro*	Human mononuclear cells from middle aged (40–59 years old) and elderly (60–80 years old) donors: 5 μM	Improved the inflammatory profile (TNF α, IL 6 and IL 10), but the effect was more obvious in middle-aged group	Santos et al. (2023)
Resveratrol	*In vivo*	Young and old male Wistar rats: 10 mg/kg for 10 weeks	Reverted age-related changes in inflammatory (INF γ and TNF α) markers in the rat heart.	Torregrosa-Munumer et al. (2021)
Resveratrol	*In vivo*	6-month-old Sprague-Dawley rats: 1.25 mg/day for 5 months	Improved age-related cognitive performance decline via increasing cerebral blood flow and downregulating several pathways (Eicosanoid signalling, MIF-mediated innate immunity, NF-κB signalling, TNFR2 signalling, and IL 6 signalling) in the brain	Garrigue et al. (2021)

(continued)

Table 9.1 (continued)

Name	Study type	Model & Treatment	Effects and Mechanisms	Ref.
Resveratrol	*In vivo*	Annual short-lived fishes: 200 μg/g food for 2, 5, and 8 months	Ameliorated age-related structure/function degeneration of thymus and kidney, which could be related to the reduction the levels of NF-κB and inflammatory factors with the up-regulation of SIRT1/ SIRT3 expression	Hou et al. (2023)
Resveratrol	*In vivo*	Old (20 months old) male Wistar rats: 20 mg/kg, intraperitoneal injection for 28 days	Improved episodic-like memory and motor coordination through modulating NF-κB neuroinflammation signalling pathways in the hippocampus	Sarubbo et al. (2023)
Structurally modified resveratrol analogues (butyrate, isobutyrate, palmitate, acetate, and diacetate)	*In vitro*	Epiderm full thickness skin cultures: 1%	Butyrate could most significantly inhibit the level of ageing biomarker S100 calcium-binding protein A8 and increase the level of anti-ageing biomarker TGFB1; Inhibited the gene expression of IL1A, IL1R2, IL 6 and IL 8	Lephart and Andrus (2017)
Pterostilbene	*In vivo*	Male and Female *Drosophila Melanogaster*: 50, 100 and 200 μM	Increased the average lifespan of both genders with a higher effect in females; Reduced the inflammatory mediators dome and egr only in female flies	Beghelli et al. (2022)
Pterostilbene	*In vitro*	PM-induced HaCaT cell line: 5–80 μM	Alleviated PM-induced skin inflammation and ageing partly through inhibiting inflammation (COX-2) and ageing (MMP-9) protein cascades	Teng et al. (2021)
Activation of SIRT1				
Resveratrol	*In vivo*	Short-lived fish: 200 μg/g food for 2, 5, and 8 months	Ameliorated age-related structure/function degeneration of thymus and kidney via up-regulating SIRT1/SIRT3 expression	Hou et al. (2023)

(continued)

Table 9.1 (continued)

Name	Study type	Model & Treatment	Effects and Mechanisms	Ref.
Resveratrol	*In vitro*	H_2O_2-induced bone marrow mesenchymal stem cells (pre-mature senescence model): 1, 5, and 10 μM	Inhibited premature senescence partially via regulating RELA/SIRT1	He et al. (2022)
Resveratrol	*In vivo*	28-week-old male ddY mice: 0.4 g/kg diet for 32 weeks	Attenuated ageing-associated adverse changes in heart and skeletal muscle via a SIRT1 activation pathway	Hosoda et al. (2023)
Resveratrol	*In vivo, in vitro*	Zebrafish: 1 μM BV2 microglia cells and PC12 rat pheochromocytoma cells: 20 μM;	Alleviated 27HC-induced senescence and locomotor behaviour disorder and ageing via regulating SIRT1-mediated STAT3 signalling	Liu et al. (2021)
Resveratrol	*In vivo*	Female Sprague Dawley rats: 20 mg/kg/day for 45 days	Delayed ovarian ageing possibly via decreasing oxidative damage and enhancing SIRT1	Wu et al. (2022a)
Resveratrol	*In vivo, in vitro*	*Nothobranchius guentheri*: 200 μg/g food for 3, 6, or 9 months; HEK293T cells: 20 mM	Activated SIRT1/Nrf2 to suppress inflammation and ER stress, and finally delayed ovarian ageing	Zhu et al. (2023)
Pterostilbene and pterostilbene nicotinate	*In vitro, ex vivo*	Ang II- or H_2O_2-treated Human umbilical vein endothelial cells: 1, 10, 100 and 1000 nM; Aortic rings from Sprague–Dawley rats: 100 and 1000 nM	Alleviated vascular endothelial senescence and induced endothelium-dependent relaxations via activation of SIRT1	Zhang et al. (2021)
IMS analogues	*In silico, in vitro*	Yeast model: 25–100 μM, measured at days 0 and 6; Human chronic myeloid leukaemia K562 cell line: 2.5–25 μM, measured at days 4 and 5; Molecular docking analysis with SIRT1	Displayed binding affinity towards SIRT1; Extended the chronological life span	Naini et al. (2019)
Regulation of Mitochondrial Function				

(continued)

Table 9.1 (continued)

Name	Study type	Model & Treatment	Effects and Mechanisms	Ref.
Resveratrol	*In vivo*	Female ICR mice: 50 mg/kg BW/day, intraperitoneal injection for 15 days	Improved mitochondria function; Protected against postovulatory oocyte ageing	Liang et al. (2018)
Resveratrol	*In vivo*	Young and old zebrafish: 20 mg/L for 1 or 10 days	Promoted mitochondrial function and prevent ageing	Wang et al. (2019)
Resveratrol	Pilot trial	Old adults with functional limitations: 500 and 1000 mg/day combined with exercise for 12 weeks	The combination improved skeletal muscle mitochondrial function and mobility-related indices of physical function	Harper et al. (2021)
Pterostilbene	*In silico*	Hierarchical cell simulation design and agent-based modelling: 100 μM	Significantly slowed the rate of natural degeneration and ageing, which was presumably related to the mitochondrial dynamics and function	Hoffman et al. (2017)
Inhibition of Apoptosis				
Resveratrol	*In vivo*	Young and old male Wistar rats: 10 mg/kg for 10 weeks	Reverted age-related markers in rat heart partly by reducing the expression levels of AIF, Bcl-2, and XIAP	Torregrosa-Munumer et al. (2021)
Resveratrol	*In vivo*	D-galactose-induced ageing C57BL/6 mice: 40 mg/kg, gavage for 8 weeks	Ameliorated AGEs-related renal dysfunction in ageing mice via improving renal cellular senescence, apoptosis, and fibrosis	Lan et al. (2023)
Resveratrol and copper	*In vivo*	C57BL/6 mice: 1 mg/kg of RSV and 0.1 μg/kg of Cu twice daily, oral gavage for 12 months	Reduced the ageing hallmarks in brain cells, involving apoptosis	Pal et al. (2022)
Resveratrol	*In vivo*	D-galactose-induced ageing male ICR mice: 25, 50, 100 mg/kg/day, gavage for 6 weeks	Reduced cellular senescence, cellular proliferation and apoptosis in the thymus via increasing the expression of FoxN1	Wei et al. (2021)

(continued)

Table 9.1 (continued)

Name	Study type	Model & Treatment	Effects and Mechanisms	Ref.
Piceatannol	*In vitro*	HaCaT cells: 4.88 μg/mL for 24 h	Showed the protective effect against UVB-induced skin ageing partially by decreasing the levels of apoptosis-related proteins (Bim, Bax, and cleaved caspase-7)	Jeong et al. (2023)
Reduction of DNA Damage				
Resveratrol	*In vitro*	Human embryonic kidney HEK293 cell line: 100 μM	Showed anti-ageing potential by protecting cells from MNNG-induced DNA damage and rescuing effects of NAD^+ precursors	Yang et al. (2021)
Resveratrol	*In vivo*	Terc−/− mice: 1 mg/kg for 3 months	Delayed ageing-associated lung degenerative changes partly through blocking DNA damage of parenchymal cells	Navarro et al. (2017)
Resveratrol and copper	*In vivo*	C57BL/6 mice: 1 mg/kg of RSV and 0.1 μg/kg of Cu twice daily, oral gavage for 12 months	Reduced the ageing hallmarks in brain cells, involving DNA damage	Pal et al. (2022)
Piceatannol	*In vitro*	HaCaT cells: 4.88 μg/mL for 24 h	Reduced the levels of CPD as well as DNA damage response proteins (such as p53, p-p53, p21), and exhibited potential in protecting from ultraviolet B-induced skin ageing	Jeong et al. (2023)
Maintenance of Telomeres				
Resveratrol	*In vivo*	Mice aged 10 months: 1 mg/kg for 10 months (combined with 0.1 mu g/kg of Cu)	Reduced the ageing hallmarks in brain cells, involving telomere attrition	Pal et al. (2022)
Resveratrol	*In vivo*	Male Wistar rats: 0.15% mg/kg diet	Delayed the vascular ageing partly through enhancing telomerase activity and increasing telomere length in the aorta	da Luz et al. (2012)
Resveratrol	*In vivo*	Female mice: 30 mg/L, for 6 or 12 months (drinking water)	Protected against ovarian ageing and age-associated infertility partially through keeping telomerase activity the telomere length resembled to young counterpart	Liu et al. (2013)

(continued)

Table 9.1 (continued)

Name	Study type	Model & Treatment	Effects and Mechanisms	Ref.
Other Mechanisms				
Resveratrol	*In vivo*	D-galactose-induced ageing male ICR mice: 25, 50, and100 mg/kg/day, gavage for 6 weeks	Reduced cellular senescence of the thymus and enhanced immunity function with the involvement of FoxN1 pathway	Wei et al. (2021)
Piceatannol	*In vitro*	Mesenchymal stromal cells: 0.001, 0.01, 0.1, 1, and 10 μM	Decreased the number of senescent cells, and promoted cell proliferation and the stemness property of MSCs	Alessio et al. (2021)
Oxyresveratrol	*In vivo, in vitro*	*Caenorhabditis elegans*: 0.5, 1.0, and 2.0 mg/mL Murine melanoma B16 cells: 1.25, 2.5, 5, 10 and 15 μM	Prolonged the lifespan of *Caenorhabditis elegans*; Reduced melanin production by inhibiting the tyrosinase activity; Inhibited age pigment production	Li et al. (2020)
Resveratrol	*In vivo*	D-galactose-induced ageing C57BL/6 mice: 40 mg/kg, gavage for 8 weeks	Decreased the accumulation of AGEs, and ameliorated renal dysfunction	Lan et al. (2023)
Pterostilbene	*In vivo*	18-month-old aged Fischer rats: 0.004% and 0.016% for 12–13 weeks	Improved age-related cognitive behavioural deficits and dopamine release; Working memory was associated with pterostilbene level in the hippocampus	Joseph et al. (2008)
Pterostilbene	Clinical trial, *in vitro*	Healthy volunteers: one fingertip unit uniformly on complete face twice daily for 8 weeks; Mouse melanoma B16F1 cells	Inhibited melanogenesis; Reduced wrinkles and fine lines, as well as improved skin hydration elasticity; Decreased the ageing markers	Majeed et al. (2020)

Abbreviation: AGE, advanced glycation end-product; AIF, apoptosis-inducing factor; Bcl-2, B-cell lymphoma 2; CAT, catalase; COX-2, cyclooxygenases 2; CPD, cyclo-butane pyrimidine dimer; ER, endoplasmic reticulum; FoxN1, transcription factors forkhead box protein N1; GPX, glutathione peroxidase; HO-1, heme oxygenase-1; IL 1β, interleukin-1β; IL1A, interleukin1 alpha; IL1R2, interleukin 1 receptor 2; IL 6, interleukin 6; IL 8, interleukin 8; IMS, imine stilbene; INF-γ, Interferon-gamma; MDA, malondialdehyde; MIF, macrophage migration inhibitory factor; MMP-9, matrix metalloproteinase 9; MNNG, N-methyl-N′-nitro-N nitrosoguanidine; MSCs, mesenchymal stromal cells; MT1H, metallothionein 1H; MT2H, metallothionein 2H; NAD^+, nicotinamide adenine dinucleotide; NF-κB, nuclear factor kappa B; NO, nitric oxide; Nrf2, nuclear factor

(continued)

Table 9.1 (continued)

erythroid 2-related factor 2; RELA, v-rel avian reticuloendotheliosis viral oncogene homolog A; p-p53, phospho-p53; PM, particulate matter; ROS, reactive oxygen species; RV, resveratrol; SIRT, sirtuin; SOD, superoxide dismutase; STAT3, signal transducer and activator of transcription 3; TGFB1, transforming growth factor beta1; TNFR2, tumour necrosis factor receptor 2; TNF α, tumour necrosis factor α; XIAP, X-linked inhibitor of apoptosis protein; 27HC, 27-hydroxycholesterol

pathway (Song et al. 2021). Another study showed that resveratrol and oxyresveratrol could significantly prolong the lifespan of *Caenorhabditis elegans* by 5% and 6.82%, respectively (Li et al. 2020). Moreover, resveratrol could significantly improve the viability of aged cells and the pathological conditions of ageing mice through inhibiting the formation of malondialdehyde (MDA) and increasing the activities of superoxide dismutase (SOD) and catalase (CAT) (Chu et al. 2021). In addition, resveratrol delayed the ovarian ageing in killifish *Nothobranchius guentheri* by activating SIRT1/nuclear factor erythroid 2-related factor 2 (Nrf2) pathway to suppress inflammation and endoplasmic reticulum (ER) stress (Zhu et al. 2023).

As for resveratrol analogues, a study found that imine stilbene analogues of resveratrol could extend the chronological lifespan of yeast and enhance cell viability of mammalian cells, which might be associated with the activity of SIRT1 protein (Naini et al. 2019). Another study found that pterostilbene reversed the negative effects of silver nanoparticles (AgNPs) on longevity and age-related function in *Drosophila*, e.g., shortened lifespan as well as the loss of intestinal integrity and stress resistance, which might be achieved by the reduction of ROS and the upregulation of antioxidant pathways (Chen et al. 2021). Beghelli et al. also evaluated the anti-ageing effects of pterostilbene in *Drosophila*, and found that pterostilbene increased the average lifespan of both male and female flies, accompanied with enhancing expressions of the longevity-related genes including SIRT2, FOXO, and Notch in both sexes but with different patterns (Beghelli et al. 2022). The study also indicated that the modulation of oxidative stress and proteins was significantly different in male and female flies treated with pterostilbene, which indicated that the anti-ageing mechanisms underlying pterostilbene intervention might be different in different sexes (Beghelli et al. 2022). In addition, a study compared the effects of resveratrol and five structurally modified resveratrol analogues (butyrate, isobutyrate, palmitate, acetate, and diacetate) on human skin gene expression, and found that butyrate could most significantly suppress the levels of ageing biomarker S100 calcium-binding protein A8 and increase the levels of anti-ageing biomarker transforming growth factor beta1 (TGFB1), indicating the potential of structurally modified resveratrol analogues (butyrate) as an anti-ageing intervention (Lephart and Andrus 2017).

Anti-ageing Mechanisms of Resveratrol and Its Analogues

As mentioned above, resveratrol and its analogues have promising potential to develop into anti-ageing agents. However, the mechanisms of resveratrol and its analogues underlying their effects on lifespan are complex due to the multi-factorial ageing process. To further elucidate the anti-ageing mechanisms of resveratrol and its analogues, we have collected and reviewed the recently published papers, and the related mechanisms are discussed in detail below (Table 9.1 and Fig. 9.2).

Amelioration of Oxidative Stress

Resveratrol and its analogues have been widely known for their strong antioxidant activities for a long time, and a plethora of studies have revealed that antioxidant activity played a crucial role in the anti-ageing effect of resveratrol and its analogues (Bohara et al. 2022; Gu et al. 2021; Zhou et al. 2021). For example, an *in vitro* study with aged oocytes indicated that resveratrol could delay postovulatory ageing through modulating oxidative stress, like increasing the gene expression of CAT, glutathione peroxidase (GPX) and SOD1 (Abbasi et al. 2021). In addition, a study showed that resveratrol increased the cell viability and recovered the kidney and brain damage in a D-galactose-induced ageing mouse model partly via decreasing MDA and enhancing SOD and CAT activities (Chu et al. 2021). For resveratrol analogues, pterostilbene reduced the production of ROS and upregulated antioxidant pathways to revert the AgNPs-induced negative effects on longevity and age-related functions in *Drosophila*, such as shortened lifespan and loss of stress resistance (Chen et al. 2021). Moreover, piceatannol alleviated behavioural disorder and brain damage in an ageing mouse model induced by D-galactose via decreasing MDA level, elevating SOD and CAT activities in the hippocampal and cortical tissues, and activating the Nrf2 pathway (Zhang et al. 2018).

Depression of Inflammation

Systemic chronic inflammation is a prominent characteristic of ageing, and resveratrol as well as its analogues could exert anti-ageing effect via targeting inflammation (Baechle et al. 2023; Saavedra et al. 2023; Zhou et al. 2021). A study found that long-term resveratrol treatment (1.25 mg/day for 5 months) could improve ageing-associated cognitive decline in the elderly male rat model, accompanied by the increase of cerebral blood flow and down-regulation of several pro-inflammatory pathways in the brain, such as eicosanoid signalling, MIF-mediated innate immunity, NF-κB signalling, dendritic cell maturation, tumour necrosis factor receptor 2 (TNFR2) signalling, interleukin (IL) 6 signalling, and production of NO and ROS

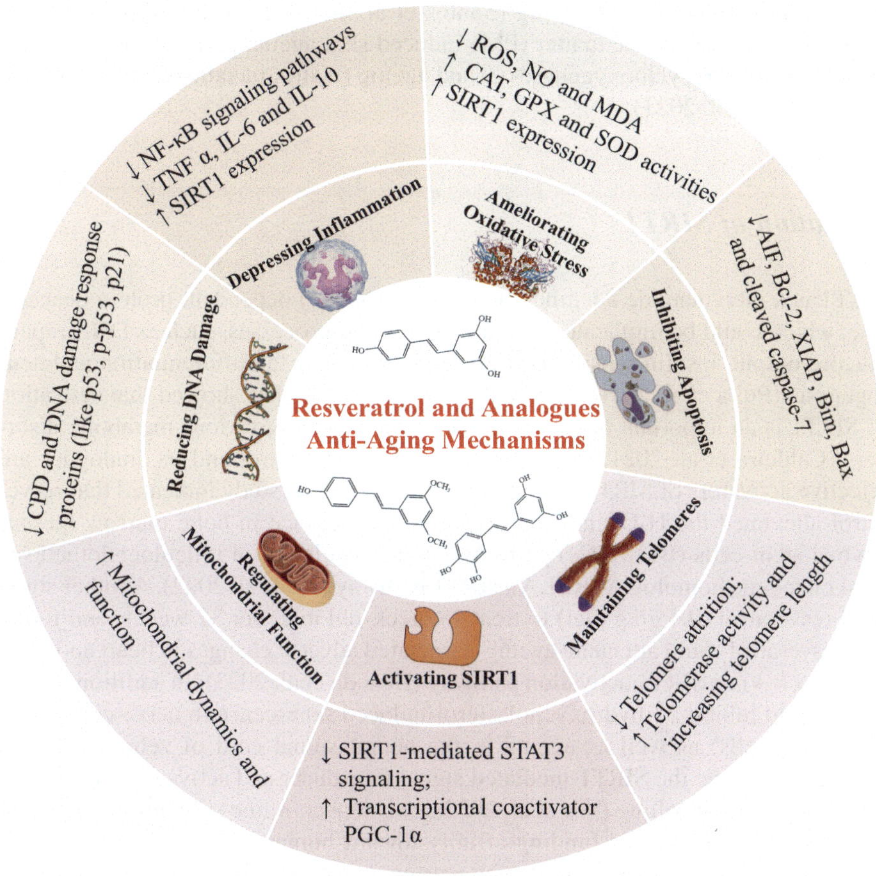

Fig. 9.2 Underlying mechanisms related to the anti-ageing effects of resveratrol and its analogues. AIF, apoptosis-inducing factor; Bax, Bcl-2-associated X protein; Bcl-2, B-cell lymphoma 2; CAT, catalase; CPD, cyclo-butane pyrimidine dimer; GPX, glutathione peroxidase; IL, interleukin; MDA, malondialdehyde; NF-κB, nuclear factor kappa B; PGC-1α, peroxisome proliferator-activated receptor-gamma coactivator-1 alpha; ROS, reactive oxygen species; SOD, superoxide dismutase; SIRT, sirtuin; STAT3, signal transducer and activator of transcription 3; TNF α, tumour necrosis factor α; XIAP, X-linked inhibitor of apoptosis protein

(Garrigue et al. 2021). Another study found that resveratrol ameliorated age-related structure/function degeneration of thymus and kidney in the short-lived killifish of the genus *Nothobranchius*, which might be achieved by the inhibition of inflammation with the up-regulation of SIRT1/SIRT3 expression (Hou et al. 2023). Moreover, a study showed that resveratrol could significantly improve the profile of inflammatory cytokines, such as tumour necrosis factor α (TNF α), IL 6 and IL 10, in human mononuclear cells (PBMCs) from both middle-aged (40–59 years old) and elderly (60–80 years old) donors. The effect was more pronounced in the middle-aged group, which indicated that it is better to consume resveratrol at an earlier time to suppress inflammation and ageing (Santos et al. 2023). In addition, pterostilbene could alleviate particulate matter (PM)-induced skin ageing partly through inhibiting inflammation (cyclooxygenases 2) and ageing (matrix metalloproteinase 9) cascades (Teng et al. 2021).

Activation of SIRT1

SIRT1 is a nicotinamide adenine dinucleotide (NAD^+) dependent protein deacetylase, which could be implicated in multiple cellular processes, such as DNA repair, glucose output, insulin sensitivity, fatty acid oxidation, fat differentiation and neurogenesis (Pulla et al. 2014; Zhu et al. 2023). Many studies showed that activation of SIRT1 is an important target for longevity and treating various metabolic disorders (Caldeira et al. 2021; Pratiwi et al. 2019). Resveratrol and its analogues are effective activators of SIRT1 to intervene with ageing. A study indicated that resveratrol alleviated the H_2O_2-induced premature senescence in bone marrow mesenchymal stem cells (BMMSCs) partially via the regulation of reticuloendotheliosis viral oncogene homolog A (RELA)/SIRT1 pathway (He et al. 2022). Another study used resveratrol (0.4 g/kg diet) to treat 28-week-old mice for 32 weeks, and found that resveratrol could attenuate ageing-associated adverse changes in heart and skeletal muscle via a SIRT1 activation pathway (Hosoda et al. 2023). In addition, resveratrol could inhibit 27-hydroxycholesterol-induced senescence in nerve cells (BV2 and PC12 cells) as well as ageing in the neural spinal cord of zebrafish larvae through activating the SIRT1-mediated signal transducer and activator of transcription 3 (STAT3) signalling (Wu et al. 2022a). Moreover, imine stilbene analogues of resveratrol could enhance binding affinity toward human SIRT1 protein *in silico* and exhibit anti-ageing activity, which was evidenced by the extension of chronological lifespan in yeast and enhancement of cell viability of mammalian cells (Naini et al. 2019).

Regulation of Mitochondrial Function

Mitochondria are among the key regulators of longevity and ageing because mitochondria are involved in various cellular or molecular activities, such as energy metabolism, oxidative stress, and cell fate decisions (Gherardi et al. 2022). Resveratrol could exert anti-ageing effect via targeting mitochondrial function and metabolism. For example, a study indicated that mitochondrial dysfunction (measured by mitochondrial DNA integrity, copy number, fusion/fission, and mitophagy/autophagy) is a critical player of ageing in zebrafish retina, and treatment of resveratrol could improve mitochondrial function and prevent ageing in zebrafish retina (Wang et al. 2019). Another study showed that the intraperitoneal injection of resveratrol (50 mg/kg/day for 15 consecutive days) to female mice aged 7–8 months could delay the apoptosis and ageing of postovulatory oocytes, which could be achieved by preserving mitochondrial function and reducing oxidative stress (Liang et al. 2018). In addition, based on the results from an *in silico* approach, pterostilbene could significantly slow the rate of natural degeneration and ageing, which was presumably related to the mitochondrial dynamics and function (Hoffman et al. 2017).

Inhibition of Apoptosis

Apoptosis is a critical mechanism to regulate cell survival and tissue/organ homeostasis and could also be an important ageing-regulatory pathway (Luo et al. 2022; Zhou et al. 2021). A study showed that resveratrol could reverse renal dysfunction in a D-galactose-induced ageing mouse model via improving the apoptosis, senescence and fibrosis of renal cells (Lan et al. 2023). Moreover, supplementation with resveratrol for 10 weeks in old male rats reverted ageing-related markers in rat heart partly by reducing the expression levels of apoptosis-inducing factor (AIF), B-cell lymphoma 2 (Bcl-2), and X-linked inhibitor of apoptosis protein (XIAP) (Torregrosa-Munumer et al. 2021). In addition, piceatannol showed protective effect against ultraviolet B-induced skin ageing partially by decreasing the levels of apoptosis-related proteins (Bim, Bax, and cleaved caspase-7) (Jeong et al. 2023).

Reduction of DNA Damage

Genomic instability is a major driver of ageing. It is reported that the accumulation of DNA damage could induce cell senescence and death, compromise tissue function, and finally result in ageing (Soto-Palma et al. 2022; Zhao et al. 2023). A study indicated that resveratrol could present anti-ageing potential by protecting HEK293 cells from DNA damage induced by N-methyl-N′-nitro-N-nitrosoguanidine (MNNG) and rescuing effects of NAD^+ precursors, and this effect was more potent when combined with quercetin (Yang et al. 2021). Another study found that resveratrol delayed ageing-associated lung degenerative changes in mouse model partly through blocking DNA damage of parenchymal cells (Navarro et al. 2017). Moreover, piceatannol exhibited potential in preventing ultraviolet B-induced skin ageing by reducing the levels of cyclo-butane pyrimidine dimer (CPD) and DNA damage response proteins (such as p53, p-p53, and p21) (Jeong et al. 2023).

Maintenance of Telomeres

Telomeres are located at the ends of chromosomes that could determine genome integrity and cell fate, and telomere attrition or shortening is a characteristic ageing marker (Benetos 2022; Verma et al. 2022). Maintaining the length and function of telomeres is important to fight against ageing. A study showed that long-term administration of resveratrol (from drinking water: 30 mg/L, for 6 or 12 months) could protect against ovarian ageing and age-associated infertility in female mice partially through keeping telomerase activity and telomere length similar to young counterparts (Liu et al. 2013). Additionally, low dose supplementation of resveratrol (0.15% mg/kg diet) to rats delayed the vascular ageing partly achieved by the increase of telomerase activity and telomere length (da Luz et al. 2012).

Other Mechanisms

Ageing is a continuous and complicated process caused by multiple factors, and the mechanisms for resveratrol to fight against ageing are also intricate (Saavedra et al. 2023). Apart from the processes and pathways mentioned above, many other mechanisms could also participate in the anti-ageing effects of resveratrol, such as regulating immunity, increasing the stemness property, inhibiting the melanogenesis, decreasing the accumulation of advanced glycation end-products (AGEs), and so on. For example, in a D-galactose-induced ageing mouse model study, resveratrol significantly reduced cellular senescence of the thymus and enhanced immunity function with the involvement of transcription factor forkhead box protein N1 (FoxN1) pathway (Wei et al. 2021). In addition, pterostilbene inhibited melanogenesis, reduced wrinkles and fine lines, and improved skin hydration elasticity, showing effect in fighting against skin ageing (Majeed et al. 2020). Moreover, piceatannol decreased the amounts of senescent cells both in senescent replicative cultures and after genotoxic stress-induced acute senescence, and promoted cell proliferation and stemness property of mesenchymal stromal cells (MSCs), suggesting that piceatannol is effective in counteracting ageing (Alessio et al. 2021). Besides, resveratrol could decrease the levels of AGEs and ameliorate the renal dysfunction in the D-galactose-induced ageing mouse (Lan et al. 2023).

In short, diverse and complex mechanisms could be involved in the anti-ageing effects of resveratrol, such as the amelioration of oxidative stress, the reduction of inflammation, the activation of SIRT1, the regulation of mitochondrial function, the inhibition of apoptosis, the reduction of DNA damage, and the maintenance of telomeres. These mechanisms might be interconnected and interdependent on each other, allowing resveratrol and its analogues to act as promising anti-ageing molecules.

Conclusion

Resveratrol and its analogues have shown promising potential in developing into anti-ageing agents. In this chapter, we summarised the anti-ageing effects of resveratrol and its analogues, such as prolonging the lifespan, and delaying the functional and structural deterioration of various tissues and organs with age. In addition, we discussed the mechanisms of action underlying their anti-ageing effects, which mainly include the suppression of oxidative stress, reduction of inflammation, activation of SIRT1 pathway, regulation of mitochondrial function, inhibition of apoptosis, amelioration of DNA damage, and maintenance of telomeres. Some structurally modified resveratrol analogues could show better stability and efficacy compared to resveratrol, which indicates that more research interest could be focused on the construction and exploration of other possible structurally modified resveratrol analogues. Moreover, the recent studies are mostly from cells, yeast, worms, flies, fish, and mouse/rat models. Therefore, more clinical studies are needed to further verify the anti-ageing effects and mechanisms of resveratrol and its analogues in the future. In summary, this chapter provides important scientific evidence for the use of resveratrol and its analogues as anti-ageing interventions.

References

Abbasi B, Dong Y, Rui R (2021) Resveratrol hinders postovulatory aging by modulating oxidative stress in porcine oocytes. Molecules 26(21):6346. https://doi.org/10.3390/molecules26216346

Alessio N, Squillaro T, Lettiero I, Galano G, De Rosa R, Peluso G, Galderisi U, Di Bernardo G (2021) Biomolecular evaluation of piceatannol's effects in counteracting the senescence of mesenchymal stromal cells: a new candidate for senotherapeutics? Int J Mol Sci 22(21):11619. https://doi.org/10.3390/ijms222111619

Baechle JJ, Chen N, Makhijani P, Winer S, Furman D, Winer DA (2023) Chronic inflammation and the hallmarks of aging. Mol Metab 74:101755. https://doi.org/10.1016/j.molmet.2023.101755

Beghelli D, Zallocco L, Barbalace MC, Paglia S, Strocchi S, Cirilli I, Marzano V, Putignani L, Lupidi G, Hrelia S, Giusti L, Angeloni C (2022) Pterostilbene promotes mean lifespan in both male and female drosophila melanogaster modulating different proteins in the two sexes. Oxidative Med Cell Longev 2022:1744408. https://doi.org/10.1155/2022/1744408

Benetos A (2022) Telomere length: from cellular senescence to trajectories of human ageing. Bulletin De L Academie Nationale De Medecine 206(7):872–882. https://doi.org/10.1016/j.banm.2022.02.018

Bohara RA, Tabassum N, Singh MP, Gigli G, Ragusa A, Leporatti S (2022) Recent overview of resveratrol's beneficial effects and its nano-delivery systems. Molecules 27(16):5154. https://doi.org/10.3390/molecules27165154

Caldeira CA, Santos MA, Araujo GR, Lara RC, Franco FN, Chaves MM (2021) Resveratrol: change of sirt 1 and ampk signaling pattern during the aging process. Exp Gerontol 146:111226. https://doi.org/10.1016/j.exger.2021.111226

Chen ZY, Su YC, Cheng FY, Yan SJ, Wang YJ (2021) Lifetime bioaccumulation of silver nanoparticles accelerates functional aging by inactivating antioxidant pathways, an effect reversed by pterostilbene. Environ Sci-Nano 8(12):3774–3791. https://doi.org/10.1039/d1en00655j

Chu SH, Yang D, Wang YP, Yang R, Qu LB, Zeng HJ (2021) Effect of resveratrol on the repair of kidney and brain injuries and its regulation on klotho gene in d-galactose-induced aging mice. Bioorg Med Chem Lett 40:127913. https://doi.org/10.1016/j.bmcl.2021.127913

da Luz PL, Tanaka L, Brum PC, Dourado PMM, Favarato D, Krieger JE, Laurindo FRM (2012) Red wine and equivalent oral pharmacological doses of resveratrol delay vascular aging but do not extend life span in rats. Atherosclerosis 224(1):136–142. https://doi.org/10.1016/j.atherosclerosis.2012.06.007

Garrigue P, Mounien L, Champion S, Mouhajir Y, Pechere L, Guillet B, Landrier JF, Seree E (2021) Long-term administration of resveratrol at low doses improves neurocognitive performance as well as cerebral blood flow and modulates the inflammatory pathways in the brain. J Nutr Biochem 97:108786. https://doi.org/10.1016/j.jnutbio.2021.108786

Gherardi G, Corbioli G, Ruzza F, Rizzuto R (2022) CoQ(10) and resveratrol effects to ameliorate aged-related mitochondrial dysfunctions. Nutrients 14(20):4326. https://doi.org/10.3390/nu14204326

Gu TY, Wang NM, Wu T, Ge Q, Chen L (2021) Antioxidative stress mechanisms behind resveratrol: a multidimensional analysis. J Food Qual 2021:5571733–5571712. https://doi.org/10.1155/2021/5571733

Guo KB, Feng YQ, Zheng XE, Sun LT, Wasan HS, Ruan SM, Shen MH (2021) Resveratrol and its analogs: potent agents to reverse epithelial-to-mesenchymal transition in tumors. Front Oncol 11:644134. https://doi.org/10.3389/fonc.2021.644134

Harper SA, Bassler JR, Peramsetty S, Yang YF, Roberts LM, Drummer D, Mankowski RT, Leeuwenburgh C, Ricart K, Patel RP, Bamman MM, Anton SD, Jaeger BC, Buford TW (2021) Resveratrol and exercise combined to treat functional limitations in late life: a pilot randomized controlled trial. Exp Gerontol 143:111111. https://doi.org/10.1016/j.exger.2020.111111

He SJ, Zhou M, Zheng HM, Wang YW, Wu SH, Gao Y, Chen JH (2022) Resveratrol inhibits the progression of premature senescence partially by regulating v-rel avian reticuloendotheliosis viral oncogene homolog a (RELA) and sirtuin 1 (SIRT1). Ren Fail 44(1):171–183. https://doi.org/10.1080/0886022x.2022.2029488

Hoffman TE, Barnett KJ, Wallis L, Hanneman WH (2017) A multimethod computational simulation approach for investigating mitochondrial dynamics and dysfunction in degenerative aging. Aging Cell 16(6):1244–1255. https://doi.org/10.1111/acel.12644

Hosoda R, Nakashima R, Yano M, Iwahara N, Asakura S, Nojima I, Saga Y, Kunimoto R, Horio Y, Kuno A (2023) Resveratrol, a sirt1 activator, attenuates aging-associated alterations in skeletal muscle and heart in mice. J Pharmacol Sci 152(2):112–122. https://doi.org/10.1016/jahs.2023.04.001

Hou YH, Li SS, Zhu HY, Qiao MX, Sun XW, Li GR (2023) Development of the thymus and kidney and effects of resveratrol on their aging in a short-lived fish. J Gerontol Ser A-Biol Sci Med Sci 78(9):1550–1557. https://doi.org/10.1093/gerona/glad089

Jeong S, Chung Y, Park S, Lee S, Choi N, Park JK (2023) Combined treatment of ginsenoside rg2 and piceatannol mixture reduces the apoptosis and DNA damage induced by uvb in hacat cells. Mol Cell Toxicol 19(1):63–70. https://doi.org/10.1007/s13273-022-00238-w

Joseph JA, Fisher DR, Cheng V, Rimando AM, Shukitt-Hale B (2008) Cellular and behavioral effects of stilbene resveratrol analogues: implications for reducing the deleterious effects of aging. J Agric Food Chem 56(22):10544–10551. https://doi.org/10.1021/jf802279h

Juarez D, Arteaga I, Cortes H, Vazquez-Roque R, Lopez-Lopez G, Flores G, Trevino S, Guevara J, Diaz A (2023) Chronic resveratrol administration reduces oxidative stress and brain cell loss and improves memory of recognition in old rats. Synapse 77(4):e22271. https://doi.org/10.1002/syn.22271

Lan KC, Peng PJ, Chang TY, Liu SH (2023) Resveratrol alleviates advanced glycation end-products-related renal dysfunction in D-galactose-induced aging mice. Meta 13(5):655. https://doi.org/10.3390/metabo13050655

Lephart ED, Andrus MB (2017) Human skin gene expression: natural (trans) resveratrol versus five resveratrol analogs for dermal applications. Exp Biol Med 242(15):1482–1489. https://doi.org/10.1177/1535370217723628

Li J, Lin ZY, Tang XD, Liu G, Chen YJ, Zhai XX, Huang QR, Cao Y (2020) Oxyresveratrol extracted fromartocarpus heterophylluslam. Inhibits tyrosinase and age pigmentsin vitroandin vivo. Food Funct 11(7):6595–6607. https://doi.org/10.1039/d0fo01193b

Liang QX, Lin YH, Zhang CH, Sun HM, Zhou L, Schatten H, Sun QY, Qian WP (2018) Resveratrol increases resistance of mouse oocytes to postovulatory aging in vivo. Aging-US 10(7):1586–1596. https://doi.org/10.18632/aging.101494

Liu MY, Yin Y, Ye XY, Zeng M, Zhao Q, Keefe DL, Liu L (2013) Resveratrol protects against age-associated infertility in mice. Hum Reprod 28(3):707–717. https://doi.org/10.1093/humrep/des437

Liu J, Jiao KL, Zhou Q, Yang J, Yang KK, Hu CY, Zhou M, Li Z (2021) Resveratrol alleviates 27-hydroxycholesterol-induced senescence in nerve cells and affects zebrafish locomotor behavior via activation of sirt1-mediated stat3 signaling. Oxidative Med Cell Longev 2021:6673343. https://doi.org/10.1155/2021/6673343

Lopez-Otin C, Blasco MA, Partridge L, Serrano M, Kroemer G (2023) Hallmarks of aging: an expanding universe. Cell 186(2):243–278. https://doi.org/10.1016/j.cell.2022.11.001

Luo ML, Mai MQ, Song WH, Yuan QH, Feng XL, Xia EQ, Guo HH (2022) The antiaging activities of phytochemicals in dark-colored plant foods: involvement of the autophagy- and apoptosis-associated pathways. Int J Mol Sci 23(19):11038. https://doi.org/10.3390/ijms231911038

Majeed M, Majeed S, Jain R, Mundkur L, Rajalakshmi HR, Lad PS, Neupane P (2020) An open-label single-arm, monocentric study assessing the efficacy and safety of natural pterostilbene (pterocarpus marsupium) for skin brightening and antiaging effects. Clin Cosmet Investig Dermatol 13:105–116. https://doi.org/10.2147/ccid.S238358

Naini R, Chikati R, Vudem DR, Kancha RK (2019) Molecular docking analysis of imine stilbene analogs and evaluation of their anti-aging activity using yeast and mammalian cell models. J Recept Signal Transduct 39(1):55–59. https://doi.org/10.1080/10799893.2019.1605529

Navarro S, Reddy R, Lee J, Warburton D, Driscoll B (2017) Inhaled resveratrol treatments slow ageing-related degenerative changes in mouse lung. Thorax 72(5):451–459. https://doi.org/10.1136/thoraxjnl-2016-208964

Pal K, Raghuram GV, Dsouza J, Shinde S, Jadhav V, Shaikh A, Rane B, Tandel H, Kondhalkar D, Chaudhary S, Mittra I (2022) A pro-oxidant combination of resveratrol and copper down-regulates multiple biological hallmarks of ageing and neurodegeneration in mice. Sci Rep 12(1):17209. https://doi.org/10.1038/s41598-022-21388-w

Pratiwi R, Prachayasittikul V, Prachayasittikul S, Nantasenamat C (2019) Rational design of novel sirtuin 1 activators via structure-activity insights from application of qsar modeling. EXCLI J 18:207–222. https://doi.org/10.17179/excli2019-1274

Pulla VK, Alvala M, Sriram DS, Viswanadha S, Sriram D, Yogeeswari P (2014) Structure-based drug design of small molecule sirt1 modulators to treat cancer and metabolic disorders. J Mol Graph Model 52:46–56. https://doi.org/10.1016/j.jmgm.2014.06.005

Saavedra D, Ane-Kouri AL, Barzilai N, Caruso C, Cho KH, Fontana L, Franceschi C, Frasca D, Ledon N, Niedernhofer LJ, Pereira K, Robbins PD, Silva A, Suarez GM, Berghe WV, von Zglinicki T, Pawelec G, Lage A (2023) Aging and chronic inflammation: highlights from a multidisciplinary workshop. Immun Ageing 20(1):25. https://doi.org/10.1186/s12979-023-00352-w

Santos MA, Franco FN, Caldeira CA, de Araujo GR, Vieira A, Chaves MM (2023) Resveratrol has its antioxidant and anti-inflammatory protective mechanisms decreased in aging. Arch Gerontol Geriatr 107:104895. https://doi.org/10.1016/j.archger.2022.104895

Sarubbo F, Ramis MR, Tejada S, Jimenez-Garcia M, Esteban S, Miralles A, Moranta D (2023) Resveratrol improves episodic-like memory and motor coordination through modulating neuroinflammation in old rats. J Funct Foods 104:105533. https://doi.org/10.1016/j.jff.2023.105533

Shen PY, Yue YR, Sun QC, Kasireddy N, Kim KH, Park Y (2017) Piceatannol extends the lifespan of caenorhabditis elegans via daf-16. Biofactors 43(3):379–387. https://doi.org/10.1002/biof.1346

Song JB, Liu L, Hao KG, Mao S, Tang YX, Tong XL, Dai FY (2021) Resveratrol elongates the lifespan and improves antioxidant activity in the silkworm bombyx mori. J Pharm Anal 11(3):374–382. https://doi.org/10.1016/j.jpha.2020.06.005

Soto-Palma C, Niedernhofer LJ, Faulk CD, Dong X (2022) Epigenetics, DNA damage, and aging. J Clin Invest 132(16):e158446. https://doi.org/10.1172/jci158446

Teng WL, Huang PH, Wang HC, Tseng CH, Yen FL (2021) Pterostilbene attenuates particulate matter-induced oxidative stress, inflammation and aging in keratinocytes. Antioxidants 10(10):1552. https://doi.org/10.3390/antiox10101552

Torregrosa-Munumer R, Vara E, Fernandez-Tresguerres JA, Gredilla R (2021) Resveratrol supplementation at old age reverts changes associated with aging in inflammatory, oxidative and apoptotic markers in rat heart. Eur J Nutr 60(5):2683–2693. https://doi.org/10.1007/s00394-020-02457-0

United Nations (2023) World population prospects 2022: summary of results. United Nations Retrieved June 2 from https://www.un.org/development/desa/pd/sites/www.un.org.development.desa.pd/files/wpp2022_summary_of_results.pdf

Verma AK, Singh P, Al-Saeed FA, Ahmed AE, Kumar S, Kumar A, Dev K, Dohare R (2022) Unravelling the role of telomere shortening with ageing and their potential association with diabetes, cancer, and related lifestyle factors. Tissue Cell 79:101925. https://doi.org/10.1016/j.tice.2022.101925

Wang N, Luo ZW, Jin M, Sheng WW, Wang HT, Long XY, Wu Y, Hu PP, Xu H, Zhang X (2019) Exploration of age-related mitochondrial dysfunction and the anti-aging effects of resveratrol in zebrafish retina. Aging-US 11(10):3117–3137. https://doi.org/10.18632/aging.101966

Wei TT, Feng YK, Cao JH, Li JH, Yuan SL, Ding Y, Chai YR (2021) Dosage effects of resveratrol on thymus involution in d-galactose-treated mice. J Food Biochem 45(5):e13709. https://doi.org/10.1111/jfbc.13709

Wu HF, Xue JC, Di HQ, Lv CT, Hao YL, Nie ZY (2022a) Resveratrol improves ovarian function in aged rat by inhibiting oxidative stress and activating the sirt1. Gen Physiol Biophys 41(1):53–61. https://doi.org/10.4149/gpb_2021040

Wu SX, Xiong RG, Huang SY, Zhou DD, Saimaiti A, Zhao CN, Shang A, Zhang YJ, Gan RY, Li HB (2022b) Effects and mechanisms of resveratrol for prevention and management of cancers: an updated review. Crit Rev Food Sci Nutr 63:12422–12440. https://doi.org/10.1080/10408398.2022.2101428

Yang Y, Liu ZW, Zhao BS, Joseph S, Ma YC, Xu H, Cheng HW (2021) Resveratrol and quercetin potentiate the cell protection and rescue effects of NAD(+) precursors in hek293 cells challenged by DNA damaging agent, N-methyl-N′-nitro-N-nitrosoguanidine. Nat Prod Commun 16(9):1934578X2110454. https://doi.org/10.1177/1934578x211045465

Zhang Y, Zhang LH, Chen X, Zhang N, Li G (2018) Piceatannol attenuates behavioral disorder and neurological deficits in aging mice via activating the Nrf2 pathway. Food Funct 9(1):371–378. https://doi.org/10.1039/c7fo01511a

Zhang LL, Zheng JW, Tie X, Lin T, Yang WQ, Li ZQ, Zou Y, Guan GM, Liu PQ, Luo WW, Li ZM (2021) Pterostilbene and its nicotinate derivative ameliorated vascular endothelial senescence and elicited endothelium-dependent relaxations via activation of sirtuin 1. Can J Physiol Pharmacol 99(9):900–909. https://doi.org/10.1139/cjpp-2020-0583

Zhao Y, Simon M, Seluanov A, Gorbunova V (2023) DNA damage and repair in age-related inflammation. Nat Rev Immunol 23(2):75–89. https://doi.org/10.1038/s41577-022-00751-y

Zhou DD, Luo M, Huang SY, Saimaiti A, Shang A, Gan RY, Li HB (2021) Effects and mechanisms of resveratrol on aging and age-related diseases. Oxidative Med Cell Longev 2021:9932218. https://doi.org/10.1155/2021/9932218

Zhou DD, Li JH, Xiong RG, Saimaiti A, Huang SY, Wu SX, Yang ZJ, Shang A, Zhao CN, Gan RY, Li HB (2022) Bioactive compounds, health benefits and food applications of grape. Food Secur 11(18):2755. https://doi.org/10.3390/foods11182755

Zhu HY, Li XW, Qiao MX, Sun XW, Li GR (2023) Resveratrol alleviates inflammation and er stress through sirt1/nrf2 to delay ovarian aging in a short-lived fish. J Gerontol Ser A-Biol Sci Med Sci 78(4):596–602. https://doi.org/10.1093/gerona/glad009

Chapter 10
Hormetic Effects of Phytochemicals with Anti-Ageing Properties

Calogero Caruso, Giulia Accardi, Anna Aiello, and Giuseppina Candore

Abstract In the fields of biology and medicine, hormesis is defined as the adaptive response of cells and organisms to moderate and usually intermittent stress. Examples include radiation, pharmaceutical agents, as well as dietary and lifestyle factors such as calorie restriction and physical exercise. However, in the present chapter, we will focus on the hormetic role of certain phytochemicals, compounds that naturally occur in plants, playing roles in plant colour, flavour, and disease resistance, with nutraceutical properties. Indeed, these compounds exhibit health-promoting, disease-preventing, or medicinal properties, mostly through a hormetic mechanism.

Keywords Hormesis · Inflammation · Oxidation · Phytochemicals · Polyphenols

Introduction

Presently, individuals in developed countries experience a significantly extended lifespan compared to earlier times. Nonetheless, they are not immune to the ageing process. Many countries witness a rise in older population, accompanied by an increase in the prevalence of age-related diseases and healthcare costs. Throughout the twentieth century, the life expectancy at birth in developed countries increased by 30 years. Initially, this was attributed to a decrease in neonatal and infant mortality, and subsequently, to a decline in mortality during middle and older age. In 1900, approximately 40% of children born in countries with reliable data had a life expectancy exceeding 65 years. Currently, in these same countries, over 88% of all newborns can expect to live beyond 65 years, and at least 44% will reach the age of 85

C. Caruso (✉) · G. Accardi · A. Aiello · G. Candore
Laboratory of Immunopathology and Immunosenescence, Department of Biomedicine, Neurosciences and Advanced Diagnostics, University of Palermo, Palermo, Italy
e-mail: calogero.caruso@unipa.it; giulia.accardi@unipa.it; anna.aiello@unipa.it; giuseppina.candore@unipa.it

© The Author(s), under exclusive license to Springer Nature Switzerland AG 2024

V. I. Korolchuk, J. R. Harris (eds.), *Biochemistry and Cell Biology of Ageing: Part V, Anti-Ageing Interventions*, Subcellular Biochemistry 107,
https://doi.org/10.1007/978-3-031-66768-8_10

or older. This heightened life expectancy is closely linked to socio-economic development, facilitating improved access to food and clean water, enhanced housing and living conditions, reduced susceptibility to infectious diseases, and availability of medical care. Additionally, it has been proposed that enhanced education plays a crucial role in empowering individuals to make healthier lifestyle choices (Accardi et al. 2019; Lutz and Kebede 2018; Caruso and Candore 2021; Christensen et al. 2009; Oeppen and Vaupel 2002).

Older individuals exhibit reduced resilience to environmental challenges and pathological stimuli, such as infections. In our society, the understanding of the ageing process involves a diminished capacity to withstand chronic diseases and the simultaneous decline in mobility, sensory, and cognitive functions. This is accompanied by a rapid escalation in healthcare expenses, attributable to the growing older population in the Western world. It is imperative to implement preventive therapeutic measures promptly to decelerate the ageing process (Longo et al. 2015). Among preventive mechanisms, a key role is played by a healthy lifestyle, which obviously includes a healthy dietary pattern. The Mediterranean Diet (MedDiet) stands out as one of the extensively researched healthful dietary patterns (Vasto et al. 2014a, b).

Phytochemicals

MedDiet constitutes a nutritional regimen featuring a low-glycaemic index and limited intake of animal proteins. This diet incorporates phytochemical compounds present in vegetables, fruits, red wine, olive oil, or nuts, imparting anti-inflammatory and antioxidant effects. The health benefits of this lifestyle pattern go beyond just being a diet, as it also derives advantages from nutraceuticals (Accardi et al. 2016a, b; Vasto et al. 2014a, b), which are described as "naturally derived bioactive compounds present in foods, dietary supplements, and herbal products, exhibiting health-promoting, disease-preventing, or medicinal properties." Coined in 1989 by Stephen De Felice, the term "nutraceuticals" originates from the fusion of nutrition and pharmaceutics (Gupta et al. 2010). Nutraceuticals may contain phytochemicals among other bioactive compounds. Nutraceuticals, in general, cover a broader spectrum of compounds, comprising vitamins, minerals, amino acids, and other bioactive substances, whereas phytochemicals specifically denote compounds derived from plants. Numerous nutraceutical products derive their health-promoting properties from the inclusion of phytochemicals. Phytochemicals, a subgroup of compounds found in plants, are incorporated into nutraceuticals, which are products designed to offer health benefits beyond basic nutrition, often incorporating phytochemicals alongside other bioactive substances. The utilisation of plants rich in phytochemicals during the development of nutraceuticals underscores the intimate relationship between these two concepts in the pursuit of enhancing health and preventing diseases (Aiello et al. 2016).

In summary, phytochemicals are compounds that naturally occur in plants, playing roles in plant colour, flavour, and disease resistance. Research has explored the potential health benefits of phytochemicals, which include antioxidant and anti-inflammatory properties. While not deemed essential for fundamental human survival, studies indicate that they might have positive effects on health. Flavonoids, carotenoids, glucosinolates, and polyphenols are among the examples of phytochemicals (Thakur et al. 2020).

Hence, polyphenols constitute a particular category within the realm of phytochemicals, characterised by the presence of multiple phenol structural units. These compounds are extensively distributed throughout the plant kingdom, present in fruits, vegetables, tea, coffee, red wine, and various other plant-based foods. The richness in polyphenols is the characteristic that makes extra virgin olive oil (EVOO), the main constituent of the MedDiet, different from all other edible oils and makes it a nutraceutical product (Gambino et al. 2018). Acknowledged for their antioxidant properties, polyphenols play a role in neutralising harmful free radicals within the body. Common subclasses of polyphenols encompass flavonoids, phenolic acids, lignans, and stilbenes, each exhibiting a distinctive chemical structure and being prevalent in different plant types. Besides EVOO, foods rich in polyphenols include berries (abundant in flavonoids), green tea (high in catechins), and red wine (containing resveratrol, a type of stilbene). To summarise, polyphenols, although a subset of phytochemicals, do not encompass all phytochemicals, representing a distinct category of plant compounds distinguished by unique chemical structures and biological activities. As secondary metabolites of plants, polyphenols commonly serve in defence against ultraviolet radiation or aggression by pathogens (Pandey and Rizvi 2009).

As briefly mentioned before, dietary polyphenols fall into four primary categories typically present in glycosidic form or as esters (Pandey and Rizvi 2009). Polyphenols represent a substantial family of molecules acting through various mechanisms. Following polyphenol treatments, the activation of Sirtuin 1 (SIRT1) and AMP-activated protein kinase (AMPK), along with the inhibition of E1A-binding protein p300 (EP300) and mTORC1, has been observed. An outcome common to these signals should be the induction of autophagy, potentially explaining the health-promoting effects of these molecules (Madeo et al. 2019).

Indeed, the deacetylases known as SIRT1 family proteins promote longevity in diverse species and could mediate many of the beneficial effects of dietary restriction. Likewise, AMPK is a conserved, energy-sensing serine/threonine kinase that is activated when cellular energy levels are low, resulting in increasing levels of AMP. AMPK activation generates insulin-sensitising effects resulting in increased glucose uptake in skeletal muscles, decreased hepatic glucose production, and enhanced fatty acid oxidation in several tissues (Longo et al. 2015). Pietrocola et al. (2015) demonstrated that the acetyltransferase EP300, negatively controls autophagy, and its inhibitors induce autophagy. Finally, mTORC1 activation leads to protein translation and cell growth, whereas its inhibition blocks growth and induces stress response pathways such as autophagy (Longo et al. 2015).

Despite the mechanisms of action of polyphenols remain incompletely elucidated, their antioxidant capacity is a significant aspect. On the other hand, polyphenols, along with other phytochemicals, are suggested to operate in a hormetic manner, stimulating various stress-response pathways that ultimately contribute to health benefits, often referred to as hormetics. Mechanisms such as the activation of nuclear factor erythroid 2 (NRF2), along with the downregulation of NF-κB, have been described for these so-called hormetics (Martel et al. 2019; Martucci et al. 2017). As detailed in the next section, hormesis denotes adaptive processes where low doses of stress alter the original homeostasis in a cell or organism, leading to adaptive responses with beneficial effects.

The effects of various hormetics within healthful diets such as the MedDiet likely contribute to the beneficial effects of healthful diets. Characterising hormetics and understanding their mechanisms of action opens the possibility of using these substances as therapeutic agents to enhance health and healthspan through pharmacological approaches (Ros and Carrascosa 2020).

Phytochemicals as Hormetic Agents

Hormesis is a biphasic dose-response observed in biological systems exposed to increasing amounts of either a condition or a substance. At high doses, exposures lead to inhibitory or toxic effects, while at low doses they induce stimulating or beneficial effects (Mattson and Calabrese 2010) (Fig. 10.1).

Mild stress, capable of disturbing homeostatic balance and thus hormetic, activates and enhances molecular and cellular responses that will bolster the organism ability to face more severe challenges (Calabrese and Mattson 2017). Receptors, intracellular messengers, and transcription factors involved in the expression of genes coding for cytoprotective proteins such as antioxidant enzymes, growth

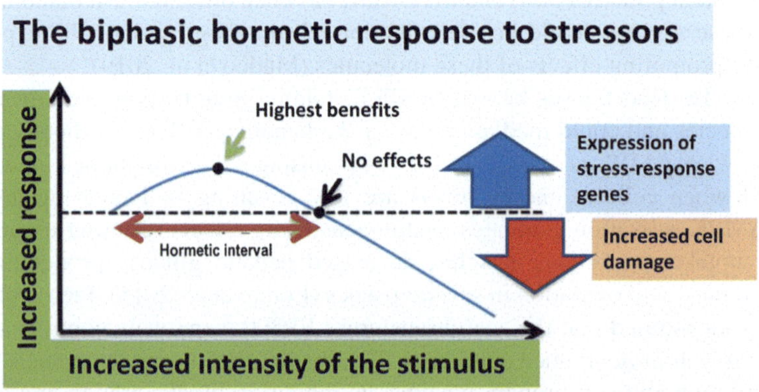

Fig. 10.1 Hormesis describes a biphasic dose-response feature to stressful stimuli (published under a Creative Commons CC-BY license from Leri et al. 2020)

factors, and chaperones constitute the pathway that mediates the response to mild stress and thus with hormetic properties (Mattson and Calabrese 2010). There are several endogenous and exogenous factors capable of eliciting a biphasic hormetic response. They range from diet (caloric restriction) to lifestyle (physical activity), from radiation to drugs: beneficial at low doses, potentially toxic at high doses (Masoro 2006; Ji et al. 2010; Mattson and Calabrese 2010). More relevant to our discussion is the fact that various phytochemicals can act hormetically, activating the aforementioned pathway at low doses (Lee et al. 2014).

Studies on experimental models, and in humans, have demonstrated how hormetic factors reduce the accumulation of senescent cells, thereby also reducing the risk of age-related functional impairments. All this, resulting in an extension of healthy life, makes hormesis considered as an anti-ageing intervention (Gems and Partridge 2008; Son et al. 2008). The positive effect of hormesis, as mentioned earlier, is attributed to the stimulation of stress response pathways, known to play a key role in the ageing process as genes positively regulated by these pathways influence ageing by regulating lifespan (Davinelli et al. 2012). The constant decline in stress response mechanisms is indeed a general characteristic of the ageing process, although the cause-and-effect relationship is not clear, in other words, whether the deterioration of stress response mechanisms is a contributing factor to ageing or a consequence of it (Kourtis and Tavernarakis 2011).

Polyphenols and other plant-derived compounds potentially facilitate their health-enhancing effects by activating one or more adaptive stress response pathways, such as NRF2, while concurrently inhibiting NF-κB. NRF2 serves as a pivotal transcription factor regulating a plethora of genes (>500) associated with cytoprotective functions. On the other hand, activation of NF-κB signalling pathways elevates the expression of genes linked to ageing, leading to tissue degradation and inflamm-ageing, and promotes the activation of genes that inhibit cell death, culminating in immune ageing and inflammation. In its inactive state, NRF2 remains bound to the cysteine-rich protein Kelch-like ECH-associated protein 1 (KEAP1) within the cytoplasm under normal conditions. KEAP1 acts as a sensor for oxidative stress, capturing NRF2 through interaction with its N-terminal NEH2 domain and promoting the degradation of NRF2 via ubiquitin-mediated proteolysis. Upon exposure to oxidative or electrophilic stress, NRF2 is liberated and relocates to the nucleus. There, it triggers the expression of antioxidants by binding to antioxidant response element sequences (Alì et al. 2021; Lee et al. 2014; Martel et al. 2019; Martucci et al. 2017).

Prolonged and excessive activation of NRF2 may result in pathophysiological consequences, whereas its natural activation prompts a beneficial response to mild stress, leading to improved health outcomes and increased lifespan in animal models. As a result, the NRF2 signalling pathway is recognized as hormetic. As for the transcriptional activity of NF-κB, it rises with ageing, correlating with several age-related diseases. This is because in aged cells NF-κB is activated by oxidative stress, but as mentioned earlier, the control of oxidative stress by hormesis curbs its pathological activation (Alì et al. 2021; Lee et al. 2014; Martel et al. 2019; Martucci et al. 2017).

In clinical studies investigating the potential benefits of antioxidants, not only has no beneficial effect of phytochemicals at high doses been demonstrated, but several studies have shown a potential harmful effect of these compounds at high doses. On the other hand, the absence of reduced metabolites of phytochemicals in both animal models and humans indicates that phytochemicals do not act as direct antioxidants (Alì et al. 2021; Riccioni et al. 2007).

On the other hand, it is well-known that only a fraction of the ingested phytochemicals reaches systemic circulation due to their significantly reduced intestinal absorption; in other words, phytochemicals have low bioavailability. However, it is possible that at least some of their positive effects are independent of their ability to be absorbed, as they depend on a prebiotic effect on the composition of the intestinal microbiota, resulting in a decrease in pathogenic bacteria and an increase in "beneficial" bacteria. This modulation might be responsible for the positive effects on organismal health (Sorrenti et al. 2020). Therefore, modulation of the intestinal microbiota contributes, at least in part, to the significant bioactivity of phytochemicals, despite their previously mentioned low bioavailability. This could also explain why they can have overlapping effects despite having different chemical structures (Martel et al. 2020). Additionally, it has been suggested that phytochemicals may also have epigenetic effects, thus contributing to achieving healthy ageing and longevity (Puca et al. 2018).

Although various clinical studies have not provided consistent results even when the same phytochemicals are used, it is undeniable that various phytochemicals can be harmful if consumed in high doses (with carcinogenic, neurotoxic, genotoxic, teratogenic, cytotoxic, nephrotoxic, and hepatotoxic effects), while at low doses they can trigger stress response pathways with positive effects on health status (Ali et al. 2008; Bode and Dong 2011; Guldiken et al. 2018; Zhang et al. 2004). There are several reasons for this inconsistency in results, including primarily different tested doses, varying levels of extract quality, and diverse study designs (Cicero and Colletti 2021). Further studies are needed to determine the optimal dosage and percentage of phytochemical presence in foods, as well as to ascertain whether there are interactions between phytochemical intake and other anti-ageing strategies, such as moderate physical activity, which may utilize the same pathways correlated with mild stress. In this regard, it has been suggested that spacing out phytochemical intake could be beneficial as it would allow for better cellular adaptation (Martel et al. 2019).

In the following two sections, we elucidate two facets of hormesis that are not directly associated with phytochemicals but nevertheless underscore the significance of hormetic processes in human health and longevity.

Hormesis and Inflammation

Neurohormesis refers to the central nervous system's (CNS) ability to respond to toxic agents, both endogenous (such as nitric oxide, carbon monoxide, glutamate, Ca^{2+}) and exogenous, representing mild stress that enhances neuronal resistance to more intense insults (Santoro et al. 2020). Conversely, a sedentary lifestyle, often accompanied by nutrient-rich and high-fat diets, can negatively impact the CNS, compromising cellular stress resistance and neuroplasticity (Raefsky and Mattson 2017). In models, a high-fat and high-sugar diet impairs hippocampal plasticity and cognitive performance (Stranahan et al. 2008). In accordance with what has been observed in models, older individuals with metabolic syndrome or diabetes exhibit lower performance in cognitive tests involving information processing speed, attention, and executive functions compared to healthy controls (van den Berg et al. 2008).

An important source of pro-inflammatory cytokines in ageing is adipose tissue, thus contributing to the typical pro-inflammatory state of old age, known as inflammageing, which is a risk factor for major age-related diseases such as cardiovascular diseases, type 2 diabetes, sarcopenia, and osteoporosis (Caruso and Candore 2021; Santoro et al. 2020). However, according to the obesity paradox theory, the low-grade inflammation produced by adipocytes could stimulate a hormetic response capable of improving glucose metabolism and damage responses, thus protecting the organism from age-related chronic diseases, improving health status, adaptation, and likely contributing to successful ageing (Santoro et al. 2020). This could be attributed to molecules such as mitokines, stress response molecules known to be produced in response to mitochondrial stress, as well as endoplasmic reticulum stress and other stresses. Mitokines can enhance energy metabolism and protect against the deleterious effects of high-fat diets and possess anti-inflammatory activity (Santoro et al. 2020).

Hormesis and Lifespan

As reviewed by Calabrese et al. (2024), various lifestyle behaviours as well as many phytochemicals and pharmaceuticals can augment the lifespan of animal models, with a hormetic dose-response effect. Lifespan extension is observed in various animal models with various types of hormetic stimuli. All the data reviewed thus far show that the hormetic process is capable of prolonging life in animal models considering both chemical and behavioural hormetics, although, in percentage terms, the extent of lifespan lengthening is quite mild.

The obvious reason for the existence of experimental models is to understand how to extract useful information for human health. In this regard, it is beyond doubt that humans are naturally exposed to numerous hormetic stimuli throughout life. Many of these hormetic stimuli, besides increasing life expectancy, could also have the property of delaying the onset and slowing the progression of acute pathological conditions such as heart attacks and strokes, and chronic pathological conditions such as type 2 diabetes and Alzheimer's disease. It is worth noting, therefore, that hormetics presumably have a dual effect, not only on extending lifespan but also on ensuring that this lifespan is healthy, by avoiding or delaying the onset of age-related diseases. We can thus speculate that this process is already underway through hormetic stimuli present in the lifestyle of at least a portion of the global population. As mentioned earlier, in models, the size of lifespan extension in response to hormetic stimuli is not excessive and reaches a higher level (25% increase) when hormetic stimuli are optimized, which obviously does not occur in the real lives of humans. However, it should be considered that unlike in experimental models, in humans, hormetic stimuli are multiple, and thus there might be a positive synergistic effect, which on the other hand may be limited by the constraints of biological plasticity. Nonetheless, in any case, the information obtained from studying animal models is relevant, both at an individual level and at a public health level, for understanding how to achieve a healthy lifespan (Calabrese et al. 2024).

Concluding Remarks

A long life in a healthy, vigorous, and youthful body has always been one of humanity greatest dreams. Recent advancements in low-calorie diets for laboratory animals promise that 1 day medicine will allow us to exert total control over ageing. This chapter has explored the role of phytochemicals in slowing ageing and delaying or preventing the onset of age-related diseases through hormesis. It is worth noting that the goal of ageing research is not to increase human longevity regardless of consequences, but to enhance active longevity, free from disability and functional dependence.

Finally, extensive experimental data demonstrate the hormetic effects of phytochemicals, which occur through the induction of cellular stress resistance mechanisms. However, to demonstrate high-quality evidence regarding the health benefits of various phytochemicals, human intervention studies must be designed taking into account the following factors: (1) phytochemicals need to be sufficiently characterised, along with the optimal dose for the hormetic effect, (2) the characteristics of the target populations, including their nutritional status, health condition, and genetic background, must be considered, (3) clinically relevant, sensitive, reproducible, and feasible endpoints must be identified, (4) the duration of the intervention must be commonly agreed upon. Nevertheless, even though experimental data have not always translated into a definitive clinical effect, the antioxidant and anti-inflammatory properties of phytochemicals have been widely accepted.

References

Accardi G, Aiello A, Gargano V, Gambino CM, Caracappa S, Marineo S, Vesco G, Carru C, Zinellu A, Zarcone M, Caruso C, Candore G (2016a) Nutraceutical effects of table green olives: a pilot study with Nocellara del Belice olives. Immun Ageing 13:11. https://doi.org/10.1186/s12979-016-0067-y

Accardi G, Aiello A, Gambino CM, Virruso C, Caruso C, Candore G (2016b) Mediterranean nutraceutical foods: strategy to improve vascular ageing. Mech Ageing Dev 159:63–70. https://doi.org/10.1016/j.mad.2016.02.007

Accardi G, Aprile S, Candore G, Caruso C, Cusimano R, Cristaldi L, Di Bona D, Duro G, Galimberti D, Gambino CM, Ligotti ME, Mazzucco W, Vasto S, Aiello A (2019) Genotypic and phenotypic aspects of longevity: results from a Sicilian survey and implication for the prevention and treatment of age-related diseases. Curr Pharm Des 25:228–235. https://doi.org/10.2174/1381612825666190313115233

Aiello A, Accardi G, Candore G, Carruba G, Davinelli S, Passarino G, Scapagnini G, Vasto S, Caruso C (2016) Nutrigerontology: a key for achieving successful ageing and longevity. Immun Ageing 13:17. https://doi.org/10.1186/s12979-016-0071-2

Ali BH, Blunden G, Tanira MO, Nemmar A (2008) Some phytochemical, pharmacological and toxicological properties of ginger (Zingiber officinale Roscoe): a review of recent research. Food Chem Toxicol 46:409–420. https://doi.org/10.1016/j.fct.2007.09.085

Alì S, Davinelli S, Accardi G, Aiello A, Caruso C, Duro G, Ligotti ME, Pojero F, Scapagnini G, Candore G (2021) Healthy ageing and Mediterranean diet: a focus on hormetic phytochemicals. Mech Ageing Dev 200:111592. https://doi.org/10.1016/j.mad.2021.111592

Bode AM, Dong Z (2011) The two faces of capsaicin. Cancer Res 71:2809–2814. https://doi.org/10.1158/0008-5472.CAN-10-3756

Calabrese EJ, Mattson MP (2017) How does hormesis impact biology, toxicology, and medicine? NPJ Aging Mech Dis 3:13. https://doi.org/10.1038/s41514-017-0013-z

Calabrese EJ, Nascarella M, Pressman P, Hayes AW, Dhawan G, Kapoor R, Calabrese V, Agathokleous E (2024) Hormesis determines lifespan. Ageing Res Rev 94:102181. https://doi.org/10.1016/j.arr.2023.102181

Caruso C, Candore G (2021) Conclusions. Slowing aging and fighting age-related diseases, from bench to bedside? In: Caruso C, Candore G (eds) Human aging. Academic, pp 341–354. https://doi.org/10.1016/B978-0-12-822569-1.00001-9

Christensen K, Doblhammer G, Rau R, Vaupel JW (2009) Ageing populations: the challenges ahead. Lancet 374:1196–1208. https://doi.org/10.1016/S0140-6736(09)61460-4

Cicero AFG, Colletti A (2021) Nutraceutical approach to age-related diseases—the clinical evidence on cognitive decline. In: Caruso C, Candore G (eds) Human aging. Academic, pp 255–272. https://doi.org/10.1016/B978-0-12-822569-1.00015-9

Davinelli S, Willcox DC, Scapagnini G (2012) Extending healthy ageing: nutrient sensitive pathway and centenarian population. Immun Ageing 9:9. https://doi.org/10.1186/1742-4933-9-9

Gambino CM, Accardi G, Aiello A, Candore G, Dara-Guccione G, Mirisola M, Procopio A, Taormina G, Caruso C (2018) Effect of extra virgin olive oil and table olives on the immuneinflammatory responses: potential clinical applications. Endocr Metab Immune Disord Drug Targets 18:14–22. https://doi.org/10.2174/1871530317666171114113822

Gems D, Partridge L (2008) Stress-response hormesis and aging: "that which does not kill us makes us stronger". Cell Metab 7:200–203. https://doi.org/10.1016/j.cmet.2008.01.001

Guldiken B, Ozkan G, Catalkaya G, Ceylan FD, Ekin Yalcinkaya I, Capanoglu E (2018) Phytochemicals of herbs and spices: health versus toxicological effects. Food Chem Toxicol 119:37–49. https://doi.org/10.1016/j.fct.2018.05.050

Gupta S, Chauhan D, Mehla K, Sood P, Nair A (2010) An overview of nutraceuticals: current scenario. J Basic Clin Pharm 1:55–62

Ji LL, Dickman JR, Kang C, Koenig R (2010) Exercise-induced hormesis may help healthy aging. Dose Response 8:73–79. https://doi.org/10.2203/dose-response.09-048.Ji

Kourtis N, Tavernarakis N (2011) Cellular stress response pathways and ageing: intricate molecular relationships. EMBO J 30:2520–2531. https://doi.org/10.1038/emboj.2011.162

Lee J, Jo DG, Park D, Chung HY, Mattson MP (2014) Adaptive cellular stress pathways as therapeutic targets of dietary phytochemicals: focus on the nervous system. Pharmacol Rev 66:815–868. https://doi.org/10.1124/pr.113.007757

Leri M, Scuto M, Ontario ML, Calabrese V, Calabrese EJ, Bucciantini M, Stefani M (2020) Healthy effects of plant polyphenols: molecular mechanisms. Int J Mol Sci 21:1250. https://doi.org/10.3390/ijms21041250

Longo VD, Antebi A, Bartke A, Barzilai N, Brown-Borg HM, Caruso C, Curiel TJ, de Cabo R, Franceschi C, Gems D, Ingram DK, Johnson TE, Kennedy BK, Kenyon C, Klein S, Kopchick JJ, Lepperdinger G, Madeo F, Mirisola MG, Mitchell JR, Passarino G, Rudolph KL, Sedivy JM, Shadel GS, Sinclair DA, Spindler SR, Suh Y, Vijg J, Vinciguerra M, Fontana L (2015) Interventions to slow aging in humans: are we ready? Aging Cell 14:497–510. https://doi.org/10.1111/acel.12338

Lutz W, Kebede E (2018) Education and health: redrawing the Preston curve. Popul Dev Rev 44(2):343–361. https://doi.org/10.1111/padr.12141

Madeo F, Carmona-Gutierrez D, Hofer SJ, Kroemer G (2019) Caloric restriction mimetics against age-associated disease: targets, mechanisms, and therapeutic potential. Cell Metab 29:592–610. https://doi.org/10.1016/j.cmet.2019.01.018

Martel J, Ojcius DM, Ko YF, Ke PY, Wu CY, Peng HH, Young JD (2019) Hormetic effects of phytochemicals on health and longevity. Trends Endocrinol Metab 30:335–346. https://doi.org/10.1016/j.tem.2019.04.001

Martel J, Ojcius DM, Ko YF, Young JD (2020) Phytochemicals as prebiotics and biological stress inducers. Trends Biochem Sci 45:462–471. https://doi.org/10.1016/j.tibs.2020.02.008

Martucci M, Ostan R, Biondi F, Bellavista E, Fabbri C, Bertarelli C, Salvioli S, Capri M, Franceschi C, Santoro A (2017) Mediterranean diet and inflammaging within the hormesis paradigm. Nutr Rev 75:442–455. https://doi.org/10.1093/nutrit/nux013

Masoro EJ (2006) Role of hormesis in life extension by caloric restriction. Dose Response 5:163–173. https://doi.org/10.2203/dose-response.06-005.Masoro

Mattson MP, Calabrese EJ (2010) Hormesis: what it is and why it matters. In: Mattson M, Calabrese E (eds) Hormesis. Humana Press. https://doi.org/10.1007/978-1-60761-495-1_1

Oeppen J, Vaupel JW (2002) Demography. Broken limits to life expectancy. Science 296:1029–1031. https://doi.org/10.1126/science

Pandey KB, Rizvi SI (2009) Plant polyphenols as dietary antioxidants in human health and disease. Oxidative Med Cell Longev 2:270–278. https://doi.org/10.4161/oxim.2.5.9498

Pietrocola F, Lachkar S, Enot DP, Niso-Santano M, Bravo-San Pedro JM, Sica V, Izzo V, Maiuri MC, Madeo F, Mariño G, Kroemer G (2015) Spermidine induces autophagy by inhibiting the acetyltransferase EP300. Cell Death Differ 22:509–516. https://doi.org/10.1038/cdd.2014.215

Puca AA, Spinelli C, Accardi G, Villa F, Caruso C (2018) Centenarians as a model to discover genetic and epigenetic signatures of healthy ageing. Mech Ageing Dev 174:95–102. https://doi.org/10.1016/j.mad.2017.10.004

Raefsky SM, Mattson MP (2017) Adaptive responses of neuronal mitochondria to bioenergetic challenges: roles in neuroplasticity and disease resistance. Free Radic Biol Med 102:203–216. https://doi.org/10.1016/j.freeradbiomed.2016.11.045

Riccioni G, Bucciarelli T, Mancini B, Di Ilio C, Capra V, D'Orazio N (2007) The role of the antioxidant vitamin supplementation in the prevention of cardiovascular diseases. Expert Opin Investig Drugs 16:25–32. https://doi.org/10.1517/13543784.16.1.25

Ros M, Carrascosa JM (2020) Current nutritional and pharmacological anti-aging interventions. Biochim Biophys Acta Mol basis Dis 1866:165612. https://doi.org/10.1016/j.bbadis.2019.165612

Santoro A, Martucci M, Conte M, Capri M, Franceschi C, Salvioli S (2020) Inflammaging, hormesis and the rationale for anti-aging strategies. Ageing Res Rev 64:101142. https://doi.org/10.1016/j.arr.2020.101142

Son TG, Camandola S, Mattson MP (2008) Hormetic dietary phytochemicals. NeuroMolecular Med 10:236–246. https://doi.org/10.1007/s12017-008-8037-y

Sorrenti V, Fortinguerra S, Caudullo G, Buriani A (2020) Deciphering the role of polyphenols in sports performance: from nutritional genomics to the gut microbiota toward phytonutritional epigenomics. Nutrients 12:1265. https://doi.org/10.3390/nu12051265

Stranahan AM, Arumugam TV, Cutler RG, Lee K, Egan JM, Mattson MP (2008) Diabetes impairs hippocampal function through glucocorticoid-mediated effects on new and mature neurons. Nat Neurosci 11:309–317. https://doi.org/10.1038/nn2055

Thakur M, Singh K, Khedkar R (2020) Phytochemicals: extraction process, safety assessment, toxicological evaluations, and regulatory issues. In: Prakash B (ed) Functional and preservative properties of phytochemicals. Academic, pp 341–361. https://doi.org/10.1016/B978-0-12-818593-3.00011-7

van den Berg E, Dekker JM, Nijpels G, Kessels RP, Kappelle LJ, de Haan EH, Heine RJ, Stehouwer CD, Biessels GJ (2008) Cognitive functioning in elderly persons with type 2 diabetes and metabolic syndrome: the Hoorn study. Dement Geriatr Cogn Disord 26:261–269. https://doi.org/10.1159/000160959

Vasto S, Buscemi S, Barera A, Di Carlo M, Accardi G, Caruso C (2014a) Mediterranean diet and healthy ageing: a Sicilian perspective. Gerontology 60:508–518. https://doi.org/10.1159/000363060

Vasto S, Barera A, Rizzo C, Di Carlo M, Caruso C, Panotopoulos G (2014b) Mediterranean diet and longevity: an example of nutraceuticals? Curr Vasc Pharmacol 12:735–738. https://doi.org/10.2174/1570161111666131219111818

Zhang H, Cifone MA, Murli H, Erexson GL, Mecchi MS, Lawlor TE (2004) Application of simplified in vitro screening tests to detect genotoxicity of aristolochic acid. Food Chem Toxicol 42:2021–2028. https://doi.org/10.1016/j.fct.2004.07.016

Chapter 11
Melatonin as a Chronobiotic and Cytoprotector in Non-communicable Diseases: More than an Antioxidant

Daniel P. Cardinali, Seithikurippu R. Pandi-Perumal, and Gregory M. Brown

Abstract A circadian disruption, manifested by disturbed sleep and low-grade inflammation, is commonly seen in noncommunicable diseases (NCDs). Cardiovascular, respiratory and renal disorders, diabetes and the metabolic syndrome, cancer, and neurodegenerative diseases are among the most common NCDs prevalent in today's 24-h/7 days Society. The decline in plasma melatonin, which is a conserved phylogenetic molecule across all known aerobic creatures, is a constant feature in NCDs. The daily evening melatonin surge synchronizes both the central pacemaker located in the hypothalamic suprachiasmatic nuclei (SCN) and myriads of cellular clocks in the periphery ("chronobiotic effect"). Melatonin is the prototypical endogenous chronobiotic agent. Several meta-analyses and consensus studies support the use of melatonin to treat sleep/wake cycle disturbances associated with NCDs. Melatonin also has cytoprotective properties, acting primarily not only as an antioxidant by buffering free radicals, but also by regulating inflammation, down-regulating pro-inflammatory cytokines, suppressing low-grade inflammation, and preventing insulin resistance, among other effects. Melatonin's phylogenetic conservation is explained by its versatility of effects. In animal models of NCDs, melatonin treatment prevents a wide range of low-inflammation-linked alterations. As a result, the therapeutic efficacy of melatonin as a chronobiotic/cytoprotective drug has been proposed. Sirtuins 1 and 3 are at the heart of melatonin's chronobiotic

D. P. Cardinali (✉)
Faculty of Medical Sciences, Pontificia Universidad Católica Argentina, Buenos Aires, Argentina
e-mail: daniel_cardinali@uca.edu.ar

S. R. Pandi-Perumal
Saveetha Medical College and Hospitals, Saveetha Institute of Medical and Technical Sciences, Saveetha University, Chennai, India

G. M. Brown
Centre for Addiction and Mental Health, Department of Psychiatry, University of Toronto, Toronto, ON, Canada
e-mail: gregory.brown@camh.ca

© The Author(s), under exclusive license to Springer Nature Switzerland AG 2024

V. I. Korolchuk, J. R. Harris (eds.), *Biochemistry and Cell Biology of Ageing: Part V, Anti-Ageing Interventions*, Subcellular Biochemistry 107,
https://doi.org/10.1007/978-3-031-66768-8_11

and cytoprotective function, acting as accessory components or downstream elements of circadian oscillators and exhibiting properties such as mitochondrial protection. Allometric calculations based on animal research show that melatonin's cytoprotective benefits may require high doses in humans (in the 100 mg/day range). If melatonin is expected to improve health in NCDs, the low doses currently used in clinical trials (i.e., 2–10 mg) are unlikely to be beneficial. Multicentre double-blind studies are required to determine the potential utility of melatonin in health promotion. Moreover, melatonin dosage and levels used should be re-evaluated based on preclinical research information.

Keywords Circadian system · Cytoprotection · Inflammation · Melatonin · Sirtuins · Sleep/wake cycle

Introduction

Noncommunicable diseases (NCDs) account for 71% of all yearly worldwide deaths (Khan 2019). Cardiovascular diseases (CVD), cancer, respiratory disorders, diabetes, and neurological diseases account for more than 80% of NCD deaths. NCDs are strongly linked to impairment, dependency, and the need for long-term care, particularly in the elderly.

According to the World Health Organization, the number of people aged 60 and up will double from one billion in 2019 to two billion in 2050, with low- and middle-income countries accounting for 80% of all older people (https://www.who.int/news-room/fact-sheets/detail/ageing-and-health). People will encounter a host of health and quality-of-life challenges as they live longer lifetimes, including a rise in the prevalence of NCDs. NCDs are characterized in the aged by a persistent low-grade pro-inflammatory state (metainflammation), often known as "inflammaging" (Franceschi et al. 2018; Fulop et al. 2018; Barbé-Tuana et al. 2020).

NCDs involve the alteration of several pro- and anti-inflammatory pathways. The nuclear factor (NF)-κB and cyclooxygenase (COX) pathways are major proinflammatory pathways, whereas the hypothalamic/pituitary-adrenal axis, the inflammatory resolution pathway (specialized pro-resolving mediators, SPMs), and the melatonin/endocannabinoid/angiotensin 1,7 axis are major anti-inflammatory pathways (García et al. 2020; Lissoni 2021; Mińczuk et al. 2022). Circadian disruption plays a major role in the dysregulation of pro- and anti-inflammatory pathways (Besedovsky et al. 2022).

Inflammation mediators are involved in almost every human disease, as well as a wide range of biological processes, metabolism, and behavioral functions (Medzhitov 2021). The goal of inflammatory responses is to restore homeostasis. Indeed, the ability of the human body to resolve inflammation reduces with age, resulting in an imbalance that favors proinflammatory processes. As a comorbidity of inflammaging, sleep cycle disruption causes a slew of pathophysiological changes that hasten NCDs (Besedovsky et al. 2022).

This review examines the several activities of melatonin as a chronobiotic and cytoprotector in the context of NCDs (Cardinali 2019a, b). Melatonin is a methoxyindole that has pleiotropic functions. It has several qualities that make it effective in the treatment of both circadian dysregulation and inflammation. Melatonin acts as a chronobiotic (i.e., synchronization, phase shifting, and amplitude enhancement of circadian rhythms). It also acts as a direct and indirect antioxidant, immunological modulator, and mitochondrial protector and modulator. Melatonin levels decline with age, and they are even lower in people with NCDs. Melatonin reduces inflammatory responses and inflammation progression (Cardinali and Hardeland 2017). The relationship between melatonin and sirtuins (SIRT), which are also known for their roles as age suppressors, accessory components, or downstream elements of circadian oscillators, will be discussed in this review (Hardeland 2019). SIRT1 and SIRT3 appear to be central to the chronobiotic and cytoprotective activities of melatonin in NCDs.

Metainflammation, Inflammaging

The term "metainflammation" refers to the low degree imbalance of inflammatory and anti-inflammatory signals that occur in NCDs, particularly as people age ("inflammaging") (Fulop et al. 2021). This imbalance contributes to the development or worsening of NCDs such as CVD, metabolic syndrome and diabetes, neurodegenerative, renal, lung, and dermatological conditions. Metainflammation is characterized by increased inflammatory indexes including tumour necrosis factor (TNF)-α, interferon (IFN)-β, interleukin (IL)-1, IL-6, IL-8, IL-12, IL-17, and IL-22, chemokines, and other inflammatory factors such as C-reactive protein and monocyte chemoattractant protein-1 (MCP-1) (Xia et al. 2016).

Macrophages play a delicate role in balancing pro- and anti-inflammatory responses. They perform critical innate immunological functions, such as clearing dying cells via phagocytosis (Lu et al. 2021). Macrophages are classified into two polarization states: conventionally activated (M1) and alternatively activated (M2). M1 macrophages significantly express TNF-α, IL-6, MCP-1, and inducible nitric oxide synthase (iNOS), which are associated with pro-inflammatory cytokines or oxidative stress, while the anti-inflammatory cytokine IL-10 is highly expressed in M2 macrophages (Lu et al. 2021).

It is worth noting that inflammatory mediator levels tend to rise with age even in the absence of NCD. However, when stress occurs, it causes inflammatory damage to components of cells including proteins, lipids, and DNA, as well as contributing further to the age-related decline in physiological functions, particularly in neural, immune, and endocrine cells that regulate homeostasis (Bulut et al. 2021). As a result, age-related functional losses include a slowly acting, long-lasting type of oxidative stress caused by the increasing production of reactive oxygen and nitrogen species (ROS and RNS), which is increased by mitochondrial damage (García et al. 2020; Bader and Winklhofer 2020).

The term "oxidative stress" refers to an increase in the production of ROS and RNS in comparison to the level of antioxidants present in the body's natural defence systems. Melatonin is a unique endogenous antioxidant because of its anti-inflammatory and antioxidant properties, combined with its role as a metabolic regulator (Hardeland et al. 2015; Majidinia et al. 2018; García et al. 2020). Melatonin seems to be useful in promoting health.

The Circadian Apparatus

The daily and seasonal variations caused by Earth's rotation around the solar orbit have a significant impact on the species that inhabit it. The most visible manifestation of this periodic pattern is the daily light-dark (LD) cycle, which over eons led to the development of the endogenous circadian timing systems to synchronize biological functions with the environment (Foster 2020). Thus, the adaptation to anticipate predictable changes in the environment, such as light and darkness, temperature, food availability, or predator activity, evolved (predictive homeostasis) (Burdakov 2019). Hence, the circadian clock has become one of the most critical biological functions for living beings, working as a multipurpose timer to adapt the homeostatic system to the 24 h LD cycle, including sleep/wakefulness, hormone secretions, immunity, and most other physiological activities.

The mammalian circadian timing system comprises numerous tissue-specific cellular clocks. To generate appropriate physiological and behavioural responses, the phases of this multitude of body clocks are controlled by a master circadian oscillator (aka. pacemaker) found in the anterior hypothalamic suprachiasmatic nuclei (SCN) (Hastings et al. 2018). Natural light is the most pervasive and prominent synchronizer among photic (e.g., natural LD cycle) as well as non-photic (e.g., food, behavioural arousal, etc.) cues ("*Zeitgebers*"). The retinohypothalamic tract (RHT) entrains the SCN via messenger neurotransmitters, controlling the differential expression of clock genes and clock-controlled genes within SCN cells and influencing observable output in the form of physiology and behaviour (Hastings et al. 2020).

Circadian clocks are based on the so-called clock genes, with key genes encoding proteins that feedback and repress transcription at the molecular level. These oscillators consist of interconnected transcriptional and post-translational feedback loops that are regulated by a small number of core clock genes (Welz and Benitah 2020). Transgenic gene deletion technology in rats and mice was used to study the negative and positive transcriptional/translational feedback loops that comprise the core clockwork. The 24-h oscillation in clock gene expression is caused by a delay in the feedback loops, regulated in part by phosphorylation of the clock proteins, affecting their stability, nuclear re-entry, and transcription complex formation (Takahashi 2017).

At least four overlapping feedback loops comprise the mammalian circadian clock. As the *Clock* and *Bmal1* genes in mammals are transcribed, the transcription factors CLOCK and BMAL1 are produced, which then form dimers via basic helix-loop-helix (bHLH) domains. The dimers subsequently stimulated the transcription of two additional genes, *Per* and *Cry*, resulting in the creation of the proteins PER and CRY, which dimerized and inhibited the expression of CLOCK and BMAL1. The cycle is restarted as PER and CRY decline over time (Takahashi 2017). *Per* and *Cry* mRNA levels in the SCN peak in the middle to late afternoon in both nocturnal and diurnal mammals (Hastings et al. 2020). *Bmal1* mRNA levels rise around midnight, but *Clock* mRNA is present in the SCN at a constant concentration all day (Lee et al. 2001). Production of PER and CRY is limited by binding to the E-box element of the promoter regions of *Bmal1*, *Clock*, *Rev-Erb* and other clock-controlled genes via the CLOCK/BMAL1 complex (Takahashi 2017). Casein kinase 1 δ/ε phosphorylates PER and CRY, which are then translocated to the nucleus (Lee et al. 2001).

The master oscillation is further modulated by four secondary regulatory loops. One of them involves the nuclear receptors REV-ERB and ROR (retinoid-related orphan receptor). REV-ERB suppresses *Bmal1*, whereas ROR promotes it via attaching to the RORE (response element-binding site) sequence in *Bmal1*'s promoter region (Preitner et al. 2003; Fontaine and Staels 2007). A second regulatory loop is given by the protein ROR that also binds to the RORE element in the promoter of CLOCK, and to NFIL3 (nuclear factor, interleukin 3 regulated), resulting in their transcription. NR1D1 (nuclear receptor subfamily 1 group D member 1) and possibly other proteins from this family inhibit ROR binding to the RORE element. A third regulatory loop includes DBP (D-box binding PAR b ZIP transcription factor), a protein whose expression is controlled by BMAL1:CLOCK from the first loop binds that binds to the D box in the promoter region of PER. NFIL3 from the second loop regulates this binding negatively (Takahashi 2017).

CRY1, like PER1/2, is regulated by a combinatorial mechanism involving both E-box and RORE, resulting in a distinct phase from DBP and REV-ERB. The newly described DEC (AKA Basic helix-loop-helix family member e40 loop) is an ancillary circadian loop characterized by the expression of DEC and other circadian-controlled genes that are controlled by BMAL1:CLOCK. DEC, in turn, inhibits BMAL1:CLOCK binding to the E-box element, regulating its expression (fourth regulatory loop) (Ono et al. 2021). Phosphorylation and ubiquitylation by the E3 ligase complex regulate PER and CRY stability, leading to proteasomal degradation.

Circadian rhythm disruptions are harmful to one's health (Welz and Benitah 2020). Chronic jet lag and shift work have been linked to CVD disease (Crnko et al. 2019), memory loss (Snider and Obrietan 2018), hormonal timing disruptions (Maierova et al. 2016), type 2 diabetes mellitus (Stenvers et al. 2019; Oosterman et al. 2020; Tsereteli et al. 2021), impaired reproductive health (Caba et al. 2018; Pan et al. 2020), and metabolic disorders (Spiegel et al. 1999; Reinke and Asher 2019; Che et al. 2021). The use of chronotherapies, such as melatonin and/or timely light exposure, to modulate the molecular elements of circadian rhythms to alleviate these negative effects is beginning to gain traction in the scientific literature (Cardinali et al. 2023).

Melatonin as a Chronobiotic

The physiological regulation of the circadian rhythm of sleep/wakefulness (the body's main circadian rhythm) includes 24-h and homeostatic components (Borbély et al. 2016). Melatonin, an important 24-h component of the circadian clock, regulates sleep timing. Circadian rhythms in both the synthesis and secretion of pineal melatonin are tightly tied to the sleep rhythm not only in normal but also in blind patients (Emens and Eastman 2017). The onset of nightly melatonin secretion begins around 2 h prior to an individual's habitual bedtime and has been connected to the onset of evening weariness. Some studies have documented that sleep proclivity is the physiologic outcome of an increase in endogenous melatonin (Auld et al. 2017; Gobbi and Comai 2019).

Melatonin plays an important role in the coordination of circadian rhythmicity. Melatonin secretion is a "hand" of the biological clock that responds to SCN signals so that the timing of the melatonin rhythm reveals the phase of the clock (i.e., internal clock time relative to external clock time) as well as amplitude (Pévet et al. 2021). Melatonin, in another sense, is a chemical signal of the darkness: the longer the night, the greater the duration of its secretion. In most mammalian species, this secretion pattern serves as a timing cue for seasonal rhythms (Clarke and Caraty 2013; Wahab et al. 2018). The sympathetic nervous system regulates pineal melatonin production through a neural pathway that travels from the hypothalamic paraventricular nucleus (PVN) and ends at the upper level of the thoracic spinal cord (Pévet et al. 2021). The superior cervical ganglion (SCG) postganglionic sympathetic nerve terminals release norepinephrine into the pineal gland, where it interacts with β- (primarily) and α-adrenoceptors on pineal cell membranes to activate melatonin synthesis. Melatonin is highly diffusible, so it is not stored in the pineal and is released as soon as it is produced (Tan et al. 2018). The SCN-melatonin loop describes a collection of components that regulates circadian rhythms. The RHT, the SCN, the PVN, the intermediolateral cell column of the cervical spinal cord, the SCG, the pineal gland, and the melatonin rhythm are all part of this loop, which includes also the melanopsin-containing retinal ganglion cells (mRGCs). The melatonin produced leads to negative feedback effects on the SCN (Tan et al. 2018).

Vertebrate species produce pineal melatonin exclusively during the dark phase of the LD cycle. Melatonin is generated constantly during the night, regardless of the species' day activity/rest cycle, and is closely related to the external photoperiod. It should be emphasized that melatonin is produced at night providing that there is no light present (Pévet et al. 2021). During the night, blue light stimulates mRGCs, resulting in the reduction of pineal sympathetic norepinephrine (NE) release, thus lowering or eliminating melatonin generation.

The effects of melatonin as an internal *zeitgeber* (time cues or time giver) of the circadian clock, like the effects of the external *zeitgeber* light, are time-dependent. Melatonin alters the phase of the circadian clock in rats, which could explain how melatonin affects sleep in humans (Pévet et al. 2021). Clinical trials with melatonin in blind subjects (who have free-running circadian rhythms) provide indirect support for such a physiological involvement (Skene and Arendt 2007). It has been shown that the phase response curve for melatonin was opposite (i.e., approximately 180 degrees out of phase) to that of light providing more concrete evidence for this theory (Lewy 2010).

Melatonin receptors are present in both the CNS and the periphery (Dubocovich et al. 2010). The G-protein coupled (GPCR) families of MT1 and MT2 receptors have been cloned. GPR50, a new melatonin receptor subfamily member, was recently added (Cecon et al. 2018). GPR50 has a structure similar to MT1 and MT2, however it does not bind to melatonin or any other known ligand. Nevertheless, these receptors' capacity to form homo- and heteromers with other GPCRs, such as the serotonin 5-HT 2C receptor, may alter receptor activity.

While melatonin's main physiological role is to control circadian and seasonal rhythmicity, its actions are not confined to receptor-rich areas. Melatonin has effects on mitochondria including modulating electron flux, the permeability transition pore, and organelle's biogenesis, as well as having anti-excitatory properties, immunomodulation (including pro- and anti-inflammatory properties), antioxidant actions, and energy metabolism (Tan and Reiter 2019). Most of these actions are not mediated by receptors.

Melatonin is loosely linked to albumin in human blood (Cardinali et al. 1972) and is metabolized by hydroxylation in the liver before conjugation with sulphate or glucuronide (Claustrat and Leston 2015). 6-sulphatoxymelatonin (aMT6s) is the primary metabolite in human urine. In the brain, melatonin is converted to kynurenine metabolites, some of which (e.g. N^1-acetyl-5-methoxykynuramine) share its well-documented antioxidant effects. Additional antioxidant metabolites of melatonin are cyclic 3-hydroxymelatonin and N^1-acetyl-N^2-formyl-5-methoxykynuramine. Thus, melatonin administration to experimental animals and humans triggers an antioxidant cascade (Reiter et al. 2017).

The pineal gland produces nearly all circulating melatonin in mammals. However, a significant portion of melatonin is produced locally in cells, organs, and tissues including lymphocytes, bone marrow, the thymus, the gastrointestinal tract, the skin, and the eyes, where it can play an autocrine or paracrine role (Acuña-Castroviejo et al. 2014). Indeed, melatonin is known to be synthesized in every animal cell with mitochondria where it can play a functional role (Tan and Reiter 2019). Although it is widely accepted that natural melatonin's chronobiotic influence is mediated by MT receptors, a chronobiotic effect can also be observed when pharmaceutical quantities of fast-release melatonin (that saturate receptors) are utilized (Fig. 11.1). Even at such high doses, melatonin ingested as a fast-release preparation at a single time of day (bedtime) maintains chronobiotic effects (Cardinali et al. 2020b). In conclusion, the case for utilizing melatonin as a preventive drug in age-related NCDs is supported by improvements in the immunoinflammatory disease as well as general improvements and the prevention of possible consequences brought on by difficulties in maintaining normal circadian rhythmicity.

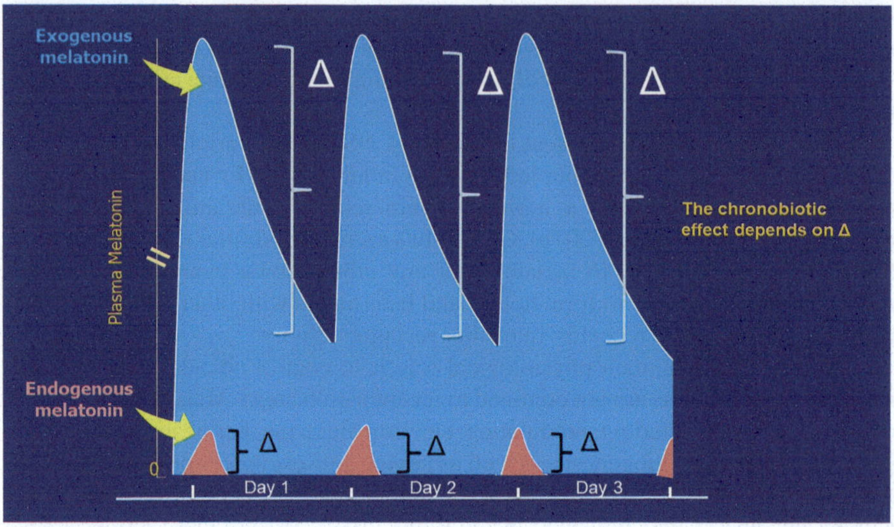

Fig. 11.1 Due to the pharmacokinetic characteristics i.e., since melatonin has a relatively short half-life in the blood when administered orally as a quick-release preparation at bedtime, it produces a chronobiotic signal regardless of the dosage given (reproduced with permission from Cardinali et al. 2020b)

Melatonin and Metainflammation

Melatonin's role in lowering inflammation has gotten a lot of attention recently, especially when it comes to therapeutic possibilities for those with low endogenous melatonin levels (Cardinali et al. 2023). Melatonin is a methoxyindole that has been shown to decrease with age and, more importantly, in various age-related NCDs (Hardeland 2012; Vasey et al. 2021). Low melatonin levels are found in people with chronic diseases, notably coronary heart disease, metabolic syndrome, and type 2 diabetes mellitus (Nagtegaal et al. 1995; Girotti et al. 2000; Altun et al. 2002; Yaprak et al. 2003; Hernández et al. 2007). Moreover, polymorphisms found in human melatonin receptor genes in prediabetes, type 2 diabetes, elevated cholesterol, and coronary heart disease suggest that melatonergic signalling deviations may favour the development of these disorders. In mice knocking out the melatonin receptor MT1 resulted in insulin resistance (Contreras-Alcantara et al. 2010).

Melatonin has anti-inflammatory properties via several physiopathogenic mechanisms. One of these is metabolic dysregulation repair, which includes preventing insulin resistance, an inflammation-promoting change that is associated with the metabolic syndrome (Cuesta et al. 2013; Lee et al. 2020). Melatonin is effective in reducing insulin resistance in a variety of animals and tissues, and with different induction methods. The reduction of serine phosphorylation of insulin receptor substrate 1 (IRS-1) is a key effect, which is frequently followed by an increase in IRS-1 expression (Du and Wei 2014). Melatonin and the melatonergic agonist piromelatine are known to reverse the inhibition of insulin signal transduction (She et al. 2009). Insulin resistance has been discovered as an early predictor of low-grade neuroinflammation in Alzheimer's and Parkinson's diseases (Verdile et al. 2015; Sun et al. 2020a).

Another useful action is to avoid processes that promote or lead to inflammation. Calcium overload, excessive nitric oxide (NO) release, which results in the formation of peroxynitrite, peroxynitrite-derived free radicals, and, eventually, tyrosine nitration, are all examples, as is a mitochondrial dysfunction due to oxidative stress (Hardeland et al. 2015; Cardinali and Hardeland 2017). Many of these alterations have been connected to low-grade inflammation in a range of organs and are associated with ageing. These include microglial activation and vicious cycles caused by overexcitation and oxidant damage in the central nervous system, resulting in decreased activity of neurons and astrocytes. Melatonin has been found in animal models to protect against these damaging processes by acting as an anti-excitatory agent, protecting mitochondria, lowering peroxynitrite-related damage, and blocking microglial activation.

The immunological effects of melatonin are a third significant aspect relevant to metainflammation. Melatonin's many immunomodulatory functions include both pro-inflammatory and anti-inflammatory effects, which result in either prooxidant or antioxidant disequilibrium (Carrillo-Vico et al. 2013; Hardeland 2019; Markus et al. 2021). In immunocompromised people, melatonin is often pro-inflammatory. The exact mechanisms by which melatonin acts pro- or anti-inflammatory are unknown, although inflammation intensity and the temporal sequence of initiation and healing processes are certainly involved. The anti-inflammatory effects of melatonin become increasingly essential as people age. Melatonin decreases pro-inflammatory cytokines such as TNF-α, IL-1, and IL-6 and it increases anti-inflammatory cytokine IL-10 in the livers of elderly, ovariectomized female rats (Kireev et al. 2008). Corresponding findings in the hippocampal dentate gyrus occur, as was an increase in SIRT1, a protein with strong anti-inflammatory properties (Kireev et al. 2014). TNF-α and IL-1 levels are lower in the liver, pancreas, and heart of the senescence-accelerated mouse strain SAMP8, whereas IL-10 levels were higher (Cuesta et al. 2010, 2011; Forman et al. 2011). Melatonin has been shown to have anti-inflammatory properties in brain injury, ischemia/reperfusion (I/R) lesions, hemorrhagic shock, and many forms of high-grade inflammation, including endotoxemia and sepsis. For example, the use of melatonin as a SARS-CoV-2 antidote has been advocated (Zhang et al. 2020; Cardinali et al. 2020a).

Distinguishing between direct and indirect anti-inflammatory effects of melatonin via changes in the phase or amplitude of local circadian oscillators is not always possible (Boivin et al. 2003; Bollinger et al. 2011; Hardeland et al. 2012). Melatonin has been demonstrated to influence various metabolic sensing factors such as peroxisome proliferator-activated receptor coactivator 1 α (PGC-1α), phosphoinositide 3-kinase, protein kinase B, and the accessory oscillator components AMP kinase, NAMPT, and SIRT1. It has been reported that melatonin will induce the induction of antioxidant enzymes in the rat liver and pancreas under inflammatory conditions including the expression and nuclear translocation of nuclear factor erythroid

2-related factor 2 (Nrf2), which mediates the upregulation of protective enzymes (Jung et al. 2010). Additionally, melatonin reduces NF-κB production by attracting a histone deacetylase (HDAC) to its promoter, reducing pro-inflammatory molecules such as TNF-, IL-1, and iNOS.

Melatonin, Sirtuins, and the Anti-Inflammatory Network

SIRT1 and SIRT3 are two sirtuins that are particularly important for the anti-inflammatory properties of melatonin (Mayo et al. 2017). SIRT1 is a multifunctional protein, which regulates gene transcription by deacetylating both histone and non-histone sites. Non-histone targets that regulate stress responses, inflammation, cellular senescence, and apoptosis include P53, FOXO transcription factor, PGC1α, and NF-κB (Watroba and Szukiewicz 2021). SIRT1 was initially thought to be an NAD^+ -dependent histone deacetylase because its activity depends on the cofactor NAD^+. Insulin sensitivity is improved by Sirt1 overexpression that deacetylates PGC-1α, a transcriptional coactivator that regulates glucose homeostasis at the transcriptional level, influencing glucose tolerance (Milne et al. 2007). Furthermore, Sirt1 overexpression in mice progeny reduces insulin resistance, improves glucose tolerance, prevents hepatic steatosis, and reduces ROS generation (Nguyen et al. 2017).

SIRT1 is also involved in signalling pathways that participate in the development of cognitive impairment, heart disease, ageing, cancer, and energy homeostasis, including lipid and glucose homeostasis (Watroba and Szukiewicz 2021). SIRT1 has been linked to increased longevity and the prevention of neurodegenerative diseases. SIRT1 overexpression in Alzheimer's disease reduces the increase in amyloid-β (Aβ) deposition (Fernando and Wijayasinghe 2021). SIRT1 overexpression is also beneficial in Parkinson's disease because it reduces acetylation of SIRT1 substrate and inhibits α-synuclein aggregation by preventing protein misfolding (Jęśko et al. 2017). SIRT1, like melatonin, has been shown in numerous studies to have antioxidant and anti-inflammatory properties (Mayo et al. 2017). This includes inhibiting TLR4 (toll-like receptor 4) signalling, suppressing NF-κB activation, upregulating Nrf2, and suppressing NLRP3 inflammasome activation. TLR4 activation is dependent on the inflammatory signalling protein HMGB1 (high mobility group box-1), which is secreted by monocytes and macrophages (Hardeland 2019). SIRT1 deacetylates HMGB1, preventing its nucleo-cytoplasmic transfer and release. Importantly, HMGB1 promotes the polarization of macrophages and microglia towards the pro-inflammatory M1 type. Melatonin has also been shown to have anti-inflammatory properties through HMGB1 inhibition (Mayo et al. 2017).

Several examples of sirtuin-mediated suppression by melatonin were discovered in more severe inflammation. This was observed in normal and diabetic rats with cardiac I/R, H9C2 cardiomyocytes with endoplasmic reticulum stress, LPS-treated microglial cell lines, and mice with cecal ligation/puncture-induced brain injury (Hardeland 2019). SIRT3 regulates the pyruvate dehydrogenase complex (PDH) and participates in ATP synthesis, making it an important factor in mitochondrial function.

Several studies have discovered that melatonin acts at the mitochondrial level via SIRT3 (Mayo et al. 2017). Switching from cytosolic aerobic glycolysis to oxidative phosphorylation enhances ATP biosynthesis, ATP production-coupled oxygen consumption rate, and lactic acid secretion. Melatonin, through its impacts on SIRT3 and PDH, boosted the mitochondrial membrane potential and the activity of complexes I and IV in the electron transport chain. Melatonin significantly improves mitochondrial energy metabolism by reversing the Warburg effect and stimulating SIRT3 (Reiter et al. 2020). Melatonin induces the transformation of pro-inflammatory glycolytic M1 macrophages into anti-inflammatory M2 macrophages (Xia et al. 2019) Melatonin stimulates mitochondrial pyruvate metabolism, the tricarboxylic acid cycle, oxidative phosphorylation, and ROS production via down-regulating hypoxia-inducible factor 1 (HIF-1), resulting in PDH disinhibition. Melatonin and its metabolites are especially efficient direct scavengers of partly reduced derivatives of oxygen under these circumstances, in addition to decreasing mitochondrial ROS generation.

Because macrophages and associated cells play an important role in inflammation, their differentiation into pro-inflammatory M1 or anti-inflammatory M2 phenotypes is essential for maintaining the pro−/anti-inflammatory balance (Fujisaka 2021). Melatonin can shift this balance towards the anti-inflammatory side by favouring M2 polarization and discouraging M1 polarization. The MT1 receptor-mediated activation of NF-κB degradation is one of the most important anti-inflammatory effects of decreasing M1 activity. Additionally, ROR has been shown to decrease NF-kB activities. Because ROR cannot bind melatonin, the effect of methoxyindole on the transcription factor must be indirect (Xia et al. 2019).

One intriguing notion is that SIRT1 acts on ROR as a partial modulator of melatonin effects. SIRT1 activation deacetylates PGC-1α, allowing ROR to attach to its response elements. SIRT1 and the circadian clock collaborate. SIRT1 influences the circadian clock in both the brain and peripheral organs (Masri 2015; Soni et al. 2021). Circadian gene expression patterns of *Per1*, *Per2*, *Cry1*, and *Cry2* in Sirt1-deficient mice are changed. Sirt1 and PER 2 work together to repress each other (Wang et al. 2016). SIRT1 deacetylates and destroys PER2 in the liver. SIRT1 also controls circadian rhythms by binding to the CLOCK-BMAL1 complex on a regular basis. As a result, the acetylation and deacetylation of its constituents have an impact on the molecular circadian clock.

Sirtuins and circadian clock components collaborate to control oxidative metabolism through NAD^+ and NADH responses (Anderson et al. 2017; Griffiths et al. 2020). The CLOCK-BMAL1 heterodimer not only activates the clock genes Per and Cry, as well as other clock-controlled genes, but it also regulates the activity of the gene *Nampt*, which encodes the rate-limiting enzyme nicotinamide phosphoribosyl transferase (NAmPRTase or NAMPT), the metabolite of which is NAD^+. The synthesis of NAD^+ has a distinct circadian cycle due to oscillations in NAMPT levels. The distribution of NAD^+ in the cytosol, nucleus, and mitochondria maintains the cellular redox status, which is required for the normal functioning of the bioenergetic enzymatic machinery (Anderson et al. 2017; Griffiths et al. 2020). The findings suggest that a complex set of regulators, including SIRT1, stabilizes the molecular circadian clock via multiple mechanisms. Melatonin regulates SIRT1 activity, which may be central to the cytoprotective and chronobiotic effects of methoxyindole (Emamgholipour et al. 2016; Bonomini et al. 2018; Stacchiotti et al. 2019; Favero et al. 2020). SIRT1 signalling in antioxidative response pathways mediates melatonin's cardioprotective action during I/R. The antioxidant enzymes manganese superoxide dismutase (MnSOD) and catalase are produced when SIRT1 is deacetylated. Acetylated FOXO1 increases apoptosis (Ac-FOXO1). SIRT1 and Ac-FOXO1 expression were drastically enhanced and lowered in melatonin-treated myocardial I/R rats, respectively. In the I/R plus vehicle group, SIRT1 expression was lowered whereas Ac-FOXO1 expression was dramatically elevated (Yu et al. 2014). Melatonin therapy boosted antiapoptotic gene Bcl-2 expression by upregulating SIRT1 and thereby reducing Ac-FOXO1.

Because it reduces oxidative stress, regulates inflammatory responses, and inhibits apoptotic pathways, SIRT1 is also the effector responsible for melatonin's protective role in kidney function in severely burned rats (Bai et al. 2016; Owczarek et al. 2020). SIRT1 contributes to melatonin's protective role after cecal ligation and puncture in a C57BL/6J mouse model of sepsis (Zhao et al. 2015). Melatonin alleviates the neuroinflammatory and oxidative stress caused by septic encephalopathy (Hu et al. 2017). A SIRT1 inhibitor reduced this benefit, implying that melatonin's beneficial effect is mediated by SIRT1 (Zhao et al. 2015). The activation of the NLRP3 inflammasome in several systems, under various conditions, and the effects of melatonin have recently been reviewed (Volt et al. 2016; Zheng et al. 2021; Sayed et al. 2021). Melatonin's modulation of NF-B signalling, which is vital in the protection of oxidative damage, was connected to these findings. Additionally, NF-κB has been found to promote pyroptosis in adipose tissue, which melatonin inhibits. TLR4 activation via the IFN- adaptor protein, a toll-receptor-associated activator of interferon (TRIF), is another pro-inflammatory pathway (Lwin et al. 2021; Feng et al. 2022). Melatonin has been demonstrated to block TRIF and TLR4 and hence lower the production of pro-inflammatory cytokines such as TNF- α, IL-1, IL-6, and IL-8. Melatonin's effects on this pathway are likely to be more widespread because TLR4 also causes pro-oxidant actions via NF-κB.

Melatonin's Therapeutic Value in Animal and Clinical Models of NCDs

As previously indicated, NCDs consistently co-exit with reduced circulating melatonin levels. A modest number of clinical trials with melatonin doses ranging from 2 to 5 mg/day yielded partially favourable results. Melatonin, on the other hand, was highly effective in reducing symptomatology in animal models of NCDs. Allometric calculations based on these animal studies predict that cytoprotective melatonin doses for humans should be in the 40–100 mg/day range, doses which are rarely used clinically (Cardinali 2019c).

Melatonin reduces obesity, type 2 diabetes, and hepatic steatosis in rats (Pan et al. 2006; Martínez Soriano et al. 2020). Melatonin infusions restored most of the observed physical abnormalities and modifications in the pro-inflammatory metabolic profile in various animal models of obesity. Streptozotocin-induced type 1 diabetic mice treated with melatonin caused pancreatic β-cell regeneration and proliferation as well as lowering blood glucose levels (Kanter et al. 2006; Hajam et al. 2021). Melatonin deficiency in the bloodstream after pinealectomy causes hyperinsulinemia and lipid build-up in the rat liver (Nishida et al. 2003). Long-term treatment with melatonin improves lipid metabolism in type 2 diabetic mice and increased insulin sensitivity. Melatonin treatment increased glycogen content in rats' livers, while intraperitoneal injection of 10 mg/kg melatonin improved glucose consumption, and insulin sensitivity, and alleviated hepatic steatosis in high-fat diet-induced diabetic mice (Shieh et al. 2009).

The causes of the drop in body weight after taking melatonin deserve to be investigated further because it occurs in the absence of major changes in food intake. The fact that melatonin influences seasonal changes in adiposity by increasing the activity of the sympathetic nervous system, which innervates white and brown fat, suggests a role for that system (Bartness et al. 2002; Ryu et al. 2018). Melatonin influences both white and brown adipose tissue recruitment and metabolic activity in mammals (Fernández Vázquez et al. 2018; de Souza et al. 2019; Halpern et al. 2019, 2020). Melatonin's hypertrophic impact and the functional stimulation of brown adipose tissue have been postulated as potential therapeutic options for human obesity.

Melatonin human equivalent dose (HED) for a 75 kg adult can be calculated by normalizing body surface area from animal doses (Reagan-Shaw et al. 2008; Blanchard and Smoliga 2015; Nair et al. 2018). Body surface area has been advocated as a factor to use when converting a dose for translation from animals to humans because it correlates well with several biological parameters such as O_2 consumption, caloric expenditure, basal metabolism, blood volume, circulating plasma proteins, and renal function across several mammalian species. It should be noted that theoretical HEDs of melatonin derived from various research studies are 2- to 3-orders-of-magnitude higher than those used in humans.

A summary of the effects of melatonin in animal models of age-related NCDs has been provided (Cardinali 2019a, b). Melatonin reduced both the area of injury and the number of injured myocardium regions in a rat model of myocardial infarction (caused by a 3-h closure of the left anterior descending coronary artery) (Castagnino et al. 2002). Melatonin has been proven in several trials in rats and mice to minimize indications of cardiac damage, improve antioxidant defences, and regulate lipid profiles. This was also seen in cardiomyopathy caused by streptozotocin or doxorubicin (Kandemir et al. 2019). Melatonin improves cardiac progenitor cell therapeutic efficacy in a mouse model of myocardial infarction treated with cardiac progenitor cells (Ma et al. 2018). The subcellular distribution of melatonin in the heart of rats discovered that at a dose of 40 mg/kg b.w., the nucleus and mitochondrion had their highest melatonin concentrations (Acuña-Castroviejo et al. 2018). It should be noted that for a 70 kg human adult, the authors calculated a HED of melatonin = 112 mg/day (Acuña-Castroviejo et al. 2018).

Melatonin-mediated mechanisms in Alzheimer's disease prevention have been identified through cell line studies. Comprehensive reviews documenting melatonin's activity in reversing disrupted signalling mechanisms in neurodegeneration, including proteostasis dysfunction, autophagic integrity disruption, and anomalies in the insulin, Notch, and Wnt/beta-catenin signalling pathways, can be found in (Shukla et al. 2019; Melhuish Beaupre et al. 2021). The findings in transgenic Alzheimer's disease models support the hypothesis that melatonin affects Aβ metabolism primarily in the early stages of the pathogenic process (Corpas et al. 2018; Jürgenson et al. 2019; Sun et al. 2020b). The HED of melatonin for a 75 kg adult ranged from 2- to 3-orders of magnitude higher than those used in humans [see for ref. Cardinali 2019b].

The mechanism by which melatonin inhibits Aβ production is being actively investigated. Melatonin inhibits progressive sheet and/or amyloid fibrils by interacting with $A\beta_{40}$ and $A\beta_{42}$ (Pappolla et al. 1998), an interaction that appears to be dependent on structural melatonin rather than on its antioxidant properties. Melatonin may aid peptide clearance by increasing proteolytic breakdown and inhibiting the formation of secondary sheets. Melatonin also effectively protects cells in vitro and in vivo against Aβ-induced neurotoxicity and cell death due to oxidative stress. Melatonin has been shown to protect against Aβ toxicity, especially at the mitochondrial level (Cardinali 2019b). Melatonin inhibits tau hyperphosphorylation in neuroblastoma cells by modulating protein kinases and phosphatases (Solís-Chagoyán et al. 2020). Melatonin boosts Aβ clearance in the glymphatic system of AD transgenic mice (Pappolla et al. 2018). As a result, sleep disturbance as a comorbidity in Alzheimer's disease may contribute to the disease's development and progression via Aβ clearance failure (Bitar et al. 2021). The activation of microglia, which leads in increased production of pro-inflammatory cytokines, is another element in the pathogenesis of Alzheimer's disease. Melatonin inhibited the production of pro-inflammatory cytokines in microglia induced by Aβ, NF-kB, and NO (Rosales-Corral et al. 2003; Baeeri et al. 2021; Zhang et al. 2021). Melatonin also inhibited the DNA binding activity of NF-kB (Hardeland 2019).

CSF melatonin levels drop even in the preclinical stages of Alzheimer's disease, when patients have no cognitive impairment, implying that CSF melatonin reduction could be an early trigger and marker for the disease (Liu et al. 1999; Colwell 2021). Although it is uncertain whether relative melatonin shortage is a result or cause of neurodegeneration, it is apparent that melatonin deficiency worsens Alzheimer's disease, and that early circadian disturbance is a crucial deficit to consider. Melatonin levels were found to differ significantly between patients with mild cognitive impairment and those with Alzheimer's disease, with a negative relationship between neuropsychological examination and melatonin levels (Şirin et al. 2015; Zhang et al. 2021). According to meta-analyses and consensus reports, melatonin therapy improves sleep in dementia patients (Xu et al. 2015; Zhang et al. 2016; Trotti and Karroum 2016; Fatemeh et al. 2022). Melatonin treatment has not always resulted in persons with fully developed Alzheimer's disease. It should be highlighted that the heterogeneity of the sample investigated is one of the concerns with these individuals with fully established disorders. A review of published research on the use of melatonin in the early stages of cognitive decline, on the other hand, indicated that taking melatonin every night before retiring improves sleep quality and cognitive function in this stage of the disease (see for reference, (Wade et al. 2014; Wang et al. 2017; Cardinali 2019b; Sumsuzzman et al. 2021; Liu et al. 2021).

In terms of clinical studies on the therapeutic value of melatonin in age-related NCDs, type 2 diabetic patients have low circulating levels of melatonin, with a concurrent and expected regulation of mRNA expression of melatonin membrane receptors (Tütüncü et al. 2005; Otamas et al. 2020; el Aghoury et al. 2020; Tanaka et al. 2021). Furthermore, melatonin receptor allelic variants have been linked to an increase in fasting blood glucose levels and/or an increased risk of type 2 diabetes (Bouatia-Naji et al. 2009; Prokopenko et al. 2009; Tam et al. 2010; Bonnefond and Froguel 2017; Bai et al. 2020) as well as the polycystic ovarian syndrome (PCOS), another component of the metabolic syndrome (Song et al. 2015; Yi et al. 2020). Melatonin secretion is reduced in patients with coronary artery disease (Brugger et al. 1995; Sakotnik et al. 1999; Girotti et al. 2000, 2003; Domínguez-Rodríguez et al. 2002; Yaprak et al. 2003; Misaka et al. 2019), and in elderly hypertensive patients, nocturnal urinary melatonin excretion was inversely associated with the non-dipper pattern of hypertensive disease (Jonas et al. 2003; Obayashi et al. 2013). Melatonin (5 mg/day) therapy reduced nocturnal blood pressure in hypertensives while also alleviating age-related cardiovascular rhythm abnormalities (Scheer 2005; Cagnacci et al. 2005; Grossman et al. 2006; Gubin et al. 2016; Imenshahidi et al. 2020; Campos et al. 2020).

Melatonin (5 mg/day) treatment improves metabolic syndrome in obese and PCOS patients, as well as in bipolar and schizophrenic patients receiving second-generation antipsychotics (Koziróg et al. 2011; Modabbernia et al. 2014; Romo-Nava et al. 2014; Agahi et al. 2018; Tagliaferri et al. 2018; Mohammadi-Sartang et al. 2018; Alizadeh et al. 2021; Duan et al. 2021). Melatonin treatment improves the enzyme profile in patients with alcoholic hepatic steatosis (Gonciarz et al. 2010; Abdi et al. 2021). Melatonin therapy has been shown in several studies to improve glycemic control in type 2 diabetes patients (Kadhim et al. 2006; Raygan et al. 2019; Pourhanifeh et al. 2020; Satari et al. 2021; Anton et al. 2021; Bazyar et al. 2021). Distinguishing core symptoms (glucose homeostasis) from diabetes-associated pathologies such as liver steatosis, cardiovascular disease, retinopathy, nephropathy, or osteoporosis is critical in human studies (Banerjee et al. 2021). Melatonin is therapeutically effective in the majority of these diabetes-associated diseases.

Concluding Remarks

NCDs associated with aging pose a significant public health challenge. CVD, cancer, respiratory diseases, diabetes, and neurological diseases account for more than 80% of NCD deaths, and NCDs are strongly linked to disability, reliance, and long-term care needs (Khan 2019). In this review, we covered two main etiopathogenic processes that lead to NCDs: inflammation and circadian disturbance, the latter produced by living in a 24/7 civilization that interrupts the sleep/wake cycle. As a result of sleep/wake cycle dysregulation, a plethora of pathophysiological alterations that accelerate ageing occur. Melatonin emerges as a feasible non-toxic chronobiotic/cytoprotective strategy in this scenario (Fig. 11.2).

It is important to note that melatonin is extremely safe. The lethal dose-50 for melatonin intraperitoneal injection was determined for rats (1168 mg/kg) and mice (1131 mg/kg), but the lethal dose for melatonin oral administration (assessed up to 3200 mg/kg in rats) and melatonin subcutaneous injection (assessed up to 1600 mg/kg in rats and mice) could not be determined (Sugden 1983).

In humans, melatonin has an excellent safety profile and is generally well tolerated (Cardinali et al. 2022; Menczel Schrire et al. 2022). As was previously mentioned, melatonin combines two characteristics that are crucial for the prevention and treatment of age-related NCDs: it is a powerful chronobiotic that helps to rectify circadian disruption and a cytoprotective drug that treats inflammaging. In addition to its well-known antioxidative and anti-inflammatory actions, which have shown efficacy in the treatment of multiple diseases/conditions in which excessive free radical-mediated oxidative damage and hyperinflammation are causative factors, the studies summarized herein support melatonin's use as a viable preventive agent in the low-grade inflammation found in age-related NCDs.

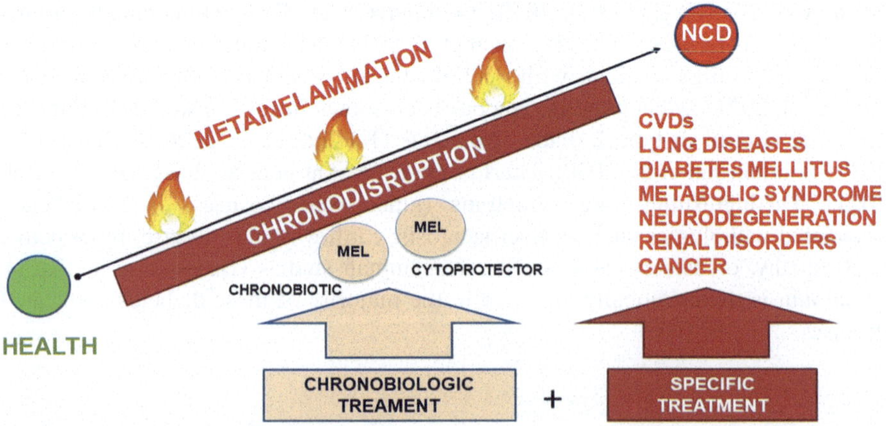

Fig. 11.2 The double effect of melatonin (MEL) in noncommunicable disease (NCD). CVDs cardiovascular disorders

Melatonin is frequently used as a dietary supplement or dietary product to treat sleep disturbances in various countries. According to the European Food Safety Authority (EFSA) melatonin reduces sleep onset delay. Melatonin may now be used orally to support "regulate sleep-wake cycles," "relaxation," and "sleep patterns" (Agostoni et al. 2011). Melatonin, melatonin-rich foods, and bio-extracts of melatonin can now be produced as dietary supplements, medications, and consumer goods for the general public, according to the EFSA.

To what degree melatonin has therapeutic efficacy in the prevention or treatment of NCDs, more study is required. Melatonin's potential and usefulness need to be explored and further investigated, which calls for multicentre double-blind trials. Because of the HED of melatonin determined from preclinical studies, melatonin dosages need to be reviewed. Indeed, given the number of scientific/medical papers that have recommended its use, melatonin's failure to garner attention as a potential treatment for healthy aging is disappointing. The fact that no significant group has advocated for its therapeutic usage in treating this condition is one of several potential causes for this. The pharmaceutical business is not motivated to promote the use of melatonin because it is not patented and affordable. Nonetheless, it would be wise for the pharmaceutical business to research the possibility of a profitable and medically effective combination of melatonin with specific medications. Due to its low cost, minimal toxicity, and ability to be taken orally, melatonin would be particularly advantageous. This is particularly true in underdeveloped nations where individuals have less money to spend on treating age-related NCDs.

Acknowledgements This research did not receive a specific grant from funding agencies in the public, commercial, or not-for-profit sectors.

Author Contributions Writing—original draft preparation, D.P.C.; writing—review and editing, S.R.P and G.M.B. All authors have read and agreed to the published version of the manuscript.

Conflict of Interest The authors declare that there are no commercial or financial relationships that could be construed as a potential conflict of interest.

References

Abdi S, Abbasinazari M, Ataei S et al (2021) Benefits and risks of melatonin in hepatic and pancreatic disorders; a review of clinical evidences. Iran J Pharm Res 20:102–109. https://doi.org/10.22037/ijpr.2020.114477.14872

Acuña-Castroviejo D, Escames G, Venegas C et al (2014) Extrapineal melatonin: sources, regulation, and potential functions. Cell Mol Life Sci 71:2997–3025

Acuña-Castroviejo D, Noguiera-Navarro MT, Reiter RJ, Escames G (2018) Melatonin actions in the heart; more than a hormone. Melatonin Res 1:21–26. https://doi.org/10.32794/MR11250002

Agahi M, Akasheh N, Ahmadvand A et al (2018) Effect of melatonin in reducing second-generation antipsychotic metabolic effects: a double blind controlled clinical trial. Diabetes Metab Syndr 12:9–15. https://doi.org/10.1016/J.DSX.2017.08.004

Agostoni C, Bresson J-L, Fairweather-Tait S et al (2011) Scientific opinion on the substantiation of a health claim related to melatonin and reduction of sleep onset latency (ID 1698, 1780, 4080) pursuant to Article 13(1) of Regulation (EC) No 1924/2006. EFSA J 9:2241. https://doi.org/10.2903/J.EFSA.2011.2241

Alizadeh M, Karandish M, Asghari Jafarabadi M et al (2021) Metabolic and hormonal effects of melatonin and/or magnesium supplementation in women with polycystic ovary syndrome: a randomized, double-blind, placebo-controlled trial. Nutr Metab (Lond) 18:57. https://doi.org/10.1186/s12986-021-00586-9

Altun A, Yaprak M, Aktoz M et al (2002) Impaired nocturnal synthesis of melatonin in patients with cardiac syndrome X. Neurosci Lett 327:143–145. https://doi.org/10.1016/S0304-3940(02)00368-3

Anderson KA, Madsen AS, Olsen CA, Hirschey MD (2017) Metabolic control by sirtuins and other enzymes that sense NAD+, NADH, or their ratio. Biochim Biophys Acta Bioenerg 1858:991–998. https://doi.org/10.1016/J.BBABIO.2017.09.005

Anton DM, Martu MA, Maris M et al (2021) Study on the effects of melatonin on glycemic control and periodontal parameters in patients with type II diabetes mellitus and periodontal disease. Medicina (Kaunas) 57:1–9. https://doi.org/10.3390/MEDICINA57020140

Auld F, Maschauer EL, Morrison I et al (2017) Evidence for the efficacy of melatonin in the treatment of primary adult sleep disorders. Sleep Med Rev 34:10–22

Bader V, Winklhofer KF (2020) Mitochondria at the interface between neurodegeneration and neuroinflammation. Semin Cell Dev Biol 99:163–171. https://doi.org/10.1016/j.semcdb.2019.05.028

Baeeri M, Didari T, Khalid M et al (2021) Molecular evidence of the inhibitory potential of melatonin against NaAsO2-induced aging in male rats. Molecules 26. https://doi.org/10.3390/molecules26216603

Bai XZ, He T, Gao JX et al (2016) Melatonin prevents acute kidney injury in severely burned rats via the activation of SIRT1. Sci Rep 6. https://doi.org/10.1038/SREP32199

Bai Y, Tang L, Li L (2020) The roles of ADIPOQ rs266729 and MTNR1B rs10830963 polymorphisms in patients with gestational diabetes mellitus: a meta-analysis. Gene 730. https://doi.org/10.1016/j.gene.2019.144302

Banerjee A, Chattopadhyay A, Bandyopadhyay D (2021) Potentially synergistic effects of melatonin and metformin in alleviating hyperglycaemia: a comprehensive review. Melatonin Res 4:522–550. https://doi.org/10.32794/MR112500110

Barbé-Tuana F, Funchal G, Schmitz CRR et al (2020) The interplay between immunosenescence and age-related diseases. Semin Immunopathol 42:545–557. https://doi.org/10.1007/S00281-020-00806-Z

Bartness TJ, Demas GE, Kay Song C (2002) Seasonal changes in adiposity: the roles of the photoperiod, melatonin and other hormones, and sympathetic nervous system. Exp Biol Med (Maywood) 227:363–376. https://doi.org/10.1177/153537020222700601

Bazyar H, Zare Javid A, Bavi Behbahani H et al (2021) Consumption of melatonin supplement improves cardiovascular disease risk factors and anthropometric indices in type 2 diabetes mellitus patients: a double-blind, randomized, placebo-controlled trial. Trials 22. https://doi.org/10.1186/S13063-021-05174-Z

Besedovsky L, Dang R, Engert LC et al (2022) Differential effects of an experimental model of prolonged sleep disturbance on inflammation in healthy females and males. PNAS Nexus 1. https://doi.org/10.1093/PNASNEXUS/PGAC004

Bitar RD, Torres-Garza JL, Reiter RJ, Phillips WT (2021) Neural glymphatic system: clinical implications and potential importance of melatonin. Melatonin Res 4:551–565. https://doi.org/10.32794/MR112500111

Blanchard OL, Smoliga JM (2015) Translating dosages from animal models to human clinical trials-revisiting body surface area scaling. FASEB J 29:1629–1634. https://doi.org/10.1096/fj.14-269043

Boivin DB, James FO, Wu A et al (2003) Circadian clock genes oscillate in human peripheral blood mononuclear cells. Blood 102:4143–4145. https://doi.org/10.1182/BLOOD-2003-03-0779

Bollinger T, Leutz A, Leliavski A et al (2011) Circadian clocks in mouse and human CD4+ T cells. PLoS One 6. https://doi.org/10.1371/JOURNAL.PONE.0029801

Bonnefond A, Froguel P (2017) Disentangling the role of melatonin and its receptor MTNR1B in type 2 diabetes: still a long way to go? Curr Diab Rep 17. https://doi.org/10.1007/s11892-017-0957-1

Bonomini F, Favero G, Rodella LF et al (2018) Melatonin modulation of sirtuin-1 attenuates liver injury in a hypercholesterolemic mouse model. Biomed Res Int 2018. https://doi.org/10.1155/2018/7968452

Borbély AA, Daan S, Wirz-Justice A, Deboer T (2016) The two-process model of sleep regulation: a reappraisal. J Sleep Res 25:131–143. https://doi.org/10.1111/JSR.12371

Bouatia Naji N, Bonnefond A, Cavalcanti-Proença C et al (2009) A variant near MTNR1B is associated with increased fasting plasma glucose levels and type 2 diabetes risk. Nat Genet 41:89–94. https://doi.org/10.1038/ng.277

Brugger P, Marktl W, Herold M (1995) Impaired nocturnal secretion of melatonin in coronary heart disease. Lancet 345:1408. https://doi.org/10.1016/S0140-6736(95)92600-3

Bulut O, Kilic G, Domínguez-Andrés J (2021) Immune memory in aging: a wide perspective covering microbiota, brain, metabolism, and epigenetics. Clin Rev Allergy Immunol. https://doi.org/10.1007/s12016-021-08905-x

Burdakov D (2019) Reactive and predictive homeostasis: roles of orexin/hypocretin neurons. Neuropharmacology 154:61–67. https://doi.org/10.1016/J.NEUROPHARM.2018.10.024

Caba M, González-Mariscal G, Meza E (2018) Circadian rhythms and clock genes in reproduction: insights from behavior and the female rabbit's brain. Front Endocrinol (Lausanne) 9. https://doi.org/10.3389/FENDO.2018.00106

Cagnacci A, Cannoletta M, Renzi A et al (2005) Prolonged melatonin administration decreases nocturnal blood pressure in women. Am J Hypertens 18:1614–1618. https://doi.org/10.1016/J.AMJHYPER.2005.05.008

Campos LA, Bueno C, Barcelos IP et al (2020) Melatonin therapy improves cardiac autonomic modulation in pinealectomized patients. Front Endocrinol (Lausanne) 11. https://doi.org/10.3389/FENDO.2020.00239

Cardinali DP (2019a) Melatonin as a chronobiotic/cytoprotector: its role in healthy aging. Biol Rhythm Res 50. https://doi.org/10.1080/09291016.2018.1491200

Cardinali DP (2019b) Melatonin: clinical perspectives in neurodegeneration. Front Endocrinol (Lausanne) 10. https://doi.org/10.3389/fendo.2019.00480

Cardinali DP (2019c) Are melatonin doses employed clinically adequate for melatonin-induced cytoprotection? Melatonin Res 2:106–132. https://doi.org/10.32794/mr11250025

Cardinali DP, Hardeland R (2017) Inflammaging, metabolic syndrome and melatonin: a call for treatment studies. Neuroendocrinology 104. https://doi.org/10.1159/000446543

Cardinali DP, Lynch HJ, Wurtman RJ (1972) Binding of melatonin to human and rat plasma proteins. Endocrinology 91:1213–1218. https://doi.org/10.1210/ENDO-91-5-1213

Cardinali DP, Brown GM, Pandi-Perumal SR (2020a) Can melatonin be a potential "silver bullet" in treating Covid-19 patients? Diseases 8. https://doi.org/10.3390/diseases8040044

Cardinali DP, Brown GM, Reiter RJ, Pandi-Perumal SR (2020b) Elderly as a high-risk group during Covid-19 pandemic: effect of circadian misalignment, sleep dysregulation and melatonin administration. Sleep Vigil 1–7. https://doi.org/10.1007/s41782-020-00111-7

Cardinali DP, Brown GM, Pandi-Perumal SR (2022) Melatonin's benefits and risks as a therapy for sleep disturbances in the elderly: current insights. Nat Sci Sleep 14:1843–1855. https://doi.org/10.2147/NSS.S380465

Cardinali DP, Pandi-Perumal SR, Brown GM (2023) Melatonin as a chronobiotic and cytoprotector in healthy aging. In: Jagota A (ed) Sleep and clocks in aging and longevity, Healthy ageing and longevity, vol 18. Springer, Cham, pp 277–312. https://doi.org/10.1007/978-3-031-22468-3_14
Carrillo-Vico A, Lardone PJ, Álvarez-Sánchez N et al (2013) Melatonin: buffering the immune system. Int J Mol Sci 14:8638–8683. https://doi.org/10.3390/IJMS14048638
Castagnino HE, Lago N, Centrella JM et al (2002) Cytoprotection by melatonin and growth hormone in early rat myocardial infarction as revealed by Feulgen DNA staining. Neuroendocrinol Lett 23:391–395
Cecon E, Oishi A, Jockers R (2018) Melatonin receptors: molecular pharmacology and signalling in the context of system bias. Br J Pharmacol 175:3263–3280. https://doi.org/10.1111/BPH.13950
Che T, Yan C, Tian D et al (2021) The association between sleep and metabolic syndrome: a systematic review and meta-analysis. Front Endocrinol (Lausanne) 12:773646. https://doi.org/10.3389/fendo.2021.773646
Clarke IJ, Caraty A (2013) Kisspeptin and seasonality of reproduction. Adv Exp Med Biol 784:411–430. https://doi.org/10.1007/978-1-4614-6199-9_19
Claustrat B, Leston J (2015) Melatonin: physiological effects in humans. Neurochirurgie 61:77–84. https://doi.org/10.1016/J.NEUCHI.2015.03.002
Colwell CS (2021) Defining circadian disruption in neurodegenerative disorders. J Clin Invest 131. https://doi.org/10.1172/JCI148288
Contreras-Alcantara S, Baba K, Tosini G (2010) Removal of melatonin receptor type 1 induces insulin resistance in the mouse. Obesity (Silver Spring) 18:1861–1863. https://doi.org/10.1038/OBY.2010.24
Corpas R, Griñán-Ferré C, Palomera-Ávalos V et al (2018) Melatonin induces mechanisms of brain resilience against neurodegeneration. J Pineal Res 65. https://doi.org/10.1111/JPI.12515
Crnko S, du Pré BC, Sluijter JPG, van Laake LW (2019) Circadian rhythms and the molecular clock in cardiovascular biology and disease. Nat Rev Cardiol 16:437–447. https://doi.org/10.1038/S41569-019-0167-4
Cuesta S, Kireev R, Forman K et al (2010) Melatonin improves inflammation processes in liver of senescence-accelerated prone male mice (SAMP8). Exp Gerontol 45:950–956. https://doi.org/10.1016/J.EXGER.2010.08.016
Cuesta S, Kireev R, García C et al (2011) Beneficial effect of melatonin treatment on inflammation, apoptosis and oxidative stress on pancreas of a senescence accelerated mice model. Mech Ageing Dev 132:573–582. https://doi.org/10.1016/J.MAD.2011.10.005
Cuesta S, Kireev R, García C et al (2013) Melatonin can improve insulin resistance and aging-induced pancreas alterations in senescence-accelerated prone male mice (SAMP8). Age (Dordr) 35:659–671. https://doi.org/10.1007/S11357-012-9397-7
de Souza CAP, Gallo CC, de Camargo LS et al (2019) Melatonin multiple effects on brown adipose tissue molecular machinery. J Pineal Res 66. https://doi.org/10.1111/JPI.12549
Domínguez-Rodríguez A, Abreu-González P, García MJ et al (2002) Decreased nocturnal melatonin levels during acute myocardial infarction. J Pineal Res 33:248–252. https://doi.org/10.1034/J.1600-079X.2002.02938.X
Du Y, Wei T (2014) Inputs and outputs of insulin receptor. Protein Cell 5:203–213. https://doi.org/10.1007/S13238-014-0030-7
Duan C, Jenkins ZM, Castle D (2021) Therapeutic use of melatonin in schizophrenia: a systematic review. World J Psychiatry 11:463–476. https://doi.org/10.5498/wjp.v11.i8.463
Dubocovich ML, Delagrange P, Krause DN et al (2010) International union of basic and clinical pharmacology. LXXV. Nomenclature, classification, and pharmacology of G protein-coupled melatonin receptors. Pharmacol Rev 62. https://doi.org/10.1124/pr.110.002832
el Aghoury AA, Elsayed ET, el Kholy NM et al (2020) Melatonin receptor 1A gene polymorphism rs13140012 and serum melatonin in atherosclerotic versus non-atherosclerotic Egyptian ESRD patients: pilot study. Heliyon 6:e04394. https://doi.org/10.1016/j.heliyon.2020.e04394

Emamgholipour S, Hossein-Nezhad A, Sahraian MA et al (2016) Evidence for possible role of melatonin in reducing oxidative stress in multiple sclerosis through its effect on SIRT1 and antioxidant enzymes. Life Sci 145:34–41. https://doi.org/10.1016/j.lfs.2015.12.014

Emens JS, Eastman CI (2017) Diagnosis and treatment of non-24-h sleep-wake disorder in the blind. Drugs 77:637–650. https://doi.org/10.1007/S40265-017-0707-3

Fatemeh G, Sajjad M, Niloufar R et al (2022) Effect of melatonin supplementation on sleep quality: a systematic review and meta-analysis of randomized controlled trials. J Neurol 269:205–216. https://doi.org/10.1007/S00415-020-10381-W

Favero G, Franco C, Stacchiotti A et al (2020) Sirtuin1 role in the melatonin protective effects against obesity-related heart injury. Front Physiol 11:103. https://doi.org/10.3389/fphys.2020.00103

Feng R, Adeniran SO, Huang F et al (2022) The ameliorative effect of melatonin on LPS-induced Sertoli cells inflammatory and tight junctions damage via suppression of the TLR4/MyD88/NF-κB signaling pathway in newborn calf. Theriogenology 179:103–116. https://doi.org/10.1016/J.THERIOGENOLOGY.2021.11.020

Fernández Vázquez G, Reiter RJ, Agil A (2018) Melatonin increases brown adipose tissue mass and function in Zücker diabetic fatty rats: implications for obesity control. J Pineal Res 64. https://doi.org/10.1111/jpi.12472

Fernando KKM, Wijayasinghe YS (2021) Sirtuins as potential therapeutic targets for mitigating neuroinflammation associated with Alzheimer's disease. Front Cell Neurosci 15. https://doi.org/10.3389/FNCEL.2021.746631

Fontaine C, Staels B (2007) The orphan nuclear receptor Rev-erbα: a transcriptional link between circadian rhythmicity and cardiometabolic disease. Curr Opin Lipidol 18:141–146. https://doi.org/10.1097/MOL.0b013e3280464ef6

Forman K, Vara E, García C et al (2011) Effect of a combined treatment with growth hormone and melatonin in the cardiological aging on male SAMP8 mice. J Gerontol A Biol Sci Med Sci 66:823–834. https://doi.org/10.1093/GERONA/GLR083

Foster RG (2020) Sleep, circadian rhythms and health. Interface Focus 10:20190098. https://doi.org/10.1098/rsfs.2019.0098

Franceschi C, Garagnani P, Parini P et al (2018) Inflammaging: a new immune-metabolic viewpoint for age-related diseases. Nat Rev Endocrinol 14:576–590. https://doi.org/10.1038/S41574-018-0059-4

Fujisaka S (2021) The role of adipose tissue M1/M2 macrophages in type 2 diabetes mellitus. Diabetol Int 12:74–79. https://doi.org/10.1007/s13340-020-00482-2

Fulop T, Witkowski JM, Olivieri F, Larbi A (2018) The integration of inflammaging in age-related diseases. Semin Immunol 40:17–35. https://doi.org/10.1016/J.SMIM.2018.09.003

Fulop T, Larbi A, Pawelec G et al (2021) Immunology of aging: the birth of inflammaging. Clin Rev Allergy Immunol. https://doi.org/10.1007/S12016-021-08899-6

García S, Giménez VMM, Marón FJM et al (2020) Melatonin and cannabinoids: mitochondrial-targeted molecules that may reduce inflammaging in neurodegenerative diseases. Histol Histopathol 35:789–800. https://doi.org/10.14670/HH-18-212

Girotti L, Lago M, Ianovsky O et al (2000) Low urinary 6-sulphatoxymelatonin levels in patients with coronary artery disease. J Pineal Res 29. https://doi.org/10.1034/j.1600-079X.2000.290302.x

Girotti L, Lago M, Ianovsky O et al (2003) Low urinary 6-sulfatoxymelatonin levels in patients with severe congestive heart failure. Endocrine 22. https://doi.org/10.1385/ENDO:22:3:245

Gobbi G, Comai S (2019) Sleep well. Untangling the role of melatonin MT1 and MT2 receptors in sleep. J Pineal Res 66. https://doi.org/10.1111/JPI.12544

Gonciarz M, Gonciarz Z, Bielanski W et al (2010) The pilot study of 3-month course of melatonin treatment of patients with nonalcoholic steatohepatitis: effect on plasma levels of liver enzymes, lipids and melatonin. J Physiol Pharmacol 61:705–710

Griffiths HBS, Williams C, King SJ, Allison SJ (2020) Nicotinamide adenine dinucleotide (NAD+): essential redox metabolite, co-substrate and an anti-cancer and anti-ageing therapeutic target. Biochem Soc Trans 48:733–744. https://doi.org/10.1042/BST20190033

Grossman E, Laudon M, Yalcin R et al (2006) Melatonin reduces night blood pressure in patients with nocturnal hypertension. Am J Med 119:898–902. https://doi.org/10.1016/j.amjmed.2006.02.002
Gubin DG, Gubin GD, Gapon LI, Weinert D (2016) Daily melatonin administration attenuates age-dependent disturbances of cardiovascular rhythms. Curr Aging Sci 9:5–13. https://doi.org/10.2174/1874609809666151130220011
Hajam YA, Rai S, Pandi-Perumal SR et al (2021) Coadministration of melatonin and insulin improves diabetes-induced impairment of rat kidney function. Neuroendocrinology. https://doi.org/10.1159/000520280
Halpern B, Mancini MC, Bueno C et al (2019) Melatonin increases brown adipose tissue volume and activity in patients with melatonin deficiency: a proof-of-concept study. Diabetes 68:947–952. https://doi.org/10.2337/DB18-0956
Halpern B, Mancini MC, Mendes C et al (2020) Melatonin deficiency decreases brown adipose tissue acute thermogenic capacity of in rats measured by 18F-FDG PET. Diabetol Metab Syndr 12:82. https://doi.org/10.1186/s13098-020-00589-1
Hardeland R (2012) Neurobiology, pathophysiology, and treatment of melatonin deficiency and dysfunction. Sci World J 2012. https://doi.org/10.1100/2012/640389
Hardeland R (2019) Aging, melatonin, and the pro-and anti-inflammatory networks. Int J Mol Sci 20:1–33. https://doi.org/10.3390/ijms20051223
Hardeland R, Madrid JA, Tan DX, Reiter RJ (2012) Melatonin, the circadian multioscillator system and health: the need for detailed analyses of peripheral melatonin signaling. J Pineal Res 52:139–166. https://doi.org/10.1111/J.1600-079X.2011.00934.X
Hardeland R, Cardinali DP, Brown GM, Pandi-Perumal SR (2015) Melatonin and brain inflammaging. Prog Neurobiol 127–128. https://doi.org/10.1016/j.pneurobio.2015.02.001
Hastings MH, Maywood ES, Brancaccio M (2018) Generation of circadian rhythms in the suprachiasmatic nucleus. Nat Rev Neurosci 19:453–469
Hastings MH, Smyllie NJ, Patton AP (2020) Molecular-genetic manipulation of the suprachiasmatic nucleus circadian clock. J Mol Biol 432:3639–3660
Hernández C, Abreu J, Abreu P et al (2007) Nocturnal melatonin plasma levels in patients with OSAS: the effect of CPAP. Eur Respir J 30:496–500. https://doi.org/10.1183/09031936.00051906
Hu W, Deng C, Ma Z et al (2017) Utilizing melatonin to combat bacterial infections and septic injury. Br J Pharmacol 174:754–768. https://doi.org/10.1111/BPH.13751
Imenshahidi M, Karimi G, Hosseinzadeh H (2020) Effects of melatonin on cardiovascular risk factors and metabolic syndrome: a comprehensive review. Naunyn Schmiedeberg's Arch Pharmacol 393:521–536. https://doi.org/10.1007/s00210-020-01822-4
Jęśko H, Wencel P, Strosznajder RP, Strosznajder JB (2017) Sirtuins and their roles in brain aging and neurodegenerative disorders. Neurochem Res 42:876–890. https://doi.org/10.1007/S11064-016-2110-Y
Jonas M, Garfinkel D, Zisapel N et al (2003) Impaired nocturnal melatonin secretion in non-dipper hypertensive patients. Blood Press 12:19–24
Jung KH, Hong SW, Zheng HM et al (2010) Melatonin ameliorates cerulein-induced pancreatitis by the modulation of nuclear erythroid 2-related factor 2 and nuclear factor-kappa B in rats. J Pineal Res 48:239–250. https://doi.org/10.1111/J.1600-079X.2010.00748.X
Jürgenson M, Zharkovskaja T, Noortoots A et al (2019) Effects of the drug combination memantine and melatonin on impaired memory and brain neuronal deficits in an amyloid-predominant mouse model of Alzheimer's disease. J Pharm Pharmacol 71:1695–1705. https://doi.org/10.1111/JPHP.13165
Kadhim HM, Ismail SH, Hussein KI et al (2006) Effects of melatonin and zinc on lipid profile and renal function in type 2 diabetic patients poorly controlled with metformin. J Pineal Res 41:189–193. https://doi.org/10.1111/J.1600-079X.2006.00353.X
Kandemir YB, Tosun V, Güntekin Ü (2019) Melatonin protects against streptozotocin-induced diabetic cardiomyopathy through the mammalian target of rapamycin (mTOR) signaling pathway. Adv Clin Exp Med 28:1171–1177. https://doi.org/10.17219/acem/103799

Kanter M, Uysal H, Karaca T, Sagmanligil HO (2006) Depression of glucose levels and partial restoration of pancreatic beta-cell damage by melatonin in streptozotocin-induced diabetic rats. Arch Toxicol 80:362–369. https://doi.org/10.1007/S00204-005-0055-Z

Khan HTA (2019) Population ageing in a globalized world: risks and dilemmas? J Eval Clin Pract 25:754–760. https://doi.org/10.1111/JEP.13071

Kireev RA, Tresguerres ACF, Garcia C et al (2008) Melatonin is able to prevent the liver of old castrated female rats from oxidative and pro-inflammatory damage. J Pineal Res 45:394–402. https://doi.org/10.1111/J.1600-079X.2008.00606.X

Kireev RA, Vara E, Viña J, Tresguerres JAF (2014) Melatonin and oestrogen treatments were able to improve neuroinflammation and apoptotic processes in dentate gyrus of old ovariectomized female rats. Age (Dordr) 36. https://doi.org/10.1007/S11357-014-9707-3

Koziróg M, Poliwczak AR, Duchnowicz P et al (2011) Melatonin treatment improves blood pressure, lipid profile, and parameters of oxidative stress in patients with metabolic syndrome. J Pineal Res 50:261–266. https://doi.org/10.1111/J.1600-079X.2010.00835.X

Lee C, Etchegaray JP, Cagampang FRA et al (2001) Posttranslational mechanisms regulate the mammalian circadian clock. Cell 107:855–867. https://doi.org/10.1016/S0092-8674(01)00610-9

Lee YH, Jung HS, Kwon MJ et al (2020) Melatonin protects INS-1 pancreatic β-cells from apoptosis and senescence induced by glucotoxicity and glucolipotoxicity. Islets 87–98. https://doi.org/10.1080/19382014.2020.1783162

Lewy A (2010) Clinical implications of the melatonin phase response curve. J Clin Endocrinol Metab 95:3158–3160. https://doi.org/10.1210/jc.2010-1031

Lissoni P (2021) How to modulate the immunoinflammatory biological response in cancer, autoimmunity and viral infections by a psychoneuroimmune approach. Nov Approaches Cancer Study 5. https://doi.org/10.31031/NACS.2021.05.000622

Liu R-Y, Zhou J-N, van Heerikhuize J et al (1999) Decreased melatonin levels in postmortem cerebrospinal fluid in relation to aging, Alzheimer's disease, and apolipoprotein E-epsilon 4/4 genotype. J Clin Endocrinol Metab 84:323–327. https://doi.org/10.1210/JCEM.84.1.5394

Liu Y-S, Wang Y-M, Zha D-J (2021) Brain functional and structural changes in Alzheimer's disease with sleep disorders: a systematic review. Front Psych 12:772068. https://doi.org/10.3389/fpsyt.2021.772068

Lu B, Huang L, Cao J et al (2021) Adipose tissue macrophages in aging-associated adipose tissue function. J Physiol Sci 71. https://doi.org/10.1186/S12576-021-00820-2

Lwin T, Yang JL, Ngampramuan S et al (2021) Melatonin ameliorates methamphetamine-induced cognitive impairments by inhibiting neuroinflammation via suppression of the TLR4/MyD88/NFκB signaling pathway in the mouse hippocampus. Prog Neuro-Psychopharmacol Biol Psychiatry 111. https://doi.org/10.1016/J.PNPBP.2020.110109

Ma W, He F, Ding F et al (2018) Pre-treatment with melatonin enhances therapeutic efficacy of cardiac progenitor cells for myocardial infarction. Cell Physiol Biochem 47:1287–1298. https://doi.org/10.1159/000490224

Maierova L, Borisuit A, Scartezzini JL et al (2016) Diurnal variations of hormonal secretion, alertness and cognition in extreme chronotypes under different lighting conditions. Sci Rep 6. https://doi.org/10.1038/SREP33591

Majidinia M, Reiter RJ, Shakouri SK, Yousefi B (2018) The role of melatonin, a multitasking molecule, in retarding the processes of ageing. Ageing Res Rev 47:198–213. https://doi.org/10.1016/J.ARR.2018.07.010

Markus RP, Sousa KS, da Silveira Cruz-Machado S et al (2021) Possible role of pineal and extra-pineal melatonin in surveillance, immunity, and first-line defense. Int J Mol Sci 22. https://doi.org/10.3390/ijms222212143

Martínez Soriano B, Güemes A, Pola G et al (2020) Effect of melatonin as an antioxidant drug to reverse hepatic steatosis: experimental model. Can J Gastroenterol Hepatol 2020. https://doi.org/10.1155/2020/7315253

Masri S (2015) Sirtuin-dependent clock control: new advances in metabolism, aging and cancer. Curr Opin Clin Nutr Metab Care 18:521–527. https://doi.org/10.1097/MCO.0000000000000219

Mayo JC, Sainz RM, González Menéndez P et al (2017) Melatonin and sirtuins: a "not-so unexpected" relationship. J Pineal Res 62. https://doi.org/10.1111/JPI.12391

Medzhitov R (2021) The spectrum of inflammatory responses. Science (1979) 374:1070–1075. https://doi.org/10.1126/SCIENCE.ABI5200

Melhuish Beaupre LM, Brown GM, Gonçalves VF, Kennedy JL (2021) Melatonin's neuroprotective role in mitochondria and its potential as a biomarker in aging, cognition and psychiatric disorders. Transl Psychiatry 11. https://doi.org/10.1038/s41398-021-01464-x

Menczel Schrire Z, Phillips CL, Chapman JL et al (2022) Safety of higher doses of melatonin in adults: a systematic review and meta-analysis. J Pineal Res 72. https://doi.org/10.1111/JPI.12782

Milne JC, Lambert PD, Schenk S et al (2007) Small molecule activators of SIRT1 as therapeutics for the treatment of type 2 diabetes. Nature 450:712–716. https://doi.org/10.1038/NATURE06261

Mińczuk K, Baranowska-kuczko M, Krzyżewska A et al (2022) Cross-talk between the (endo)cannabinoid and renin-angiotensin systems: basic evidence and potential therapeutic significance. Int J Mol Sci 23. https://doi.org/10.3390/IJMS23116350

Misaka T, Yoshihisa A, Yokokawa T et al (2019) Plasma levels of melatonin in dilated cardiomyopathy. J Pineal Res 66. https://doi.org/10.1111/jpi.12564

Modabbernia A, Heidari P, Soleimani R et al (2014) Melatonin for prevention of metabolic side-effects of olanzapine in patients with first-episode schizophrenia: randomized double-blind placebo-controlled study. J Psychiatr Res 53:133–140. https://doi.org/10.1016/j.jpsychires.2014.02.013

Mohammadi-Sartang M, Ghorbani M, Mazloom Z (2018) Effects of melatonin supplementation on blood lipid concentrations: a systematic review and meta-analysis of randomized controlled trials. Clin Nutr 37:1943–1954. https://doi.org/10.1016/J.CLNU.2017.11.003

Nagtegaal E, Smits M, Swart W et al (1995) Melatonin secretion and coronary heart disease. Lancet 346:1299. https://doi.org/10.1016/S0140-6736(95)91901-5

Nair A, Morsy MA, Jacob S (2018) Dose translation between laboratory animals and human in preclinical and clinical phases of drug development. Drug Dev Res 79:373–382. https://doi.org/10.1002/ddr.21461

Nguyen LT, Chen H, Pollock C, Saad S (2017) SIRT1 reduction is associated with sex-specific dysregulation of renal lipid metabolism and stress responses in offspring by maternal high-fat diet. Sci Rep 7. https://doi.org/10.1038/S41598-017-08694-4

Nishida S, Sato R, Murai I, Nakagawa S (2003) Effect of pinealectomy on plasma levels of insulin and leptin and on hepatic lipids in type 2 diabetic rats. J Pineal Res 35:251–256. https://doi.org/10.1034/J.1600-079X.2003.00083.X

Obayashi K, Saeki K, Iwamoto J et al (2013) Nocturnal urinary melatonin excretion is associated with non-dipper pattern in elderly hypertensives. Hypertens Res 36:736–740. https://doi.org/10.1038/HR.2013.20

Ono D, Honma K-I, Schmal C, et al (2021) CHRONO and DEC1/DEC2 compensate for lack of CRY1/CRY2 in expression of coherent circadian rhythm but not in generation of circadian oscillation in the neonatal mouse SCN. Sci Rep 11. doi:https://doi.org/10.1038/S41598-021-98532-5

Oosterman JE, Wopereis S, Kalsbeek A (2020) The circadian clock, shift work, and tissue-specific insulin resistance. Endocrinology 161. https://doi.org/10.1210/endocr/bqaa180

Otamas A, Grant PJ, Ajjan RA (2020) Diabetes and atherothrombosis: the circadian rhythm and role of melatonin in vascular protection. Diab Vasc Dis Res 17. https://doi.org/10.1177/1479164120920582

Owczarek A, Gieczewska KB, Polanska M et al (2020) Melatonin lowers HIF-1α content in human proximal tubular cells (HK-2) due to preventing its deacetylation by sirtuin 1. Front Physiol 11:572911. https://doi.org/10.3389/fphys.2020.572911

Pan M, Song YL, Xu JM, Gan HZ (2006) Melatonin ameliorates nonalcoholic fatty liver induced by high-fat diet in rats. J Pineal Res 41:79–84. https://doi.org/10.1111/J.1600-079X.2006.00346.X

Pan X, Taylor MJ, Cohen E et al (2020) Circadian clock, time-restricted feeding and reproduction. Int J Mol Sci 21. https://doi.org/10.3390/ijms21030831

Pappolla M, Bozner P, Soto C et al (1998) Inhibition of Alzheimer beta-fibrillogenesis by melatonin. J Biol Chem 273:7185–7188. https://doi.org/10.1074/JBC.273.13.7185

Pappolla MA, Matsubara E, Vidal R et al (2018) Melatonin treatment enhances Aβ lymphatic clearance in a transgenic mouse model of amyloidosis. Curr Alzheimer Res 15:637–642. https://doi.org/10.2174/1567205015666180411092551

Pévet P, Challet E, Felder-Schmittbuhl MP (2021) Melatonin and the circadian system: keys for health with a focus on sleep. Handb Clin Neurol 179:331–343. https://doi.org/10.1016/B978-0-12-819975-6.00021-2

Pourhanifeh MH, Hosseinzadeh A, Dehdashtian E et al (2020) Melatonin: new insights on its therapeutic properties in diabetic complications. Diabetol Metab Syndr 12:30. https://doi.org/10.1186/s13098-020-00537-z

Preitner N, Brown S, Ripperger J et al (2003) Orphan nuclear receptors, molecular clockwork, and the entrainment of peripheral oscillators. Novartis Found Symp 253:89–101. https://doi.org/10.1002/0470090839.ch7

Prokopenko I, Langenberg C, Florez JC et al (2009) Variants in MTNR1B influence fasting glucose levels. Nat Genet 41:77–81. https://doi.org/10.1038/NG.290

Raygan F, Ostadmohammadi V, Bahmani F et al (2019) Melatonin administration lowers biomarkers of oxidative stress and cardio-metabolic risk in type 2 diabetic patients with coronary heart disease: a randomized, double-blind, placebo-controlled trial. Clin Nutr 38:191–196. https://doi.org/10.1016/j.clnu.2017.12.004

Reagan-Shaw S, Nihal M, Ahmad N (2008) Dose translation from animal to human studies revisited. FASEB J 22:659–661. https://doi.org/10.1096/FJ.07-9574LSF

Reinke H, Asher G (2019) Crosstalk between metabolism and circadian clocks. Nat Rev Mol Cell Biol 20:227–241. https://doi.org/10.1038/s41580-018-0096-9

Reiter RJ, Rosales-Corral S, Tan DX et al (2017) Melatonin as a mitochondria-targeted antioxidant: one of evolution's best ideas. Cell Mol Life Sci 74:3863–3881

Reiter RJ, Sharma R, Ma Q et al (2020) Melatonin inhibits Warburg-dependent cancer by redirecting glucose oxidation to the mitochondria: a mechanistic hypothesis. Cell Mol Life Sci 77:2527–2542. https://doi.org/10.1007/S00018-019-03438-1

Romo-Nava F, Alvarez-Icaza González D, Fresán-Orellana A et al (2014) Melatonin attenuates antipsychotic metabolic effects: an eight-week randomized, double-blind, parallel-group, placebo-controlled clinical trial. Bipolar Disord 16:410–421. https://doi.org/10.1111/BDI.12196

Rosales-Corral S, Tan DX, Reiter RJ et al (2003) Orally administered melatonin reduces oxidative stress and proinflammatory cytokines induced by amyloid-beta peptide in rat brain: a comparative, in vivo study versus vitamin C and E. J Pineal Res 35:80–84. https://doi.org/10.1034/J.1600-079X.2003.00057.X

Ryu V, Zarebidaki E, Albers HE et al (2018) Short photoperiod reverses obesity in Siberian hamsters via sympathetically induced lipolysis and Browning in adipose tissue. Physiol Behav 190:11–20. https://doi.org/10.1016/j.physbeh.2017.07.011

Sakotnik A, Liebmann PM, Stoschitzky K et al (1999) Decreased melatonin synthesis in patients with coronary artery disease. Eur Heart J 20:1314–1317. https://doi.org/10.1053/EUHJ.1999.1527

Satari M, Bahmani F, Reiner Z et al (2021) Metabolic and anti-inflammatory response to melatonin administration in patients with diabetic nephropathy. Iran J Kidney Dis 1:22–30

Sayed RKA, Fernández-Ortiz M, Rahim I et al (2021) The impact of melatonin supplementation and nlrp3 inflammasome deletion on age-accompanied cardiac damage. Antioxidants (Basel) 10. https://doi.org/10.3390/antiox10081269

Scheer FAJL (2005) Potential use of melatonin as adjunct antihypertensive therapy. Am J Hypertens 18:1619–1620. https://doi.org/10.1016/j.amjhyper.2005.07.013

She M, Deng X, Guo Z et al (2009) NEU-P11, a novel melatonin agonist, inhibits weight gain and improves insulin sensitivity in high-fat/high-sucrose-fed rats. Pharmacol Res 59:248–253. https://doi.org/10.1016/J.PHRS.2009.01.005

Shieh JM, Wu HT, Cheng KC, Cheng JT (2009) Melatonin ameliorates high fat diet-induced diabetes and stimulates glycogen synthesis via a PKCzeta-Akt-GSK3beta pathway in hepatic cells. J Pineal Res 47:339–344. https://doi.org/10.1111/J.1600-079X.2009.00720.X

Shukla M, Chinchalongporn V, Govitrapong P, Reiter RJ (2019) The role of melatonin in targeting cell signaling pathways in neurodegeneration. Ann N Y Acad Sci 1443:75–96. https://doi.org/10.1111/NYAS.14005

Şirin FB, Kumbul Doğuc D, Vural H et al (2015) Plasma 8-isoPGF2α and serum melatonin levels in patients with minimal cognitive impairment and Alzheimer disease. Turk J Med Sci 45:1073–1077. https://doi.org/10.3906/SAG-1406-134

Skene DJ, Arendt J (2007) Circadian rhythm sleep disorders in the blind and their treatment with melatonin. Sleep Med 8:651–655. https://doi.org/10.1016/j.sleep.2006.11.013

Snider KH, Obrietan K (2018) Modulation of learning and memory by the genetic disruption of circadian oscillator populations. Physiol Behav 194:387–393. https://doi.org/10.1016/j.physbeh.2018.06.035

Solís-Chagoyán H, Domínguez-Alonso A, Valdés-Tovar M et al (2020) Melatonin rescues the dendrite collapse induced by the pro-oxidant toxin okadaic acid in organotypic cultures of rat hilar hippocampus. Molecules 25. https://doi.org/10.3390/MOLECULES25235508

Song X, Sun X, Ma G et al (2015) Family association study between melatonin receptor gene polymorphisms and polycystic ovary syndrome in Han Chinese. Eur J Obstet Gynecol Reprod Biol 195:108–112. https://doi.org/10.1016/J.EJOGRB.2015.09.043

Soni SK, Basu P, Singaravel M et al (2021) Sirtuins and the circadian clock interplay in cardioprotection: focus on sirtuin 1. Cell Mol Life Sci 78:2503–2515. https://doi.org/10.1007/s00018-020-03713-6

Spiegel K, Leproult R, van Cauter E (1999) Impact of sleep debt on metabolic and endocrine function. Lancet 354:1435–1439. https://doi.org/10.1016/S0140-6736(99)01376-8

Stacchiotti A, Grossi I, García-Gómez R et al (2019) Melatonin effects on non-alcoholic fatty liver disease are related to microRNA-34a-5p/Sirt1 axis and autophagy. Cells 8. https://doi.org/10.3390/cells8091053

Stenvers DJ, Scheer FAJL, Schrauwen P et al (2019) Circadian clocks and insulin resistance. Nat Rev Endocrinol 15:75–89. https://doi.org/10.1038/S41574-018-0122-1

Sugden D (1983) Psychopharmacological effects of melatonin in mouse and rat. J Pharmacol Exp Ther 227(3):587–591

Sumsuzzman DM, Choi J, Jin Y, Hong Y (2021) Neurocognitive effects of melatonin treatment in healthy adults and individuals with Alzheimer's disease and insomnia: a systematic review and meta-analysis of randomized controlled trials. Neurosci Biobehav Rev 127:459–473. https://doi.org/10.1016/J.NEUBIOREV.2021.04.034

Sun C, Qiu X, Wang Y et al (2020a) Long-term oral melatonin alleviates memory deficits, reduces amyloid-β deposition associated with downregulation of BACE1 and mitophagy in APP/PS1 transgenic mice. Neurosci Lett 735. https://doi.org/10.1016/j.neulet.2020.135192

Sun Y, Ma C, Sun H, et al (2020b) Metabolism: a novel shared link between diabetes mellitus and Alzheimer's disease. J Diabetes Res 2020. doi:https://doi.org/10.1155/2020/4981814

Tagliaferri V, Romualdi D, Scarinci E et al (2018) Melatonin treatment may be able to restore menstrual cyclicity in women with PCOS: a pilot study. Reprod Sci 25:269–275. https://doi.org/10.1177/1933719117711262

Takahashi JS (2017) Transcriptional architecture of the mammalian circadian clock. Nat Rev Genet 18:164–179

Tam CHT, Ho JSK, Wang Y et al (2010) Common polymorphisms in MTNR1B, G6PC2 and GCK are associated with increased fasting plasma glucose and impaired beta-cell function in Chinese subjects. PLoS One 5. https://doi.org/10.1371/journal.pone.0011428

Tan D-X, Reiter RJ (2019) Mitochondria: the birth place, battle ground and the site of melatonin metabolism in cells. Melatonin Res 2:44–66. https://doi.org/10.32794/mr11250011

Tan DX, Xu B, Zhou X, Reiter RJ (2018) Pineal calcification, melatonin production, aging, associated health consequences and rejuvenation of the pineal gland. Molecules 23. https://doi.org/10.3390/MOLECULES23020301

Tanaka K, Okada Y, Maiko H et al (2021) Associations between urinary 6-sulfatoxymelatonin excretion and diabetic vascular complications or arteriosclerosis in patients with type 2 diabetes. J Diabetes Investig 12:601–609. https://doi.org/10.1111/jdi.13374

Trotti LM, Karroum EG (2016) Melatonin for sleep disorders in patients with neurodegenerative diseases. Curr Neurol Neurosci Rep 16. https://doi.org/10.1007/s11910-016-0664-3

Tsereteli N, Vallat R, Fernandez-Tajes J et al (2021) Impact of insufficient sleep on dysregulated blood glucose control under standardised meal conditions. Diabetologia. https://doi.org/10.1007/s00125-021-05608-y

Tütüncü NB, Batur MK, Yildirir A et al (2005) Melatonin levels decrease in type 2 diabetic patients with cardiac autonomic neuropathy. J Pineal Res 39:43–49. https://doi.org/10.1111/J.1600-079X.2005.00213.X

Vasey C, McBride J, Penta K (2021) Circadian rhythm dysregulation and restoration: the role of melatonin. Nutrients 13. https://doi.org/10.3390/NU13103480

Verdile G, Keane KN, Cruzat VF et al (2015) Inflammation and oxidative stress: the molecular connectivity between insulin resistance, obesity, and Alzheimer's disease. Mediat Inflamm 2015. https://doi.org/10.1155/2015/105828

Volt H, García JA, Doerrier C et al (2016) Same molecule but different expression: aging and sepsis trigger NLRP3 inflammasome activation, a target of melatonin. J Pineal Res 60:193–205. https://doi.org/10.1111/JPI.12303

Wade AG, Farmer M, Harari G et al (2014) Add-on prolonged-release melatonin for cognitive function and sleep in mild to moderate Alzheimer's disease: a 6-month, randomized, placebo-controlled, multicenter trial. Clin Interv Aging 9:947–961. https://doi.org/10.2147/CIA.S65625

Wahab F, Atika B, Ullah F et al (2018) Metabolic impact on the hypothalamic kisspeptin-Kiss1r signaling pathway. Front Endocrinol (Lausanne) 9:123. https://doi.org/10.3389/fendo.2018.00123

Wang RH, Zhao T, Cui K et al (2016) Negative reciprocal regulation between Sirt1 and Per2 modulates the circadian clock and aging. Sci Rep 6. https://doi.org/10.1038/SREP28633

Wang YY, Zheng W, Ng CH et al (2017) Meta-analysis of randomized, double-blind, placebo-controlled trials of melatonin in Alzheimer's disease. Int J Geriatr Psychiatry 32:50–57. https://doi.org/10.1002/GPS.4571

Watroba M, Szukiewicz D (2021) Sirtuins at the service of healthy longevity. Front Physiol 12. https://doi.org/10.3389/FPHYS.2021.724506

Welz PS, Benitah SA (2020) Molecular connections between circadian clocks and aging. J Mol Biol 432:3661–3679. https://doi.org/10.1016/J.JMB.2019.12.036

Xia S, Zhang X, Zheng S et al (2016) An update on inflamm-aging: mechanisms, prevention, and treatment. J Immunol Res 2016. https://doi.org/10.1155/2016/8426874

Xia Y, Chen S, Zeng S et al (2019) Melatonin in macrophage biology: current understanding and future perspectives. J Pineal Res 66

Xu J, Wang LL, Dammer EB et al (2015) Melatonin for sleep disorders and cognition in dementia: a meta-analysis of randomized controlled trials. Am J Alzheimers Dis Other Dement 30:439–447. https://doi.org/10.1177/1533317514568005

Yaprak M, Altun A, Vardar A et al (2003) Decreased nocturnal synthesis of melatonin in patients with coronary artery disease. Int J Cardiol 89:103–107. https://doi.org/10.1016/S0167-5273(02)00461-8

Yi S, Xu J, Shi H et al (2020) Association between melatonin receptor gene polymorphisms and polycystic ovarian syndrome: a systematic review and meta-analysis. Biosci Rep 40. https://doi.org/10.1042/BSR20200824

Yu L, Sun Y, Cheng L et al (2014) Melatonin receptor-mediated protection against myocardial ischemia/reperfusion injury: role of SIRT1. J Pineal Res 57:228–238. https://doi.org/10.1111/JPI.12161

Zhang W, Chen XY, Su SW et al (2016) Exogenous melatonin for sleep disorders in neurodegenerative diseases: a meta-analysis of randomized clinical trials. Neurol Sci 37:57–65. https://doi.org/10.1007/S10072-015-2357-0

Zhang R, Wang X, Ni L et al (2020) COVID-19: melatonin as a potential adjuvant treatment. Life Sci 250

Zhang H, Zhang Y, Li Y et al (2021) Bioinformatics and network pharmacology identify the therapeutic role and potential mechanism of melatonin in AD and rosacea. Front Immunol 12. https://doi.org/10.3389/FIMMU.2021.756550

Zhao L, An R, Yang Y et al (2015) Melatonin alleviates brain injury in mice subjected to cecal ligation and puncture via attenuating inflammation, apoptosis, and oxidative stress: the role of SIRT1 signaling. J Pineal Res 59:230–239. https://doi.org/10.1111/JPI.12254

Zheng R, Ruan Y, Yan Y et al (2021) Melatonin attenuates neuroinflammation by down-regulating NLRP3 Inflammasome via a SIRT1-dependent pathway in MPTP-induced models of Parkinson's disease. J Inflamm Res 14:3063–3075. https://doi.org/10.2147/JIR.S317672

Chapter 12
Role of Vitamin B in Healthy Ageing and Disease

Kathleen Mikkelsen, Maria Trapali, and Vasso Apostolopoulos

Abstract B vitamin complex consist of vitamins B1, B2, B5, B6, B9, B12 and is pivotal for overall health, influencing vital functions such as, energy metabolism, DNA maintenance, and healthy immune system. Inadequate B vitamin levels are associated with various health issues, including neurocognitive problems, immune imbalances, and inflammation. In ageing individuals, deficiencies in B vitamins increase the risk of cardiovascular ailments, stroke, cognitive disorders, neurodegeneration, mental health issues, and methylation-related disorders. These result primarily due to changes in glycation, mitochondria, and oxidative stress. Thus, ensuring optimal vitamin B levels in the ageing population may be beneficial in preventing such age-related diseases. In this chapter we discuss the extensive role of B vitamins in the ageing process.

Keywords Vitamin B · Ageing · Niacin · Thiamine · Folate · Cobalamin · Riboflavin · Pyridoxine

Introduction

In society today, the concept of successful ageing is commonly defined as maximizing one's lifespan while minimizing the physical and mental decline and disability that often come with advancing age (Bowling and Dieppe 2005). Achieving successful ageing, however, can depend on a multitude of factors, encompassing

---

K. Mikkelsen · V. Apostolopoulos (✉)
Institute for Health and Sport, Victoria University, Werribee, VIC, Australia
e-mail: kathleen.mikkelsen@vu.edu.au; vasso.apostolopoulos@vu.edu.au

M. Trapali
Laboratory of Chemistry and Biochemistry and Cosmetic Science, Department of Biomedical Medicine, University of West Attica, Egaleo, Attiki, Greece
e-mail: ymaria@uniwa.gr

© The Author(s), under exclusive license to Springer Nature Switzerland AG 2024

V. I. Korolchuk, J. R. Harris (eds.), *Biochemistry and Cell Biology of Ageing: Part V, Anti-Ageing Interventions*, Subcellular Biochemistry 107,
https://doi.org/10.1007/978-3-031-66768-8_12

genetic predisposition, as well as sociocultural and lifestyle choices (Trapali 2021). These choices may involve decisions regarding exercise, smoking habits, alcohol consumption, and dietary preferences. In addition to various approaches such as dietary modifications, pharmaceutical interventions, and lifestyle adjustments aimed at promoting health and extending longevity, specific nutritional strategies may also play a role in delaying the onset or preventing certain degenerative diseases and cognitive decline among the elderly population (Porter et al. 2016).

An expanding body of research studies demonstrate a pivotal role that B vitamins play in maintaining overall health and ensuring proper physiological and biochemical functions (Mikkelsen and Apostolopoulos 2018). These functions encompass vital processes such as energy metabolism, the production of monoamine oxidase, DNA methylation, DNA synthesis, DNA repair, phospholipid maintenance, immune system support, and regulation of steroid hormone activity (Mikkelsen et al. 2016a). There is a connection between vitamin B deficiency and its role in the development of several diseases, including pellagra, beri-beri, Wernicke-Korsakoff syndrome, and pernicious anaemia. Moreover, vitamin B deficiency plays a part in neurocognitive disorders, encompassing conditions like Alzheimer's disease, dementia, depression and anxiety disorders (Pan et al. 2016; Mikkelsen et al. 2016b; Jerneren et al. 2015; Mitchell et al. 2014), as well as mitochondrial dysfunction (Fu et al. 2014; Du et al. 2014; Abdou and Hazell 2014), immune dysfunction, inflammatory conditions (Mikkelsen et al. 2017b; Pariante 2015; Kiykim et al. 2015; Slavich and Irwin 2014; Eyre and Baune 2012), insulin sensitivity, peripheral neuropathy, lethargy, liver damage, anaemia and fatigue (Mikkelsen et al. 2016a). Reduced vitamin B levels in the elderly have been specifically associated with cardiovascular issues, cognitive impairment, osteoporosis, and methylation-related disorders. Such deficiencies can heighten the susceptibility to degenerative conditions, notably cardiovascular disease, cognitive disorders, and osteoporosis. Additionally, when deficiencies in B vitamins coincide with genetic variations (e.g. MTHFR 677C-T), it can lead to complications related to one-carbon metabolism (Porter et al. 2016). Enhancing vitamin B levels in the elderly could potentially offer advantages in averting degenerative diseases (Fig. 12.1). At present, 8% of the global population is aged 65 and older, a proportion projected to rise to 16% by 2050 (Beard and Bloom 2015). One quarter of the global cost of disease, is carried by the elderly which means the maintenance of health during ageing becomes a public health priority. Improving nutritional outcomes has the potential to prevent or delay the onset of age-related health deterioration.

Vitamin B

The B vitamins constitute a cluster of eight vital water-soluble vitamins that collaborate both independently and collectively to support the body's physiological functions. This group encompasses B1, B2, B3, B5, B6, B7, B9, and B12 (Fig. 12.1). Notably, choline, sometimes referred to as B4, falls outside the traditional B

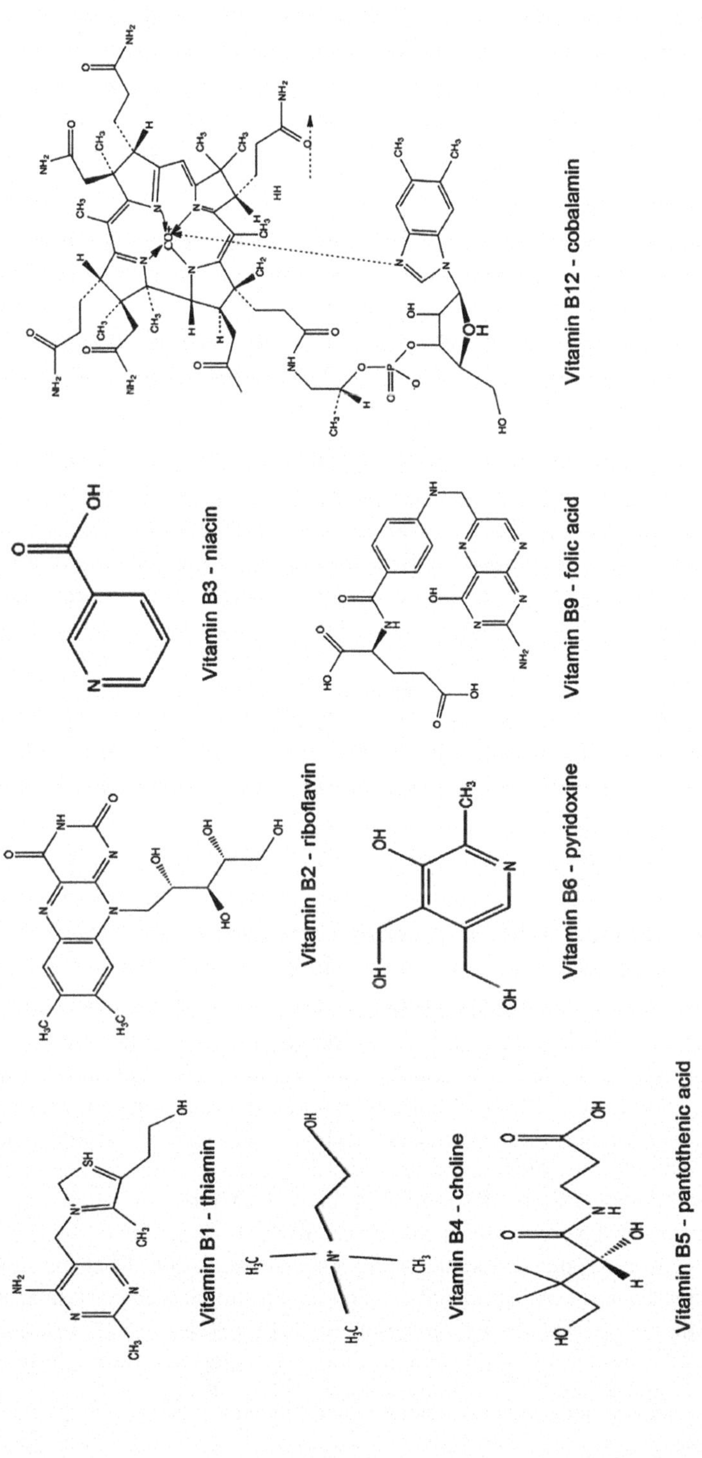

Fig. 12.1 Chemical structures of vitamin B1, B2, B3, B5, B6, B9 and B12

vitamins category, yet it gained official recognition as an essential nutrient by the Institute of Medicine in 1998 (Zeisel 1992). B vitamins are necessary for effective functioning of the methylation cycle and the synthesis and repair of DNA, RNA, protein, and phospholipids. They are important in supporting immune system function, regulating inflammation, aiding in cell metabolism and repair, and facilitating energy production. In addition, they are important for the maintenance of a well-functioning nervous system, mood stability, cognitive performance, and serve as essential building blocks for neurotransmitters. B vitamins act as co-enzymes in several enzymatic processes and collectively participate in the metabolism of carbohydrates, proteins, fats, as well as the processing of other vitamins, minerals, and drugs.

B vitamins are present in animal and dairy products like meat, poultry, fish, eggs, and milk. However, they are also plentiful in plant-based diets, with fresh fruits, vegetables, nuts, seeds, grains, legumes, soy products, and fortified cereals serving as good sources. Vitamin B12, on the other hand, is produced by bacteria and archaea and is found in limited amounts in plants and soil, often via bacterial contamination, typically originating from faecal matter (Herbert 1988). The main source of B12 in the human diet is from animal origin and the presence of B12 in animal flesh is mostly dependent on the process of bio magnification through food chains (Rizzo et al. 2016). Yeast based spreads represent a rich source vitamin B, and some spreads are fortified with vitamin B12 (Mikkelsen et al. 2018). Nutritional yeast stands out as another abundant source of vitamin B12, frequently chosen by individuals adhering to vegetarian or vegan diets as a non-animal alternative for B12 supplementation. In the elderly population, vitamin B deficiency is prevalent, notably affecting metabolically linked B vitamins such as B6, B9, and B12. This deficiency often arises from issues like poor absorption, inadequate dietary intake, or heightened nutritional demands.

Vitamin B Deficiency in the Ageing Population

There are several causes for vitamin B deficiency in the ageing population. As such, a common cause of B2/B9 deficiency is low intake, B12 deficiency primarily results from malabsorption, while B6 deficiency is often due to elevated needs associated with ageing (Porter et al. 2016). Additional factors contributing to B vitamin deficiency include interactions between drugs and nutrients, genetic disorders, and specific medical conditions. The natural ageing process can also impact the absorption, transport, and metabolism of B vitamins in the body. Dietary elements play a substantial role in the development of numerous ailments experienced by the ageing population, with micronutrient status being a significant factor. Micronutrient deficiency frequently arises from reduced food intake or limited dietary variety, as highlighted by the World Health Organization in 2017 (WHO 2017). Micronutrient malabsorption is a prevalent issue, and the elderly population is particularly susceptible to B12 deficiency due to malabsorption. The absorption of vitamin B12 involves a series of intricate steps that rely on various factors for successful assimilation. Initially, B12, often bound to animal protein in the diet, is ingested and undergoes breakdown in the stomach due to the action of pepsin and hydrochloric acid. Subsequently, it is liberated as free B12 and forms a complex with R protein, which is secreted from bile. Pancreatic enzymes then dismantle this complex, releasing B12 once more. In the duodenum, parietal cells release intrinsic factor, which binds to the free B12, allowing it to travel unimpeded into the ileum, where it attaches to mucosal cell receptors. Ultimately, it is transported through the portal system by transcobalamin, a transport protein, which delivers it to all somatic cells (Wong 2015). There are a few pathophysiological changes that can occur with ageing which prevents B12 absorption along this pathway. Hypochloridria, which can occur due to atrophic gastritis, or the use of certain drugs such as proton pump inhibitors, histamine H2 blockers, can interfere with the release of B12 from binding protein. Pernicious anaemia, gastrectomy and atrophic gastritis can interfere with the release of intrinsic factor and prevents B12 from travelling to the ileum. Exocrine pancreatic insufficiency impairs the secretion of pancreatic enzyme which prevents the release of B12 from B12-R protein. Within the ileum, Crohn's disease, ileum resection, bacterial overgrowth and the use of certain drugs including metformin and cholestyramine, can all cause interference with B12 absorption. Finally, transcobalamin deficiency can prevent B12 from being carried within the plasma to somatic cells (Wong 2015).

Vitamin B1: Thiamine

Vitamin B1 (also referred as, thiamine) is part of the co-enzyme thiamine pyrophosphate. It plays an important role in generating nerve impulses and facilitating the synthesis of various vital compounds, including neurotransmitters, fatty acids, nucleic acids, steroids, and complex carbohydrates. It is required for the proper functioning of the nervous system, and its deficiency can result in anxiety, mood disorders and depression, in addition to, severe cardiovascular issues and heart failure (Mikkelsen et al. 2016a, b, 2017a, b, 2018; Nemazannikova et al. 2017). Thiamine deficiency can result in neurological damage leading to the generation of free radicals and increased oxidative stress, resulting in damage to axons, incorrect myelin production, and excitotoxicity mediated by glutamate. Deficiency in thiamine in the body has immune-related consequences, resulting in heightened neuroinflammation, infiltration of T cells, and an upsurge in the production of pro-inflammatory cytokines such as IL-1, TNF-alpha, and IL-6. Further, microglia and astrocytes in the brain overproduce CD40 and CD40 ligand. Thiamine primarily binds to serum proteins, such as albumin within the bloodstream. Its uptake in the small intestine occurs via passive transport at lower concentrations and through active transport mechanisms at higher concentrations. Following absorption, thiamine and its metabolites are excreted via the urine (Abdou and Hazell 2014; Carney et al. 1979; Read and Harrington 1981; Zhang et al. 2013). Thiamine serves as an essential nutrient for the metabolic and cellular processes in the brain, as well as for

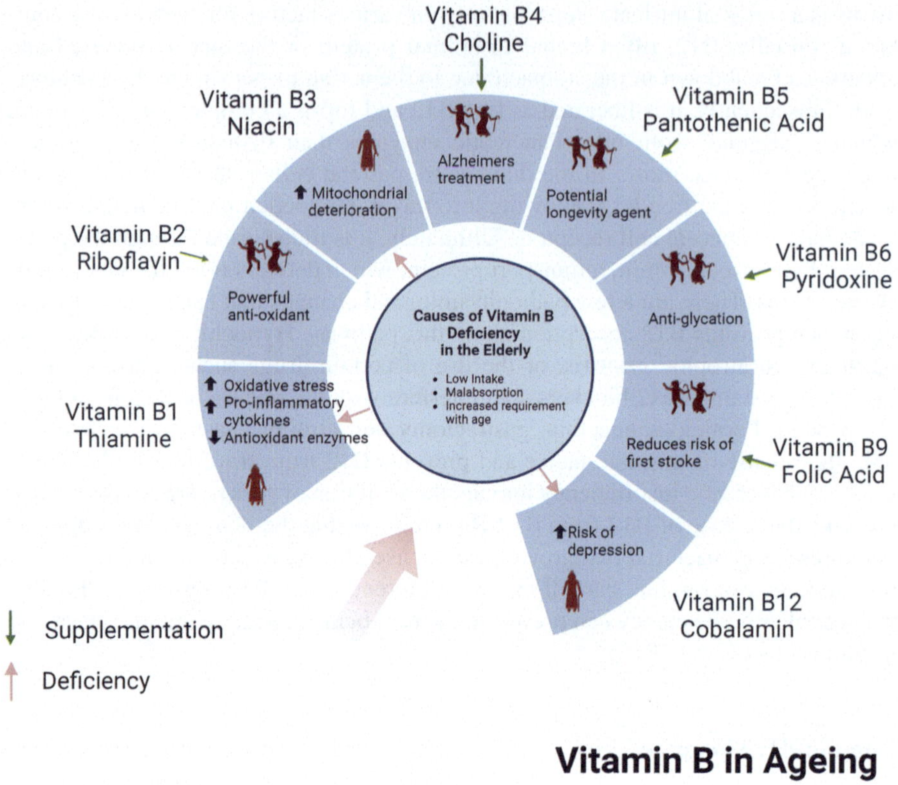

Fig. 12.2 Summary of vitamin B complex contributing to ageing

the synthesis of neurotransmitters (i.e., acetylcholine). Its deficiency can contribute to the disruption of oxidative metabolism, cause endoplasmic reticulum stress, induce neuroinflammation and autophagy dysfunction, and ultimately, neurodegeneration (Fig. 12.2). These factors play a role in the pathogenesis of age-related diseases such as dementia, Parkinson's, Alzheimer's, and Huntington's disease.

Thiamine Deficiency: Neurodegeneration Mechanisms

Evidence from both *in vitro* and *in vivo* studies indicates that thiamine deficiency leads to elevated production of reactive oxygen species (ROS), leading to oxidative stress in brain and neuronal tissues (Liu et al. 2017; Wang et al. 2017). Several factors may contribute to this, including dysfunction of mitochondria, a significant source of ROS production and a key focus of thiamine deficiency-related effects. Notably, mitochondrial dysfunction can result in an increase in binding site densities for the translocator protein, and is noted in the brains of animals with thiamine deficiency (Leong et al. 1994). In addition, it has been suggested that thiamine deficiency can interfere with the antioxidant system within the brain. Thiamine deficiency has been demonstrated to have a significant impact on antioxidant enzymes like superoxide dismutase, catalase, and glutathione peroxidase in animal models, as shown by Sharma and colleagues (Sharma et al. 2013). Thiamine deficiency leads to an increase in pro-inflammatory cytokines, chemokines, and transcription factors, contributing to neuro-inflammation. This neuro-inflammation, in turn, can disrupt mitochondrial function and increase oxidative stress levels. Additionally, autophagy, a vital process responsible for maintaining cell health by breaking down, transporting, and recycling damaged organelles, is often observed to be impaired in neurodegenerative diseases (Shibata et al. 2006). Furthermore, deficiency of thiamine stimulates accumulation of autophagosomes and upregulates autophagy marker expression; however, thiamine supplementation can reverse these changes (Liu et al. 2017). Endoplasmic reticulum stress represents another potential mechanism that may contribute to neurodegeneration. Among its several functions, the endoplasmic reticulum is responsible for protein folding, protein modification, and transport. When the endoplasmic reticulum encounters stress, it triggers a response known as the unfolded protein response, which is an adaptive mechanism aimed at restoring homeostasis in the face of this stress. However, if the unfolded protein response fails to alleviate the stress in the endoplasmic reticulum, it can lead to apoptotic cell death. In conditions such as Huntington's disease, amyotrophic lateral sclerosis, Parkinson's disease and, Alzheimer's disease, the presence of protein aggregates in the brain is a common indicator of neuronal loss. Existing evidence strongly suggests that endoplasmic reticulum stress plays a role in contributing to damage within the central nervous system (Begum et al. 2013; Mota et al. 2015; Wang et al. 2017). Although the strong evidence linking thiamine deficiency with aspects of cognitive decline exists, there are limited studies linking thiamine supplementation with improvement of cognitive symptoms in the ageing population (Fig. 12.2). However, one study showed that high dose of thiamine supplementation in newly diagnosed patients with Parkinson's disease, was effective in reversing symptoms (Costantini et al. 2013).

Vitamin B2: Riboflavin

Vitamin B2, (also referred as, riboflavin), is a potent antioxidant and acts as a co-enzyme in several reactions, particularly in the realms of energy and iron metabolism (Fig. 12.2). Riboflavin exists in two active forms: flavin mononucleotide and flavin adenine dinucleotide, both of which play essential roles as co-factors in vital metabolic processes. Flavin adenine dinucleotide is a component of complex II within the electron transport chain, participating in processes like phosphorylation, fatty acid metabolism and pyruvate oxidation. Flavin mononucleotide is necessary for the conversion of tryptophan into niacin (vitamin B3), as well as for the reduction of glutathione, via glutathione reductase within complex II. B2 deficiency can lead to anaemia, inflammation, cognitive impairment, depression, and, during embryonic development, the risk of congenital malformations (Mikkelsen et al. 2016a, b).

In older individuals, a deficiency in vitamin B2 has been associated with reduced cognitive function, increased risk of depression, adverse personality changes including heightened aggression and notable alterations within the central nervous system (Mahabadi et al. 2023). Within the immune system riboflavin plays an important role in the function of mucosal associated invariant T (MAIT) cells. MAIT cells are the only T cells activated by MR1-bound riboflavin metabolite derivatives. Free riboflavin is bound to albumin and specific immunoglobulins in the serum and gets absorbed in the upper region of the small intestine. It is eliminated through urine as riboflavin or other metabolites (Sheraz et al. 2014; Murakami et al. 2008; Miyake et al. 2006; Powers 2003; Foraker et al. 2003; Massey 2000). In addition, vitamin B2 has been shown to have anti-inflammatory and anti-cancer properties (Mikkelsen et al. 2019).

Riboflavin: Oxidative Damage

Oxidative stress or oxidative stress represents a disturbance in the balance between the production of reactive oxygen species (ROS) and the ability of a biological system to inactivate these toxic molecules and repair the damage they cause. ROS damage all cell components, including proteins, lipids, and DNA. ROS are classified into the following four categories: (1) free radicals, such as the hydroxyl radical (•OH), (2) ions, such as the hypochlorite anion (ClO-), resulting from dissociation of hypochlorous acid (HClO), (3) combinations of free radicals and ions, such as superoxide anion (•O2-), and (4) molecules, such as hydrogen peroxide (H_2O_2). The free radical theory of ageing suggests that oxidative damage accumulated in cells and tissues over time, due to aerobic metabolism is a key factor (Wickens 2001). If this theory holds true, it implies that antioxidants may be able to extend lifespan. Consequently, there has been extensive research into the role of endogenous antioxidants and their relationship with the longevity of various organisms (Sadowska-Bartosz and Bartosz 2014). Some well-known and frequently studied antioxidants, i.e., vitamin C, vitamin E, resveratrol, curcumin, tocopherol, and coenzyme Q, have all shown the ability to prolong the lifespan of model organisms (Sadowska-Bartosz and Bartosz 2014).

Riboflavin, despite being reported for its antioxidant activities, has been relatively overlooked in studies exploring the connection between antioxidants and lifespan. Riboflavin acts as an antioxidant through two mechanisms: (1) it prevents lipid peroxidation (Naziroglu et al. 2015), and, (2) it mitigates reperfusion oxidative injury (Ashoori and Saedisomeolia 2014). In a study of 400 fruit flies, the lifespan of the group supplemented with riboflavin was extended by 14.1% compared non-supplemented control group, which were subjected to hydrogen peroxide to induce oxidative stress (Zou et al. 2017). Additionally, the riboflavin group exhibited enhanced levels of superoxide dismutase SOD1, an antioxidant enzyme that naturally decreases with age, when compared to the control group (Zou et al. 2017). It also reduced ROS levels in keratoconus via determination the total collagen protein in the keratoconic stroma (Cheung et al. 2014). In diabetic male Sprague-Dawley rats, Riboflavin treatment significantly reduced lipid peroxidation, increased SOD, while increased glucose uptake in skeletal muscles and white adipose tissue was noted (Alam et al. 2015). It is clear that further research is warranted to establish a potential link between riboflavin and its ability to slow down the ageing process in humans.

Vitamin B3: Niacin

Vitamin B3 or niacin (nicotinic acid and nicotinamide) are both precursors of the coenzymes nicotinamide adenine dinucleotide (NAD) and nicotinamide adenine dinucleotide phosphate (NADPH). The enzyme NAD+ kinase phosphorylates NAD to produce NADPH. NAD is an essential component in various metabolic processes, including the breakdown of fats, carbohydrates, proteins, and alcohol, as well as playing a crucial role in cell signalling and DNA repair. On the other hand, NADPH is necessary for the synthesis of fatty acids and cholesterol. Further, in a 2 × 6-week randomized, double-blind, placebo-controlled, crossover clinical trial, it was shown that supplementation with the NAD+ precursor, nicotinamide riboside, was well-tolerated and successfully increased NAD+ metabolism in healthy middle-aged and older adults. This study also hinted at potential advantages of supplementation, emphasizing the need for further clinical trials to investigate its potential in reducing blood pressure and arterial stiffness in this demographic (Martens et al. 2018). Deficiency of niacin, a form of vitamin B3, can manifest in cognitive symptoms such as tension, anxiety, depression, as well as bipolar disorder and schizophrenia. Further, its deficiency has also been associated with pellagra, a condition characterized by dermatitis, inflamed skin, dementia, diarrhoea, and sores around the mouth (Mikkelsen et al. 2016a, b). Niacin also decreases pro-inflammatory cytokines and mitigates the impact of inflammation, as evidenced in conditions like atherosclerosis (Lipszyc et al. 2013; Digby et al. 2012). It is absorbed in the stomach and the small intestine and stored in the liver.

NAD and Mitochondria

Energy metabolism is one the of the key duties of NAD. Ageing has been linked to decreased levels of NAD and this in turn has been implicated in mitochondrial deterioration (Lanza and Nair 2010), and can be reversed with dietary supplements that increase cellular levels of NAD (Gomes et al. 2013) (Fig. 12.2). Nicotinamide riboside is a precursor to NAD and effective in boosting NAD levels. In fact, treatment with nicotinamide riboside for 1 week is adequate to boost NAD levels to restore muscle health and mitochondrial homeostasis in mice (Gomes et al. 2013). Furthermore, nicotinamide riboside has been found to replenish NAD+ levels in the mitochondria of mice genetically engineered with Cockayne syndrome, a neurodegenerative disorder associated with accelerated ageing (Scheibye-Knudsen et al. 2014). In human studies, supplementation with nicotinamide riboside demonstrated the ability to elevate NAD levels. NAD synergises with a group of enzymes known as sirtuins, which are implicated as significant contributors to longevity and overall health benefits. NAD and sirtuins work in tandem to stimulate the formation of new mitochondria and ensure the proper functioning of existing ones. When NAD levels decline, a common occurrence with ageing, it can lead to reduced activity of Sirt1. This decline disrupts communication between the cell nuclei and mitochondria, potentially causing mitochondrial dysfunction and contributing to neurodegenerative diseases. Further research may uncover methods to combine the actions of sirtuins and NAD, as promising anti-ageing intervention that regulates health and extends life (Imai and Guarente 2014).

Vitamin B4: Choline

Choline, while sharing a similar chemical structure to other vitamin B's, does not fall under their strict classification. Choline plays a multifaceted role in the body, contributing to various functions such as cell signalling and membrane function. It is a key component of phospholipids like phosphatidylcholine and sphingomyelin, serving as a fundamental element in lecithin and playing an important role in the structure of cell membranes and plasma lipoproteins. In tissues, it serves to prevent accumulation of fat deposits in the liver and aids in the transport of fat into cells. Additionally, choline is responsible for the synthesis of acetylcholine, an important molecule involved in several functions, including memory and muscular control (Nemazannikova et al. 2017).

Choline: Alzheimer's Disease

Choline plays a significant role in the management of patients with Alzheimer's disease. Previous research suggests that cholinergic neurons are important in processes related to learning, memory, and cognitive functions. Both muscarinic and nicotinic acetylcholine receptors are instrumental in the formation of new memories. The proper functioning of cholinergic brain processes is dependent on the neurotransmitter acetylcholine, and a decrease in acetylcholine levels results in a corresponding reduction in cholinergic function (Fu et al. 2004; Bird et al. 1983). In individuals suffering from Alzheimer's disease, a multitude of pathological changes occur within the brain. These changes include the loss of brain tissue and the presence of neurofibrillary tangles and harmful deposits referred to as senile amyloid-beta plaques. Much of this damage has been attributed to the impact of oxidative stress and inflammation. Choline acetyltransferase (ChAT) is the enzyme responsible for facilitating the synthesis of acetylcholine in cholinergic neurons. In Alzheimer's disease patients, there is a distinct loss of cholinergic neurons in the basal forebrain, and this reduction in ChAT activity is closely correlated with the severity of the disease (Fu et al. 2004). In mice, recombinant TAT-choline acetyltransferase fusion protein injected into the brains was shown to improve memory and cognition (Fu et al. 2004). One of the most widely accepted treatment approaches for Alzheimer's disease involves the use of medications that inhibits the action of acetylcholinesterase, an enzyme responsible for breaking down acetylcholine. By blocking the activity of this enzyme, these drugs effectively increase the availability of acetylcholine for neurons in the brain. Of note, these drugs do not cure Alzheimer's disease, but temporarily slow down progression of disease or offer some relief. However, this treatment can be costly and may lead to unpleasant side effects. A safer and more natural supplement currently under investigation in Alzheimer's studies is CDP-choline or citicoline. CDP-choline serves as a precursor to acetylcholine and has the potential to enhance acetylcholine levels in the brain. Additionally, it shows promise in supporting neural repair processes, particularly in Alzheimer's disease (Arenth et al. 2011) and other conditions involving degenerative and vascular cognitive decline (Gareri et al. 2015). A recent study found that a lack of dietary choline resulted in changes in the hippocampal networks, disruptions in protein networks, disturbances in mitochondrial functions, increased inflammation, and altered metabolic processing in mice. Consequently, maintaining an adequate dietary intake of choline is essential to prevent widespread organ damage and mitigate the key pathological features associated with Alzheimer's disease (Dave et al. 2023).

Vitamin B5: Pantothenic Acid

Vitamin B5 or pantothenic acid (pantothenate) is part of the chemical structure of co-enzyme A and a crucial component of the Krebs cycle. Pantothenic acid is involved in energy and amino acid metabolism, fatty acid and glycogen synthesis, and the production of steroid hormones melatonin and acetylcholine. Pantothenic acid contributes to the normal development of the central nervous system and although rare, its deficiency can result in fatigue, irritability, insomnia, nausea, stomach cramps, depression, and hypoglycaemia (Mikkelsen et al. 2016a, b; Mitchell et al. 2014). Pantothenic acid binds to proteins such as acyl carrier protein and must be converted to free pantothenic acid in the intestine prior to absorption in the small intestine. It is excreted intact via the urine (Spry et al. 2013; Jansen et al. 2013).

Pantothenic Acid: Longevity

Throughout history, humans have sought remedies to extend life. Some evidence suggests that pantothenic acid, may have a role in this quest. An early intervention dating back to 1948 involved the feeding of royal jelly to fruit flies which significantly increased lifespan compared to those without it. Of interest, royal jelly is one of the richest sources of pantothenic acid. This finding hinted at the potential significance of pantothenic acid in the context of longevity (Gardner 1948). Fifty years later, interest in the actions of pantothenic acid resurfaced, where it was noted that mice fed high doses of royal jelly increased their lifespan by 25% compared to control mice. This was accompanied with reduced DNA damage; possibly as a result of a reduction in oxidative damage (Inoue et al. 2003). In addition, royal jelly has been shown to extend the lifespan of *C. elegans* (Honda et al. 2011). It is largely unknown however which constituent of royal jelly is responsible for its longevity effects, although it is assumed that pantothenic acid may play an important role. In fact, mice fed with calcium pantothenate (B5) (added to the drinking water) lived an extra 104 days compared to control mice. However, a study has proposed that the longevity-promoting effects of royal jelly might be attributed to its proteins, specifically the major royal jelly proteins (Xin et al. 2016). As such, further research is required to ascertain the longevity properties of pantothenic acid (Fig. 12.2).

Vitamin B6: Pyridoxamine

Vitamin B6 is found in three naturally occurring forms, pyridoxine, pyridoxal and pyridoxamine, and has several essential functions within the endocrine, neurological and immune systems (Mikkelsen et al. 2017b). The active co-enzyme form, pyridoxal-5 phosphate is the biologically active form of vitamin B6 and, aids the synthesis of neurotransmitters (i.e., epinephrine, serotonin, dopamine, GABA). Pyridoxal-5 phosphate aids in the conversion of tryptophan to niacin or serotonin and is involved in steroid hormone activity (Mikkelsen et al. 2016a, b). Pyridoxal-5 phosphate serves multiple functions, including the breakdown of amino acids and the transport of amine groups. Additionally, it plays a role in glucose and lipid metabolism, and it facilitates the synthesis and breakdown of sphingolipids. Vitamin B6 is primarily linked to the development of depression in terms of its pathogenesis (Mikkelsen et al. 2016a). Deficiency states of vitamin B6 can lead to elevated levels of homocysteine, which, in turn, can disrupt methylation process and has been linked to seizures, confusion, migraines, irritability, and depression (Mikkelsen et al. 2016a).

In regard to the benefits of vitamin B6 in the immune system, it has been shown to reduce excessive inflammation by decreasing the accumulation of sphingosine-1-phosphate (S1P) in macrophages through the action of S1P lyase (SPL). Vitamin B6 achieves this by suppressing pro-inflammatory signals like nuclear factor-κB and mitogen-activated protein kinases pathways, and by enhancing SPL activity (Du et al. 2020). Interestingly, in conjunction with deficiency of vitamin B12, it can amplify the inflammatory response of dendritic cells (Mikkelsen et al. 2017b). It was more recently noted that vitamin B6 in addition to other vitamins and minerals have anti-inflammatory properties and immune boosting abilities in patients with COVID-19 (Shakoor et al. 2021a, b, c, d). These findings reveal a novel anti-inflammatory mechanism for vitamin B6 and offer insights into its clinical applications.

The conversion of the inactive forms of vitamin B6 (pyridoxal, pyridoxine, pyridoxamine) into the active form, pyridoxal phosphate, is facilitated by pyridoxal kinase, and this process requires the presence of zinc for full activation. Vitamin B6 is absorbed in the jejunum and ileum through passive diffusion and is eliminated from the body via urine primarily as 4-pyridoxic acid. When vitamin B6 levels are elevated, a small amount may also be excreted in faeces (Richard 2014; Wu and Lu 2012; Murakami et al. 2008; Wang and Kuo 2007; Balk et al. 2007; Miyake et al. 2006; Malouf and Grimley Evans 2003).

Pyridoxine and Glycation

Vitamin B6 in the form of pyridoxamine has been recognized for its potential role in combatting the ageing process. Among the various theories of ageing, the cross-linking theory addresses how proteins interact with sugars to create a complex of protein and glucose, often displaying a yellow-brown colour, encompassing several cross-linked structures. Initially, food chemists were the pioneers in studying this phenomenon, known as the Maillard reaction. This reaction, also referred to as browning or caramelization, is commonly used in cooking to enhance flavour, colour, and texture. Interestingly, in the 1970s and 1980s, it was revealed that this same process occurs within the body and is referred to as advanced glycation (Tamanna and Mahmood 2015). The outcomes of advanced glycation, often referred to advanced glycation end products (AGEs), have been associated with heightened production of free radicals and are believed to contribute to age-related complications by causing tissue damage. These AGEs can accumulate in various body tissues, including the eyes (Jansirani and Anathanaryanan 2004), skin (Bailey 2001), blood vessels, kidneys, lungs (King 2001) and, brain (Shimizu et al. 2013). Accumulations of AGEs in any of these regions can disrupt the regular functioning of the respective organ or tissue. For instance, when AGEs amass in the eyes, they have the potential to trigger cataract formation. In joint tissues, their presence may contribute to arthritis, while in the kidneys, they can be linked to nephrosis and kidney failure. Within the vascular system, AGEs can lead to complications in blood vessels, as it can lead to the build-up of plaques, thickening of the basement membrane, loss of vessel elasticity, ultimately contributing to heart disease. Importantly, vitamin B6, in all its forms, possesses the capability to act as an anti-glycation agent, with pyridoxamine being particularly potent among the three forms (Voziyan and Hudson 2005). Pyridoxamine effectively prevents AGEs formation by capturing reactive carbonyl groups produced during sugar and lipid breakdown. It also inhibits specific stages of the glycation process by binding to catalytic redox metal ions and trapping reactive oxygen species. The potential of vitamin B6 as an inhibitor of AGEs is promising, however further research is required to fully understand this association.

Vitamin B9: Folic Acid

Vitamin B9 (folic acid, folate) plays an important role in many functions within the body. Folate, primarily in its co-enzyme form known as tetrahydrofolate, plays a crucial role in facilitating a series of one-carbon transfer reactions within metabolism. Together with vitamins B12 and B6, it is essential for methylation, a process that recycles homocysteine into methionine. Additionally, folate aids in the synthesis, repair, and methylation of DNA, making it an important component for cell division. In fact, cells require folate to function and divide properly. Furthermore, folate, in the form of 5-methyl tetrahydrofolate, contributes to the regulation of monoamine neurotransmission and plays a role in the breakdown of norepinephrine and dopamine, as well as the synthesis of these neurotransmitters, along with serotonin (Mikkelsen et al. 2016a). In addition, vitamin B9 helps convert B12 to a co-enzyme form whilst B12 is required to convert B9 to a co-enzyme form. Deficiency in folate has been associated with neural tube defects in the foetus, and in adults, it is associated with, anaemia, neurocognitive effects such as behaviour disorders, cognitive decline and depression (Mikkelsen et al. 2016a). In addition, low levels of folate have been linked to dementia and Alzheimer's disease (Mikkelsen et al. 2016b). Symptoms of folate deficiency include mental confusion, weakness, fatigue, shortness of breath, irritability, headache, a smooth red tongue, depressive symptoms, and elevated homocysteine levels (Fava and Mischoulon 2009; Loria-Kohen et al. 2013; Walker et al. 2012). Individuals experiencing megaloblastic anaemia, the clinical manifestation of folate deficiency, often have compromised immune responses, which can be reversed through supplementation with folic acid. Folates exist in food as polyglutamates and must undergo intestinal breakdown to folate monoglutamates before absorption can take place. The excretion of vitamin B9 happens primarily through urine (Jerneren et al. 2015; Mitchell et al. 2014; Du et al. 2014; Loria-Kohen et al. 2013; Walker et al. 2012; Fava and Mischoulon 2009; Balk et al. 2007).

Folic Acid: Stroke and Neurodegenerative Diseases

Folic acid is well-known for its role in preventing neural tube defects during foetal development (Pitkin 2007). To address this issue, some countries have implemented folic acid food fortification policies, with approximately 80 countries worldwide now having mandatory fortification in place. An intriguing and unexpected outcome of this fortification has been a decrease in stroke-related mortality (Fig. 12.2). In fact, it was noted that the reduction in stroke mortality was evident after the implementation of mandatory folic acid fortification policies in Canada and the USA. In contrast, during the same period, England, and Wales, which did not implement folic acid food fortification, did not experience a similar improvement in stroke outcomes (Yang et al. 2006). Another study investigated whether a combination therapy of enalapril (a medication used to treat high blood pressure) and folic acid was more effective in reducing primary strokes compared to enalapril alone among

Chinese adults with hypertension. The results of this study demonstrated that supplementation with folic acid significantly lowered the risk of a first stroke (Huo et al. 2015). Further, in a meta-analysis of randomized controlled trials of folic acid supplementation in countries without mandatory folic acid food fortification, was shown to be linked to reduced incidences of stroke (Hsu et al. 2018). More recently, in a cross-sectional study involving 8371 adults it was noted that associations existed between lower folate and incidence of stroke and it was concluded that higher levels of folate is required for stroke prevention, although underlying mechanisms require further investigation (Zhang et al. 2023). In a cost-effectiveness analysis of the use of folic acid as a treatment for primary prevention of stroke in hypertensive patients, it was noted that the combination of enalapril and folic acid for hypertension treatment in China was cost-effective, particularly for males aged 55–65 with higher stroke risk. Some lower-risk subgroups may not benefit, but with reduced drug costs, more patients could benefit. These findings have broad implications for healthcare policies in countries with similar demographics and low folate intake (Zhang et al. 2022).

It is believed that the decrease in stroke risk may be due to its ability to reduce homocysteine levels as elevated homocysteine is correlated with increased risk of stroke (Holmes et al. 2011; Casas et al. 2005), whilst, lower blood homocysteine levels are associated with reduced risk for stroke (Saposnik et al. 2009). Homocysteine is a by-product that arises from the methylation cycle, a crucial process the body uses to manage stress, combat infections, and process toxic metabolites. Methylation reactions are integral to nearly all internal chemical processes. On occasion, these reactions may not proceed efficiently, resulting in elevated homocysteine levels. This increase can potentially lead to metabolic dysfunction and may be associated with certain neurodegenerative conditions, including, depression, dementia, Alzheimer's, and Parkinson's disease (Bottiglieri et al. 2000). To ensure effective methylation, an adequate dietary intake of folate is crucial. Folate plays a pivotal role in the methylation cycle, serving as an essential component for the conversion of homocysteine into methionine and in the synthesis of S-adenosylmethionine, a key molecule required for the methylation of DNA, proteins, and lipids (Mikkelsen et al. 2016b; Saposnik et al. 2009). It is important to recognize that the methylation cycle relies not only on folate but also on the synergistic action of vitamins B1, B2, B6, and B12. Inadequate levels of these vitamins due to factors such as low dietary intake, poor absorption, or increased requirements can result in elevated homocysteine levels, contributing to the development of numerous age-associated neurocognitive disorders (Tucker et al. 2005; Xiu et al. 2012).

Vitamin B12: Cobalamin

Vitamin B12, also referred to as cobalamin, is characterized by a cobalt centred corrin nucleus. Cyanocobalamin is the metabolically active form of vitamin B12, but other forms found naturally in biological systems include methyl cobalamin, cob(I) alamin, 5′-deoxyadenosylcobalamin and hydroxy-cobalamin. The active C-Co bonds found in vitamin B12 play crucial roles in facilitating isomerase and methyltransferase reactions. These reactions are vital for extracting energy from proteins and fats and are integral to the process of methylation. Vitamin B12 is essential for the proper functioning of the nervous system, maintenance of nerve cells, cellular synthesis, as well as the breakdown of fatty acids and amino acids, as reviewed in (Mikkelsen et al. 2016a, b, 2017b). A close relationship exists between B12 and folate as each depends on the other for activation. Vitamin B12 is synthesized by bacteria and interestingly, human colon bacteria make large amounts of vitamin B12, but it is not absorbed through the colon. An important study back in the 1950s showed that when vegan patients suffering from vitamin B12 deficiency were fed water extracts of their own stools were subsequently cured of their B12 deficiency. Clinical symptoms of B12 deficiency includes, fatigue, loss of appetite, weight loss, weakness, constipation, peripheral tingling and soreness of the mouth and tongue (Mikkelsen et al. 2016a, b, 2017b; Nemazannikova et al. 2017). In a recent retrospective chart review from the Dhulikhel hospital in Nepal, of 1454 patients, it was noted that a significant proportion (33.5%) were vitamin B12 deficient with, 27.9% being borderline deficient (Gyawali et al. 2023). Vitamin B12 deficiency has also been noted in several elderly cultures, such as the Slovenian (58.3% deficient) (Lavrisa et al. 2022), Taiwanese (Chen et al. 2005) and Indian populations (61.7% deficient) (Malik and Trilok-Kumar 2020). Vitamin B12 deficiency is therefore a great concern, especially among the ageing population, as it can lead to significant impact on overall wellbeing. In fact, in a cross-sectional study (n = 100), those with low levels of vitamin B12 correlated to high patient healthcare questionnaire-9 scores which is an indicator of depression (Fatima et al. 2023).

Vitamin B12: Depression

A deficiency in vitamin B12 has been linked to a range of neuropsychiatric symptoms, including, mental fatigue, low mood, severe depression, manic episodes, psychosis, suicidal tendencies, reduced cognitive function, and heightened agitation. Further, B12 deficiency can disrupt the production of high turnover cells like red blood cells, potentially leading to megaloblastic or pernicious anaemia. When B12 deficiency occurs, it can trap folate in its inactive form, impeding DNA production. In relation to the immune system, a deficiency in vitamin B12 has been associated with a decrease in CD8 T cells and natural killer cells, and an increase of TNF-alpha by macrophages. Interestingly, deviations in the levels of IL-6 and TNF-alpha were corrected with vitamin B12 supplementation (Fava and Mischoulon 2009; Loria-Kohen et al. 2013; Walker et al. 2012). In terms of absorption, vitamin B12 in food is bound to proteins. Its release is aided by hydrochloric acid and proteases in the stomach. Parietal cells in the stomach secrete intrinsic factor, which forms a complex with B12, enabling its absorption in the ileum. This absorption process relies on the presence of intrinsic factor. After absorption, B12 is transported into cells via the plasma transporter transcobalamin II. Inside the cells, it undergoes degradation through lysosomal activity, and free B12 enters the cytoplasm. Most of the B12 is excreted through bile and then reabsorbed and stored in the liver, with a small amount excreted in the faeces. Larger quantities of vitamin B12 are excreted in the urine when consumed in excess. Within the body, vitamin B12 is initially bound to proteins in food. The release of B12 is facilitated by the presence of hydrochloric acid and proteases in the stomach. Parietal cells located in the stomach mucosa secrete intrinsic factor, forming a complex with B12, which is essential for its absorption in the ileum. This absorption process is highly dependent on the capacity of intrinsic factor. Once absorbed, vitamin B12 is transported into cells through the plasma transporter transcobalamin II. Inside the cells, it undergoes degradation by lysosomal activity, and the free B12 is then transferred into the cytoplasm. The majority of B12 is excreted via bile, where it is subsequently reabsorbed and stored in the liver. A minor amount of B12 exits the body through faeces. Larger doses of B12 taken are excreted in the urine (Jerneren et al. 2015; Mitchell et al. 2014; Du et al. 2014; Loria-Kohen et al. 2013; Walker et al. 2012; Fava and Mischoulon 2009; Balk et al. 2007).

The normal function of the brain and nervous system is reliant on adequate levels of vitamin B12 being obtained in the diet or via supplementation. Vitamin B12 deficiency can result in development of severe depression, irritability, mania, psychosis, and suicidal behaviours (Petridou et al. 2016; Almeida et al. 2015; Sengul et al. 2014; Mitchell et al. 2014; Syed et al. 2013; Seppala et al. 2013). The link between vitamin B12 and neurotransmitter synthesis sheds light on why vitamin B12 deficiency can lead to depression. Low vitamin B12 levels are particularly common in the elderly, often stemming from age-related disruptions in absorption, transfer, and metabolism. Within the brain methylation reactions including one-carbon metabolism are responsible for generating monoamine neurotransmitters like dopamine, serotonin, and norepinephrine. A deficiency of vitamin B12 can reduce the activity of methionine synthase, which hampers the formation of tetrahydrofolate and leads to the entrapment of folate as 5-methyl tetrahydrofolate. The interconnectedness of the methionine pathway with the purine and thymidylate cycles plays a critical role in neurotransmitter synthesis. Disturbances in any of these pathways can disrupt the balance within others, potentially affecting neurotransmitter production and contributing to depression (Dayon et al. 2017; Mitchell et al. 2014).

In a recent 4-year longitudinal study in 3849 individuals aged over 50 years, and folate and vitamin B12 plasma levels analysed, it was clear that individuals with lower vitamin B12 status at baseline had a significantly higher likelihood (51% higher) of incident depression 4 years later; there was no association with folate status (Laird et al. 2023). In addition, in a case report of a 66-year-old having sad mood, lack of motivation- sleep- and energy, hallucinations and suicidal ideas, in the preceding 6-months. There was no cognitive decline or neurologic deficits, and she was treated for major depressive disorder without psychotic symptoms as an initial diagnosis. However, her vitamin B12 levels were extremely low and after a series of vitamin B12 injections symptoms improved significantly and medications for major depressive disorder were removed (Hanna et al. 2009). Likewise, a 64-year-old individual with no prior history of mental illness who was hospitalized for confusion, delusions, depression in the preceding 2 months was found to have severe vitamin B12 deficiency, and the mental health status improved significantly within a few days of vitamin B12 injection (Durand et al. 2003). Worldwide there has been a rise in in bariatric surgeries like gastric bypass and sleeve gastrectomy which has led to a growing concern of micronutrient deficiencies. In particular in the United Arab Emirates, people undergoing these procedures are at a heightened risk of developing neurological, cognitive, and cardiovascular issues due to vitamin B especially, B12 deficiency. It was concluded that dietary interventions and supplements are crucial post-bariatric surgery to prevent adverse outcomes (Al Mansoori et al. 2021).

Acknowledgments KM was supported by the Victoria University Postgraduate Scholarship and the Vice Chancellors top-up Scholarship. VA was supported by the Vice Chancellors Distinguished Fellow scheme, VU Research, Victoria University, Australia.

References

Abdou E, Hazell AS (2014) Thiamine deficiency: an update of pathophysiologic mechanisms and future therapeutic considerations. Neurochem Res. https://doi.org/10.1007/s11064-014-1430-z

Al Mansoori A, Shakoor H, Ali HI, Feehan J, Al Dhaheri AS, Cheikh Ismail L, Bosevski M, Apostolopoulos V, Stojanovska L (2021) The effects of bariatric surgery on vitamin B status and mental health. Nutrients 13(4). https://doi.org/10.3390/nu13041383

Alam MM, Iqbal S, Naseem I (2015) Ameliorative effect of riboflavin on hyperglycemia, oxidative stress and DNA damage in type-2 diabetic mice: mechanistic and therapeutic strategies. Arch Biochem Biophys 584:10–19. https://doi.org/10.1016/j.abb.2015.08.013

Almeida OP, Ford AH, Flicker L (2015) Systematic review and meta-analysis of randomized placebo-controlled trials of folate and vitamin B12 for depression. Int Psychogeriatr 27(5):727–737. https://doi.org/10.1017/S1041610215000046

Arenth PM, Russell KC, Ricker JH, Zafonte RD (2011) CDP-choline as a biological supplement during neurorecovery: a focused review. PM R 3(6 Suppl 1):S123–S131. https://doi.org/10.1016/j.pmrj.2011.03.012

Ashoori M, Saedisomeolia A (2014) Riboflavin (vitamin B(2)) and oxidative stress: a review. Br J Nutr 111(11):1985–1991. https://doi.org/10.1017/S0007114514000178

Bailey AJ (2001) Molecular mechanisms of ageing in connective tissues. Mech Ageing Dev 122(7):735–755

Balk EM, Raman G, Tatsioni A, Chung M, Lau J, Rosenberg IH (2007) Vitamin B6, B12, and folic acid supplementation and cognitive function: a systematic review of randomized trials. Arch Intern Med 167(1):21–30. https://doi.org/10.1001/archinte.167.1.21

Beard JR, Bloom DE (2015) Towards a comprehensive public health response to population ageing. Lancet 385(9968):658–661. https://doi.org/10.1016/S0140-6736(14)61461-6

Begum G, Harvey L, Dixon CE, Sun D (2013) ER stress and effects of DHA as an ER stress inhibitor. Transl Stroke Res 4(6):635–642. https://doi.org/10.1007/s12975-013-0282-1

Bird TD, Stranahan S, Sumi SM, Raskind M (1983) Alzheimer's disease: choline acetyltransferase activity in brain tissue from clinical and pathological subgroups. Ann Neurol 14(3):284–293. https://doi.org/10.1002/ana.410140306

Bottiglieri T, Laundy M, Crellin R, Toone BK, Carney MW, Reynolds EH (2000) Homocysteine, folate, methylation, and monoamine metabolism in depression. J Neurol Neurosurg Psychiatry 69(2):228–232

Bowling A, Dieppe P (2005) What is successful ageing and who should define it? BMJ 331(7531):1548–1551. https://doi.org/10.1136/bmj.331.7531.1548

Carney MW, Williams DG, Sheffield BF (1979) Thiamine and pyridoxine lack newly-admitted psychiatric patients. Br J Psychiatry J Ment Sci 135:249–254

Casas JP, Bautista LE, Smeeth L, Sharma P, Hingorani AD (2005) Homocysteine and stroke: evidence on a causal link from mendelian randomisation. Lancet 365(9455):224–232. https://doi.org/10.1016/S0140-6736(05)17742-3

Chen KJ, Pan WH, Yang FL, Wei IL, Shaw NS, Lin BF (2005) Association of B vitamins status and homocysteine levels in elderly Taiwanese. Asia Pac J Clin Nutr 14(3):250–255

Cheung IM, McGhee CN, Sherwin T (2014) Beneficial effect of the antioxidant riboflavin on gene expression of extracellular matrix elements, antioxidants and oxidases in keratoconic stromal cells. Clin Exp Optom 97(4):349–355. https://doi.org/10.1111/cxo.12138

Costantini A, Pala MI, Compagnoni L, Colangeli M (2013) High-dose thiamine as initial treatment for Parkinson's disease. BMJ Case Rep 2013. https://doi.org/10.1136/bcr-2013-009289

Dave N, Judd JM, Decker A, Winslow W, Sarette P, Villarreal Espinosa O, Tallino S, Bartholomew SK, Bilal A, Sandler J, McDonough I, Winstone JK, Blackwood EA, Glembotski C, Karr T, Velazquez R (2023) Dietary choline intake is necessary to prevent systems-wide organ pathology and reduce Alzheimer's disease hallmarks. Aging Cell 22(2):e13775. https://doi.org/10.1111/acel.13775

Dayon L, Guiraud SP, Corthesy J, Da Silva L, Migliavacca E, Tautvydaite D, Oikonomidi A, Moullet B, Henry H, Metairon S, Marquis J, Descombes P, Collino S, Martin FJ, Montoliu I, Kussmann M, Wojcik J, Bowman GL, Popp J (2017) One-carbon metabolism, cognitive impairment and CSF measures of Alzheimer pathology: homocysteine and beyond. Alzheimers Res Ther 9(1):43. https://doi.org/10.1186/s13195-017-0270-x

Digby JE, Martinez F, Jefferson A, Ruparelia N, Chai J, Wamil M, Greaves DR, Choudhury RP (2012) Anti-inflammatory effects of nicotinic acid in human monocytes are mediated by GPR109A dependent mechanisms. Arterioscler Thromb Vasc Biol 32(3):669–676. https://doi.org/10.1161/ATVBAHA.111.241836

Du J, Zhu M, Bao H, Li B, Dong Y, Xiao C, Zhang GY, Henter I, Rudorfer M, Vitiello B (2014) The role of nutrients in protecting mitochondrial function and neurotransmitter signaling: implications for the treatment of depression, PTSD, and suicidal behaviors. Crit Rev Food Sci Nutr 56(15):2560–2578. https://doi.org/10.1080/10408398.2013.876960

Du X, Yang Y, Zhan X, Huang Y, Fu Y, Zhang Z, Liu H, Zhang L, Li Y, Wen Q, Zhou X, Zuo D, Zhou C, Li L, Hu S, Ma L (2020) Vitamin B6 prevents excessive inflammation by reducing accumulation of sphingosine-1-phosphate in a sphingosine-1-phosphate lyase-dependent manner. J Cell Mol Med 24(22):13129–13138. https://doi.org/10.1111/jcmm.15917

Durand C, Mary S, Brazo P, Dollfus S (2003) Psychiatric manifestations of vitamin B12 deficiency: a case report. Encéphale 29(6):560–565

Eyre H, Baune BT (2012) Neuroplastic changes in depression: a role for the immune system. Psychoneuroendocrinology 37(9):1397–1416. https://doi.org/10.1016/j.psyneuen.2012.03.019

Fatima A, Shah F, Khan HS, Rauf K, Ali S, Khan SA (2023) Frequency of depression in patients with vitamin B12 deficiency. JRMC [Internet] 27(2)

Fava M, Mischoulon D (2009) Folate in depression: efficacy, safety, differences in formulations, and clinical issues. J Clin Psychiatry 70(Suppl 5):12–17. https://doi.org/10.4088/JCP.8157su1c.03

Foraker AB, Khantwal CM, Swaan PW (2003) Current perspectives on the cellular uptake and trafficking of riboflavin. Adv Drug Deliv Rev 55(11):1467–1483

Fu AL, Li Q, Dong ZH, Huang SJ, Wang YX, Sun MJ (2004) Alternative therapy of Alzheimer's disease via supplementation with choline acetyltransferase. Neurosci Lett 368(3):258–262. https://doi.org/10.1016/j.neulet.2004.05.116

Fu L, Doreswamy V, Prakash R (2014) The biochemical pathways of central nervous system neural degeneration in niacin deficiency. Neural Regen Res 9(16):1509–1513. https://doi.org/10.4103/1673-5374.139475

Gardner TS (1948) The use of Drosophila melanogaster as a screening agent for longevity factors; the effects of biotin, pyridoxine, sodium yeast nucleate, and pantothenic acid on the life span of the fruit fly. J Gerontol 3(1):9–13

Gareri P, Castagna A, Cotroneo AM, Putignano S, De Sarro G, Bruni AC (2015) The role of citicoline in cognitive impairment: pharmacological characteristics, possible advantages, and doubts for an old drug with new perspectives. Clin Interv Aging 10:1421–1429. https://doi.org/10.2147/CIA.S87886

Gomes AP, Price NL, Ling AJ, Moslehi JJ, Montgomery MK, Rajman L, White JP, Teodoro JS, Wrann CD, Hubbard BP, Mercken EM, Palmeira CM, de Cabo R, Rolo AP, Turner N, Bell EL, Sinclair DA (2013) Declining NAD(+) induces a pseudohypoxic state disrupting nuclear-mitochondrial communication during aging. Cell 155(7):1624–1638. https://doi.org/10.1016/j.cell.2013.11.037

Gyawali P, Bhatt RD, Karmacharya RM, Pant V, Khadka A (2023) High burden of vitamin B12 deficiency among adults and elderly. J Nepal Health Res Counc 20(3):702–707. https://doi.org/10.33314/jnhrc.v20i3.4407

Hanna S, Lachover L, Rajarethinam RP (2009) Vitamin b(1)(2) deficiency and depression in the elderly: review and case report. Prim Care Companion J Clin Psychiatry 11(5):269–270. https://doi.org/10.4088/PCC.08l00707

Herbert V (1988) Vitamin B-12: plant sources, requirements, and assay. Am J Clin Nutr 48(3 Suppl):852–858

Holmes MV, Newcombe P, Hubacek JA, Sofat R, Ricketts SL, Cooper J, Breteler MM, Bautista LE, Sharma P, Whittaker JC, et al. (2011) Effect modification by population dietary folate on the association between MTHFR genotype, homocysteine, and stroke risk: a meta-analysis of genetic studies and randomised trials. Lancet 378(9791):584–594. https://doi.org/10.1016/S0140-6736(11)60872-6

Honda Y, Fujita Y, Maruyama H, Araki Y, Ichihara K, Sato A, Kojima T, Tanaka M, Nozawa Y, Ito M, Honda S (2011) Lifespan-extending effects of royal jelly and its related substances on the nematode Caenorhabditis elegans. PLoS One 6(8):e23527. https://doi.org/10.1371/journal.pone.0023527

Hsu CY, Chiu SW, Hong KS, Saver JL, Wu YL, Lee JD, Lee M, Ovbiagele B (2018) Folic acid in stroke prevention in countries without mandatory folic acid food fortification: a meta-analysis of randomized controlled trials. J Stroke 20(1):99–109. https://doi.org/10.5853/jos.2017.01522

Huo Y, Li J, Qin X, Huang Y, Wang X, Gottesman RF, Tang G, Wang B, Chen D, He M, Fu J, Cai Y, Shi X, Zhang Y, Cui Y, Sun N, Li X, Cheng X, Wang J, Yang X, Yang T, Xiao C, Zhao G, Dong Q, Zhu D, Wang X, Ge J, Zhao L, Hu D, Liu L, Hou FF, Investigators C (2015) Efficacy of folic acid therapy in primary prevention of stroke among adults with hypertension in China: the CSPPT randomized clinical trial. JAMA 313(13):1325–1335. https://doi.org/10.1001/jama.2015.2274

Imai S, Guarente L (2014) NAD+ and sirtuins in aging and disease. Trends Cell Biol 24(8):464–471. https://doi.org/10.1016/j.tcb.2014.04.002

Inoue S, Koya-Miyata S, Ushio S, Iwaki K, Ikeda M, Kurimoto M (2003) Royal Jelly prolongs the life span of C3H/HeJ mice: correlation with reduced DNA damage. Exp Gerontol 38(9):965–969

Jansen PA, Hermkens PH, Zeeuwen PL, Botman PN, Blaauw RH, Burghout P, van Galen PM, Mouton JW, Rutjes FP, Schalkwijk J (2013) Combination of pantothenamides with vanin inhibitors as a novel antibiotic strategy against gram-positive bacteria. Antimicrob Agents Chemother 57(10):4794–4800. https://doi.org/10.1128/AAC.00603-13

Jansirani, Anathanaryanan PH (2004) A comparative study of lens protein glycation in various forms of cataract. Indian J Clin Biochem 19(1):110–112. https://doi.org/10.1007/BF02872403

Jerneren F, Elshorbagy AK, Oulhaj A, Smith SM, Refsum H, Smith AD (2015) Brain atrophy in cognitively impaired elderly: the importance of long-chain omega-3 fatty acids and B vitamin status in a randomized controlled trial. Am J Clin Nutr 102(1):215–221. https://doi.org/10.3945/ajcn.114.103283

King RH (2001) The role of glycation in the pathogenesis of diabetic polyneuropathy. Mol Pathol 54(6):400–408

Kiykim E, Kiykim A, Cansever MS, Zeybek CA (2015) Biotinidase deficiency mimicking primary immune deficiencies. BMJ Case Rep 2015. https://doi.org/10.1136/bcr-2014-209275

Laird EJ, O'Halloran AM, Molloy AM, Healy M, Hernandez B, O'Connor DMA, Kenny RA, Briggs R (2023) Low vitamin B(12) but not folate is associated with incident depressive symptoms in community-dwelling older adults: a 4-year longitudinal study. Br J Nutr 130(2):268–275. https://doi.org/10.1017/S0007114521004748

Lanza IR, Nair KS (2010) Mitochondrial function as a determinant of life span. Pflugers Arch 459(2):277–289. https://doi.org/10.1007/s00424-009-0724-5

Lavrisa Z, Hristov H, Hribar M, Zmitek K, Kusar A, Korousic Seljak B, Gregoric M, Blaznik U, Gregoric N, Zaletel K, Oblak A, Osredkar J, Pravst I (2022) Dietary intake and status of vitamin B12 in Slovenian population. Nutrients 14(2). https://doi.org/10.3390/nu14020334

Leong DK, Le O, Oliva L, Butterworth RF (1994) Increased densities of binding sites for the "peripheral-type" benzodiazepine receptor ligand [3H]PK11195 in vulnerable regions of the rat brain in thiamine deficiency encephalopathy. J Cereb Blood Flow Metab 14(1):100–105. https://doi.org/10.1038/jcbfm.1994.14

Lipszyc PS, Cremaschi GA, Zorrilla-Zubilete M, Bertolino ML, Capani F, Genaro AM, Wald MR (2013) Niacin modulates pro-inflammatory cytokine secretion. A potential mechanism involved in its anti-atherosclerotic effect. Open Cardiovasc Med J 7:90–98. https://doi.org/10.2174/1874192401307010090

Liu D, Ke Z, Luo J (2017) Thiamine deficiency and neurodegeneration: the interplay among oxidative stress, endoplasmic reticulum stress, and autophagy. Mol Neurobiol 54(7):5440–5448. https://doi.org/10.1007/s12035-016-0079-9

Loria-Kohen V, Gomez-Candela C, Palma-Milla S, Amador-Sastre B, Hernanz A, Bermejo LM (2013) A pilot study of folic acid supplementation for improving homocysteine levels, cognitive and depressive status in eating disorders. Nutr Hosp 28(3):807–815. https://doi.org/10.3305/nh.2013.28.3.6335

Mahabadi N, Bhusal A, Banks SW (2023) Riboflavin deficiency. StatPearls, Treasure Island

Malik A, Trilok-Kumar G (2020) Status of vitamin B(12) among healthy adult and elderly population in India: a review. J Nutr Sci Vitaminol (Tokyo) 66(Supplement):S361–S368. https://doi.org/10.3177/jnsv.66.S361

Malouf R, Grimley Evans J (2003) The effect of vitamin B6 on cognition. Cochrane Database Syst Rev 4:CD004393. https://doi.org/10.1002/14651858.CD004393

Martens CR, Denman BA, Mazzo MR, Armstrong ML, Reisdorph N, McQueen MB, Chonchol M, Seals DR (2018) Chronic nicotinamide riboside supplementation is well-tolerated and elevates NAD(+) in healthy middle-aged and older adults. Nat Commun 9(1):1286. https://doi.org/10.1038/s41467-018-03421-7

Massey V (2000) The chemical and biological versatility of riboflavin. Biochem Soc Trans 28(4):283–296

Mikkelsen K, Apostolopoulos V (2018) B vitamins and ageing. Subcell Biochem 90:451–470. https://doi.org/10.1007/978-981-13-2835-0_15

Mikkelsen K, Stojanovska L, Apostolopoulos V (2016a) The effects of vitamin B in depression. Curr Med Chem 23(38):4317–4337

Mikkelsen K, Stojanovska L, Tangalakis K, Bosevski M, Apostolopoulos V (2016b) Cognitive decline: a vitamin B perspective. Maturitas 93:108–113. https://doi.org/10.1016/j.maturitas.2016.08.001

Mikkelsen K, Stojanovska L, Polenakovic M, Bosevski M, Apostolopoulos V (2017a) Exercise and mental health. Maturitas 106:48–56. https://doi.org/10.1016/j.maturitas.2017.09.003

Mikkelsen K, Stojanovska L, Prakash M, Apostolopoulos V (2017b) The effects of vitamin B on the immune/cytokine network and their involvement in depression. Maturitas 96:58–71. https://doi.org/10.1016/j.maturitas.2016.11.012

Mikkelsen K, Hallam K, Stojanovska L, Apostolopoulos V (2018) Yeast based spreads improve anxiety and stress. J Funct Foods 40:471–476. https://doi.org/10.1016/j.jff.2017.11.034

Mikkelsen K, Prakash MD, Kuol N, Nurgali K, Stojanovska L, Apostolopoulos V (2019) Anti-tumor effects of vitamin B2, B6 and B9 in promonocytic lymphoma cells. Int J Mol Sci 20(15). https://doi.org/10.3390/ijms20153763

Mitchell ES, Conus N, Kaput J (2014) B vitamin polymorphisms and behavior: evidence of associations with neurodevelopment, depression, schizophrenia, bipolar disorder and cognitive decline. Neurosci Biobehav Rev 47C:307–320. https://doi.org/10.1016/j.neubiorev.2014.08.006

Miyake Y, Sasaki S, Tanaka K, Yokoyama T, Ohya Y, Fukushima W, Saito K, Ohfuji S, Kiyohara C, Hirota Y, Osaka M, Child Health Study Group (2006) Dietary folate and vitamins B12, B6, and B2 intake and the risk of postpartum depression in Japan: the Osaka Maternal and Child Health Study. J Affect Disord 96(1–2):133–138. https://doi.org/10.1016/j.jad.2006.05.024

Mota SI, Costa RO, Ferreira IL, Santana I, Caldeira GL, Padovano C, Fonseca AC, Baldeiras I, Cunha C, Letra L, Oliveira CR, Pereira CM, Rego AC (2015) Oxidative stress involving changes in Nrf2 and ER stress in early stages of Alzheimer's disease. Biochim Biophys Acta 1852(7):1428–1441. https://doi.org/10.1016/j.bbadis.2015.03.015

Murakami K, Mizoue T, Sasaki S, Ohta M, Sato M, Matsushita Y, Mishima N (2008) Dietary intake of folate, other B vitamins, and omega-3 polyunsaturated fatty acids in relation to depressive symptoms in Japanese adults. Nutrition 24(2):140–147. https://doi.org/10.1016/j.nut.2007.10.013

Naziroglu M, Celik O, Uguz AC, Butun A (2015) Protective effects of riboflavin and selenium on brain microsomal Ca2+-ATPase and oxidative damage caused by glyceryl trinitrate in a rat headache model. Biol Trace Elem Res 164(1):72–79. https://doi.org/10.1007/s12011-014-0199-x

Nemazannikova N, Mikkelsen K, Stojanovska L, Blatch GL, Apostolopoulos V (2017) Is there a link between vitamin B and multiple sclerosis? Med Chem. https://doi.org/10.2174/1573406413666170906123857

Pan X, Fei G, Lu J, Jin L, Pan S, Chen Z, Wang C, Sang S, Liu H, Hu W, Zhang H, Wang H, Wang Z, Tan Q, Qin Y, Zhang Q, Xie X, Ji Y, Cui D, Gu X, Xu J, Yu Y, Zhong C (2016) Measurement of blood thiamine metabolites for Alzheimer's disease diagnosis. EBioMedicine 3:155–162. https://doi.org/10.1016/j.ebiom.2015.11.039

Pariante CM (2015) Neuroscience, mental health and the immune system: overcoming the brain-mind-body trichotomy. Epidemiol Psychiatr Sci 25:101–105. https://doi.org/10.1017/S204579601500089X

Petridou ET, Kousoulis AA, Michelakos T, Papathoma P, Dessypris N, Papadopoulos FC, Stefanadis C (2016) Folate and B12 serum levels in association with depression in the aged: a systematic review and meta-analysis. Aging Ment Health 20(9):965–973. https://doi.org/10.1080/13607863.2015.1049115

Pitkin RM (2007) Folate and neural tube defects. Am J Clin Nutr 85(1):285S–288S

Porter K, Hoey L, Hughes CF, Ward M, McNulty H (2016) Causes, consequences and public health implications of low B-vitamin status in ageing. Nutrients 8(11). https://doi.org/10.3390/nu8110725

Powers HJ (2003) Riboflavin (vitamin B-2) and health. Am J Clin Nutr 77(6):1352–1360

Read DH, Harrington DD (1981) Experimentally induced thiamine deficiency in beagle dogs: clinical observations. Am J Vet Res 42(6):984–991

Richard FE (2014) Pyridoxine deficiency workup. http://emedicine.medscape.com/article/124947-workup. Accessed 18 Mar 2015

Rizzo G, Lagana AS, Rapisarda AM, La Ferrera GM, Buscema M, Rossetti P, Nigro A, Muscia V, Valenti G, Sapia F, Sarpietro G, Zigarelli M, Vitale SG (2016) Vitamin B12 among vegetarians: status, assessment and supplementation. Nutrients 8(12). https://doi.org/10.3390/nu8120767

Sadowska-Bartosz I, Bartosz G (2014) Effect of antioxidants supplementation on aging and longevity. Biomed Res Int 2014:404680. https://doi.org/10.1155/2014/404680

Saposnik G, Ray JG, Sheridan P, McQueen M, Lonn E, Heart Outcomes Prevention Evaluation 2 Investigators (2009) Homocysteine-lowering therapy and stroke risk, severity, and disability:

additional findings from the HOPE 2 trial. Stroke 40(4):1365–1372. https://doi.org/10.1161/STROKEAHA.108.529503

Scheibye-Knudsen M, Mitchell SJ, Fang EF, Iyama T, Ward T, Wang J, Dunn CA, Singh N, Veith S, Hasan-Olive MM, Mangerich A, Wilson MA, Mattson MP, Bergersen LH, Cogger VC, Warren A, Le Couteur DG, Moaddel R, Wilson DM 3rd, Croteau DL, de Cabo R, Bohr VA (2014) A high-fat diet and NAD(+) activate Sirt1 to rescue premature aging in Cockayne syndrome. Cell Metab 20(5):840–855. https://doi.org/10.1016/j.cmet.2014.10.005

Sengul O, Uygur D, Gulec M, Dilbaz B, Simsek EM, Goktolga U (2014) The comparison of folate and vitamin B12 levels between depressive and nondepressive postmenopausal women. Turk J Med Sci 44(4):611–615

Seppala J, Koponen H, Kautiainen H, Eriksson JG, Kampman O, Leiviska J, Mannisto S, Mantyselka P, Oksa H, Ovaskainen Y, Viikki M, Vanhala M, Seppala J (2013) Association between vitamin b12 levels and melancholic depressive symptoms: a Finnish population-based study. BMC Psychiatry 13:145. https://doi.org/10.1186/1471-244X-13-145

Shakoor H, Feehan J, Al Dhaheri AS, Ali HI, Platat C, Ismail LC, Apostolopoulos V, Stojanovska L (2021a) Immune-boosting role of vitamins D, C, E, zinc, selenium and omega-3 fatty acids: could they help against COVID-19? Maturitas 143:1–9. https://doi.org/10.1016/j.maturitas.2020.08.003

Shakoor H, Feehan J, Al Dhaheri AS, Cheikh Ismail L, Ali HI, Alhebshi SH, Apostolopoulos V, Stojanovska L (2021b) Role of vitamin D supplementation in aging patients with COVID-19. Maturitas 152:63–65. https://doi.org/10.1016/j.maturitas.2021.03.006

Shakoor H, Feehan J, Apostolopoulos V, Platat C, Al Dhaheri AS, Ali HI, Ismail LC, Bosevski M, Stojanovska L (2021c) Immunomodulatory effects of dietary polyphenols. Nutrients 13(3). https://doi.org/10.3390/nu13030728

Shakoor H, Feehan J, Mikkelsen K, Al Dhaheri AS, Ali HI, Platat C, Ismail LC, Stojanovska L, Apostolopoulos V (2021d) Be well: a potential role for vitamin B in COVID-19. Maturitas 144:108–111. https://doi.org/10.1016/j.maturitas.2020.08.007

Sharma A, Bist R, Bubber P (2013) Thiamine deficiency induces oxidative stress in brain mitochondria of Mus musculus. J Physiol Biochem 69(3):539–546. https://doi.org/10.1007/s13105-013-0242-y

Sheraz MA, Kazi SH, Ahmed S, Anwar Z, Ahmad I (2014) Photo, thermal and chemical degradation of riboflavin. Beilstein J Org Chem 10:1999–2012. https://doi.org/10.3762/bjoc.10.208

Shibata M, Lu T, Furuya T, Degterev A, Mizushima N, Yoshimori T, MacDonald M, Yankner B, Yuan J (2006) Regulation of intracellular accumulation of mutant Huntingtin by Beclin 1. J Biol Chem 281(20):14474–14485. https://doi.org/10.1074/jbc.M600364200

Shimizu F, Sano Y, Tominaga O, Maeda T, Abe MA, Kanda T (2013) Advanced glycation end-products disrupt the blood-brain barrier by stimulating the release of transforming growth factor-beta by pericytes and vascular endothelial growth factor and matrix metalloproteinase-2 by endothelial cells in vitro. Neurobiol Aging 34(7):1902–1912. https://doi.org/10.1016/j.neurobiolaging.2013.01.012

Slavich GM, Irwin MR (2014) From stress to inflammation and major depressive disorder: a social signal transduction theory of depression. Psychol Bull 140(3):774–815. https://doi.org/10.1037/a0035302

Spry C, Macuamule C, Lin Z, Virga KG, Lee RE, Strauss E, Saliba KJ (2013) Pantothenamides are potent, on-target inhibitors of Plasmodium falciparum growth when serum pantetheinase is inactivated. PLoS One 8(2):e54974. https://doi.org/10.1371/journal.pone.0054974

Syed EU, Wasay M, Awan S (2013) Vitamin B12 supplementation in treating major depressive disorder: a randomized controlled trial. Open Neurol J 7:44–48. https://doi.org/10.2174/1874205X01307010044

Tamanna N, Mahmood N (2015) Food processing and Maillard reaction products: effect on human health and nutrition. Int J Food Sci 2015:526762. https://doi.org/10.1155/2015/526762

Trapali MJ (2021) Biomarkers of aging: a mini review. Rev Clin Pharmacol Pharmacokinet 35(2):55–58

Tucker KL, Qiao N, Scott T, Rosenberg I, Spiro A 3rd (2005) High homocysteine and low B vitamins predict cognitive decline in aging men: the veterans affairs normative aging study. Am J Clin Nutr 82(3):627–635

Voziyan PA, Hudson BG (2005) Pyridoxamine: the many virtues of a Maillard reaction inhibitor. Ann N Y Acad Sci 1043:807–816. https://doi.org/10.1196/annals.1333.093

Walker JG, Batterham PJ, Mackinnon AJ, Jorm AF, Hickie I, Fenech M, Kljakovic M, Crisp D, Christensen H (2012) Oral folic acid and vitamin B-12 supplementation to prevent cognitive decline in community-dwelling older adults with depressive symptoms--the Beyond Ageing Project: a randomized controlled trial. Am J Clin Nutr 95(1):194–203. https://doi.org/10.3945/ajcn.110.007799

Wang HS, Kuo MF (2007) Vitamin B6 related epilepsy during childhood. Chang Gung Med J 30(5):396–401

Wang X, Xu M, Frank JA, Ke ZJ, Luo J (2017) Thiamine deficiency induces endoplasmic reticulum stress and oxidative stress in human neurons derived from induced pluripotent stem cells. Toxicol Appl Pharmacol 320:26–31. https://doi.org/10.1016/j.taap.2017.02.009

WHO (2017) Ageing and nutrition: a growing global challenge. http://www.who.int/nutrition/topics/ageing/en/index1.html

Wickens AP (2001) Ageing and the free radical theory. Respir Physiol 128(3):379–391

Wong CW (2015) Vitamin B12 deficiency in the elderly: is it worth screening? Hong Kong Med J 21(2):155–164. https://doi.org/10.12809/hkmj144383

Wu XY, Lu L (2012) Vitamin B6 deficiency, genome instability and cancer. Asian Pac J Cancer Prev 13(11):5333–5338

Xin XX, Chen Y, Chen D, Xiao F, Parnell LD, Zhao J, Liu L, Ordovas JM, Lai CQ, Shen LR (2016) Supplementation with major royal-jelly proteins increases lifespan, feeding, and fecundity in Drosophila. J Agric Food Chem 64(29):5803–5812. https://doi.org/10.1021/acs.jafc.6b00514

Xiu LL, Lee MS, Wahlqvist ML, Chen RC, Huang YC, Chen KJ, Li D (2012) Low and high homocysteine are associated with mortality independent of B group vitamins but interactive with cognitive status in a free-living elderly cohort. Nutr Res 32(12):928–939. https://doi.org/10.1016/j.nutres.2012.09.005

Yang Q, Botto LD, Erickson JD, Berry RJ, Sambell C, Johansen H, Friedman JM (2006) Improvement in stroke mortality in Canada and the United States, 1990 to 2002. Circulation 113(10):1335–1343. https://doi.org/10.1161/CIRCULATIONAHA.105.570846

Zeisel SH (1992) Choline: an important nutrient in brain development, liver function and carcinogenesis. J Am Coll Nutr 11(5):473–481

Zhang G, Ding H, Chen H, Ye X, Li H, Lin X, Ke Z (2013) Thiamine nutritional status and depressive symptoms are inversely associated among older Chinese adults. J Nutr 143(1):53–58. https://doi.org/10.3945/jn.112.167007

Zhang T, Liang Z, Lin T, Cohen DJ, Arrieta A, Wang X, Qin X, Wang B, Huo Y, Liu GG, Jiang J, Zhang Z (2022) Cost-effectiveness of folic acid therapy for primary prevention of stroke in patients with hypertension. BMC Med 20(1):407. https://doi.org/10.1186/s12916-022-02601-z

Zhang P, Xie X, Zhang Y (2023) Associations between homocysteine and B vitamins and stroke: a cross-sectional study. Front Neurol 14:1184141. https://doi.org/10.3389/fneur.2023.1184141

Zou YX, Ruan MH, Luan J, Feng X, Chen S, Chu ZY (2017) Anti-aging effect of riboflavin via endogenous antioxidant in fruit fly Drosophila melanogaster. J Nutr Health Aging 21(3):314–319. https://doi.org/10.1007/s12603-016-0752-8

Chapter 13
Mineral Supplements in Ageing

Simon Welham, Peter Rose, Charlotte Kirk, Lisa Coneyworth, and Amanda Avery

Abstract With advancing age, achievement of dietary adequacy for all nutrients is increasingly difficult and this is particularly so for minerals. Various factors impede mineral acquisition and absorption including reduced appetite, depressed gastric acid production and dysregulation across a range of signalling pathways in the intestinal mucosa. Minerals are required in sufficient levels since they are critical for the proper functioning of metabolic processes in cells and tissues, including energy metabolism, DNA and protein synthesis, immune function, mobility, and skeletal integrity. When uptake is diminished or loss exceeds absorption, alternative approaches are required to enable individuals to maintain adequate mineral levels. Currently, supplementation has been used effectively in populations for the restoration of levels of some minerals like iron, zinc, and calcium, but these may not be without inherent challenges. Therefore, in this chapter we review the current understanding around the effectiveness of mineral supplementation for the minerals most clinically relevant for the elderly.

Keywords Ageing · Mineral · Supplement · Nutrition · Micronutrient · Iron · Zinc · Calcium · Phosphorus · Selenium · Magnesium

Introduction

Life expectancy has increased over several decades with human longevity rising significantly between 2000 to 2016 by approximately 5 years (Bruins et al. 2019). Globally, people of diverse cultural backgrounds are now reaching ages of more than 80 years in many countries (Riley 2001), these gains being attributed to better

S. Welham (✉) · P. Rose · C. Kirk · L. Coneyworth · A. Avery
Division of Food, Nutrition and Dietetics, School of Biosciences, University of Nottingham, Sutton Bonington Campus, Sutton Bonington, Leicestershire, UK
e-mail: simon.welham@nottingham.ac.uk; Peter.Rose@nottingham.ac.uk; lisa.coneyworth@nottingham.ac.uk; amanda.avery@nottingham.ac.uk

© The Author(s), under exclusive license to Springer Nature Switzerland AG 2024

V. I. Korolchuk, J. R. Harris (eds.), *Biochemistry and Cell Biology of Ageing: Part V, Anti-Ageing Interventions*, Subcellular Biochemistry 107,
https://doi.org/10.1007/978-3-031-66768-8_13

nutrition, wealth, behavioural changes, education, and access to adequate health care provisions (Bruins et al. 2019). Unfortunately, gains in human longevity are mirrored by parallel increases in several non-communicable diseases (NCDs) including diabetes, cardiovascular diseases, arthritic conditions, some cancers, and neurodegenerative conditions (Budreviciute et al. 2020; Watkins et al. 2022; WHO 2019). Key drivers of these NCDs are unhealthy diets, low physical activity, tobacco and alcohol use, and point to the fact that these diseases are preventable or could be better managed due to the strong lifestyle component (WHO 2019). Consequently, many researchers across the world are focusing their efforts on the development of better nutritional strategies that maintain health in the elderly and can be developed as a strategy to prevent and/or reduce the relative risk of developing NCDs. These efforts support the World Health Organization (WHO) and United Nations (UN) plans to tackle aged-related declines in health and the promotion of health span in the general population (Figueira et al. 2016). This recognition is now informing the development of programmes targeting future sustainable development goals (Watkins et al. 2022). These goals are aimed at reducing premature mortality caused by NCDs, coupled to the targeting of better mental health and wellbeing policies. Combined these approaches will be critical in the future development of new management strategies aimed at elderly populations.

For many years, nutritionists have pointed to the fact that many micronutrients are limiting in elderly individuals due to poor diet, physiological limitations, and polypharmacy. Indeed, in the elderly, vitamins and minerals are often limiting, since intakes and bioavailability, are often diminished. Moreover, limitations in the methodologies needed to assess nutrient status are still being developed, and there are still significant gaps in our understanding of biochemical cycling mechanisms linked to the retention of key nutrients in tissues and organs. To date, it is estimated that 5–10% of community-dwelling adults >70 years of age are undernourished. Moreover, this figure increasing significantly to between 30% to 65% for institutionalized elderly patients (NICE 2006). At present, it is widely recognised that many nutrients including various B vitamins (Joshi 2015; Zhu et al. 2020), and many minerals including calcium (Ca^{2+}), iodine (I_2), magnesium (Mg^{2+}), iron (Fe^{2+}), zinc (Zn^{2+}), and selenium (Se) are diminished in older people. These deficiencies are often associated with many NCDs since micronutrient insufficiencies reduce the proper function of the immune system, general metabolism, and increase the risk of frailty among others (Ames et al. 1993; High 1999; Lorenzo-López et al. 2017; Maggini et al. 2018; Semba et al. 2006; Tyrovola et al. 2023). Consequently, intervention using micronutrient supplementation is often seen as a potential strategy to manage micronutrient deficiencies in aged populations and thereby prevent NCDs. In the UK for example, the National Institute for Health and Care Excellence (NICE) guidelines recommend the use of supplementation for patients at risk of malnutrition or who are already malnourished (NICE 2006). Furthermore, in recent times there has been a significant expansion in the commercial availability of multivitamins, minerals, and other supplements, with several products being specifically developed for use by the elderly. Such is the growth in this market that business forecasts point to growth estimated to be in the region of US$185 billion dollars by

2025, with additional contributions from online sales of 6.9% per annum as a further driver of product growth (vitality-pro.com). In the United Kingdom (UK), market predictors show similar patterns of expansion, with values in the order of £520 million in 2022, with sales growth of 17% since 2017. Current estimates show that in the UK alone 38% of all adults are using these products (Mintel 2022). In the general population, increased popularity stems from the need to compensate for perceived poor diets and the reported health promoting benefits linked to the marketing of various commercial products. However, despite the availability of these items many questions remain as to their benefits in the elderly. For example, many human intervention studies are difficult to interpret since the administered supplements are typically composed of more than one micronutrient, vitamin complex or bioactive. This makes it difficult to pinpoint a specific link between an age-related ailment with that of an individual nutrient. Moreover, research is still trying to better understand the impact of changes in the bioavailability, tissue distribution, and utilisation of supplements in elderly people, and this area requires further study. This is especially important since there is considerable variation in the types and amounts of foods consumed by elderly populations both between and within countries. In addition, suboptimal nutrient-density of diets and excess amounts of salt, sugar, and saturated fat are common in processed or pre-prepared foods and this can mask the need for diets with adequate supplies of essential nutrients.

While there is a need for future research to provide a better understanding of these areas, we attempt here to summarise the current data for mineral specific supplementation where available.

Iron

Iron is an essential dietary component required as a cofactor for a wide range of enzymes and other proteins. Its importance is conferred through its capacity to readily undergo redox cycling between the Fe^{2+} (ferrous) and Fe^{3+} (ferric) states (Wallace 2016). The biological roles of iron containing proteins are broad, including energy generation, oxygen transport, immune function, and detoxification of ingested and circulating substances. The haem molecule, which has iron at its core, is the most abundant iron repository, with haemoglobin containing approximately 65% of the body's iron. Because of the ease with which free iron can result in the generation of oxygen radicals (Koskenkorva-Frank et al. 2013), its distribution is very tightly controlled with little free iron in any tissue compartment. Iron status is maintained by intake as there is no specific excretory pathway for iron. Whilst recycling accounts for the majority of the internal supply, iron must be consumed in the diet as it is adventitiously lost through blood loss, sloughing of intestinal cells, shed skin and hair, urine, and sweat in addition to that used for reproduction and growth. Daily losses reach approximately 2 mg per day in healthy individuals not experiencing blood loss (Sharp and Srai 2007) and this must therefore be replaced from dietary sources.

Iron Sources and Digestion

Iron recycling provides the majority [around 90%; (Slusarczyk and Mleczko-Sanecka 2021)] of daily iron requirements, while the remainder comes from food sources. Recommended iron intake for elderly females and males is 8.7 mg day^{-1} (SACN 2011). Principal dietary sources include fortified cereals (including bread), meat and seafood, meat products and vegetables with some from nuts. However, the ability to absorb dietary iron depends upon the form in which it is consumed. The most bioavailable form of iron is heme iron (iron as a component of a haem molecule), derived from animal products, with the remainder being non-haem and which is predominantly in the ferric ($^{3+}$) state. Iron is solubilised in the stomach including that associated with ingested ferritin [whether plant or animal derived; (Hoppler et al. 2008)] and subsequently absorbed either as haem iron or as non-haem Fe^{2+}. Haem iron can be directly absorbed through the haem carrier protein 1 [HCP1 or SLC46A1; (Sharp and Srai 2007)], whilst Fe^{2+} is taken up into the enterocyte via the divalent mineral transporter 1 (SLC11A2, previously DMT1). However, since the predominant form of iron in the diet is Fe^{3+}, which cannot be transported through SLC11A2, then for it to be absorbed it must be reduced to Fe^{2+} both in the stomach through reaction with ascorbic acid (Lane and Richardson 2014) and other reductive dietary components [e.g. histidine and cysteine; (Glahn and Van Campen 1997; Swain et al. 2002)], and in the duodenum via the ferric reductase enzyme, Cytochrome B Reductase 1 (CYBRD1) located on the apical membrane of the enterocyte (Wallace 2016). Once reduced, Fe^{2+} is stabilised by the low pH of the stomach and upper duodenum. This acidic environment further aids uptake through SLC11A2 by virtue of its action as a metal-proton cotransporter. Bioavailability of dietary iron is impeded, however, by certain antinutritional factors present predominantly in the plant components of the diet (Samtiya et al. 2020). Binding of phytic acid (from seeds, legumes, wholegrains, and nuts), calcium and phenolic compounds from a range of beverages (including tea and coffee) directly impedes availability for transport and results in an increased load reaching the large intestine.

Iron is absorbed into the enterocyte as part of a haem molecule or as Fe^{2+}. The haem molecule is degraded by cytoplasmic haem oxygenase 1 (HMOX1) to release its iron as Fe^{2+}. Once free in the cytoplasm Fe^{2+} presents an oxidative threat to the cell and is either oxidised to form Fe^{3+} and bound by intracellular ferritin, or transported as Fe^{2+} through the basolateral transporter, ferroportin [SLC11A3; (Sharp and Srai 2007)]. In the circulation, again, Fe^{2+} represents an oxidative threat, so is itself oxidised to Fe^{3+} by the membrane bound ferroxidase hephaestin (Petrak and Vyoral 2005) or circulating ceruloplasmin (Eid et al. 2014) and bound by transferrin. Cells requiring iron express transferrin receptors on their surface which bind circulating transferrin and internalise it along with its iron cargo via endocytosis. Transferrin-containing vesicles fuse with lysosomes which pump in protons to reduce intralysosomal pH leading to dissociation of iron from transferrin. This Fe^{3+} iron is reduced back to Fe^{2+} and exported into the cytoplasm through SLC11A2, where it is re-oxidised and bound to ferritin from which it can be removed when needed (Steere et al. 2012).

Causes of Iron Deficiency

Iron deficiency impacts many molecular processes, but the earliest and clearest clinical manifestation is anaemia. A range of markers are used to assess iron status with each being indicative of different stages or levels (Pfeiffer and Looker 2017). Short term limitations of provision are met by release from intracellular ferritin bound iron stores, predominantly stored in hepatocytes and some macrophage populations. Circulating ferritin concentrations are reflective of the intracellular iron storage capacity and therefore is used as an indicator of iron stores. As iron status diminishes, the reduced capacity to adequately supply the needs for erythropoiesis becomes apparent. This is initially reflected in the transport of iron in the circulation and is indicated by the saturation of serum transferrin (transferrin saturation or TSAT), which diminishes with iron deficiency, the levels of erythrocyte protoporphyrin (reflects iron availability during erythropoiesis) which drops, and the concentration of serum transferrin receptor (sTfR) which becomes elevated. Finally, as erythropoiesis is compromised this is reflected in reduced circulating haemoglobin (Hb) levels.

Anaemia poses a significant threat to health in the elderly, hampering physical and cognitive function, increasing risk of frailty, recovery time from illness and results in increased length of hospital stay. Inadequate intake is a significant cause of iron deficiency in the elderly in the developed world [approximately 16%; (Guralnik et al. 2004)], but there are numerous other factors that additionally contribute to iron status and may result in iron deficiency even in the face of adequate intakes. As a result, estimates of the prevalence of anaemia in the elderly vary widely, being dependent on the mode of assessment, and are highly context specific (Wawer et al. 2018) ranging from <10% in healthy community dwelling individuals (Yildirim et al. 2015) to >60% among some hospitalized groups (Zilinski et al. 2014).

Poor dietary intakes can be caused by reduced appetite due to limited physical activity, consumption of iron poor diets, consumption of monotonous diets and, particularly in the case of institutionalised individuals, dependence on others for assisted eating. A reduced rate of gastric emptying and altered taste contributes to the volume and choice of foods consumed (Drewnowski and Shultz 2001) often towards a smaller quantity of relatively low nutrient density foods. Consumption of diets containing a predominance of ferric iron (e.g. plant based diets) or those with a relative abundance of anti-nutritional factors, such as polyphenols or phytic acid (Petry et al. 2010), may further diminish the bioavailability of dietary iron, thus impairing absorption.

In addition to dietary factors and the simple consequences of ageing, there are numerous clinical factors which contribute to the ability to absorb dietary iron. Inflammation, both acute and chronic, can impact iron status via upregulation of hepcidin. Hepcidin is a liver-synthesized circulating factor which directly impacts iron absorption and recycling by inhibiting ferroportin activity (Ganz 2005). Circulating hepcidin interacts with ferroportin to induce its internalization and subsequent degradation. Inflammation enhances hepcidin expression (Fraenkel 2017; Nemeth et al. 2004, 2003) with the result that both intestinal iron absorption and iron recycled from macrophage degradation of senescent RBCs are reduced. Chronic consumption of certain medications may additionally impact iron bioavailability. Proton pump inhibitors prescribed for the reduction of gastric acid secretion have been found to impair absorption as a result of the diminished ability to reduce ferric iron and stabilise it as Fe^{2+} (Ito and Jensen 2010). Additionally, blood loss, particularly from occult and overt intestinal bleeding (Rockey 2005), is a major contributor to iron deficiency in elderly groups.

Iron Supplementation

The requirement to restore iron levels is evident, however the method for achieving this must be carefully considered in light of the underlying cause. In some instances, oral supplementation is appropriate, however, in situations where either the underlying condition or the medications being consumed impede absorption of iron from the diet, then alternative mechanisms must be sought, and barriers addressed if possible. Alternative routes of supplementation include intravenous provision of iron and, as a last resort, blood transfusion (Goodnough and Schrier 2014; Silverstein et al. 2008).

Anaemia is categorised either as being due to inadequate intake of iron (iron deficiency anaemia; IDA) or other critical nutrients (folate and vitamin B12), functional iron deficiency classified as anaemia of chronic disease (ACD) and anaemia of inflammation [AI; Nemeth and Ganz 2014], or unexplained anaemia (UA). ACD/AI may result from a range of different conditions, which cause a pro-inflammatory phenotype and result in elevated hepcidin levels alongside inflammatory cytokine mediated suppression of erythropoiesis and enhancement of macrophage erythrocyte destruction.

Oral supplementation, when appropriate, is typically provided as ferrous sulphate, ferrous fumarate and ferrous gluconate and frequently given at very high levels [~200 mg; (Tay and Soiza 2015)], being more than 20 times higher than the age specific recommendations (SACN 2011). Evidence indicates that supplementation of these forms of iron at such levels, while able to elevate Hb concentration, nonetheless are associated with adverse outcomes and may be no more effective than supplementation with much lower quantities [e.g. 15 mg; (Lindblad et al. 2015; Prentice et al. 2017)]. Adverse effects of oral supplementation including upper gastrointestinal erosive mucosal injury, nausea, diarrhoea and constipation are frequent and become more severe with increasing dose (Cancelo-Hidalgo et al. 2013). This has led to the generation of modified forms which slow iron release, such as ferrous sulphate with mucoprotease, which exhibits much better tolerability. Other alternative forms such as Iron(III)-hydroxide polymaltose complex (IPC) provides non-ionic iron stably complexed with polymaltose which limits its scope for engaging in redox reactions in the intestinal lumen if not absorbed and, as a consequence, induces far fewer adverse effects (Toblli and Brignoli 2007).

The effectiveness of oral supplementation is compromised by the fact that a large proportion (around half) of the elderly population either consume proton pump inhibitors or suffer from atrophic gastritis [from 21% among those 61–69 to 37% in those over 80; (Kassarjian and Russell 1989)], resulting in either reduced or absent stomach acid secretion (hypochlorhydria or achlorhydria respectively). In the case of the latter group (atrophic gastritis), this is an irreversible part of ageing as it is associated with loss of acid secreting parietal cells. As iron must be stabilised in the ferrous state by a low pH environment, the reduction of acid secretion greatly diminishes its bioavailability regardless of the source from which it is derived. Consequently, there is little or no further beneficial impact of iron supplementation with increasing quantities in individuals with either PPI or gastritis induced hypochlorhydria and simply results in a greater iron load reaching the lower intestine.

Since inflammation elevates hepcidin levels, diminishing ferroportin and consequently iron uptake from the diet, oral supplementation in people with ACD/AI is far less effective and intravenous (iv) provision may be required. In general, it is preferable where possible to treat the underlying cause and reduce inflammation, but iv supplementation can be extremely effective if well tolerated. Individuals suffering from inflammatory bowel disease (IBD) are frequently anaemic due to impaired iron absorption, gastrointestinal bleeding and chronic inflammation (Semrin et al. 2006). Intravenous iron supplementation is recommended under such conditions although some will respond to oral iron (Gasche et al. 2007). Iron deficiency is similarly common in those suffering from renal disease and has a significant effect on morbidity in those progressing to chronic kidney disease (CKD) and end stage renal failure [ESRD; (Gutiérrez 2021)]. Hepcidin levels are typically elevated in CKD rendering oral supplementation less effective although they have some utility in sensitive individuals (Pergola et al. 2019). However, iv supplementation is often preferred as it may be better tolerated and achieve more effective outcomes (Shepshelovich et al. 2016).

Because of the impaired absorptive capacity of older adults, it is suggested that an oral iron absorption test be carried out in iron deficient individuals before the mode of supplementation is decided in order to determine the likelihood of success of oral strategies before their adoption. This would help to select a more effective approach to restoring iron status and reduce the clinical burden of adverse outcomes (Silay et al. 2015).

Zinc

Zinc is a divalent cation (Zn^{2+}), essential for optimal human health. Ubiquitous throughout the cells of the human body, it is present in its greatest concentration within skeletal muscle (60%) and bone (30%), with further notable concentrations located in the skin, liver and hair (Clarys et al. 1984; Forbes et al. 1953; Molina and DiMaio 2012; Yamanaka et al. 2007). The adult human body contains approximately 1–3 g zinc (Maret and Sandstead 2006). Zinc plays a vital role in many physiological and biochemical functions which can largely be classified according to catalytic, structural and regulatory functions. Such functions include but are not limited to the normal functioning of the immune system, DNA and protein synthesis, wound healing and cell division (Lowe et al. 2009; Mocchegiani et al. 1998; Prasad 2008). Predominantly intracellular, zinc is essential in the structure and function of thousands of proteins including over 300 enzymes spanning all six enzyme classes (Andreini et al. 2006; Vallee and Falchuk 1993). The vast number of zinc dependent biological processes underpin the significant impact of zinc deficiency on the normal functioning of the human body.

Zinc is primarily absorbed in the small intestine via a saturable carrier-mediated mechanism (Cousins 1985; Lee et al. 1989). Free zinc ions (Zn^{2+}) are liberated from the food matrix during digestion and absorbed from the intestinal lumen via specialist transporters located on the apical and basolateral membranes of the enterocyte. There are two families of zinc transporters. The ZnT family (SLC30A) of transporters typically promote the reduction of intracellular zinc availability by enhancing the efflux of zinc from cells into extracellular space or intracellular vesicles. The Zrt-Irt-related protein transporters (ZIP, SLC39A) increase intracellular zinc availability by functioning in the uptake of zinc into the cytosol of cells from either the extracellular space or intracellular vesicles (Cousins et al. 2006; Devergnas et al. 2004; Sekler et al. 2007). Both families of the ZnT and ZIP transporter proteins exhibit unique tissue specific expression and responsiveness to dietary zinc status (Liuzzi and Cousins 2004). Dietary zinc is transported from the lumen into the enterocytes predominantly via the ZIP4 transporter located on the apical brush border and then exported from the enterocyte into the portal blood via ZnT1 (Küry et al. 2002; McMahon and Cousins 1998; Wang et al. 2002). The majority of zinc found in the plasma is protein bound, predominantly to albumin with smaller amounts bound to alpha-macroglobulin and transferrin for transportation around the body to the site of use (Scott and Bradwell 1983). Zinc binding metallothioneins (MT) bind free zinc, acting as a buffer in the redistribution of intracellular zinc by transporting the cation to the membrane bound transporters for sequestration into organelles to reduce free cytoplasmic zinc ions (Colvin et al. 2010; Maret and Krezel 2007).

Zinc homeostasis is tightly regulated and synergistic adjustments to zinc absorption and endogenous excretion are the principal mechanism of such regulation. Zinc is lost from the body via body fluids including faeces, urine, perspiration, semen and menstruation in addition to sloughed skin cells, hair loss and nail loss (Brown et al. 2004). These losses, combined with the lack of zinc storage within the body leads to a requirement for daily dietary zinc intake.

Dietary Zinc

In the UK, red meat and animal foods are a significant source of dietary zinc with red meat and processed meat contributing approximately 32% and 27% of men and women's dietary zinc intake respectively (SACN Iron and Health Report 2011). Globally, plant-based foods including pulses and cereals are a major source of dietary zinc although the presence of dietary inhibitors including phytate and tannins may reduce the accessibility of zinc and subsequent absorption in the gastrointestinal tract. The source of dietary zinc and potential impact on zinc status should be a consideration for older adults choosing to consume a diet low in or void of animal-based products.

The recommended dietary intake of zinc in adults is 7.0 mg/day and 9.5 mg/day for females and males respectively in the UK. In the US, recommended dietary intake levels are set higher at 12 mg/day and 15 mg/day for females and males, respectively. Although dietary zinc recommendations increase throughout childhood, there are no additional recommendations for increased intake for older adults.

Dietary factors may influence zinc absorption in the GI tract. Phytate, the major phosphorus containing compound present in plant-based foods binds to zinc and limits zinc accessibility in the lumen (Maares and Haase 2020). This impact on bioavailability may be in part, predicted by the molar ratios of phytate:zinc with ratios from 6:1 beginning to impact absorption of zinc (Fredlund et al. 2006) and further influenced by the zinc content of the meal and the protein source itself (Sandstrom et al. 1989). Food preparation techniques including soaking, fermentation and sprouting may encourage the enzymatic breakdown of phytate by phytase (Hurrell 2004) whereas mechanical impact of milling may reduce phytate content further (Schlemmer et al. 2009). High dietary intakes of both heme iron and inorganic iron are associated with a negative impact on dietary zinc absorption. High intakes of dietary zinc are associated with impaired copper absorption (Sandstead 1995). Citrate is a ligand, present in dairy milk which binds to zinc and enhances bioavailability (Wegmuller et al. 2014). Dietary protein intake is associated with increased zinc absorption with protein from animal sources enhancing absorption to a greater extent than plant based proteins (Lonnerdal 2000). Older adults excluding animal products from their diets may need to consider the impact of such exclusion on zinc intake and subsequent accessibility and absorption.

Measuring Zinc Status

Typical symptoms of zinc deficiency including compromised growth and impaired immune response may be multifactorial and indicative of general malnourishment. The lack of a unique clinical health outcome, reliable indicator or biomarker specifically attributed to inadequate zinc status renders the diagnosis of marginal zinc deficiency challenging. Plasma zinc is the biomarker most frequently used to evaluate the likelihood of zinc deficiency and is currently recommended by the

International Zinc Nutrition Consultative Group (iZiNCG), the World Health Organisation (WHO) and UNICEF. Urinary and hair zinc concentrations are also noted to be reflective of recent dietary intake and are responsive to supplementation in healthy individuals (Lowe et al. 2009). Concerns have been raised relating to the physiological insensitivity of such measures due to diurnal and post-prandial fluctuations. Further concerns have been raised with using plasma zinc measures in older adults (Mocchegiani et al. 2008) as serum zinc levels may appear adequate (12–16 μM) even following periods of reduced dietary zinc intake and reduced intracellular zinc ion availability (Mocchegiani et al. 2006). Further research is required to adequately establish a reliable method to evaluate zinc status and risk of deficiency in all population subgroups.

Zinc Deficiency and Supplementation

Zinc deficiency manifests in a multitude of health outcomes including stunted growth (Prasad 1985), immune dysfunction (Mocchegiani et al. 1998; Prasad 2008), skin lesions and impaired wound healing (Evans 1986), gastrointestinal disturbances (Hambidge 1992) and neurological/psychological changes (Marcellini et al. 2006; Sandstead 2000). Malnutrition is the main cause of zinc deficiency worldwide with more than 25% of the global population estimated to be at risk of deficiency (Maret and Sandstead 2006). Although severe deficiency is uncommon in western populations, marginal deficiency may have a negative impact on health outcomes (Haase et al. 2006). It is well reported that older adults are at increased risk of malnourishment and therefore at increased risk of zinc deficiency. This risk may be increased due to an age-related reduction in energy requirements and/or sensory impairment leading to decreased appetite and reduced food intake (Stewart-Knox et al. 2005); alterations in the function of zinc transporter proteins reducing absorption in the GI tract; poor dentition leading to inadequate mastication and/or interactions with medications (Mocchegiani et al. 2013). Mild zinc deficiency in adults over the age of 65 years has been identified to be of clinical concern with studies from around the world suggesting almost half of adults in this age group may be considered to have inadequate zinc intake (Andriollo-Sanchez et al. 2005; Briefel et al. 2000; Kogirima et al. 2007; Mocchegiani et al. 2008). Living arrangements may also impact the risk of zinc deficiency with older adults living dependently in institutions including care homes, nursing homes and retirement homes at increased risk of inadequate zinc intake compared to older adults residing independently in the community (Vural et al. 2020).

Zinc supplementation initiatives have significantly reduced the risk of all-cause mortality in children under the age of 5 years (Rouhani et al. 2022). Pharmacological doses of zinc supplementation have long been used in the treatment of the autosomal recessive disease, Acrodermatitis Enteropathica (AE). AE is characterised by severe zinc deficiency because of impaired zinc absorption in the GI tract due to reduced function of the ZIP4 transporter. Supplementation of AE patients to a level high enough to ward off deficiency symptoms is deemed adequate. Pharmacological supplementation of zinc has also been used in the treatment of Wilsons disease; a genetic disorder characterised by the accumulation of copper in the human body. Wilsons disease patients consume high level zinc supplementation (5 mg zinc acetate, three times daily) to avoid this copper accumulation (Brewer 2001).

In the general population, zinc supplements are most frequently consumed orally as single nutrient supplements or in combination with other micronutrients. Zinc is supplemented primarily in the form of zinc sulphate, zinc acetate or zinc gluconate. Older adults prescribed oral nutrition supplement (ONS) drinks may consume a significant proportion of recommended intake via this route. Many human intervention trials have been carried out which explore the potential impact of zinc supplementation on health outcomes in older adults. Such benefits have suggested potential enhancements in immune response (Mocchegiani et al. 2013), reduction in oxidative stress (Mousavi et al. 2020), inflammation (Ceylan et al. 2021) and decreasing the risk of age-related degenerative diseases (Mocchegiani et al. 2013). However, additional attention to the impact of supplementation above the recommended intakes and potential impacts on health outcomes is required. Inconsistencies in the efficacy of zinc supplementation have been reported often due to heterogeneity in study design particularly relating to the dose, duration, and form of zinc supplementation. Current recommendations fail to consider the risk of toxicity from high dose supplementation (Maret and Sandstead 2006) and some concerns have been raised about possible accumulation of intracellular zinc and potential toxicity following high dose or extended duration of zinc treatment (Mocchegiani et al. 2013). Further inconsistencies in response of older adults to zinc supplementation may be in part explained by genetic variation in interleukin-6 -174 G/C locus. Higher levels of IL-6 production have been observed in C-carriers with a subsequent observation in impaired immune response, increased metallothionein, low intracellular zinc ion availability and increased zinc deficiency compared to C+ carriers (Franceschi et al. 2005; Giacconi et al. 2004; Mocchegiani et al. 2006). Further studies are required in order to elucidate if C-carriers may benefit more from zinc supplementation compared to C+ carriers and therefore if a personalised approach to zinc supplementation in older adults would be beneficial. It must also be noted that the habitual consumption of zinc rich foods in the diet of older adults may prove advantageous in the maintenance of adequate zinc status in this population subgroup and further research is required to establish the impact of zinc supplementation on health outcomes when baseline zinc status is deemed adequate.

Selenium

Selenium (Se) is an essential trace element for animals and humans first discovered in 1817 by Swedish scientist Jacob Berzelius (Bodnar et al. 2012; Reilly 2006). Se exists in multiple forms represented by inorganic species like selenide (Se^{2-}), elemental selenium (Se^0), selenite (SeO_3^{2-}), and selenate (SeO_4^{2-}), and organic species such as selenomethionine and selenocysteine. Se is typically found in soils with levels ranging between 0.005 to 1200 μg/g soil but most commonly between 0.1 and 10 μg/g (Alexander 2015). In the UK total Se in soil ranges from 0.1–4 mg Se/kg, however Broadley and White estimate that greater than 95% of UK soils contain less than 1 mg Se/kg (Broadley et al. 2006). Soil concentrations of Se are governed by geochemical conditions, such as rock type, pH and soil water content which ultimately influence the levels of Se in edible plants (Sors et al. 2005). Because of this the accumulation of Se in foods is known to vary widely and is dependent on regional differences in Se soil conditions. A study on bread-making wheat in 1982 and 1992 found that major cereal growing regions in the UK (e.g. South, Midlands and North), had low Se levels in soil (Adams et al. 2002).

The main form of Se in food is Selenomethionine (SeMet) (Weekley and Harris 2013) which can be incorporated into proteins and can replaces the sulphur atom of the amino acid methionine and cysteine (Schubert et al. 1987). Inorganic Se (selenite and selenate) is rarely found in foods apart from in cereals (Cubadda et al. 2010), mushrooms (Stefánka et al. 2001) and certain vegetables (Pedrero et al. 2006). Brazil nuts contain the highest amount of Se mostly in the form SeMet (Thomson et al. 2008), followed by sesame seeds (Kápolna et al. 2007) and several cereals crops (Thiry et al. 2012). Vegetables of Alliums and Brassicaceae families are sources of methylselenocysteine (Pyrzynska 2009). The major source of Se in the diets of humans are meats, fish, eggs which typically contain between 0.01–0.3 μg g^{-1} fresh weight, and various nuts [brazil nuts 5300 μg/100 g reported (Thorn et al. 1978)]. Cereals and grains typically contain between (0.01–0.55 μg g^{-1} fresh weight) whereas fruits and vegetables typically contain much lower levels of Se 0.001–0.022 μg g^{-1} fresh weight (Reilly 1993). A recent study on foodstuffs produced in Scotland classified 90% of samples (including wheat, potato, broccoli, beef-steak, milk) as deficient in Se (<0.6 mg Se/kg) (Fordyce et al. 2010). In recent times, efforts have been made to enhance Se levels in foods using soil fertilisation strategies or plant breeding procedures [reviewed in Broadley et al. (2006)].

Individual Se intakes range from 3–7000 μg Se/d worldwide although most are at the lower end of the spectrum (Broadley et al. 2006; Fordyce 2005). Suboptimal Se status is widespread epidemic across Europe, UK and the middle east (Stoffaneller and Morse 2015). Interestingly eastern Europe has a lower intake of Se compared to western Europe (Stoffaneller and Morse 2015). The middle east has varying Se intakes which have been attributed to varying good habits, food imports and difference in religions and cultures (Stoffaneller and Morse 2015). Selenium intake and status had declined from >60 μg Se/d in 1974 to 29–39 μg Se/d in just 30 years (Rayman 1997, 2000, 2002, 2004) a systematic review paper concluded that the UK had a 'consistent overall suboptimal serum Se status' across a broad range of population age groups (Stoffaneller and Morse 2015). Indeed, dietary intake of Se in the UK are less than half the RNI of 1 μg Se/kg of body weight (Brown and Arthur 2001) (60 μg/day for women and 75 μg/day for men), hence the UK was listed as a country of low Se intake by Scientific Advisory Committee on Nutrition Report 2013 (SACN 2013). The decline in Se intake of the UK population has been attributed to multiple factors including the switch in sourcing wheat from high Se soils of

North America to UK grown wheat (Adams et al. 2002; Broadley et al. 2006), and decreased Se input to soils resulting from a reduction in coal combustion due to the 'Clean Air Act 1956' (Broadley et al. 2006) and the increased use of triple superphosphate fertilisers (White et al. 2004) which contains less than <4 mg Se/kg compared to single superphosphate ≤55 mg Se/kg (Broadley et al. 2006) as well as the behavioural shift in eating less meat (Fordyce et al. 2010) which increases reliance on Se poor vegetables.

Once ingested, Se is incorporated into the aforementioned seleno-amino acids (Bodnar et al. 2012), and subsequently used in the synthesis of various selenoproteins. Of the 25 selenoproteins characterised to date most are peroxidases and include proteins like glutathione peroxidase and thioredoxin reductase. These proteins play critical roles in cellular redox reactions linked to cytoprotection, inflammatory processes, and defences against oxidative stress (Papp et al. 2007; Roman et al. 2014; Wrobel et al. 2016). It is worth noting that in extreme cases, chronic toxicity has been reported and known to occur at Se levels >400 μg/day (Winkel et al. 2012), and the European Food Safety Authority (EFSA) has recently recommended a upper tolerable limit of 233 μg/day for both men and women (EFSA Panel on Nutrition et al. 2023). Acute toxicity includes hypotension, tachycardia, and symptoms like diarrhoea, fatigue, hair loss, joint pain, nail discoloration or brittleness, nausea, and headaches (MacFarquhar et al. 2010). People who have been exposed to excess selenium often have the characteristic of 'garlic breath'.

Selenium Deficiency

Se deficiency occurs due to suboptimal levels of Se intakes *viz.* <40 μg/day and is associated with a wealth of symptoms such as growth retardation and impaired bone metabolism (Moreno-Reyes et al. 2001), as well as abnormalities in thyroid function (Reeves and Hoffmann 2009). Se deficiency can also predispose people to Keshan disease (associated with heart enlargement and cardiac arrhythmia), and Kashin-Beck disease (enlarged joints and osteoarthritis) (Roman et al. 2014).

The impact of selenium deficiency has not been well characterized in older populations and is rare in the general population due to robust biofortification strategies. However, it is widely reported that serum selenium concentrations decrease with age. Potential causes of selenium deficiency are linked to:

- Malnutrition resulting from poor appetite, poorly fitting dentures, food insecurity,
- Malabsorption in the gastrointestinal tract, Crohn's disease,
- Those undergoing total parenteral nutrition, and gastrointestinal bypass surgery,
- Renal disorders and in those undergoing dialysis,
- Drug usage—cholesterol lowering medication viz. statins. Anti-inflammatory medications—inhalant, systemic, and topical corticosteroids, Beclomethasone, Budesonide, Dexamethasone, Hydrocortisone, Fluticasone, Mometasone

Furoate, Prednisone, Methylprednisolone and Triamcinolone. Anticonvulsant Medications including derivatives of valproic acid.

Selenium Supplementation

The therapeutic role of selenium as a micronutrient has been highlighted numerous times, with Se being shown to aid immune and endocrine function, metabolic cycling, and redox cellular homeostasis. Typically, Se is available in tablet form containing either one or a combination of sodium selenate, sodium selenite and selenium-enriched yeast (Puccinelli et al. 2017). Se-enriched yeast products show wide variability in Se content and Se speciation (Rayman 2004). However, these products appears popular as a commercially available health food supplement (Khabbaz Koche Ghazi et al. 2021; Ravn-Haren et al. 2008) reviewed in Rayman (2004). Since 2002 in the EU, only inorganic forms of Se are allowed as food supplements (Puccinelli et al. 2017). Numerous human intervention trials have been conducted assessing the potential impact of Se supplementation on key heath parameters in the elderly. However, most of these studies assess Se heath impacts in combination with other compounds such as various dietary antioxidants like vitamin E, β-carotene, and coenzyme Q(10) respectively. These combined interventions make extrapolation of key impacts of Se supplementation difficult or when studies are investigated in replete participants (Perri et al. 2024).

Calcium

Calcium is required as a structural component of teeth and bones as well as for a wide range of biological roles including mediation of intracellular signalling, nerve signal transduction, muscle contraction and vascular tone (Institute of Medicine Committee to Review Dietary Reference Intakes for Vitamin 2011). The calcium in bones, which comprises around 99% of that present in the body, in addition to providing structural support, represents a storage reservoir enabling release of ionised calcium into the circulation when required. The circulating calcium concentration is maintained within tightly controlled limits, mediated by altering uptake efficiency, rate of excretion and release of calcium from bone.

The regulation of calcium homeostasis is coordinated by parathyroid hormone (PTH) acting both directly on target tissues and indirectly through vitamin D signalling. Declining calcium levels are detected by the parathyroid gland through activation of the calcium sensing receptor (CasR), resulting in the release of PTH into the circulation. Direct effects of PTH mediate calcium release into the circulation and reabsorption from filtrate through actions on bone and kidney respectively. In bone, PTH binds its receptor on the surface of osteoblasts (Jüppner et al. 1991) inducing recruitment and differentiation of osteoclasts, which in turn embark upon bone resorption leading to calcium release into the circulation (Raggatt and Partridge 2010). In the distal tubule and collecting duct of the kidney, PTH increases reabsorption of calcium while impeding reabsorption of phosphate in the proximal tubule, thereby enabling elevated ionic calcium to be returned to the circulation. PTH additionally enhances renal production of active vitamin D, 1,25 dihydroxy vitamin D (1,25[OH]$_2$D) through activation of the enzyme 25 hydroxyvitamin D-1α hydroxylase [CYP27B1; (Fleet 2022)]. At the same time, activity of the 1,25[OH]$_2$D degrading enzyme 25-Hydroxyvitamin D-24-hydroxylase (CYP24A1) is reduced. Circulating 1,25[OH]$_2$D exerts various effects on a range of target tissues to increase plasma calcium concentrations through activation of the vitamin D receptor (VDR). In the intestine, calcium absorption is stimulated via two separate routes, a saturable and a non-saturable pathway. The saturable pathway involves transcriptional upregulation of calcium channels [TRPV6 and potentially Ca_v1.3; (Cui et al. 2012; Kellett 2011)] and is active in the duodenum, jejunum, caecum and colon [not active in the ileum; (Pansu et al. 1983)]. Calcium enters the cell via these channels and is expelled at the basolateral membrane via the energy dependent PMCA1b and NCX1 (Khanal and Nemere 2008; Wongdee and Charoenphandhu 2015). The calcium binding protein calbindin-d is believed to play a buffering role regulating intracellular concentrations in intestinal and renal epithelia which both are exposed to large calcium fluxes during uptake and resorption respectively (Cui et al. 2012). As circulating calcium levels rise, this activates CASR receptors resulting in a rapid and dramatic decrease in PTH secretion (Chen and Goodman 2004).

Calcium is lost in urine and faeces with a small amount lost in sweat. To replace losses, calcium must be consumed in the diet, with recommended intakes being similar for elderly females and males [700 mg day^{-1}; (SACN 2011)]. The largest contribution to calcium intake in omnivores is from dairy products (mostly milk, cheese and yoghurt), providing up to 70% of daily intake (Institute of Medicine Committee to Review Dietary Reference Intakes for Vitamin 2011). Fortified cereals additionally provide a significant proportion of intake, while vegetables, grains, fruit, meat and poultry contribute relatively little (BDA 2021). For those not consuming dairy products, there is a need to obtain calcium from alternative sources including supplements and fortified foods and drinks. Where plant-based alternative "milk" products are fortified, the bioaccessibility of calcium (and other minerals) is generally good (Silva et al. 2022) but currently only around 70% are adequately fortified (Drewnowski 2022). This potentially poses a significant problem for those excluding dairy and requires consideration in the nutritional management of such elderly individuals.

Causes and Consequences of Low Calcium Status

Low calcium status can result from dietary deficiency and from various clinical outcomes which impact uptake and homeostasis. Calcium absorption declines with advancing age being particularly marked in women after the onset of menopause, at which point there is an initial drop of around 2% at the time of oestrogen loss, followed by a relatively steady decline of ~0.02% per year subsequently (Heaney et al. 1989). While calcium absorption diminishes, bone mineral loss also accelerates with age up to around 1% per year depending on age and clinical context (Burger et al. 1998; Jones et al. 1994; Peacock et al. 2000). The progressive loss of bone mineral, along with other factors, ultimately results in the weakening of bones, increased incidence of osteoporosis and increased risk of fracture (NOGG 2021; Weaver et al. 2016). While this loss to the structural components of bone is occurring, it tends to do so whilst maintaining circulating calcium concentrations within the range necessary to support the broader calcium functions of the body.

Approximately half of all women and one fifth of men will sustain a fracture due to a fall from standing height (fragility fracture), over the course of their lifetime (van Staa et al. 2001). Such fractures significantly reduce the quality of life resulting in pain, physical disability, mental health issues, and elevated mortality (NOGG 2021). Some types of fracture, such as hip fractures, are associated with particularly high mortality rates, reaching 10% within 30 days of fracture and around 25% after 1 year (Hawley et al. 2016). The current estimated cost to the NHS of fragility fractures if £4.7 billion per annum comprising both immediate post-fall support and requirement for longer term institutional care.

Acute changes in calcium status provide different clinical challenges. When circulating calcium levels decline this is of much more immediate clinical significance. Hypocalcaemia is characterised by serum total calcium concentrations corrected for albumin of <2.12 mmol [8.5 mg dl^{-1}; (Pepe et al. 2020)]. Disorders resulting in hypocalcaemia are typically divided into PTH dependent and independent. PTH dependent mechanisms are driven either by reduced secretion of parathyroid hormone (hypoparathyroidism) or impairment of its function (pseudohypoparathyroidism). Hypoparathyroidism is predominantly a consequence of genetic disorders of which there are several, autoimmune parathyroid damage, and most commonly as a result of neck surgery. Additionally, hypomagnesaemia blocks PTH secretion (Bove-Fenderson and Mannstadt 2018), thought to be due to aberrant activation of the calcium sensing receptor CASR. PTH independent mechanisms include specific organ dysfunction, such as chronic kidney disease with concomitant reduction in 1,25$[OH]_2$D synthesis, or may be drug induced. Such pharmacological outcomes are seen with the use of loop diuretics (leading to excessive calcium loss), inhibitors of bone resorption which limit release into the circulation, and anticonvulsants which can reduce vitamin D levels by increasing their metabolism (Stamp et al. 1972).

Calcium Supplementation

Acute hypocalcaemia constitutes a medical emergency and is treated by intravenous infusion of calcium gluconate or, less commonly due to venous irritation, calcium carbonate (Turner et al. 2016), alongside treatment of the underlying cause. Chronic hypocalcaemia, however, requires supplementation. Whilst recommended intakes of 1 g day^{-1} would be sufficient in healthy subjects, those suffering from hypoparathyroidism require much larger quantities due to the inefficiency of absorption. Supplement doses in the region of up to 3 g in multiple daily doses (3–4 per day) are necessary to enable sufficient acquisition of calcium. In addition, the form in which the calcium is delivered is also important as is the time of consumption, in many cases being most effective when consumed between meals. Oral supplementation is generally delivered as calcium carbonate, calcium phosphate, calcium citrate or calcium citrate malate (Weaver and Peacock 2011).

Calcium supplements are commonly consumed in an attempt to reduce the incidence and severity of osteoporosis and are recommended for those not obtaining adequate quantities from the diet (NOGG 2021; Peacock et al. 2000). Osteoporosis is characterised by a low bone mineral mass alongside a progressive breakdown of the bone microarchitecture, ultimately resulting in greater fragility and propensity to fracture (Amin et al. 2023). Numerous pharmacological treatments are available which reduce bone loss and support mineralisation. The efficacy of these treatments is contingent upon an adequate calcium supply and so daily calcium consumption is required to maintain status, as they may result in depression of circulating calcium concentrations. However, global recommendations vary widely, from 700 mg day^{-1} (COMA 1991) to 1300 mg day^{-1} (NHMRC 2006), indicating a relative lack of consensus on precise requirements. Depending on the country of origin therefore, variable supplementation levels may be recommended. While many studies of calcium, vitamin D and combined supplementation have been conducted, evidence to support their usefulness for prevention of bone loss, particularly those without a diagnosed bone disorder, is not clear (Bolland et al. 2015; Bristow et al. 2022; Lips et al. 2010; Manoj et al. 2023; Weaver et al. 2016) unless underlying deficiency is already present and further, it has been suggested that they may increase the risk of kidney stones (Kahwati et al. 2018). Difficulty in reaching definitive conclusion may reflect inherent challenges posed by the study of supplement outcomes, in that short term trials are hampered by a limited ability to effectively assess bone mineral changes and physical consequence (e.g. fracture), while longer trials face challenges such as the necessity to assume adherence to supplementation regimens which cannot be guaranteed.

Several studies have raised concerns regarding the potential for calcium and vitamin D supplementation to increase the risk of cardiovascular disease. However, the findings are again equivocal, and a consensus statement released to the effect that there is only moderate quality evidence of any relationship between vitamin D and/or calcium supplementation and cardiovascular outcomes (Kopecky et al. 2016). Ultimately, the recommendation has been made that only patients at risk of calcium and vitamin D deficiency would be likely to benefit from calcium and vitamin D supplementation (Harvey et al. 2017).

Phosphorus

Phosphorus (P) is an important mineral found in bones, teeth and the genetic building blocks, DNA and RNA (Carpenter 2000). In humans, phosphorus makes up approximately 1–1.4% of the fat-free mass, of which around 85% is in bones and teeth, with the remaining 15% distributed throughout the blood and soft tissues. Phosphorus is needed for healthy bone mineralisation and the maintenance and repair of all tissues and cells. Phosphorus has an important role in carbohydrate and energy metabolism being required for the synthesis of the body's key energy source, adenosine triphosphate (ATP) which provides energy to drive and support most cellular processes such as muscle contraction and nerve impulses. In the form of phospholipids, phosphorus is also a key component of cell membrane structure. Many proteins and sugars in the body are phosphorylated allowing them to be used appropriately—this includes the activation and deactivation of enzymes. In addition, phosphorus plays key roles in the maintenance of normal pH in extracellular fluid (Carpenter 2000). As it is one of the most abundant components of all tissues, disturbances in phosphorus homeostasis may influence any organ system.

Inorganic phosphorus levels, principally phosphate (PO_4), are highly regulated through intestinal and renal transport systems. Phosphate is absorbed in the intestinal tract. In healthy people the majority of phosphate filtered at the glomerulus is reabsorbed by the renal tubular cells. Vitamin D increases uptake and regulates homeostasis, maintaining an appropriate bone content of phosphate. Renal excretion is controlled by vitamin D, parathyroid hormone (PTH) and fibroblast growth factor 23 [FGF23; (Peacock 2021)]. However, there are some concerns about excessive P intakes given similar homeostatic mechanisms are also involved in maintaining a healthy calcium homeostasis. The mechanisms regulating serum Ca and P are complex and concentrations may be significantly compromised by chronically high intakes of P. A high P intake inhibits the elevation in serum 1,25-dihydroxyvitamin D [1,25$(OH)_2$D] concentration in response to low dietary Ca intake (Calvo 1993). A sustained increase in PTH after the intake of P has been found in animal and human studies (Takeda et al. 2014). Repetitive hormonal adaptations relating to elevated serum phosphate may lead to bone resorption and increase the risk of fractures (Calvo and Uribarri 2013), and may also result in cellular calcification in damaged arterial and heart valvular walls (Kemi et al. 2009).

The normal reference range for serum phosphate in adults is commonly regarded as 0.8–1.5 mmol/L. People with chronic or mild hypophosphataemia (phosphate levels 0.6–0.8 mmol/L) may be asymptomatic or have symptoms that are vague, such as fatigue, which may overlap with other potential nutritional deficiencies. Otherwise, bone pain, muscle weakness (cardiac, diaphragmatic, and skeletal), rhabdomyolysis and impaired mental status (confusion, hallucinations) are the most common presenting symptoms associated with lower phosphate levels (Amanzadeh and Reilly Jr. 2006; Imel and Econs 2012). Disorientation, congestive heart failure, seizures, some focal neurological deficits, and muscular pain are all potential symptoms associated with more severe acute hypophosphataemia [<0.3 mmol/L; (Leung and Crook 2019)].

Phosphorus is abundant in most diets through meat, dairy products, eggs, nuts and grains. The recommended daily intake for all adults is 700 mg. In many countries, dietary intake of phosphorus is higher (Adatorwovor et al. 2015; Calvo et al. 2014) than the recommended daily allowance including in the diet of most older adults. Indeed, for many countries the median intake is just above 1000 mg (Moshfegh et al. 2009). In addition to naturally occurring dietary sources of phosphorus, the food industry commonly uses phosphorus additives resulting in increased intake (Kalantar-Zadeh et al. 2010). The phosphorus from inorganic additives is almost completely absorbed (Bell et al. 1977) and may have different effects to those of natural dietary phosphorus (Kemi et al. 2009). It has been suggested that phosphorus additives may add as much as 1 g of phosphorus to the diet in industrialised countries where the intake of processed foods and drinks containing phosphoric acid is high and ever increasing (Takeda et al. 2002).

Causes of Hypophosphataemia in Older Adults

Whilst rare in the general population, hypophosphataemia is more common in hospitalised patients (2.2–3.1%). This is particularly so in in patients in intensive care (45% of those hospitalised), those with sepsis (65–80%) and people with chronic obstructive pulmonary disease (20%). Patients suffering from a chronic alcohol dependency are also more frequently hypophosphataemic [30–40%; (Geerse et al. 2010; Zhao et al. 2016)]. Significant hypophosphataemia (below 0.4 mmol/L) is predominantly due to decreased intake, renal losses, redistribution into cells or to conditions affecting circulating PTH levels. Many of the patients with low phosphate concentrations frequently exhibit further electrolyte imbalances.

Despite phosphate being naturally present in many foods and thus plentiful in the normal diet, there are several reasons why older people may be at risk of deficiency. These may include:

- *Poor dietary intake*: due to difficulties with chewing and swallowing; poor appetite which may arise due to low mood, polypharmacotherapy, chronic alcohol dependency; restrictive diets such as a solely plant-based diet where high phytate intakes may reduce the bioavailability of minerals such as P; food insecurity limiting access to a balanced diet
- *Malnutrition* may be due to malabsorption resulting from damage, inflammation or surgery to the small intestine, or persistent diarrhoea/vomiting
- *Vitamin D deficiency* due to lack of sun exposure if limited mobility or medical conditions resulting in malabsorption of fat-soluble vitamins e.g. coeliac disease
- *Inappropriate use of gut phosphate binders* such as aluminium hydroxide which is commonly used in Antacid preparations used to manage gastric reflux and heartburn
- Low dietary phosphorus intake and large calcium supplement doses as a treatment for osteoporosis (Heaney 2004)

- *Increased renal excretion*: this may be due to hyperparathyroidism resulting in reduced absorption in the proximal renal tubules, a high salt intake leading not only to reduced reabsorption of sodium but also of phosphate, renal tubular disorders which may be due to certain conditions or chronic heavy metal exposure, long periods of renal dialysis, use of medications resulting in renal phosphate leak such as aminoglycosides (bactericidal antibiotics) or cisplatin used in the management of certain cancers, chronic need for intravenous iron therapy which may result in increased excretion of the hormone required to maintain normal phosphate homeostasis, tumour induced osteomalacia leading to excessive renal phosphate loss
- *Movement of phosphate from extracellular to the intracellular compartment*: this may arise due to diabetic ketoacidosis when on treatment, insulin rapidly drives phosphate into the cells, as a consequence of refeeding syndrome where carbohydrate is given after a period of fasting and would be associated with low potassium, low calcium, low magnesium and abnormal glucose metabolism or as a result of acute respiratory alkalosis resulting in an increased demand for phosphate intracellularly
- *Chronic infections*: hypermetabolism during the acute phase of any infection increases the requirement for phosphate
- Intravenous iron: contemporary formulations permit the delivery of high doses of elemental iron to enable correction of total iron deficiency in one or two infusions, but these can then increase the secretion of FGF23 (Schaefer et al. 2022)
- *Parathyroidectomy*
- *Leukaemia or lymphomas*

Use of Phosphorus Supplements

Mild hypophosphataemia often resolves without treatment and it is important to look at the underlying cause(s) and the severity. Clinical investigations should also consider serum Ca and Mg levels, renal function, liver function tests (alkaline phosphatase), PTH & Vit D. levels and bone density dependent on other presenting symptoms and medical history. Treatment should ideally address the underlying cause of the deficiency before oral supplements are recommended.

Phosphate supplements should be given when hypophosphataemia may be anticipated such as after periods of chronically poor dietary intake or starvation and especially in the refeeding period. Chronic alcohol dependency may also warrant supplementation as may induce hypermetabolic states where demands for P are higher. There is limited guidance as to the amount of P that should be given in these situations with variations in guidance for emergency medicine.

Given the significance of having a balanced intake of both calcium and phosphorus and the importance of Vit D, it is not recommended that a person takes phosphorus as a single supplement, unless there are indications for doing so. There are no suggestions that phosphate supplements may reduce mortality and morbidity and indeed some evidence of the opposite effect if taken inappropriately (Chang et al. 2014).

It is much better that healthy older adults get the P that they require from their diet together with a Vit D supplement of 10 µg/day (400 IU/day) as recommended in some countries for older adults. In the US an upper limit of 4000 mg/day for men and women aged 51–70 years and an upper limit of 3000 mg/day for those aged 71 years or older is recommended (Kemi et al. 2009). More consideration could be given to the diet and given the importance of having an appropriate calcium to phosphorus ratio in the diet, increasing dairy intake may be beneficial. Dairy intake has been positively linked to bone health in interventions studies, observational and retrospective studies (Rizzoli et al. 2010). Milk and cheese, excluding processed cheese, are high in natural phosphate (albeit free from phosphate additives). Work has shown that elevated consumption of milk and cheese results in a lower mean serum PTH concentration (Kemi et al. 2009). This is due to the high calcium content of milk and cheese which, as a consequence of its PTH lowering effect, slows the rate of absorption of phosphorus in the intestine (Calvo 1993). It has been suggested that a combined calcium-phosphorus salt may have long-term benefits on fracture prevention through a reduction in bone resorption. This was observed in health postmenopausal women after short-term consumption of milk or dairy products and was seen both in community dwelling (Bonjour et al. 2008) and institutionalised older adults (Bonjour et al. 2009, 2011).

Magnesium

Magnesium (Mg^{2+}) is the second most abundant element in the body present in its divalent ion form. This element is the fourth most abundant mineral in the human body with around 60% of Mg^{2+} partitioned into the skeleton and around 40% in soft tissues. Only <1% is found in the blood.

The majority of all metabolic process in cells and tissues of humans require magnesium since it is widely used as a co-factor in enzyme catalysed reactions. The main function of Mg^{2+} is in energy production, protein synthesis (Reddy et al. 2018) and Vitamin D metabolism (Dai et al. 2018), although this element is also important for the proper function of the renal, hepatic and brain metabolic and physiological processes. In addition, Mg^{2+} is needed in DNA repair reactions. For many important enzymatic reactions, Mg^{2+} acts as a cofactor providing a structural or catalytic component of the protein, and for others, is associated with specific substrates. At present, over 600 enzymes are known to require Mg^{2+} as a cofactor and a further 200 as an enzyme activator (Caspi et al. 2012). Critically, Mg^{2+} is required to form the Mg-ATP complex that permits enzymatic engagement with ATP, and therefore is the rate-limiting factor for ATP dependent enzymatic reactions involved in carbohydrate, energy metabolism, protein and nucleic acid synthesis. Key enzymes of the glucose oxidation pathways are Mg^{2+} dependent, and the activation of vitamin B1 into thiamine diphosphate (TDP) requires Mg^{2+}. TDP is an essential coenzyme required for the oxidative metabolism of glucose in the brain (Sang et al. 2018). Finally, several studies show important roles of Mg^{2+} in cytochrome P450 (CYP) mediated reactions especially those involved in vitamin D activation (Campbell and Hayes 1974).

Magnesium participates in the active transport of elements like calcium and potassium across cell membranes and therefore functions in mediating nerve conduction, muscle contraction, and heart rhythm muscle contraction and relaxation (Topf and Murray 2003). In addition, magnesium is needed for the synthesis of the cellular antioxidant glutathione (GSH), a tripeptide, that is needed for healthy ageing and to mitigate free radical mediated damage to proteins lipids and DNA. GSH also plays important roles in drug detoxification reactions. Other biological effects of Mg^{2+} include roles in inflammation. Moderate or subclinical Mg^{2+} deficiency induces a chronic low-grade inflammation that correlates with elevated levels of inflammatory cytokines and the production of free radicals in cells and tissues (Nielsen 2014). As such, Mg^{2+} deficiency is a known risk factor for several pro-inflammatory conditions like obesity, hypertension, and type 2 diabetes (T2DM). It is also widely recognised that Mg^{2+} intake is inversely associated with levels of C reactive protein (Piuri et al. 2021) and magnesium depletion can lead to chronic inflammation both directly (Mazidi et al. 2018; Nielsen 2018) and indirectly. The latter through modifications to the intestinal microbiota (Lobionda et al. 2019) which can have further negative consequences for the older person.

Magnesium Deficiency

Traditionally serum magnesium levels have been routinely used in clinical practice to determine Mg^{2+} status because it is a feasible and inexpensive laboratory test (EFSA Panel on Dietetic Products 2015). However, serum levels do not correlate well with overall status given serum Mg^{2+} occurs as a very small fraction of the total tissue pools. This means that Mg^{2+} deficiency tends to be underestimated and there is a high risk of undiagnosed hypomagnesemia even in high-middle income countries (Gröber et al. 2015). The clinical reference range for serum Mg^{2+} is between 0.70–1.05 mmol/L (Nielsen 2016; Razzaque 2018) with hypomagnesemia occurring when serum Mg^{2+} concentrations fall below 0.7 mmol/L (Topf and Murray 2003). Recently, it has been proposed that urinary Mg^{2+} secretion should be used in parallel to serum measures to determine Mg^{2+}status when serum level < 0.82 mmol/L. Typically, a Mg^{2+} urinary excretion rate of 40–80 mg/day is deemed indicative of Mg^{2+} deficiency (normal excretion would be approx. 120 mg/day) (Costello et al. 2016). Another factor impacting on Mg^{2+} status is the propensity of this mineral to bind to proteins. About 25% of plasma magnesium is bound to albumin so fluctuations in albumin concentrations will additionally affect magnesium levels (Kroll and Elin 1985).

Subclinical hypomagnesemia can lead to a number of different symptoms that are non-specific and may overlap with issues linked to other nutritional imbalances

(Al Alawi et al. 2018). Common end-point symptoms include depression and fatigue however, early signs of deficiency include loss of appetite, nausea, vomiting, fatigue, and weakness. As the deficiency worsens, a range of clinical outcomes ensue, including muscle contractions and cramps, seizures, abnormal heart rhythms, numbness, tingling, personality changes, and coronary spasms. Severe deficiency is defined by serum Mg^{2+} concentration falling below 0.3–0.4 mmol/L. This can precipitate other mineral imbalances and drive hypocalcaemia or hypokalaemia leading to increased risk of cardiac arrhythmias (de Baaij et al. 2015). A chronic low Mg^{2+} status has been associated with an increased risk of osteoporosis and sarcopenia thus increasing the risk of frailty in older people (Beaudart et al. 2019).

Causes of Hypomagnesemia in Older Adults

There are several potential causes of hypomagnesemia including an inadequate dietary intake and researchers have indicated that in western countries the average dietary magnesium intakes are often below those recommended (DiNicolantonio et al. 2018). In America, at least 60% of US adults do not meet their RDA for magnesium (US Department of Agriculture, 2019). Magnesium status is generally impaired in older people (Barbagallo et al. 2021). Data from the UK NDNS suggests adults aged 65 years and over have low intakes with an increase in women aged >65 years having intakes below the lower reference nutrient intake (LRNI) observed over recent years. Intake is also related to income (NDNS trend data). The UK recommended intake is 300 mg for men and 270 mg for women with recommended intakes slightly higher in other countries (US: 360 mg/day). The main dietary sources are green leafy vegetables, legumes, nuts, seeds, wholegrains, dairy and water. Poor dietary intakes are linked to a high intake of processed foods and low ingestion or incorporation of vegetables and legumes in diets. Additionally, changes in agricultural practices and soil erosion are driving reductions in magnesium levels in vegetables and grains (Workinger et al. 2018).

Other potential causes of magnesium deficiency particularly relevant to older people include:

- Malnutrition resulting from poor appetite, poorly fitting dentures, food insecurity
- Gastrointestinal loss related to gastrointestinal disorders including any level of malabsorption, inflammatory bowel disease, coeliac disease
- Where part of the small intestine has been removed (region of the intestine responsible for greatest magnesium absorption)
- Poorly controlled type 2 diabetes
- Renal disorders that lead to an increased excretion
- Various drugs are known to impact on Mg^{2+} status in humans. Several medicines including proton pump inhibitors (reduce gastric acid production), diuretics (aid water loss), antibiotics, some chemotherapeutic drugs and immunosuppressants
- Chronic alcoholism

The wide use of proton pump inhibitors (PPIs) to treat gastric acid reflux like omeprazole can lead to hypomagnesemia in up to 13% of people taking them (Chrysant 2019). The UK Medicines & Healthcare products Regulator Agency (MHRA) recognise that prolonged use of PPIs is linked to hypomagnesaemia. In this instance, current recommendations are that magnesium levels are measured before and during PPI intervention. This is especially important in those taking PPI concomitantly with digoxin or other medication that may cause hypomagnesaemia.

Magnesium Supplementation

Magnesium supplementation has been shown to improve vascular function in older adults with type 2 diabetes (Barbagallo et al. 2010), and is a useful strategy to support non-pharmacotherapy for atrial and/or ventricular arrhythmias (McLean 1994). Oral Mg supplements, versus placebo, have been found to help improve clinical symptoms and survival outcomes in people with severe congestive heart failure (Stepura and Martynow 2009). Moreover, recent studies highlight an inverse association between Mg intakes and risk of stroke (Adebamowo et al. 2015) and higher Mg intake, including supplementation, and decreased risk of stroke (Veronese et al. 2020). It is widely recognised that the prevalence of T2DM increases with ageing. A meta-analysis of 18 randomised controlled trials reported that Mg supplementation led to improved glycaemic control in people with T2DM and improved insulin-sensitivity in those at high risk of T2DM (Kirkland et al., 2018). There has been increasing interest in the use of Mg supplements to help in the management of diabetic foot ulcers [DFU; (Karakousis et al. 2023)]. In a clinical trial including 70 patients with grade 3 DFU, randomly divided into two groups with one group receiving 250 mg Mg oxide and the other placebo daily. After 12 weeks, those supplemented with Mg had a significant increase in serum Mg, significant reductions in ulcer size and depth and improved glycaemic control (Razzaghi et al. 2018).

Magnesium supplements have been suggested to be helpful in the treatment of depression (Eby 3rd and Eby 2010) with higher intakes being associated with reduced symptoms of depression (Derom et al. 2013). The role of Mg in the brain is complex but it is known that this mineral influence multiple biochemical processes involved in cognitive functions. Studies using animal models have suggested that Mg supplementation in the early stages of cognitive impairment can decrease the rate of memory loss and cognitive decline (Xu et al. 2014). In humans, only a handful of clinical trials determined the role of Mg in cognitive health. In healthy adult men assessed over an 8 year period it was found that higher dietary Mg intakes were associated with a reduced risk of developing mild cognitive impairment (Cherbuin et al. 2014). An additional study which focused on more than 1000 community-dwelling participants (aged +60 years) it was found that for those consuming more than 200 mg/day of Mg the chance of developing dementia (any form) was 37% lower than those consuming below this quantity. They additionally saw a 74%

reduction in the chance of developing vascular dementia (Ozawa et al. 2012). An RCT suggested that taking Mg supplements, in the form of magnesium threonate, for 12 weeks may help in improving cognitive function in older people with impaired memory (Liu et al. 2016).

Magnesium is needed for both protein synthesis and vitamin D metabolism and hence is likely to play a pivotal role in maintaining healthy bones and muscle function. In the Health, Ageing and Body Composition Study, it was reported that in healthy Caucasians, aged 70–79 years, higher Mg intakes were associated with higher bone mineral density (BMD) (Ryder et al. 2005). Conversely inadequate dietary Mg intakes have been linked to an increased rate of bone loss in postmenopausal women with osteoporosis (New et al. 2000; Tucker 2014). Those women with higher Mg intakes, taking part in a separate large cohort study and followed for 8 years, were found to have a 27% reduced risk of future fractures (Veronese et al. 2017). In studies involving older people, there has been found to be a strong and independent relationship between serum Mg levels and muscle performance (Dominguez et al. 2006). Oral Mg supplementation (300 mg/day) was able to improve the physical performance of older participants, particularly in those with a low baseline dietary intake, suggesting that Mg supplementation may help in preventing or delaying age-related decline in physical performance (Veronese et al. 2014).

In the main, the studies that have looked at the benefits of Mg supplementation in older adults have used doses between 250–600 mg/day except for the study looking at improving cognitive function where slightly higher doses were used. There is also huge variation in the length of time that Mg supplements have been used for in the trials with intervention duration of between 7 days and 6 months so no definitive recommendation for the length of time that Mg supplements would need to be taken for benefits to be seen.

Limited experimental evidence is available that points to the most efficient supplement for use in the delivery of Mg to individuals. Currently, magnesium sulphate, oxide, carbonate, chloride, citrate, malate, acetate, gluconate, lactate, aspartate, fumarate, acetyl taurate, bis-glycinate, and threonate are all available and have been widely used in clinical trials. Organic Mg^{2+} salts appear to be more bioavailable than inorganic Mg^{2+} salts however, both forms are known to have beneficial clinical outcomes. The choice of the type of Mg^{2+} salt used may be influenced by any side effects and tolerability of a particular preparation may limit the dosage with diarrhoea a common side effect. In the UK, oral magnesium glycerophosphate is licensed and can be prescribed for treating hypomagnesaemia and oral magnesium aspartate and magnesium citrate are licensed for the treatment and prevention of magnesium deficiency. Magnesium sulphate is the choice in the UK for the treatment of symptomatic hypomagnesemia. Because Mg is excreted mainly by the kidneys the use of Mg supplements in older adults with significant renal impairment needs to be monitored to prevent hypermagnesemia which may also lead to muscle weakness and arrhythmias (van de Wal-Visscher et al. 2018).

References

Adams ML, Lombi E, Zhao FJ, McGrath SP (2002) Evidence of low selenium concentrations in UK bread-making wheat grain. J Sci Food Agric 82(10):1160–1165

Adatorwovor R, Roggenkamp K, Anderson JJ (2015) Intakes of calcium and phosphorus and calculated calcium-to-phosphorus ratios of older adults: NHANES 2005-2006 data. Nutrients 7(11):9633–9639. https://doi.org/10.3390/nu7115492

Adebamowo SN, Spiegelman D, Flint AJ, Willett WC, Rexrode KM (2015) Intakes of magnesium, potassium, and calcium and the risk of stroke among men. Int J Stroke 10(7):1093–1100. https://doi.org/10.1111/ijs.12516

Al Alawi AM, Majoni SW, Falhammar H (2018) Magnesium and human health: perspectives and research directions. Int J Endocrinol 2018:9041694. https://doi.org/10.1155/2018/9041694

Alexander J (2015) Chapter 52 - Selenium. In: Nordberg GF, Fowler BA, Nordberg M (eds) Handbook on the toxicology of metals, 4th edn. Academic, San Diego, pp 1175–1208

Amanzadeh J, Reilly RF Jr (2006) Hypophosphatemia: an evidence-based approach to its clinical consequences and management. Nat Clin Pract Nephrol 2(3):136–148. https://doi.org/10.1038/ncpneph0124

Ames BN, Shigenaga MK, Hagen TM (1993) Oxidants, antioxidants, and the degenerative diseases of aging. Proc Natl Acad Sci USA 90(17):7915–7922. https://doi.org/10.1073/pnas.90.17.7915

Amin U, McPartland A, O'Sullivan M, Silke C (2023) An overview of the management of osteoporosis in the aging female population. Womens Health (Lond) 19:17455057231176655. https://doi.org/10.1177/17455057231176655

Andreini C, Banci L, Bertini I, Rosato A (2006) Counting the zinc-proteins encoded in the human genome. J Proteome Res 5(1):196–201. https://doi.org/10.1021/pr050361j

Andriollo-Sanchez M, Hininger-Favier I, Meunier N, Toti E, Zaccaria M, Brandolini-Bunlon M et al (2005) Zinc intake and status in middle-aged and older European subjects: the ZENITH study. Eur J Clin Nutr 59(Suppl 2):S37–S41. https://doi.org/10.1038/sj.ejcn.1602296

Barbagallo M, Dominguez LJ, Galioto A, Pineo A, Belvedere M (2010) Oral magnesium supplementation improves vascular function in elderly diabetic patients. Magnes Res 23:131–137

Barbagallo M, Veronese N, Dominguez LJ (2021) Magnesium in aging, health and diseases. Nutrients 13(2). https://doi.org/10.3390/nu13020463

BDA (2021) British Dietetic Association - Calcium: Food Fact Sheet. https://www.bda.uk.com/resource/calcium.html

Beaudart C, Locquet M, Touvier M, Reginster JY, Bruyère O (2019) Association between dietary nutrient intake and sarcopenia in the SarcoPhAge study. Aging Clin Exp Res 31:815–824

Bell RR, Draper HH, Tzeng DYM, Shin HK, Schmidt GR (1977) Physiological responses of human adults to foods containing phosphate additives. J Nutr 107(1):42–50. https://doi.org/10.1093/jn/107.1.42

Bodnar M, Konieczka P, Namiesnik J (2012) The properties, functions, and use of selenium compounds in living organisms. J Environ Sci Health C 30(3):225–252. https://doi.org/10.1080/10590501.2012.705164

Bolland MJ, Leung W, Tai V, Bastin S, Gamble GD, Grey A, Reid IR (2015) Calcium intake and risk of fracture: systematic review. BMJ 351:h4580. https://doi.org/10.1136/bmj.h4580

Bonjour JP, Brandolini-Bunlon M, Boirie Y, Morel-Laporte F, Braesco V, Bertière MC, Souberbielle JC (2008) Inhibition of bone turnover by milk intake in postmenopausal women. Br J Nutr 100(4):866–874. https://doi.org/10.1017/S0007114508937429

Bonjour JP, Benoit V, Pourchaire O, Ferry M, Rousseau B, Souberbielle JC (2009) Inhibition of markers of bone resorption by consumption of vitamin D and calcium-fortified soft plain cheese by institutionalised elderly women. Br J Nutr 102:962–966

Bonjour JP, Benoit V, Pourchaire O, Rousseau B, Souberbielle JC (2011) Nutritional approach for inhibiting bone resorption in institutionalized elderly women with vitamin D insufficiency and high prevalence of fracture. J Nutr Health Aging 15(5):404–409. https://doi.org/10.1007/s12603-011-0003-y

Bove-Fenderson E, Mannstadt M (2018) Hypocalcemic disorders. Best Pract Res Clin Endocrinol Metab 32(5):639–656. https://doi.org/10.1016/j.beem.2018.05.006

Brewer GJ (2001) Zinc acetate for the treatment of Wilson's disease. Expert Opin Pharmacother 2(9):1473–1477. https://doi.org/10.1517/14656566.2.9.1473

Briefel RR, Bialostosky K, Kennedy-Stephenson J, McDowell MA, Ervin RB, Wright JD (2000) Zinc intake of the U.S. population: findings from the third National Health and Nutrition Examination Survey, 1988-1994. J Nutr 130(5S Suppl):1367S–1373S. https://doi.org/10.1093/jn/130.5.1367S

Bristow SM, Bolland MJ, Gamble GD, Leung W, Reid IR (2022) Dietary calcium intake and change in bone mineral density in older adults: a systematic review of longitudinal cohort studies. Eur J Clin Nutr 76(2):196–205. https://doi.org/10.1038/s41430-021-00957-8

Broadley MR, White PJ, Bryson RJ, Meacham MC, Bowen HC, Johnson SE et al (2006) Biofortification of UK food crops with selenium. Proc Nutr Soc 65(2):169–181. https://doi.org/10.1079/PNS2006490

Brown KM, Arthur JR (2001) Selenium, selenoproteins and human health: a review. Public Health Nutr 4(2b):593–599. https://doi.org/10.1079/PHN2001143

Brown KH, Rivera JA, Bhutta Z, Gibson RS, King JC, Lönnerdal B et al (2004) International Zinc Nutrition Consultative Group (IZiNCG) technical document #1. Assessment of the risk of zinc deficiency in populations and options for its control. Food Nutr Bull 25(1 Suppl 2):S99–S203. https://www.ncbi.nlm.nih.gov/pubmed/18046856

Bruins MJ, Van Dael P, Eggersdorfer M (2019) The role of nutrients in reducing the risk for noncommunicable diseases during aging. Nutrients 11(1). https://doi.org/10.3390/nu11010085

Budreviciute A, Damiati S, Sabir DK, Onder K, Schuller-Goetzburg P, Plakys G et al (2020) Management and prevention strategies for non-communicable diseases (NCDs) and their risk factors. Front Public Health 8:574111. https://doi.org/10.3389/fpubh.2020.574111

Burger H, de Laet CE, van Daele PL, Weel AE, Witteman JC, Hofman A, Pols HA (1998) Risk factors for increased bone loss in an elderly population: the Rotterdam Study. Am J Epidemiol 147(9):871–879. https://doi.org/10.1093/oxfordjournals.aje.a009541

Calvo MS (1993) Dietary phosphorus, calcium metabolism and bone. J Nutr 123(9):1627–1633. https://doi.org/10.1093/jn/123.9.1627

Calvo MS, Uribarri J (2013) Public health impact of dietary phosphorus excess on bone and cardiovascular health in the general population. Am J Clin Nutr 98(1):6–15. https://doi.org/10.3945/ajcn.112.053934

Calvo MS, Moshfegh AJ, Tucker KL (2014) Assessing the health impact of phosphorus in the food supply: issues and considerations. Adv Nutr 5(1):104–113. https://doi.org/10.3945/an.113.004861

Campbell TC, Hayes JR (1974) Role of nutrition in the drug-metabolizing enzyme system. Pharmacol Rev 26(3):171–197

Cancelo-Hidalgo MJ, Castelo-Branco C, Palacios S, Haya-Palazuelos J, Ciria-Recasens M, Manasanch J, Pérez-Edo L (2013) Tolerability of different oral iron supplements: a systematic review. Curr Med Res Opin 29(4):291–303. https://doi.org/10.1185/03007995.2012.761599

Carpenter TO (2000) Primary disorders of phosphate metabolism [Updated 2022 Jun 8]. In: Feingold KR, Anawalt B, Blackman MR (eds) Endotext [Internet]. MDText.com, Inc., South Dartmouth

Caspi R, Altman T, Dreher K, Fulcher CA, Subhraveti P, Keseler IM et al (2012) The MetaCyc database of metabolic pathways and enzymes and the BioCyc collection of pathway/genome databases. Nucleic Acids Res 40(D1):D742–D753. https://doi.org/10.1093/nar/gkr1014

Ceylan MN, Akdas S, Yazihan N (2021) The effects of zinc supplementation on C-reactive protein and inflammatory cytokines: A meta-analysis and systematical review. J Interf Cytokine Res 41(3):81–101. https://doi.org/10.1089/jir.2020.0209

Chang AR, Lazo M, Appel LJ, Gutiérrez OM, Grams ME (2014) High dietary phosphorus intake is associated with all-cause mortality: results from NHANES III. Am J Clin Nutr 99(2):320–327. https://doi.org/10.3945/ajcn.113.073148

Chen RA, Goodman WG (2004) Role of the calcium-sensing receptor in parathyroid gland physiology. Am J Physiol Renal Physiol 286(6):F1005–F1011. https://doi.org/10.1152/ajprenal.00013.2004

Cherbuin N, Kumar R, Sachdev PS, Anstey KJ (2014) Dietary mineral intake and risk of mild cognitive impairment: the PATH through life project. Front Aging Neurosci 6:4. https://doi.org/10.3389/fnagi.2014.00004

Chrysant SG (2019) Proton pump inhibitor-induced hypomagnesemia complicated with serious cardiac arrhythmias. Expert Rev Cardiovasc Ther 17(5):345–351. https://doi.org/10.1080/14779072.2019.1615446

Clarys JP, Martin AD, Drinkwater DT (1984) Gross tissue weights in the human body by cadaver dissection. Hum Biol 56(3):459–473. https://www.ncbi.nlm.nih.gov/pubmed/6489991

Colvin RA, Holmes WR, Fontaine CP, Maret W (2010) Cytosolic zinc buffering and muffling: their role in intracellular zinc homeostasis. Metallomics 2(5):306–317. https://doi.org/10.1039/b926662c

COMA (1991) Committee on Medical Aspects of Food Policy (1991). Dietary reference values for food energy and nutrients for the United Kingdom. https://assets.publishing.service.gov.uk/government/uploads/system/uploads/attachment_data/file/743790/Dietary_Reference_Values_-_A_Guide__1991_.pdf

Costello RB, Elin RJ, Rosanoff A, Wallace TC, Guerrero-Romero F, Hruby A et al (2016) Perspective: the case for an evidence-based reference interval for serum magnesium: the time has come. Adv Nutr 7(6):977–993. https://doi.org/10.3945/an.116.012765

Cousins RJ (1985) Absorption, transport, and hepatic metabolism of copper and zinc: special reference to metallothionein and ceruloplasmin. Physiol Rev 65(2):238–309. https://doi.org/10.1152/physrev.1985.65.2.238

Cousins RJ, Liuzzi JP, Lichten LA (2006) Mammalian zinc transport, trafficking, and signals. J Biol Chem 281(34):24085–24089. https://doi.org/10.1074/jbc.R600011200

Cubadda F, Aureli F, Ciardullo S, D'Amato M, Raggi A, Acharya R et al (2010) Changes in selenium speciation associated with increasing tissue concentrations of selenium in wheat grain. J Agric Food Chem 58(4):2295–2301

Cui M, Li Q, Johnson R, Fleet JC (2012) Villin promoter-mediated transgenic expression of transient receptor potential cation channel, subfamily V, member 6 (TRPV6) increases intestinal calcium absorption in wild-type and vitamin D receptor knockout mice. J Bone Miner Res 27(10):2097–2107. https://doi.org/10.1002/jbmr.1662

Dai Q, Zhu X, Manson JE, Song Y, Li X, Franke AA et al (2018) Magnesium status and supplementation influence vitamin D status and metabolism: results from a randomized trial. Am J Clin Nutr 108(6):1249–1258. https://doi.org/10.1093/ajcn/nqy274

de Baaij JH, Hoenderop JG, Bindels RJ (2015) Magnesium in man: implications for health and disease. Physiol Rev 95(1):1–46. https://doi.org/10.1152/physrev.00012.2014

Derom ML, Sayón-Orea C, Martínez-Ortega JM, Martínez-González MA (2013) Magnesium and depression: a systematic review. Nutr Neurosci 16(5):191–206. https://doi.org/10.1179/1476830512y.0000000044

Devergnas S, Chimienti F, Naud N, Pennequin A, Coquerel Y, Chantegrel J et al (2004) Differential regulation of zinc efflux transporters ZnT-1, ZnT-5 and ZnT-7 gene expression by zinc levels: a real-time RT-PCR study. Biochem Pharmacol 68(4):699–709. https://doi.org/10.1016/j.bcp.2004.05.024

DiNicolantonio JJ, O'Keefe JH, Wilson W (2018) Subclinical magnesium deficiency: a principal driver of cardiovascular disease and a public health crisis. Open Heart 5(1):e000668. https://doi.org/10.1136/openhrt-2017-000668

Dominguez LJ, Barbagallo M, Lauretani F, Bandinelli S, Bos A, Corsi AM et al (2006) Magnesium and muscle performance in older persons: the InCHIANTI study. Am J Clin Nutr 84(2):419–426. https://doi.org/10.1093/ajcn/84.1.419

Drewnowski A (2022) Food insecurity has economic root causes. Nat Food 3(8):555–556. https://doi.org/10.1038/s43016-022-00577-w

Drewnowski A, Shultz JM (2001) Impact of aging on eating behaviors, food choices, nutrition, and health status. J Nutr Health Aging 5(2):75–79. https://www.ncbi.nlm.nih.gov/pubmed/11426286

Eby GA 3rd, Eby KL (2010) Magnesium for treatment-resistant depression: a review and hypothesis. Med Hypotheses 74(4):649–660. https://doi.org/10.1016/j.mehy.2009.10.051

EFSA Panel on Dietetic Products, Nutrition and Allergies (2015) Scientific opinion on dietary reference values for magnesium. EFSA J 13(7):4186. https://doi.org/10.2903/j.efsa.2015.4186

EFSA Panel on Nutrition, Novel Foods and Food Allergens, Turck D, Bohn T, Castenmiller J, de Henauw S et al (2023) Scientific opinion on the tolerable upper intake level for selenium. EFSA J 21(1):e07704. https://doi.org/10.2903/j.efsa.2023.7704

Eid C, Hémadi M, Ha-Duong NT, El Hage Chahine JM (2014) Iron uptake and transfer from ceruloplasmin to transferrin. Biochim Biophys Acta 1840(6):1771–1781. https://doi.org/10.1016/j.bbagen.2014.01.011

Evans GW (1986) Zinc and its deficiency diseases. Clin Physiol Biochem 4(1):94–98. https://www.ncbi.nlm.nih.gov/pubmed/3514057

Figueira I, Fernandes A, Mladenovic Djordjevic A, Lopez-Contreras A, Henriques CM, Selman C et al (2016) Interventions for age-related diseases: shifting the paradigm. Mech Ageing Dev 160:69–92. https://doi.org/10.1016/j.mad.2016.09.009

Fleet JC (2022) Vitamin D-mediated regulation of intestinal calcium absorption. Nutrients 14(16):3351. https://doi.org/10.3390/nu14163351

Forbes RM, Cooper AR, Mitchell HH (1953) The composition of the adult human body as determined by chemical analysis. J Biol Chem 203(1):359–366. https://www.ncbi.nlm.nih.gov/pubmed/13069519

Fordyce FM (2005) Selenium deficiency and toxicity in the environment. In: Selinus O (ed) Essentials of medical geology. Springer, Dordrecht, pp 375–416

Fordyce FM, Brereton N, Hughes J, Luo W, Lewis J (2010) An initial study to assess the use of geological parent materials to predict the Se concentration in overlying soils and in five staple foodstuffs produced on them in Scotland. Sci Total Environ 408(22):5295–5305. https://doi.org/10.1016/j.scitotenv.2010.08.007

Fraenkel PG (2017) Anemia of inflammation: a review. Med Clin North Am 101(2):285–296. https://doi.org/10.1016/j.mcna.2016.09.005

Franceschi C, Olivieri F, Marchegiani F, Cardelli M, Cavallone L, Capri M et al (2005) Genes involved in immune response/inflammation, IGF1/insulin pathway and response to oxidative stress play a major role in the genetics of human longevity: the lesson of centenarians. Mech Ageing Dev 126(2):351–361. https://doi.org/10.1016/j.mad.2004.08.028

Fredlund K, Isaksson M, Rossander-Hulthén L, Almgren A, Sandberg AS (2006) Absorption of zinc and retention of calcium: dose-dependent inhibition by phytate. J Trace Elem Med Biol 20(1):49–57. https://doi.org/10.1016/j.jtemb.2006.01.003

Ganz T (2005) Hepcidin--a regulator of intestinal iron absorption and iron recycling by macrophages. Best Pract Res Clin Haematol 18(2):171–182. https://doi.org/10.1016/j.beha.2004.08.020

Gasche C, Berstad A, Befrits R, Beglinger C, Dignass A, Erichsen K et al (2007) Guidelines on the diagnosis and management of iron deficiency and anemia in inflammatory bowel diseases. Inflamm Bowel Dis 13(12):1545–1553. https://doi.org/10.1002/ibd.20285

Geerse DA, Bindels AJ, Kuiper MA, Roos AN, Spronk PE, Schultz MJ (2010) Treatment of hypophosphatemia in the intensive care unit: a review. Crit Care 14(4):R147. https://doi.org/10.1186/cc9215

Giacconi R, Cipriano C, Albanese F, Boccoli G, Saba V, Olivieri F et al (2004) The -174G/C polymorphism of IL-6 is useful to screen old subjects at risk for atherosclerosis or to reach successful ageing. Exp Gerontol 39(4):621–628. https://doi.org/10.1016/j.exger.2003.12.013

Glahn RP, Van Campen DR (1997) Iron uptake is enhanced in Caco-2 cell monolayers by cysteine and reduced cysteinyl glycine. J Nutr 127(4):642–647. https://doi.org/10.1093/jn/127.4.642

Goodnough LT, Schrier SL (2014) Evaluation and management of anemia in the elderly. Am J Hematol 89(1):88–96. https://doi.org/10.1002/ajh.23598

Gröber U, Schmidt J, Kisters K (2015) Magnesium in prevention and therapy. Nutrients 7(9):8199–8226. https://www.mdpi.com/2072-6643/7/9/5388

Guralnik JM, Eisenstaedt RS, Ferrucci L, Klein HG, Woodman RC (2004) Prevalence of anemia in persons 65 years and older in the United States: evidence for a high rate of unexplained anemia. Blood 104(8):2263–2268. https://doi.org/10.1182/blood-2004-05-1812

Gutiérrez OM (2021) Treatment of iron deficiency anemia in CKD and end-stage kidney disease. Kidney Int Rep 6(9):2261–2269. https://doi.org/10.1016/j.ekir.2021.05.020

Haase H, Mocchegiani E, Rink L (2006) Correlation between zinc status and immune function in the elderly. Biogerontology 7(5–6):421–428. https://doi.org/10.1007/s10522-006-9057-3

Hambidge KM (1992) Zinc and diarrhea. Acta Paediatr Suppl 381:82–86. https://doi.org/10.1111/j.1651-2227.1992.tb12377.x

Harvey NC, Biver E, Kaufman JM, Bauer J, Branco J, Brandi ML et al (2017) The role of calcium supplementation in healthy musculoskeletal ageing: an expert consensus meeting of the European Society for Clinical and Economic Aspects of Osteoporosis, Osteoarthritis and Musculoskeletal Diseases (ESCEO) and the International Foundation for Osteoporosis (IOF). Osteoporos Int 28(2):447–462. https://doi.org/10.1007/s00198-016-3773-6

Hawley S, Javaid MK, Prieto-Alhambra D, Lippett J, Sheard S, Arden NK et al (2016) Clinical effectiveness of orthogeriatric and fracture liaison service models of care for hip fracture patients: population-based longitudinal study. Age Ageing 45(2):236–242. https://doi.org/10.1093/ageing/afv204

Heaney RP (2004) Phosphorus nutrition and the treatment of osteoporosis. Mayo Clin Proc 79(1):91–97. https://doi.org/10.4065/79.1.91

Heaney RP, Recker RR, Stegman MR, Moy AJ (1989) Calcium absorption in women: relationships to calcium intake, estrogen status, and age. J Bone Miner Res 4(4):469–475. https://doi.org/10.1002/jbmr.5650040404

High KP (1999) Micronutrient supplementation and immune function in the elderly. Clin Infect Dis 28(4):717–722. https://doi.org/10.1086/515208

Hoppler M, Schönbächler A, Meile L, Hurrell RF, Walczyk T (2008) Ferritin-iron is released during boiling and in vitro gastric digestion. J Nutr 138(5):878–884. https://doi.org/10.1093/jn/138.5.878

Hurrell RF (2004) Phytic acid degradation as a means of improving iron absorption. Int J Vitam Nutr Res 74(6):445–452. https://doi.org/10.1024/0300-9831.74.6.445

Imel EA, Econs MJ (2012) Approach to the hypophosphatemic patient. J Clin Endocrinol Metab 97(3):696–706. https://doi.org/10.1210/jc.2011-1319

Institute of Medicine (US) Committee to Review Dietary Reference Intakes for Vitamin D and Calcium (2011) The National Academies Collection: reports funded by National Institutes of Health. In: Ross AC, Taylor CL, Yaktine AL, Del Valle HB (eds) Dietary reference intakes for calcium and vitamin D. National Academies Press, Washington (DC). Copyright © 2011, National Academy of Sciences

Ito T, Jensen RT (2010) Association of long-term proton pump inhibitor therapy with bone fractures and effects on absorption of calcium, vitamin B12, iron, and magnesium. Curr Gastroenterol Rep 12(6):448–457. https://doi.org/10.1007/s11894-010-0141-0

Jones G, Nguyen T, Sambrook P, Kelly PJ, Eisman JA (1994) Progressive loss of bone in the femoral neck in elderly people: longitudinal findings from the Dubbo osteoporosis epidemiology study. BMJ 309(6956):691–695. https://www.ncbi.nlm.nih.gov/pubmed/7950520

Joshi S (2015) Vitamin supplementation in the elderly. Clin Geriatr Med 31(3):355–366. https://doi.org/10.1016/j.cger.2015.04.004

Jüppner H, Abou-Samra AB, Freeman M, Kong XF, Schipani E, Richards J et al (1991) A G protein-linked receptor for parathyroid hormone and parathyroid hormone-related peptide. Science 254(5034):1024–1026. https://doi.org/10.1126/science.1658941

Kahwati LC, Weber RP, Pan H, Gourlay M, LeBlanc E, Coker-Schwimmer M, Viswanathan M (2018) Vitamin D, calcium, or combined supplementation for the primary prevention of fractures in community-dwelling adults: evidence report and systematic review for the

US Preventive Services Task Force. JAMA 319(15):1600–1612. https://doi.org/10.1001/jama.2017.21640

Kalantar-Zadeh K, Gutekunst L, Mehrotra R, Kovesdy CP, Bross R, Shinaberger CS et al (2010) Understanding sources of dietary phosphorus in the treatment of patients with chronic kidney disease. Clin J Am Soc Nephrol 5(3):519–530. https://doi.org/10.2215/cjn.06080809

Kápolna E, Gergely V, Dernovics M, Illés A, Fodor P (2007) Fate of selenium species in sesame seeds during simulated bakery process. J Food Eng 79(2):494–501

Karakousis ND, Pyrgioti EE, Georgakopoulos PN, Apergi K, Papanas N (2023) Magnesium and diabetic foot ulcers: a mini review. Int J Low Extrem Wounds 15347346231176117. https://doi.org/10.1177/15347346231176117

Kassarjian Z, Russell RM (1989) Hypochlorhydria: a factor in nutrition. Annu Rev Nutr 9:271–285. https://doi.org/10.1146/annurev.nu.09.070189.001415

Kellett GL (2011) Alternative perspective on intestinal calcium absorption: proposed complementary actions of Ca(v)1.3 and TRPV6. Nutr Rev 69(7):347–370. https://doi.org/10.1111/j.1753-4887.2011.00395.x

Kemi VE, Rita HJ, Kärkkäinen MU, Viljakainen HT, Laaksonen MM, Outila TA, Lamberg-Allardt CJ (2009) Habitual high phosphorus intakes and foods with phosphate additives negatively affect serum parathyroid hormone concentration: a cross-sectional study on healthy premenopausal women. Public Health Nutr 12(10):1885–1892. https://doi.org/10.1017/s1368980009004819

Khabbaz Koche Ghazi M, Ghaffari S, Naemi M, Salehi R, Taban Sadeghi M, Barati M et al (2021) Effects of sodium selenite and selenium-enriched yeast on cardiometabolic indices of patients with atherosclerosis: a double-blind randomized clinical trial study. J Cardiovasc Thorac Res 13(4):314–319. https://doi.org/10.34172/jcvtr.2021.51

Khanal RC, Nemere I (2008) Regulation of intestinal calcium transport. Annu Rev Nutr 28:179–196. https://doi.org/10.1146/annurev.nutr.010308.161202

Kirkland AE, Sarlo GL, Holton KF. (2018) The Role of Magnesium in Neurological Disorders. Nutrients. Jun 6;10(6):730. https://doi.org/10.3390/nu10060730. PMID: 29882776; PMCID: PMC6024559

Kogirima M, Kurasawa R, Kubori S, Sarukura N, Nakamori M, Okada S et al (2007) Ratio of low serum zinc levels in elderly Japanese people living in the central part of Japan. Eur J Clin Nutr 61(3):375–381. https://doi.org/10.1038/sj.ejcn.1602520

Kopecky SL, Bauer DC, Gulati M, Nieves JW, Singer AJ, Toth PP et al (2016) Lack of evidence linking calcium with or without vitamin D supplementation to cardiovascular disease in generally healthy adults: a clinical guideline from the National Osteoporosis Foundation and the American Society for Preventive Cardiology. Ann Intern Med 165(12):867–868. https://doi.org/10.7326/M16-1743

Koskenkorva-Frank TS, Weiss G, Koppenol WH, Burckhardt S (2013) The complex interplay of iron metabolism, reactive oxygen species, and reactive nitrogen species: insights into the potential of various iron therapies to induce oxidative and nitrosative stress. Free Radic Biol Med 65:1174–1194. https://doi.org/10.1016/j.freeradbiomed.2013.09.001

Kroll MH, Elin RJ (1985) Relationships between magnesium and protein concentrations in serum. Clin Chem 31(2):244–246

Küry S, Dréno B, Bézieau S, Giraudet S, Kharfi M, Kamoun R, Moisan JP (2002) Identification of SLC39A4, a gene involved in acrodermatitis enteropathica. Nat Genet 31(3):239–240. https://doi.org/10.1038/ng913

Lane DJ, Richardson DR (2014) The active role of vitamin C in mammalian iron metabolism: much more than just enhanced iron absorption! Free Radic Biol Med 75:69–83. https://doi.org/10.1016/j.freeradbiomed.2014.07.007

Lee HH, Prasad AS, Brewer GJ, Owyang C (1989) Zinc absorption in human small intestine. Am J Phys 256(1 Pt 1):G87–G91. https://doi.org/10.1152/ajpgi.1989.256.1.G87

Leung J, Crook M (2019) Disorders of phosphate metabolism. J Clin Pathol 72(11):741–747. https://doi.org/10.1136/jclinpath-2018-205130

Lindblad AJ, Cotton C, Allan GM (2015) Iron deficiency anemia in the elderly. Can Fam Physician 61(2):159. https://www.ncbi.nlm.nih.gov/pubmed/25676647

Lips P, Bouillon R, van Schoor NM, Vanderschueren D, Verschueren S, Kuchuk N et al (2010) Reducing fracture risk with calcium and vitamin D. Clin Endocrinol 73(3):277–285. https://doi.org/10.1111/j.1365-2265.2009.03701.x

Liu G, Weinger JG, Lu ZL, Xue F, Sadeghpour S (2016) Efficacy and safety of MMFS-01, a synapse density enhancer, for treating cognitive impairment in older adults: A randomized, double-blind, placebo-controlled trial. J Alzheimers Dis 49(4):971–990. https://doi.org/10.3233/jad-150538

Liuzzi JP, Cousins RJ (2004) Mammalian zinc transporters. Annu Rev Nutr 24:151–172. https://doi.org/10.1146/annurev.nutr.24.012003.132402

Lobionda S, Sittipo P, Kwon HY, Lee YK (2019) The role of gut microbiota in intestinal inflammation with respect to diet and extrinsic stressors. Microorganisms 7(8). https://doi.org/10.3390/microorganisms7080271

Lonnerdal B (2000) Dietary factors influencing zinc absorption. J Nutr 130(5S Suppl):1378S–1383S. https://doi.org/10.1093/jn/130.5.1378S

Lorenzo-López L, Maseda A, de Labra C, Regueiro-Folgueira L, Rodríguez-Villamil JL, Millán-Calenti JC (2017) Nutritional determinants of frailty in older adults: a systematic review. BMC Geriatr 17(1):108. https://doi.org/10.1186/s12877-017-0496-2

Lowe NM, Fekete K, Decsi T (2009) Methods of assessment of zinc status in humans: a systematic review. Am J Clin Nutr 89(6):2040S–2051S. https://doi.org/10.3945/ajcn.2009.27230G

Maares M, Haase H (2020) A guide to human zinc absorption: general overview and recent advances of in vitro intestinal models. Nutrients 12(3). https://doi.org/10.3390/nu12030762

MacFarquhar JK, Broussard DL, Melstrom P, Hutchinson R, Wolkin A, Martin C et al (2010) Acute selenium toxicity associated with a dietary supplement. Arch Intern Med 170(3):256–261. https://doi.org/10.1001/archinternmed.2009.495

Maggini S, Pierre A, Calder PC (2018) Immune function and micronutrient requirements change over the life course. Nutrients 10(10):1531. https://doi.org/10.3390/nu10101531

Manoj P, Derwin R, George S (2023) What is the impact of daily oral supplementation of vitamin D3 (cholecalciferol) plus calcium on the incidence of hip fracture in older people? A systematic review and meta-analysis. Int J Older People Nursing 18(1):e12492. https://doi.org/10.1111/opn.12492

Marcellini F, Giuli C, Papa R, Gagliardi C, Dedoussis G, Herbein G et al (2006) Zinc status, psychological and nutritional assessment in old people recruited in five European countries: Zincage study. Biogerontology 7(5–6):339–345. https://doi.org/10.1007/s10522-006-9048-4

Maret W, Krezel A (2007) Cellular zinc and redox buffering capacity of metallothionein/thionein in health and disease. Mol Med 13(7–8):371–375. https://doi.org/10.2119/2007-00036.Maret

Maret W, Sandstead HH (2006) Zinc requirements and the risks and benefits of zinc supplementation. J Trace Elem Med Biol 20(1):3–18. https://doi.org/10.1016/j.jtemb.2006.01.006

Mazidi M, Rezaie P, Banach M (2018) Effect of magnesium supplements on serum C-reactive protein: a systematic review and meta-analysis. Arch Med Sci 14(4):707–716. https://doi.org/10.5114/aoms.2018.75719

McLean RM (1994) Magnesium and its therapeutic uses: a review. Am J Med 96(1):63–76. https://doi.org/10.1016/0002-9343(94)90117-1

McMahon RJ, Cousins RJ (1998) Regulation of the zinc transporter ZnT-1 by dietary zinc. Proc Natl Acad Sci USA 95(9):4841–4846. https://doi.org/10.1073/pnas.95.9.4841

Mintel (2022) UK vitamins and supplements market report 2022

Mocchegiani E, Muzzioli M, Cipriano C, Giacconi R (1998) Zinc, T-cell pathways, aging: role of metallothioneins. Mech Ageing Dev 106(1–2):183–204. https://doi.org/10.1016/s0047-6374(98)00115-8

Mocchegiani E, Costarelli L, Giacconi R, Cipriano C, Muti E, Tesei S, Malavolta M (2006) Nutrient-gene interaction in ageing and successful ageing. A single nutrient (zinc) and some target genes related to inflammatory/immune response. Mech Ageing Dev 127(6):517–525. https://doi.org/10.1016/j.mad.2006.01.010

Mocchegiani E, Burkle A, Fulop T (2008) Zinc and ageing (ZINCAGE Project). Exp Gerontol 43(5):361–362. https://doi.org/10.1016/j.exger.2008.03.009

Mocchegiani E, Romeo J, Malavolta M, Costarelli L, Giacconi R, Diaz LE, Marcos A (2013) Zinc: dietary intake and impact of supplementation on immune function in elderly. Age (Dordr) 35(3):839–860. https://doi.org/10.1007/s11357-011-9377-3

Molina DK, DiMaio VJ (2012) Normal organ weights in men: part II-the brain, lungs, liver, spleen, and kidneys. Am J Forensic Med Pathol 33(4):368–372. https://doi.org/10.1097/PAF.0b013e31823d29ad

Moreno-Reyes R, Egrise D, Nève J, Pasteels JL, Schoutens A (2001) Selenium deficiency-induced growth retardation is associated with an impaired bone metabolism and osteopenia. J Bone Miner Res 16(8):1556–1563. https://doi.org/10.1359/jbmr.2001.16.8.1556

MoshfeghA, Goldman J, Ahuja JK, Rhodes D, La Comb R (2009) What we eat in America. NHANES 2005–2006. Usual nutrient intakes from food and water compared to 1997 Dietary Reference Intake for vitamin D, calcium, phosphorus and magnesium. Washington, D.C. https://www.ars.usda.gov/ARSUserFiles/80400530/pdf/0506/usual_nutrient_intake_vitD_ca_phos_mg_2005-06.pdf

Mousavi SM, Hajishafiee M, Clark CCT, Borges do Nascimento IJ, Milajerdi A, Amini MR, Esmaillzadeh A (2020) Clinical effectiveness of zinc supplementation on the biomarkers of oxidative stress: a systematic review and meta-analysis of randomized controlled trials. Pharmacol Res 161:105166. https://doi.org/10.1016/j.phrs.2020.105166

Nemeth E, Ganz T (2014) Anemia of inflammation. Hematol Oncol Clin North Am 28(4):671–681., vi. https://doi.org/10.1016/j.hoc.2014.04.005

Nemeth E, Valore EV, Territo M, Schiller G, Lichtenstein A, Ganz T (2003) Hepcidin, a putative mediator of anemia of inflammation, is a type II acute-phase protein. Blood 101(7):2461–2463. https://doi.org/10.1182/blood-2002-10-3235

Nemeth E, Tuttle MS, Powelson J, Vaughn MB, Donovan A, Ward DM et al (2004) Hepcidin regulates cellular iron efflux by binding to ferroportin and inducing its internalization. Science 306(5704):2090–2093. https://doi.org/10.1126/science.1104742

New SA, Robins SP, Campbell MK, Martin JC, Garton MJ, Bolton-Smith C et al (2000) Dietary influences on bone mass and bone metabolism: further evidence of a positive link between fruit and vegetable consumption and bone health? Am J Clin Nutr 71(1):142–151. https://doi.org/10.1093/ajcn/71.1.142

NHMRC (2006) National Health and Medical Research Council, Australian Government Department of Health and Ageing, New Zealand Ministry of Health. Nutrient Reference Values for Australia and New Zealand. https://www.nhmrc.gov.au/about-us/publications/nutrient-reference-values-australia-and-new-zealand-including-recommended-dietary-intakes

NICE (2006) National Institute for Healthcare Excellence. Nutrition support for adults: oral nutrition support, enteral tube feeding and parenteral nutrition. https://www.nice.org.uk/guidance/cg32

Nielsen FH (2014) Effects of magnesium depletion on inflammation in chronic disease. Curr Opin Clin Nutr Metab Care 17(6):525–530. https://doi.org/10.1097/mco.0000000000000093

Nielsen FH (2016) Guidance for the determination of status indicators and dietary requirements for magnesium. Magnes Res 29(4):154–160. https://doi.org/10.1684/mrh.2016.0416

Nielsen FH (2018) Magnesium deficiency and increased inflammation: current perspectives. J Inflamm Res 11:25–34. https://doi.org/10.2147/jir.s136742

NOGG (2021) National Osteoporosis Guideline Group clinical guidelines. https://www.sheffield.ac.uk/NOGG/mainrecommendations.html

SACN Iron and Health Report (2011). https://www.gov.uk/government/publications/sacn-iron-and-health-report

Ozawa M, Ninomiya T, Ohara T, Hirakawa Y, Doi Y, Hata J et al (2012) Self-reported dietary intake of potassium, calcium, and magnesium and risk of dementia in the Japanese: the Hisayama Study. J Am Geriatr Soc 60(8):1515–1520. https://doi.org/10.1111/j.1532-5415.2012.04061.x

Pansu D, Bellaton C, Roche C, Bronner F (1983) Duodenal and ileal calcium absorption in the rat and effects of vitamin D. Am J Phys 244(6):G695–G700. https://doi.org/10.1152/ajpgi.1983.244.6.G695

Papp LV, Lu J, Holmgren A, Khanna KK (2007) From selenium to selenoproteins: synthesis, identity, and their role in human health. Antioxid Redox Signal 9(7):775–806. https://doi.org/10.1089/ars.2007.1528

Peacock M (2021) Phosphate metabolism in health and disease. Calcif Tissue Int 108(1):3–15. https://doi.org/10.1007/s00223-020-00686-3

Peacock M, Liu G, Carey M, McClintock R, Ambrosius W, Hui S, Johnston CC (2000) Effect of calcium or 25OH vitamin D3 dietary supplementation on bone loss at the hip in men and women over the age of 60. J Clin Endocrinol Metab 85(9):3011–3019. https://doi.org/10.1210/jcem.85.9.6836

Pedrero Z, Madrid Y, Cámara C (2006) Selenium species bioaccessibility in enriched radish (Raphanus sativus): a potential dietary source of selenium. J Agric Food Chem 54(6):2412–2417

Pepe J, Colangelo L, Biamonte F, Sonato C, Danese VC, Cecchetti V et al (2020) Diagnosis and management of hypocalcemia. Endocrine 69(3):485–495. https://doi.org/10.1007/s12020-020-02324-2

Pergola PE, Fishbane S, Ganz T (2019) Novel oral iron therapies for iron deficiency anemia in chronic kidney disease. Adv Chronic Kidney Dis 26(4):272–291. https://doi.org/10.1053/j.ackd.2019.05.002

Perri G, Mathers JC, Martin-Ruiz C, et al. (2024) Selenium status and its determinants in very old adults: insights from the Newcastle 85+ Study. British Journal of Nutrition. 131(5):901–910. https://doi.org/10.1017/S0007114523002398

Petrak J, Vyoral D (2005) Hephaestin--a ferroxidase of cellular iron export. Int J Biochem Cell Biol 37(6):1173–1178. https://doi.org/10.1016/j.biocel.2004.12.007

Petry N, Egli I, Zeder C, Walczyk T, Hurrell R (2010) Polyphenols and phytic acid contribute to the low iron bioavailability from common beans in young women. J Nutr 140(11):1977–1982. https://doi.org/10.3945/jn.110.125369

Pfeiffer CM, Looker AC (2017) Laboratory methodologies for indicators of iron status: strengths, limitations, and analytical challenges. Am J Clin Nutr 106(Suppl 6):1606S–1614S. https://doi.org/10.3945/ajcn.117.155887

Piuri G, Zocchi M, Della Porta M, Ficara V, Manoni M, Zuccotti GV et al (2021) Magnesium in obesity, metabolic syndrome, and type 2 diabetes. Nutrients 13(2). https://doi.org/10.3390/nu13020320

Prasad AS (1985) Clinical manifestations of zinc deficiency. Annu Rev Nutr 5:341–363. https://doi.org/10.1146/annurev.nu.05.070185.002013

Prasad AS (2008) Zinc in human health: effect of zinc on immune cells. Mol Med 14(5–6):353–357. https://doi.org/10.2119/2008-00033.Prasad

Prentice AM, Mendoza YA, Pereira D, Cerami C, Wegmuller R, Constable A, Spieldenner J (2017) Dietary strategies for improving iron status: balancing safety and efficacy. Nutr Rev 75(1):49–60. https://doi.org/10.1093/nutrit/nuw055

Puccinelli M, Malorgio F, Pezzarossa B (2017) Selenium enrichment of horticultural crops. Molecules 22(6):933. https://www.mdpi.com/1420-3049/22/6/933

Pyrzynska K (2009) Selenium speciation in enriched vegetables. Food Chem 114(4):1183–1191

Raggatt LJ, Partridge NC (2010) Cellular and molecular mechanisms of bone remodeling. J Biol Chem 285(33):25103–25108. https://doi.org/10.1074/jbc.R109.041087

Ravn-Haren G, Bügel S, Krath BN, Hoac T, Stagsted J, Jørgensen K et al (2008) A short-term intervention trial with selenate, selenium-enriched yeast and selenium-enriched milk: effects on oxidative defence regulation. Br J Nutr 99(4):883–892. https://doi.org/10.1017/S0007114507825153

Rayman MP (1997) Dietary selenium: time to actIntakes and blood levels falling. British Medical Journal Publishing Group

Rayman MP (2000) The importance of selenium to human health. Lancet 356(9225):233–241. https://doi.org/10.1016/S0140-6736(00)02490-9

Rayman MP (2002) The argument for increasing selenium intake. Proc Nutr Soc 61(2):203–215

Rayman MP (2004) The use of high-selenium yeast to raise selenium status: how does it measure up? Br J Nutr 92(4):557–573
Razzaghi R, Pidar F, Momen-Heravi M, Bahmani F, Akbari H, Asemi Z (2018) Magnesium supplementation and the effects on wound healing and metabolic status in patients with diabetic foot ulcer: a randomized, double-blind, placebo-controlled trial. Biol Trace Elem Res 181(2):207–215. https://doi.org/10.1007/s12011-017-1056-5
Razzaque MS (2018) Magnesium: are we consuming enough? Nutrients 10(12). https://doi.org/10.3390/nu10121863
Reddy ST, Soman SS, Yee J (2018) Magnesium balance and measurement. Adv Chronic Kidney Dis 25(3):224–229. https://doi.org/10.1053/j.ackd.2018.03.002
Reeves MA, Hoffmann PR (2009) The human selenoproteome: recent insights into functions and regulation. Cell Mol Life Sci 66(15):2457–2478. https://doi.org/10.1007/s00018-009-0032-4
Reilly C (1993) Selenium in health and disease. Aust J Nutr Diet 50:136–136
Reilly C (2006) Selenium in food and health, 2nd edn. Springer, New York
Riley J (2001) Rising life expectancy: a global history. Cambridge University Press, Cambridge
Rizzoli R, Bianchi ML, Garabédian M, McKay HA, Moreno LA (2010) Maximizing bone mineral mass gain during growth for the prevention of fractures in the adolescents and the elderly. Bone 46(2):294–305. https://doi.org/10.1016/j.bone.2009.10.005
Rockey DC (2005) Occult gastrointestinal bleeding. Gastroenterol Clin N Am 34(4):699–718. https://doi.org/10.1016/j.gtc.2005.08.010
Roman M, Jitaru P, Barbante C (2014) Selenium biochemistry and its role for human health. Metallomics 6(1):25–54. https://doi.org/10.1039/c3mt00185g
Rouhani P, Rezaei Kelishadi M, Saneei P (2022) Effect of zinc supplementation on mortality in under 5-year children: a systematic review and meta-analysis of randomized clinical trials. Eur J Nutr 61(1):37–54. https://doi.org/10.1007/s00394-021-02604-1
Ryder KM, Shorr RI, Bush AJ, Kritchevsky SB, Harris T, Stone K et al (2005) Magnesium intake from food and supplements is associated with bone mineral density in healthy older white subjects. J Am Geriatr Soc 53(11):1875–1880. https://doi.org/10.1111/j.1532-5415.2005.53561.x
SACN (2011) Dietary reference values for energy. https://www.gov.uk/government/publications/sacn-dietary-reference-values-for-energy
SACN (2013) SACN statement on selenium and health. https://www.gov.uk/government/publications/sacn-statement-on-selenium-and-health-2013
Samtiya M, Aluko RE, Dhewa T (2020) Plant food anti-nutritional factors and their reduction strategies: an overview. Food Prod Process Nutr 2(1):6. https://doi.org/10.1186/s43014-020-0020-5
Sandstead HH (1995) Requirements and toxicity of essential trace elements, illustrated by zinc and copper. Am J Clin Nutr 61(3 Suppl):621S–624S. https://doi.org/10.1093/ajcn/61.3.621S
Sandstead HH (2000) Causes of iron and zinc deficiencies and their effects on brain. J Nutr 130(2S Suppl):347S–349S. https://doi.org/10.1093/jn/130.2.347S
Sandstrom B, Almgren A, Kivisto B, Cederblad A (1989) Effect of protein level and protein source on zinc absorption in humans. J Nutr 119(1):48–53. https://doi.org/10.1093/jn/119.1.48
Sang S, Pan X, Chen Z, Zeng F, Pan S, Liu H et al (2018) Thiamine diphosphate reduction strongly correlates with brain glucose hypometabolism in Alzheimer's disease, whereas amyloid deposition does not. Alzheimers Res Ther 10(1):26. https://doi.org/10.1186/s13195-018-0354-2
Schaefer B, Tobiasch M, Wagner S, Glodny B, Tilg H, Wolf M, Zoller H (2022) Hypophosphatemia after intravenous iron therapy: comprehensive review of clinical findings and recommendations for management. Bone 154:116202. https://doi.org/10.1016/j.bone.2021.116202
Schlemmer U, Frølich W, Prieto RM, Grases F (2009) Phytate in foods and significance for humans: food sources, intake, processing, bioavailability, protective role and analysis. Mol Nutr Food Res 53(Suppl 2):S330–S375. https://doi.org/10.1002/mnfr.200900099
Schubert A, Holden J, Wolf W (1987) Selenium content of a core group of foods based on a critical evaluation of published analytical data. J Am Diet Assoc 87(3):285–299

Scott BJ, Bradwell AR (1983) Identification of the serum binding proteins for iron, zinc, cadmium, nickel, and calcium. Clin Chem 29(4):629–633. https://www.ncbi.nlm.nih.gov/pubmed/6831689

Sekler I, Sensi SL, Hershfinkel M, Silverman WF (2007) Mechanism and regulation of cellular zinc transport. Mol Med 13(7–8):337–343. https://doi.org/10.2119/2007-00037.Sekler

Semba RD, Bartali B, Zhou J, Blaum C, Ko CW, Fried LP (2006) Low serum micronutrient concentrations predict frailty among older women living in the community. J Gerontol A Biol Sci Med Sci 61(6):594–599. https://doi.org/10.1093/gerona/61.6.594

Semrin G, Fishman DS, Bousvaros A, Zholudev A, Saunders AC, Correia CE et al (2006) Impaired intestinal iron absorption in Crohn's disease correlates with disease activity and markers of inflammation. Inflamm Bowel Dis 12(12):1101–1106. https://doi.org/10.1097/01.mib.0000235097.86360.04

Sharp P, Srai SK (2007) Molecular mechanisms involved in intestinal iron absorption. World J Gastroenterol 13(35):4716–4724. https://doi.org/10.3748/wjg.v13.i35.4716

Shepshelovich D, Rozen-Zvi B, Avni T, Gafter U, Gafter-Gvili A (2016) Intravenous versus oral iron supplementation for the treatment of anemia in CKD: an updated systematic review and meta-analysis. Am J Kidney Dis 68(5):677–690. https://doi.org/10.1053/j.ajkd.2016.04.018

Silay K, Akinci S, Yalcin A, Guney T, Ulas A, Ozer O et al (2015) The status of iron absorption in older patients with iron deficiency anemia. Eur Rev Med Pharmacol Sci 19(17):3142–3145. https://www.ncbi.nlm.nih.gov/pubmed/26400514

Silva JGS, Rebellato AP, Abreu JS, Greiner R, Pallone JAL (2022) Impact of the fortification of a rice beverage with different calcium and iron sources on calcium and iron bioaccessibility. Food Res Int 161:111830. https://doi.org/10.1016/j.foodres.2022.111830

Silverstein SB, Gilreath JA, Rodgers GM (2008) Intravenous iron therapy: a summary of treatment options and review of guidelines. J Pharm Pract 21(6):431–443. https://doi.org/10.1177/0897190008318916

Slusarczyk P, Mleczko-Sanecka K (2021) The multiple facets of iron recycling. Genes (Basel) 12(9):1364. https://doi.org/10.3390/genes12091364

Sors TG, Ellis DR, Salt DE (2005) Selenium uptake, translocation, assimilation and metabolic fate in plants. Photosynth Res 86(3):373–389. https://doi.org/10.1007/s11120-005-5222-9

Stamp TC, Round JM, Rowe DJ, Haddad JG (1972) Plasma levels and therapeutic effect of 25-hydroxycholecalciferol in epileptic patients taking anticonvulsant drugs. Br Med J 4(5831):9–12. https://doi.org/10.1136/bmj.4.5831.9

Steere AN, Byrne SL, Chasteen ND, Mason AB (2012) Kinetics of iron release from transferrin bound to the transferrin receptor at endosomal pH. Biochim Biophys Acta 1820(3):326–333. https://doi.org/10.1016/j.bbagen.2011.06.003

Stefánka Z, Ipolyi I, Dernovics M, Fodor P (2001) Comparison of sample preparation methods based on proteolytic enzymatic processes for Se-speciation of edible mushroom (Agaricus bisporus) samples. Talanta 55(3):437–447

Stepura OB, Martynow AI (2009) Magnesium orotate in severe congestive heart failure (MACH). Int J Cardiol 134:145–147

Stewart-Knox BJ, Simpson EE, Parr H, Rae G, Polito A, Intorre F et al (2005) Zinc status and taste acuity in older Europeans: the ZENITH study. Eur J Clin Nutr 59(Suppl 2):S31–S36. https://doi.org/10.1038/sj.ejcn.1602295

Stoffaneller R, Morse NL (2015) A review of dietary selenium intake and selenium status in Europe and the Middle East. Nutrients 7(3):1494–1537. https://doi.org/10.3390/nu7031494

Swain JH, Tabatabai LB, Reddy MB (2002) Histidine content of low-molecular-weight beef proteins influences nonheme iron bioavailability in Caco-2 cells. J Nutr 132(2):245–251. https://doi.org/10.1093/jn/132.2.245

Takeda E, Sakamoto K, Yokota K, Shinohara M, Taketani Y, Morita K et al (2002) Phosphorus supply per capita from food in Japan between 1960 and 1995. J Nutr Sci Vitaminol (Tokyo) 48(2):102–108. https://doi.org/10.3177/jnsv.48.102

Takeda E, Yamamoto H, Yamanaka-Okumura H, Taketani Y (2014) Increasing dietary phosphorus intake from food additives: potential for negative impact on bone health. Adv Nutr 5(1):92–97. https://doi.org/10.3945/an.113.004002

Tay HS, Soiza RL (2015) Systematic review and meta-analysis: what is the evidence for oral iron supplementation in treating anaemia in elderly people? Drugs Aging 32(2):149–158. https://doi.org/10.1007/s40266-015-0241-5

Thiry C, Ruttens A, De Temmerman L, Schneider Y-J, Pussemier L (2012) Current knowledge in species-related bioavailability of selenium in food. Food Chem 130(4):767–784. https://doi.org/10.1016/j.foodchem.2011.07.102

Thomson CD, Chisholm A, McLachlan SK, Campbell JM (2008) Brazil nuts: an effective way to improve selenium status. Am J Clin Nutr 87(2):379–384

Thorn J, Robertson J, Buss DH, Bunton NG (1978) Trace nutrients. Selenium in British food. Br J Nutr 39(2):391–396. https://doi.org/10.1079/BJN19780049

Toblli JE, Brignoli R (2007) Iron(III)-hydroxide polymaltose complex in iron deficiency anemia / review and meta-analysis. Arzneimittelforschung 57(6A):431–438. https://doi.org/10.1055/s-0031-1296692

Topf JM, Murray PT (2003) Hypomagnesemia and hypermagnesemia. Rev Endocr Metab Disord 4(2):195–206. https://doi.org/10.1023/a:1022950321817

Tucker KL (2014) Vegetarian diets and bone status. Am J Clin Nutr 100(Suppl 1):329s–335s. https://doi.org/10.3945/ajcn.113.071621

Turner J, Gittoes N, Selby P, Society for Endocrinology Clinical Committee (2016) Society for Endocrinology Endocrine Emergency Guidance: emergency management of acute hypocalcaemia in adult patients. Endocr Connect 5(5):G7–G8. https://doi.org/10.1530/EC-16-0056

Tyrovola D, Soulaidopoulos S, Tsioufis C, Lazaros G (2023) The role of nutrition in cardiovascular disease: current concepts and trends. Nutrients 15(5):1064. https://doi.org/10.3390/nu15051064

US Department of Agriculture, A. R. S. (2019) Usual nutrient intake from food and beverages, by gender and age, what we eat in America, NHANES 2013–2016. https://www.ars.usda.gov/ARSUserFiles/80400530/pdf/usual/Usual_Intake_gender_WWEIA_2013_2016.pdf

Vallee BL, Falchuk KH (1993) The biochemical basis of zinc physiology. Physiol Rev 73(1):79–118. https://doi.org/10.1152/physrev.1993.73.1.79

van de Wal-Visscher ER, Kooman JP, van der Sande FM (2018) Magnesium in chronic kidney disease: should we care? Blood Purif 45(1–3, 173):–178. https://doi.org/10.1159/000485212

van Staa TP, Dennison EM, Leufkens HG, Cooper C (2001) Epidemiology of fractures in England and Wales. Bone 29(6):517–522. https://doi.org/10.1016/s8756-3282(01)00614-7

Veronese N, Berton L, Carraro S, Bolzetta F, De Rui M, Perissinotto E et al (2014) Effect of oral magnesium supplementation on physical performance in healthy elderly women involved in a weekly exercise program: a randomized controlled trial. Am J Clin Nutr 100(3):974–981. https://doi.org/10.3945/ajcn.113.080168

Veronese N, Stubbs B, Solmi M, Noale M, Vaona A, Demurtas J, Maggi S (2017) Dietary magnesium intake and fracture risk: data from a large prospective study. Br J Nutr 117(11):1570–1576. https://doi.org/10.1017/s0007114517001350

Veronese N, Demurtas J, Pesolillo G, Celotto S, Barnini T, Calusi G et al (2020) Magnesium and health outcomes: an umbrella review of systematic reviews and meta-analyses of observational and intervention studies. Eur J Nutr 59(1):263–272. https://doi.org/10.1007/s00394-019-01905-w

Vural Z, Avery A, Kalogiros DI, Coneyworth LJ, Welham SJM (2020) Trace mineral intake and deficiencies in older adults living in the community and institutions: a systematic review. Nutrients 12(4). https://doi.org/10.3390/nu12041072

Wallace DF (2016) The regulation of iron absorption and homeostasis. Clin Biochem Rev 37(2):51–62. https://www.ncbi.nlm.nih.gov/pubmed/28303071

Wang K, Zhou B, Kuo YM, Zemansky J, Gitschier J (2002) A novel member of a zinc transporter family is defective in acrodermatitis enteropathica. Am J Hum Genet 71(1):66–73. https://doi.org/10.1086/341125

Watkins DA, Msemburi WT, Pickersgill SJ, Kawakatsu Y, Gheorghe A, Dain K et al (2022) NCD countdown 2030: efficient pathways and strategic investments to accelerate progress towards the sustainable development goal target 3.4 in low-income and middle-income countries. Lancet 399(10331):1266–1278. https://doi.org/10.1016/s0140-6736(21)02347-3

Wawer AA, Jennings A, Fairweather-Tait SJ (2018) Iron status in the elderly: a review of recent evidence. Mech Ageing Dev 175:55–73. https://doi.org/10.1016/j.mad.2018.07.003

Weaver CM, Peacock M (2011) Calcium. Adv Nutr 2(3):290–292. https://doi.org/10.3945/an.111.000463

Weaver CM, Alexander DD, Boushey CJ, Dawson-Hughes B, Lappe JM, LeBoff MS et al (2016) Calcium plus vitamin D supplementation and risk of fractures: an updated meta-analysis from the National Osteoporosis Foundation. Osteoporos Int 27(1):367–376. https://doi.org/10.1007/s00198-015-3386-5

Weekley CM, Harris HH (2013) Which form is that? The importance of selenium speciation and metabolism in the prevention and treatment of disease. Chem Soc Rev 42(23):8870–8894

Wegmuller R, Tay F, Zeder C, Brnic M, Hurrell RF (2014) Zinc absorption by young adults from supplemental zinc citrate is comparable with that from zinc gluconate and higher than from zinc oxide. J Nutr 144(2):132–136. https://doi.org/10.3945/jn.113.181487

White PJ, Bowen HC, Parmaguru P, Fritz M, Spracklen W, Spiby R et al (2004) Interactions between selenium and sulphur nutrition in Arabidopsis thaliana. J Exp Bot 55(404):1927–1937

WHO (2019) World Health Organization Noncommunicable Diseases (NCD). https://www.who.int/data/gho/data/themes/topics/topic-details/GHO/ncd-mortality

Wongdee K, Charoenphandhu N (2015) Vitamin D-enhanced duodenal calcium transport. Vitam Horm 98:407–440. https://doi.org/10.1016/bs.vh.2014.12.010

Workinger JL, Doyle RP, Bortz J (2018) Challenges in the diagnosis of magnesium status. Nutrients 10(9). https://doi.org/10.3390/nu10091202

Xu ZP, Li L, Bao J, Wang ZH, Zeng J, Liu EJ et al (2014) Magnesium protects cognitive functions and synaptic plasticity in streptozotocin-induced sporadic Alzheimer's model. PLoS One 9(9):e108645. https://doi.org/10.1371/journal.pone.0108645

Winkel LHE, Johnson CA, Lenz M, Grundl T, Leupin OX, Amini M, Charlet L (2012) Environmental selenium research: from microscopic processes to global understanding. Environ Sci Technol. 46(2):571–9. https://doi.org/10.1021/es203434d

Wrobel JK, Power R, Toborek M (2016) Biological activity of selenium: Revisited. IUBMB Life. 68(2):97–105. https://doi.org/10.1002/iub.1466

Yamanaka H, Nakajima M, Katoh M, Yokoi T (2007) Glucuronidation of thyroxine in human liver, jejunum, and kidney microsomes. Drug Metab Dispos 35(9):1642–1648. https://doi.org/10.1124/dmd.107.016097

Yildirim T, Yalcin A, Atmis V, Cengiz OK, Aras S, Varlı M, Atli T (2015) The prevalence of anemia, iron, vitamin B12, and folic acid deficiencies in community dwelling elderly in Ankara, Turkey. Arch Gerontol Geriatr 60(2):344–348. https://doi.org/10.1016/j.archger.2015.01.001

Zhao Y, Li Z, Shi Y, Cao G, Meng F, Zhu W, Yang GE (2016) Effect of hypophosphatemia on the withdrawal of mechanical ventilation in patients with acute exacerbations of chronic obstructive pulmonary disease. Biomed Rep 4(4):413–416. https://doi.org/10.3892/br.2016.605

Zhu Y, Minović I, Dekker LH, Eggersdorfer ML, van Zon SKR, Reijneveld SA et al (2020) Vitamin status and diet in elderly with low and high socioeconomic status: the lifelines-MINUTHE study. Nutrients 12(9):2659. https://doi.org/10.3390/nu12092659

Zilinski J, Zillmann R, Becker I, Benzing T, Schulz RJ, Roehrig G (2014) Prevalence of anemia among elderly inpatients and its association with multidimensional loss of function. Ann Hematol 93(10):1645–1654. https://doi.org/10.1007/s00277-014-2110-4

Chapter 14
Gut Microbiome as a Target for Anti-ageing Interventions

Tetiana R. Dmytriv and Volodymyr I. Lushchak

Abstract Trillions of various microorganisms inhabit the human intestine whilst having myriads of effects on the body. They participate in the metabolism of nutrients, support the work of the immune system, regulate operation of the nervous system, and produce vitamins, short-chain fatty acids, and a number of other compounds necessary for the host. An imbalance or disruption in the normal microbial community is called dysbacteriosis or dysbiosis. This condition is often associated with the occurrence of various pathologies including chronic low-intensity inflammation. The latter is one of the key signs of ageing. In this chapter, we consider the gut microbiome as a target for anti-ageing interventions. In particular, we describe the main functions of the gut microbiome, its changes with ageing, and discuss dysbacteriosis as a trigger of accelerated ageing. We also present anti-ageing interventions such as a diet, nutritional supplements (probiotics, prebiotics, antioxidants), and exercise and how they may affect the microbiome and enable or impede healthy longevity.

Keywords Healthy ageing · Inflammation · Intestinal permeability · Diet · Dietary supplements · Exercise

Abbreviations

GI tract	Gastrointestinal tract
HSP	Heat shock proteins

T. R. Dmytriv
Department of Biochemistry and Biotechnology, Vasyl Stefanyk Precarpathian National University, Ivano-Frankivsk, Ukraine

V. I. Lushchak (✉)
Department of Biochemistry and Biotechnology, Vasyl Stefanyk Precarpathian National University, Ivano-Frankivsk, Ukraine

Research and Development University, Ivano-Frankivsk, Ukraine
e-mail: volodymyr.lushchak@pnu.edu.ua

© The Author(s), under exclusive license to Springer Nature Switzerland AG 2024

V. I. Korolchuk, J. R. Harris (eds.), *Biochemistry and Cell Biology of Ageing: Part V, Anti-Ageing Interventions*, Subcellular Biochemistry 107,
https://doi.org/10.1007/978-3-031-66768-8_14

LPS	Lipopolysaccharides
PAMPs	Pathogen-associated molecular patterns
PD	Parkinson's disease
PRRs	Pathogen recognition receptors
ROS	Reactive oxygen species
SAD	Standard American diet
SCFA	Short-chain fatty acids
TNFα	Tumor necrosis factor α

Gut Microbiome and Its Role

Microbiota is the entire population of microorganisms that colonise a certain place. It includes not only bacteria but also viruses, fungi, archaea, and protozoa (Jandhyala et al. 2015). The human body is inhabited by approximately 10–100 trillions of symbiotic microbes with a total weight from 0.5 to 2.0 kg (Walker and Hoyles 2023). The majority of the human microbiota is located in the gastrointestinal (GI) tract (Ursell et al. 2012). It is estimated that the genomic content of the intestinal microbiome is about 100 times greater than the human genome (Thursby and Juge 2017).

The gut microbiota maintains a mutually beneficial symbiotic relationship with its host. At the same time, it forms a dynamic microbial system that performs a number of functions (Fig. 14.1). In particular, modulation of the host immune

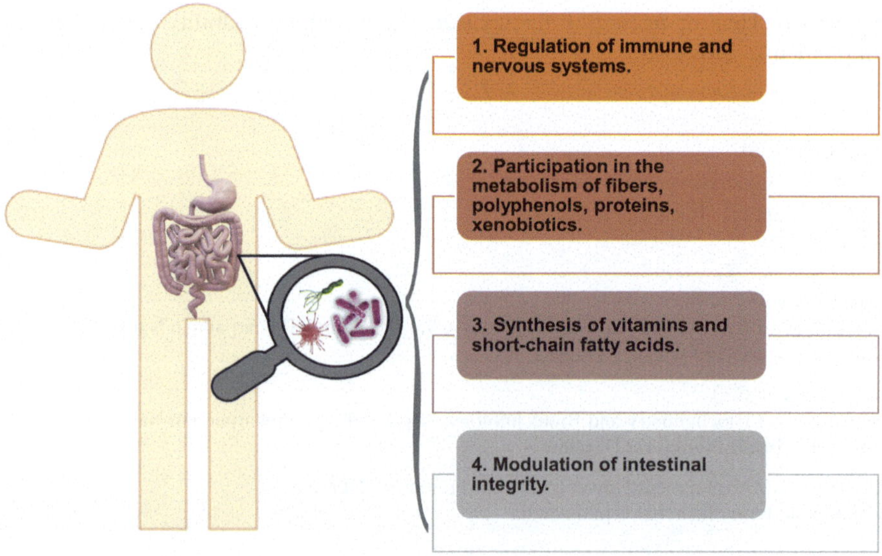

Fig. 14.1 Main functions of the gut microbiome

system is one of the most important of them. Structural components of cell walls of microorganisms, such as lipopolysaccharides (LPS), peptidoglycans, and flagella can interact with receptors of immune cells. This may stimulate of the host's immune response (Wang et al. 2022). Thus, the microbiome plays a critical role in the learning and development of response of the innate and adaptive immune systems. Microbial colonisation of the mucosal surfaces of the host is one of the key stages in the normal development of a newborn infant (Gensollen et al. 2016). The diversity of gut microbes is a necessary factor in the formation of a complex dynamic immunoregulatory host-microbiota network. Such a close relationship between the gut microbiome and the host requires the proper functioning of the host's immune system. This is necessary in order to prevent excessive exploitation of the host's resources and, at the same time, the need to maintain immune tolerance to harmless irritants of microorganisms should be met. Interaction between the immune system and the microbiota of the host contributes to immune homeostasis and reduces susceptibility to infectious and inflammatory diseases. That is especially true for so-called autoimmune diseases. However, excessive susceptibility of the immune system to the gut microbiota can lead to an abnormal immune response and the development of various pathologies (Zheng et al. 2020). Therefore, it is clear that the proper balanced interaction between microbiome and the host should be maintained to provide good health for both the partners in this symbiosis.

Another important function of the gut microbiota is its participation in the metabolism of the host. The diverse enzyme repertoire of microorganisms both differs from and complements the host organism enzymes. In this way, microorganisms provide the host with enzymes that are not encoded by its genome and enable processing of food components which human organism cannot utilise (Rowland et al. 2018). For example, microorganisms may digest polysaccharides that are not digested by mammalian enzymes, in particular, inulin and fibres. Also, in anaerobic conditions they may produce short-chain fatty acids (SCFA) from carbohydrates indigested by humans (Liu et al. 2018). Fundamentally, they have numerous regulatory functions in the body of mammals. In addition, intestinal microbiota participates in the metabolism of plant polyphenols consumed with food (Vernocchi et al. 2020). It is also known that microorganisms provide an essential capacity for proteolysis which allows the host to hydrolyse food proteins to shorter peptides, amino acids, and their derivatives. Some intestinal microbes synthesize vitamin K and an entire set of B vitamins such as biotin, cobalamin, folates, nicotinic acid, pantothenic acid, pyridoxine, riboflavin, and thiamine making them available to the host. However, to a greater extent, the vitamins are used by other bacteria and fungi which also need vitamins but they themselves are not able to synthesize them (Rowland et al. 2018). In addition to being involved in host nutrient metabolism, the gut microbiome also plays an important role in the metabolism of xenobiotics transforming them into inactive metabolites or prodrugs (Collins and Patterson 2020).

Gut microbiota also serves as an additional protective layer of the intestine that preserves its integrity. First, they prevent the adhesion of pathogens. Second, they inhibit their reproduction by releasing antimicrobial substances. Third, they secrete mucous substances that make translocation difficult due to the dense environment.

In this way, the intestinal microbiota participates in maintaining the integrity of the host's intestinal barrier (Kocot et al. 2022).

Gut microbiota also regulates the functioning of the nervous system. This bidirectional relationship between the two systems is often called "microbiome-gut-brain axis". For example, gut microorganisms influence the feeding behaviour of the host. In particular, it was found that germ-free mice depleted of their resident microbiota had a lower appetite than mice with a normal microbiome (Morais et al. 2021). More and more recent studies have demonstrated the connection between microbiota and sensation of pain. In particular, the intestinal microbiota plays a key role in the development of visceral and abdominal pain. The use of probiotics proved to be an effective means of treatment. It also controls the development of anxiety, depression, stress response, and violation of the microbiome-gut-brain axis and may lead to the development of various neuropathologies (Morreale et al. 2022).

Age-Related Changes in Gut Microbiome

Each person has a unique gut microbiome responsible for many specific functions. Its main part consists of bacteria of various species, which are classified by genus, family, order, and phyla. The composition of human microbiome depends on the age, lifestyle, diet, exposure to antibiotics, host genotype, diseases, and many other factors and *vice versa* microbiome influences many features of host organism modifying its physiological state and biological age (Rinninella et al. 2019).

Chronological age is the time elapsed from birth to a certain date. It does not account for the heterogeneity with which people age. Biological age takes into account not only the time that has passed from the birth, but also functional or physiological age related to the influence of variable effects of time on individuals (Diebel and Rockwood 2021; Maltoni et al. 2022). Changes in the composition and diversity of the gut microbiome are believed to be affected more significantly by the biological rather than chronological age (Kim and Jazwinski 2018). For example, the module of common abundance which includes the genera *Ruminococcus, Coprobacillus*, and *Eggerthella,* increases with the advance of biological age. At the same time, it does not depend on chronological age. The first two genera belong to the *Firmicutes* phylum, and the last one to *Actinobacteria*. Changes in the gut microbiota with increasing biological age can be very different from changes with chronological age (Kim and Jazwinski 2018).

Examination of the faecal samples of people of different age groups shows that the composition and variety of their microorganisms are different (Kim and Jazwinski 2018; Walker and Hoyles 2023). Scientists distinguish four main age stages of microbiome development: birth (infancy), adolescence, adulthood, and old age. Figure 14.2 shows changes in the composition and diversity of the gut microbiome with age. Babies who are born naturally are dominated by *Lactobacillus*, *Prevotella*, and *Atopobium* from the mother's vaginal microflora. In those born by caesarean section, *Staphylococcus* predominates (from the microflora of the

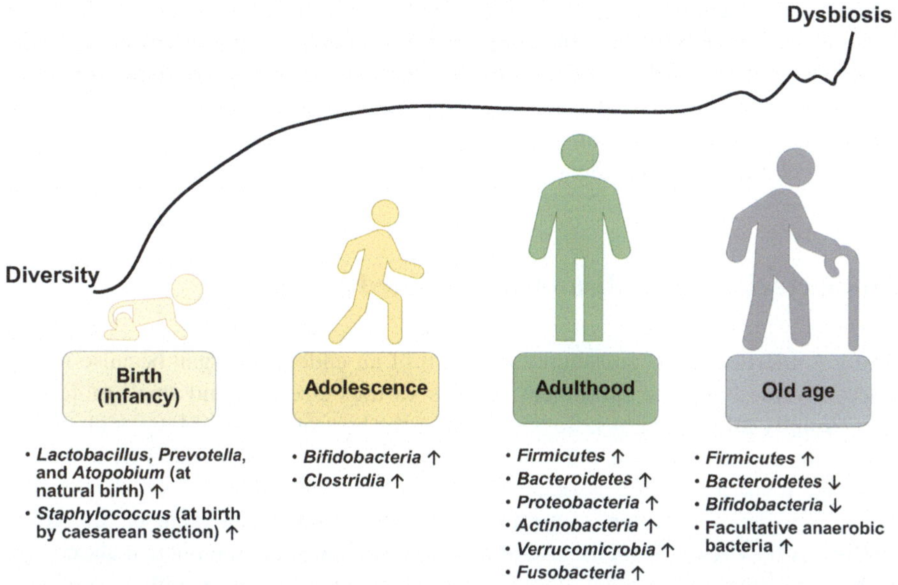

Fig. 14.2 Changes in the composition and diversity of the gut microbiome with age. From birth, the composition and diversity of the gut microbiome increases with age. In adolescence and adulthood, the diversity of microorganisms mostly remains stable. During the transition to old age, the composition and diversity of the gut microbiome is unstable. It undergoes changes, the number of facultative bacteria increases and age-related dysbacteriosis occurs. ↑—increasing; ↓—decreasing

mother's skin). During adolescence and adulthood periods, the diversity of the gut microbiome is relatively constant. In adulthood, the dominant types of gut microbes are the phyla *Firmicutes*, *Bacteroidetes*, *Proteobacteria*, *Actinobacteria*, *Verrucomicrobia*, and *Fusobacteria*. The first two of them usually represent 90% of the intestinal microbiota (Rinninella et al. 2019). With ageing, the number of anaerobic bacteria *Bifidobacteria* and *Clostridia* increases in the gut (Novakovic et al. 2020). After 60 years, changes in its composition are usually observed and the ratio of *Firmicutes* to *Bacteroidetes* is of most interest to researchers (Brooks et al. 2023). In particular, the relative abundance of *Firmicutes* increases with age, while *Bacteroides*, on the contrary, decreases (Meng et al. 2022). The number of *Bifidobacteria* also noticeably decreases and the proportion of facultative anaerobes increases, as the function of the immune system declines (Novakovic et al. 2020).

Changes in the composition of the gut microbiota with ageing are associated not only with the deterioration of immune function but also with changes in the operation of the gastrointestinal (GI) tract. This organ directly determines the supply of nutrients for the human body and therefore it affects the composition and functionality of the microbiota (An et al. 2018). With ageing, more and more gut microbiota participates in the development of inflammation. Thus, a number of processes driven by ageing are responsible for the increase in the barrier permeability of the intestine. The bacteria themselves, fragments/components of their cells, or the products of bacterial activity enter the lumen and cause inflammation (Meng et al. 2022). When the stability of the gut microbiome is lost, age-related dysbacteriosis begins. It is largely correlated with the biological age and is a sign of ageing. Preventing the age-related changes in the intestinal ecosystem mitigates microbial dysbiosis, prevents accelerated ageing, extends lifespan and life expectancy (DeJong et al. 2020).

Dysbacteriosis as a Risk of Accelerated Ageing

The gut microbiome is considered the form of an endocrine organ because it produces diverse bioactive substances such as various enzymes and small molecules that strongly affect the physiological state of the host organism (Meng et al. 2020). The state of imbalance in the natural intestinal microflora is called dysbiosis or dysbacteriosis. It is tightly associated with many age-related diseases—obesity, diabetes, various autoimmune, or inflammatory diseases (Kageyama et al. 2020). Dysbiosis may result from the use of antibiotics, changes in the relative abundance of bacteria caused by the diet, immunodeficiency, infection, or the influence of various drugs and toxins (Zeng et al. 2017).

The intestine is the intermediary and an insulator between the microbiota and immune system. The intestinal barrier consists of three main layers. Mucous layer of the intestinal lumen is the first one, whereas a continuous sheet of epithelial cells forms the second layer. The third inner layer of the intestinal barrier is formed by the mucosal immune system. Normally, the intestinal barrier prevents the interaction of gut microorganisms and harmful products of their vital activity with the immune system (Ghosh et al. 2021). Intestinal barrier dysfunction, i.e. increasing in permeability, is evolutionary conserved feature of ageing organisms (Salazar et al. 2023). Figure 14.3 shows the intestinal barrier at healthy and unhealthy ageing. Normally, the barrier is selectively permeable. Violation of the integrity with age leads to an excessive activation of the immune system by the gut microorganisms. In this case, immune system recognises pathogen-associated molecular patterns (PAMPs) such as bacterial lipopolysaccharides, peptidoglycan, and oligodeoxynucleotides (Ren et al. 2023). They are recognised by pathogen recognition receptors (PRRs) at immune (dendritic) cells and contribute to the production of pro-inflammatory cytokines such as IL-1β, IL-6, IL-8, and TNFα (Asiamah et al. 2019; Alsegiani and Shah 2022; Ren et al. 2023). This triggers the development of the inflammatory process and damage to the intestinal barrier. As a result, low-grade chronic inflammation develops with age, which is widely believed to be a sign of ageing (Ren et al. 2023).

It has been established that a change in the gut microbiota may cause many diseases. In the most general terms, in the elderly the gut microbiota represents a pro-inflammatory phenotype. In particular, low-grade chronic systemic inflammation contributes to the age-related neuroinflammation and even neurodegenerative diseases, which are closely related to chronic peripheral inflammation (Alsegiani and

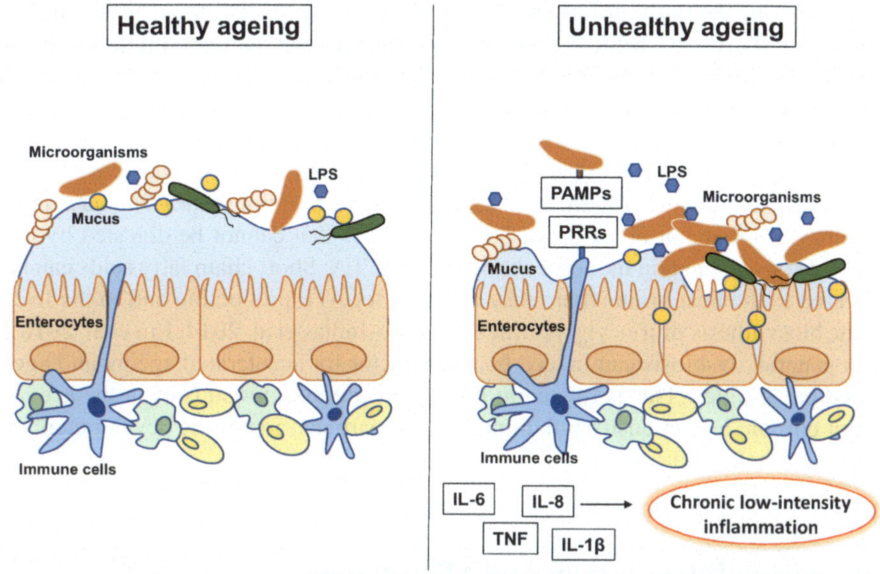

Fig. 14.3 Intestinal barrier in healthy and unhealthy aging. The intestinal barrier consists of three layers: a mucous layer, a layer of epithelial cells (enterocytes), and a layer of immune cells. Microorganisms form an additional layer on the mucus of the intestinal lumen. Optimal composition of the gut microbiome and the integrity of the intestinal barrier are the keys to healthy aging. In unhealthy aging, dysbacteriosis, a decrease in mucus secretion and a violation of the integrity of the intestinal barrier are observed. As a result, PRRs on the surface of immune cells (dendritic cells) recognize PAMPs on the surface of intestinal microorganisms leading to the production of pro-inflammatory cytokines IL-1b, IL-6, IL-8, TNF, and the development of chronic low-intensity inflammation, which is a sign of aging. Abbreviations: PAMPs pathogen-associated molecular patterns, PRRs pattern recognition receptors, IL-6/8 interleukin 6/8, TNF tumour necrosis factor

Shah 2022). Some intestinal microorganisms are quite often associated with the occurrence of neurodegenerative diseases such as Alzheimer disease (AD) and Parkinson's disease (PD). Moreover, changes in the gut microbiome can influence the development and progression of these pathologies. Bacteria such as *B. subtilis* and *E. coli* can produce amyloid proteins (Alkasir et al. 2017). Due to the increased permeability of the intestinal barrier with age, bacterial amyloids may enter the bloodstream and further may cross the blood-brain barrier and accumulate in the brain. Thus, via this mechanism bacterial amyloid can disrupt the functioning of the brain. In addition, in the elderly patients with cerebral amyloidosis and cognitive impairment, a lower number of gut bacteria with anti-inflammatory activity (*E. rectale* and *B. fragilis*) was found, whereas the number of bacteria with pro-inflammatory activity (*Escherichia/Shigella*) was increased (Mancuso and Santangelo 2018).

PD, the second most common neurodegenerative disorder after AD, is characterised by a dopamine deficiency due to the loss of dopaminergic neurons in the substantia nigra of the brain. Accumulation of aggregates of alpha-synuclein (α-syn) is the main marker of PD (Zhu et al. 2021). In general, the gut microbiome of healthy individuals substantially differs from that of patients with PD. Interestingly, transplanting of microbiota from PD patients into mice causes increased motor deficits and this effect was mediated by α-syn (Li et al. 2023).

Metabolic diseases such as obesity and diabetes are also related to the composition and diversity of the gut microbiome. For example, in obese people proportion of *Firmicutes* is increased, whereas proportion of *Bacteroidetes* is decreased relatively to healthy people (Magne et al. 2020). It is worth noting that *Firmicutes* possess enzymes that can break down polysaccharides that cannot be digested by the host and convert them in monosaccharides and SCFA. Short chain fatty acids can be delivered to the host brain where they are used for ATP production and to the liver for the biosynthesis of triacylglycerides (Moreno-Indias et al. 2014; Liu et al. 2018).

In general, dysbiosis can indeed be a risk factor for accelerated ageing and age-related pathologies. This is related to the role of gut microorganisms in the activation of chronic low-intensity inflammation and participation in the development of various age-related disorders.

Anti-ageing Interventions and Microbiome

Ageing has traditionally been considered a natural and inevitable process. In the recent years scientists refocused their research from attempting to combat ageing in general to interventions promoting healthy ageing (Liu 2022). Healthy ageing can be defined as: 1—health state evaluated through, physical, cognitive, psychological, and social aspects with the absence of diseases; and 2—the body's ability to adapt to diverse challenges throughout the life (Menassa et al. 2023). In the past few decades, scientists have shown that ageing can be slowed down and lifespan extended in model organisms. Anti-ageing interventions are usually divided into two main categories: 1—lifestyle changes and 2—pharmaceutical or genetic interventions (Liu 2022). Here we will focus more on the first group, i.e. on the interventions involving nutrition and exercise. Calorie restriction and intermittent fasting are among the most popular approaches in this direction (Ros and Carrascosa 2020). It has been established that the number of different genetic pathways control ageing and lifespan. In fact, to some extent virtually all genes are potential anti-ageing targets (Liu 2022). In recent years, the study of various anti-ageing compounds indicated that caloric restriction mimetics can induce autophagy and enhance regeneration (Chung and Chung 2019). In general, anti-ageing interventions can delay ageing and prevent the onset of many chronic diseases. In this way, they extend the duration of a healthy life (Liu 2022). Both lifestyle interventions and the use of various compounds affect the diversity and functionality of gut microorganisms. Probability of developing dysbacteriosis and related complications increases with age. A healthy microbiome can increase the duration of a healthy life, whereas an unhealthy microbiome and dysbiosis can contribute to increased intestinal permeability, the development of inflammation and various diseases ultimately leading to unhealthy ageing (Donati Zeppa et al. 2023). Thus, the gut microbiome has become a target for anti-ageing interventions. Below, we examine how diet, exercise, and the use of probiotics, prebiotics, and antioxidants affect the gut microbiome as an anti-ageing target.

Diet

Diet and nutrition are extremely important for maintaining good health. By the diet, we mean the total amount and type of food consumed by a person. Depending on the amount of nutrients that a person needs for normal metabolism, growth, and well-being in general, they are divided into macronutrients and micronutrients (Zohoori 2019). Macronutrients include proteins, fats, and carbohydrates, whereas micronutrients include trace minerals and vitamins (Costa-Pinto and Gantner 2020). Collectively they satisfy the body's energy needs and should be consumed in a balanced way. Commonly, human adults consume 45–65% of their total calories from carbohydrates, 20–35% from fat, and 10–35% from protein. Diets with a deficiency of certain macronutrients or, on the contrary, with an excess of them can be harmful (Ryan-Harshman and Aldoori 2006). Their balanced combination is necessary to preserve healthy longevity. A large number of different diets has been proposed to date, each characterised by the differences in relative proportions of micro- and macronutrients. Despite this, the combination of macronutrients that provides optimal health is still under dispute (Costa-Pinto and Gantner 2020).

The food consumed by a person is not only used by human body but some portion of it is used by gut microorganisms. The latter mainly digest components that are indigestible for human hydrolytic enzymes. Thus, what we eat directly affects the composition and functions of the gut microbiome which maintains its homeostasis in a healthy state and interacts with the immune system of the intestinal mucosal barrier (Zhang 2022). For example, the Standard American diet (SAD), also called the Western diet, contributes to metabolic endotoxemia (Horn et al. 2022). It occurs as a result of a disruption of the intestinal barrier, which allows contact between components of the intestinal microbial cell wall and Toll-like receptors located on immune cells ultimately leading to excessive activation of the immune system and development of inflammation (Kim et al. 2019). It is possible that to a large extent SAD is responsible for the ongoing epidemic of obesity, metabolic syndrome, diabetes and other so-called metabolic diseases in the USA and other Western countries. Another common diet, a so-called Western diet, is characterised by multicultural variations and excessive caloric intake. It is particularly negatively affected by the excessive consumption of refined carbohydrates, fatty meats, and fats, and a deficiency in the consumption of vegetables and fruits. It is commonly believed that such a diet contributes to obesity, as well as the development of type 2 diabetes and heart disease (Grotto and Zied 2010).

Meta-analysis shows that switching from a typical Western diet to an optimal diet at age 20 would increase life expectancy by more than a decade for women and men in the United States (Fadnes et al. 2022). The optimal diet includes the consumption of significantly more fruits, vegetables, whole grains, fish, legumes, and nuts with lower consumption of red and processed meat, sweetened beverages, and refined grains. Thus, what is eaten determines not only the state of health but also the life expectancy in general. At the same time, composition and functionality of the gut microbiome also changes, which depends on age and is often involved in the development of age-dependent chronic inflammation.

Dietary components that increase the relative abundance of anti-inflammatory and other health-promoting bacteria, such as oligosaccharides and polyphenols, may reduce systemic immune activation (Horn et al. 2022). Polyphenols are mainly found in fruits, vegetables, and cereals. They are known for their anti-inflammatory and antioxidant effects (Gill et al. 2022). Since they are components of plant foods, their consumption in especially large quantities is common for vegetarian and vegan diets. In particular, polyphenols increase the number of *Bifidobacterium* and *Lactobacillus* which have general antipathogenic and anti-inflammatory effects along with beneficial effects on the cardiovascular system (Tomova et al. 2019). It is also worth noting that plant food rich in fibres indigestible to humans may provide key substrates for the development of intestinal microorganisms. A high-fibre diet has been shown to increase gut microbial diversity and which is especially important for bacteria producing SCFA (Zhang et al. 2022). These fatty acids are produced by anaerobic fermentation of dietary fibres in the intestine: *Bacteroidetes* produce acetate and propionate, while *Firmicutes* produce butyrate. The latter one is the main energy source of colonocytes (colon cells) and the brain (Liu et al. 2018). Colonocytes are known to consume more than 70% of oxygen due to the oxidation of butyrate and thus maintain an anaerobic environment (Venegas et al. 2019). In addition, there is a correlation between SCFA level and intestinal barrier integrity. For example, butyrate can induce the expression of genes encoding tight junction proteins. SCFA deficiency leads to a weakening of the intestinal protective barrier, the functionality of epithelial cells and the production of mucus. As a result, excessive growth of microorganisms occurs eventually leading to an inflammatory reaction (Gieryńska et al. 2022). Calorie restriction is one of the most popular dietary interventions against ageing. This intervention may enhance both, average life expectancy and in some cases, maximum life expectancy (Sohal and Weindruch 1996). When the microbiome of obese females was transplanted into germ-free mice before and after a very low-calorie diet, three groups of response were noted with the very low-calorie diet. First, it improved the state of metabolism, in particular, glucose tolerance increased, and fat accumulation decreased. Second, overall higher alpha-diversity (richness) and restructuring of intestinal microbiota were observed. Third, immune ageing was slowed down (Sbierski-Kind et al. 2022). Perhaps consistent with the role of caloric restriction in longevity, women in the United States consume 25% fewer calories than men and live 5 years longer on average. Okinawans, the longest-lived on earth people, consume 40% fewer calories than Americans and live 4 years longer than Americans (Everitt and Le Couteur 2007).

In general, the intake and transformation of nutrients strongly affect the composition of the intestinal microbiota. In this regard, it is believed that anti-ageing dietary interventions may affect gut microbial communities and thereby promote healthy ageing (Vaiserman et al. 2017). However, individual, cultural, religious, climatic, traditional, and other peculiarities have to be taken into consideration when designing interventional dietary approaches.

Food Supplements: Probiotics, Prebiotics, and Antioxidants

Utilisation of probiotics and prebiotics is considered as a potential anti-ageing intervention. These supplements are believed to be able to improve the homeostasis of intestinal microflora and prevent unhealthy ageing (Jayanama and Theou 2019). Probiotics are viable microorganisms that, when ingested, supplement the host's flora. At the same time, they limit the concentration of potentially pathogenic flora in the digestive tract via competition (Chow 2002). *Bifidobacterium* and *Lactobacillus* are mostly often used as probiotics. Their use with food has many beneficial effects. In addition to inhibiting pathogenic (bad) microflora, probiotics stimulate the growth of beneficial microorganisms, improve digestion and the functioning of the immune system. For example, *Bifidobacterium* produces glutamine, which preserves the integrity of the intestinal mucosa and improves its protection (Bodke and Jogdand 2022) and butyrate described above (Liu et al. 2018).

Prebiotics are non-digestible food ingredients that selectively stimulate growth and activity of one or a limited number of beneficial (good) bacteria (Chow 2002). This definition was popular until 2008. In 2008, the sixth Meeting of the International Scientific Association of Probiotics and Prebiotics (ISAPP) defined dietary prebiotics as "a selectively fermented ingredient that results in specific changes in the composition and/or activity of the gastrointestinal microbiota, thus conferring benefit(s) upon host health" (Gibson et al. 2010). Thus, it has a positive effect on its composition and activity. Compounds of the carbohydrate group, such as short- and long-chain β-fructans and lactulose are examples of prebiotics (Davani-Davari et al. 2019). Prebiotics are used to target the composition and function of gut microbiota because they are selectively metabolised by specific microorganisms. They mainly stimulate growth the *Bifidobacterium* and *Lactobacillus* groups of bacteria, which are often called probiotics (Bedu-Ferrari et al. 2022). Thus, prebiotics and probiotics collectively regulate the composition and functionality of the gut microbiota and can ensure healthy ageing.

It is also worth mentioning the role of such dietary supplements as antioxidants, i.e. compounds that inhibit modification of different compounds by reactive oxygen species (ROS) (Lushchak 2015). A complex system of antioxidant protection operates in the cell to maintain oxidative homeostasis. It involves endogenously produced enzymes (such as catalase and superoxide dismutase) or dietary antioxidants such as vitamins C and E (Salehi et al. 2018). Pro- and antioxidant processes in living organism are well balanced but under some circumstances ROS levels may be increased leading to so-called oxidative stress which may be intensified by ageing and diseases (Lushchak 2015, 2021; Lushchak and Storey 2021). Prolonged exposure to ROS can cause the development of dysbacteriosis. In turn, this will affect the host's health, including intestinal damage, intestinal infections, and inflammatory bowel disease. Considering this, antioxidant dietary supplements potentially can protect the host from intestinal oxidative stress and influence the composition of the microbiota (Rajoka et al. 2021).

Among the intestinal microorganisms, antioxidant probiotics are singled out. It was shown that a strain of *Lactobacillus plantarum* 124 (LP124) possessed strong antioxidant activity, produced L-ascorbic acid, and promoted health and longevity (Wu et al. 2022). Oxidative stress is supposed to be related to ageing (Garaschuk et al. 2018). In particular, one of the ageing hypotheses suggests that age-related functional losses are caused by the accumulation of oxidatively modified biomolecules and their complexes (Lushchak 2021). In addition, oxidative stress accompanies a number of age-related diseases, such as cardiovascular and neurodegenerative diseases, chronic kidney disease, and cancer (Liguori et al. 2018; Garaschuk et al. 2018). Thus, the use of antioxidants to prevent oxidation processes may contribute to healthy longevity. For example, phenolic substances found in plants foods have a strong antioxidant and anti-inflammatory effects. Unabsorbed phenolic compounds undergo biotransformation by intestinal bacteria. As a result, they are absorbed, circulate in the body and produce their beneficial effects (Matsumura et al. 2023). Polyphenols are also believed to have a prebiotic effect. They stimulate growth and activity of specific microorganisms that are able to digest polyphenols and convert them into forms utilised by the host organism. Also, these compounds suppress the growth of pathogenic bacteria and inhibit the pro-inflammatory immune response in the intestine leading to the reduction of the risk of developing inflammatory bowel diseases (Mithul Aravind et al. 2021). Consumption of vitamins with pronounced antioxidant capacity, such as vitamin C and E, also modulates the composition and metabolic activity of intestinal microorganisms. Vitamin C causes a more pronounced effect. In particular, it significantly increased microbial alpha diversity and the concentration of faecal SCFA (Pham et al. 2021).

Thus, dietary supplements in the form of probiotics, prebiotics, and antioxidants are strongly interconnected and complement each other. They may be used as anti-ageing interventions due to their effects on the gut microbiome and associated beneficial effects.

Exercise

It has long been known that daily exercise is the key to healthy longevity. The ancient physician Hippocrates said: "All parts of the body, if used in moderation and exercised in labours to which each is accustomed, become thereby healthy and well developed and age slowly; but if they are unused and left idle, they become liable to disease, defective in growth and age quickly." However, lack of exercise related to sedentary style of life is now a serious problem of public health and a trigger for many diseases (Ruegsegger and Booth 2018).

Regular aerobic exercise decreases the risk of age-related diseases and prevents hypertension, cardiovascular disorders, type 2 diabetes, dementia, and others. At least 150 min/week of moderate-intensity aerobic exercise or 75 min/week of vigorous-intensity aerobic exercise is recommended for healthy ageing and longevity. The first, for example, includes walking and cycling, and the second—fitness or exhausting running (Pedersen 2019).

Again, exercise affects the composition and metabolic activity of microorganisms in the GI tract. Moderate physical activity causes positive changes in the intestinal microbiota. Intensive physical exercises, on the contrary, can lead to a decrease in the thickness of the intestinal mucus and result in increased permeability of the intestinal barrier. This allows pathogens or their constituents/fragments to enter the bloodstream and trigger inflammation (Clauss et al. 2021). Higher intestinal permeability occurs briefly during acute physical exertion. In a state of rest, contact between microbes and the immune system is weak (Lira et al. 2010). Figure 14.4 shows changes in the permeability of the intestinal barrier resulting from excessive intense exercise and regular moderate exercise. It is known that physical activity increases the body temperature and leads to the development of heat stress. Trained athletes have more efficient heat shock protein (HSP) response than sedentary people. In particular, HSPs prevent the destruction of the proteins of the tight junction between the epithelial cells of the intestinal mucosa. Athletes also have a lower circulating level of bacterial lipopolysaccharide. This indicates that regular moderate exercise has a hormetic effect. In particular, it promotes adequate adaptive response of the intestinal microbiota and, as a result, improves the stability of the intestinal barrier (Mailing et al. 2019). However, not all exercises have a positive effect on intestinal microbiota and health in general. For example, a systematic review by Bonomini-Gnutzmann et al. (2022) mentions 13 studies that showed a negative effect of aerobic exercise on the intestinal microbiota. Much of this is associated with endurance exercise, which decreases microbial diversity. Strength training has the least benefit for the microbiota. Moreover, research on athletes showed more negative consequences than on subjects who did not practice sports. Clearly, that physical activity has a positive effect on the gut microbiome of young people and in the elderly but due to discrepancies in studies, it is impossible to establish which taxa are most susceptible to the effects of exercise in the elderly (Ramos et al. 2022).

Conclusions

The gut microbiome is undoubtedly closely related to human health. It regulates the work of the immune and nervous systems, participates in the metabolism of various nutrients and xenobiotics, produces vitamins and short-chain fatty acids, and ultimately affects the integrity of the intestines. With age, its composition and functionality changes depending on many factors, in particular, lifestyle and nutrition. Development of dysbacteriosis, which contributes to age-related chronic

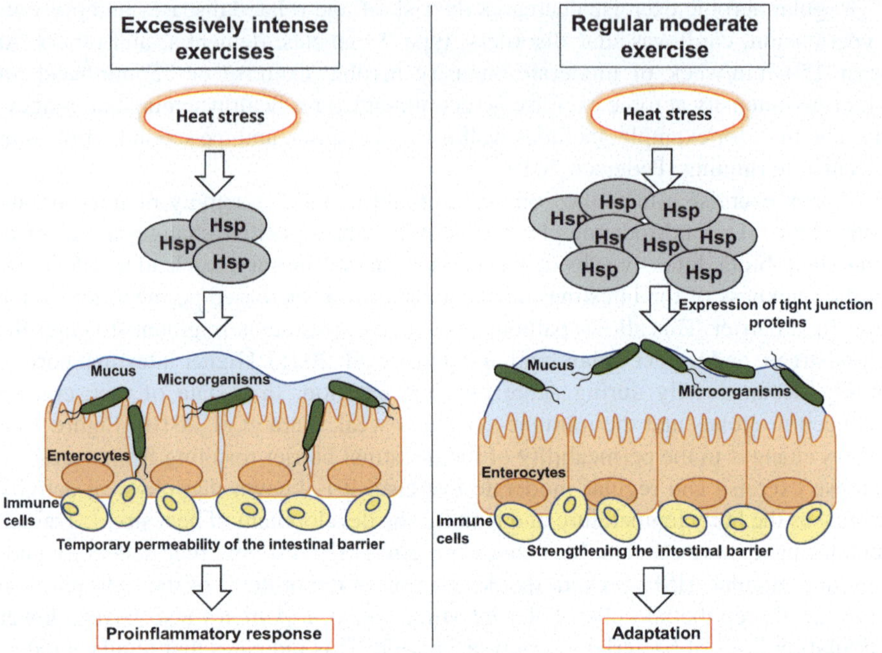

Fig. 14.4 Changes in the permeability of the intestinal barrier during excessive intense exercise and regular moderate exercise. Physical exertion leads to the development of heat stress triggering biosynthesis of HSPs which play the role of chaperones and maintain the structure of proteins in their native conformation. In case of excessive intensive exercises, temporary permeability of the intestinal barrier occurs due to heat stress resulting in intestinal microorganisms triggering of a pro-inflammatory response. Regular moderate exercise has a hormetic effect and leads to an adaptive response. In this case, more HSPs are formed, which activate the expression of tight junction proteins and helps the cells effectively cope with heat stress during regular exercise, and the normal integrity and permeability of the intestinal barrier is preserved. Abbreviations: HSP heat shock proteins

inflammation of low intensity and neurodegenerative disorders is one of the signs of ageing. Thus, regulation of gut microbiota composition may be a target for anti-ageing interventions. In particular, a balanced diet, the use of probiotics, prebiotics, and antioxidants, and a regular moderate exercise can influence the microbiome, prevent inflammation, age-related disorders and generally promote healthy ageing.

It is worth noting that the intestine is the main intermediary between the microbiota and its host due to which its integrity strongly affects the microbiome-dependent regulation of the immune, nervous and the entire body systems of the host's organism. In the intestine, all substances consumed are absorbed which directly affects their health. In addition, food components indigestible for humans, such as fibres, are substrates for the development of intestinal microorganisms. In this way, the integrity and normal homeostasis of the intestine can contribute to healthy ageing and longevity.

Maintenance and stabilisation of intestinal barrier functions may be the key to slowing down or even reversing age-related decline in function. For this reason, removing of damaging challenges and specific feeding of "good" microorganisms in human GI tract may be beneficial. Implantation of good microorganisms, or the transfer of whole microbiome from healthy subjects, may also be promising avenues for the extension of healthspan.

Acknowledgments This work was partially supported by grants from Ministry of Education and Science of Ukraine (#0122 U000894) and (#0123 U101790) to VL.

References

Alkasir R, Li J, Li X et al (2017) Human gut microbiota: the links with dementia development. Protein Cell 8:90–102. https://doi.org/10.1007/2Fs13238-016-0338-6

Alsegiani A, Shah Z (2022) The influence of gut microbiota alteration on age-related neuroinflammation and cognitive decline. Neural Regen Res 17:2407–2412. https://doi.org/10.4103/2F1673-5374.335837

An R, Wilms E, Masclee AAM et al (2018) Age-dependent changes in GI physiology and microbiota: time to reconsider? Gut 67:2213–2222. https://doi.org/10.1136/gutjnl-2017-315542

Asiamah EK, Ekwemalor K, Adjei-Fremah S et al (2019) Natural and synthetic pathogen associated molecular patterns modulate galectin expression in cow blood. J Anim Sci Technol 61:245–253. https://doi.org/10.5187/jast.2019.61.5.245

Bedu-Ferrari C, Biscarrat P, Langella P et al (2022) Prebiotics and the human gut microbiota: from breakdown mechanisms to the impact on metabolic health. Nutrients 14:2096. https://doi.org/10.3390/nu14102096

Bodke H, Jogdand S (2022) Role of probiotics in human health. Cureus 14(11):e31313. https://doi.org/10.7759/cureus.31313

Bonomini-Gnutzmann R, Plaza-Díaz J, Jorquera-Aguilera C et al (2022) Effect of intensity and duration of exercise on gut microbiota in humans: a systematic review. Int J Environ Res Public Health 19:9518. https://doi.org/10.3390/ijerph

Brooks CN, Wight ME, Azeez OE et al (2023) Growing old together: what we know about the influence of diet and exercise on the aging host's gut microbiome. Front Sports Act Living 5:1168731. https://doi.org/10.3389/fspor.2023.1168731

Chow JM (2002) Probiotics and prebiotics: a brief overview. J Ren Nutr 12:76–86. https://doi.org/10.1053/jren.2002.31759

Chung KW, Chung HY (2019) The effects of calorie restriction on autophagy: role on aging intervention. Nutrients 11:2923. https://doi.org/10.3390/2Fnu11122923

Clauss M, Gérard P, Mosca A et al (2021) Interplay between exercise and gut microbiome in the context of human health and performance. Front Nutr 8:637010. https://doi.org/10.3389/fnut.2021.637010

Collins SL, Patterson AD (2020) The gut microbiome: an orchestrator of xenobiotic metabolism. Acta Pharm Sin B 10:19–32. https://doi.org/10.1016/j.apsb.2019.12.001

Costa-Pinto R, Gantner D (2020) Macronutrients, minerals, vitamins and energy. Anaesthesia Intensive Care Med 21:157–161. https://doi.org/10.1016/j.mpaic.2019.12.006

Davani-Davari D, Negahdaripour M, Karimzadeh I et al (2019) Prebiotics: definition, types, sources, mechanisms, and clinical applications. Food Secur 8:92. https://doi.org/10.3390/2Ffoods8030092

DeJong EN, Surette MG, Bowdish DME (2020) The gut microbiota and unhealthy aging: disentangling cause from consequence. Cell Host Microbe 28:180–189. https://doi.org/10.1016/j.chom.2020.07.013

Diebel LWM, Rockwood K (2021) Determination of biological age: geriatric assessment vs biological biomarkers. Curr Oncol Rep 23:104. https://doi.org/10.1007/s11912-021-01097-9

Donati Zeppa S, Agostini D, Ferrini F et al (2023) Interventions on gut microbiota for healthy aging. Cells 12:34. https://doi.org/10.3390/cells12010034

Everitt AV, Le Couteur DG (2007) Life extension by calorie restriction in humans. Ann N Y Acad Sci 1114:428–433. https://doi.org/10.1196/annals.1396.005

Fadnes LT, Økland JM, Haaland ØA et al (2022) Estimating impact of food choices on life expectancy: a modeling study. PLoS Med 19(2):e1003889. https://doi.org/10.1371/journal.pmed.1003889

Garaschuk O, Semchyshyn HM, Lushchak VI (2018) Healthy brain aging: interplay between reactive species, inflammation and energy supply. Ageing Res Rev 43:26–45. https://doi.org/10.1016/j.arr.2018.02.003

Gensollen T, Iyer SS, Kasper DL et al (2016) How colonization by microbiota in early life shapes the immune system. Science 352:539–544. https://doi.org/10.1126/2Fscience.aad9378

Ghosh S, Whitley CS, Haribabu B et al (2021) Regulation of intestinal barrier function by microbial metabolites. Cell Mol Gastroenterol Hepatol 11:1463–1482. https://doi.org/10.1016/j.jcmgh.2021.02.007

Gibson GR, Scott KP, Rastall RA et al (2010) Dietary prebiotics: current status and new definition. Food Sci Technol Bull Funct Foods 7:1–19. https://doi.org/10.1616/1476-2137.15880

Gieryńska M, Szulc-Dąbrowska L, Struzik J et al (2022) Integrity of the intestinal barrier: the involvement of epithelial cells and microbiota—a mutual relationship. Animals 12:145. https://doi.org/10.3390/2Fani12020145

Gill PA, Inniss S, Kumagai T et al (2022) The role of diet and gut microbiota in regulating gastrointestinal and inflammatory disease. Front Immunol 13:866059. https://doi.org/10.3389/fimmu.2022.866059

Grotto D, Zied E (2010) The standard American diet and its relationship to the health status of Americans. Nutr Clin Pract 25:603–612. https://doi.org/10.1177/0884533610386234

Horn J, Mayer DE, Chen S et al (2022) Role of diet and its effects on the gut microbiome in the pathophysiology of mental disorders. Transl Psychiatry 12:164. https://doi.org/10.1038/s41398-022-01922-0

Jandhyala SM, Talukdar R, Subramanyam C et al (2015) Role of the normal gut microbiota. World J Gastroenterol 21:8836–8847. https://doi.org/10.3748/wjg.v21.i29.8787

Jayanama K, Theou O (2019) Effects of probiotics and prebiotics on frailty and ageing: a narrative review. Curr Clin Pharmacol 15:183–192. https://doi.org/10.2174/1574884714666191120124548

Kageyama T, Matsuo T, Kurakake R et al (2020) Relationship between T cells and microbiota in health and disease. Prog Mol Biol Transl Sci 171:95–129. https://doi.org/10.1016/bs.pmbts.2020.03.007

Kim S, Jazwinski SM (2018) The gut microbiota and healthy aging: a mini-review. Gerontology 64:513–520. https://doi.org/10.1159/000490615

Kim HI, Kim J-K, Kim J-Y et al (2019) *Lactobacillus plantarum LC27* and *Bifidobacterium longum LC67* simultaneously alleviate high-fat diet-induced colitis, endotoxemia, liver steatosis, and obesity in mice. Nutr Res 67:78–89. https://doi.org/10.1016/j.nutres.2019.03.008

Kocot AM, Jarocka-Cyrta E, Drabińska N (2022) Overview of the importance of biotics in gut barrier integrity. Int J Mol Sci 23:2896. https://doi.org/10.3390/ijms23052896

Li Z, Liang H, Hu Y et al (2023) Gut bacterial profiles in Parkinson's disease: a systematic review. CNS Neurosci Ther 29:140–157. https://doi.org/10.1111/cns.13990

Liguori I, Russo G, Curcio F et al (2018) Oxidative stress, aging, and diseases. Clin Interv Aging 13:757–772. https://doi.org/10.2147/2FCIA.S158513

Lira FS, Rosa JC, Pimentel GD et al (2010) Endotoxin levels correlate positively with a sedentary lifestyle and negatively with highly trained subjects. Lipids Health Dis 9:82. https://doi.org/10.1186/1476-511x-9-82

Liu JK (2022) Antiaging agents: safe interventions to slow aging and healthy life span extension. Nat Prod Bioprospect 12:18. https://doi.org/10.1007/s13659-022-00339-y

Liu H, Wang J, He T et al (2018) Butyrate: a double-edged sword for health? Adv Nutr 9:1. https://doi.org/10.1093/advances/nmx009

Lushchak V (2015) Free radicals, reactive oxygen species, oxidative stresses and their classifications. Ukr Biochem J 87:11–18. https://doi.org/10.15407/ubj87.06.011

Lushchak V (2021) Interplay between bioenergetics and oxidative stress at normal brain aging. Aging as a result of increasing disbalance in the system oxidative stress-energy provision. Pflugers Arch 473:713–722. https://doi.org/10.1007/s00424-021-02531-4

Lushchak V, Storey K (2021) Oxidative stress concept updated: definitions, classifications, and regulatory pathways implicated. EXCLI J 20:956–967. https://doi.org/10.17179/2Fexcli2021-3596

Magne F, Gotteland M, Gauthier L et al (2020) The *Firmicutes/Bacteroidetes* ratio: a relevant marker of gut dysbiosis in obese patients? Nutrients 12:1474. https://doi.org/10.3390/2Fnu12051474

Mailing LJ, Allen JM, Buford TW et al (2019) Exercise and the gut microbiome: a review of the evidence, potential mechanisms, and implications for human health. Exerc Sport Sci Rev 47:75–85. https://doi.org/10.1249/jes.0000000000000183

Maltoni R, Ravaioli S, Bronte G et al (2022) Chronological age or biological age: what drives the choice of adjuvant treatment in elderly breast cancer patients? Transl Oncol 15:101300. https://doi.org/10.1016/j.tranon.2021.101300

Mancuso C, Santangelo R (2018) Alzheimer's disease and gut microbiota modifications: the long way between preclinical studies and clinical evidence. Pharmacol Res 129:329–336. https://doi.org/10.1016/j.phrs.2017.12.009

Matsumura Y, Kitabatake M, Kayano SI et al (2023) Dietary phenolic compounds: their health benefits and association with the gut microbiota. Antioxidants 12:880. https://doi.org/10.3390/antiox12040880

Menassa M, Stronks K, Khatmi F et al (2023) Concepts and definitions of healthy ageing: a systematic review and synthesis of theoretical models. EClinicalMedicine 56:101821. https://doi.org/10.1016/j.eclinm.2022.101821

Meng X, Zhang G, Cao H et al (2020) Gut dysbacteriosis and intestinal disease: mechanism and treatment. J Appl Microbiol 129:787–805. https://doi.org/10.1111/jam.14661

Meng C, Feng S, Hao Z et al (2022) Changes in gut microbiota composition with age and correlations with gut inflammation in rats. PLoS One 17(3):e0265430. https://doi.org/10.1371/journal.pone.0265430

Mithul Aravind S, Wichienchot S, Tsao R et al (2021) Role of dietary polyphenols on gut microbiota, their metabolites and health benefits. Food Res Int 142:110189. https://doi.org/10.1016/j.foodres.2021.110189

Morais LH, Schreiber HL, Mazmanian SK (2021) The gut microbiota–brain axis in behaviour and brain disorders. Nat Rev Microbiol 19:241–255. https://doi.org/10.1038/s41579-020-00460-0

Moreno-Indias I, Cardona F, Tinahones FJ et al (2014) Impact of the gut microbiota on the development of obesity and type 2 diabetes mellitus. Front Microbiol 5:190. https://doi.org/10.3389/fmicb.2014.00190

Morreale C, Bresesti I, Bosi A et al (2022) Microbiota and pain: save your gut feeling. Cells 11:971. https://doi.org/10.3390/cells11060971

Novakovic M, Rout A, Kingsley T et al (2020) Role of gut microbiota in cardiovascular diseases. World J Cardiol 12:110–122. https://doi.org/10.4330/wjc.v12.i4.110

Pedersen BK (2019) Which type of exercise keeps you young? Curr Opin Clin Nutr Metab Care 22:167–173. https://doi.org/10.1097/mco.0000000000000546

Pham VT, Fehlbaum S, Seifert N et al (2021) Effects of colon-targeted vitamins on the composition and metabolic activity of the human gut microbiome – a pilot study. Gut Microbes 13:1–20. https://doi.org/10.1080/19490976.2021.1875774

Rajoka MSR, Thirumdas R, Mehwish HM et al (2021) Role of food antioxidants in modulating gut microbial communities: novel understandings in intestinal oxidative stress damage and their impact on host health. Antioxidants 10:1563. https://doi.org/10.3390/antiox10101563

Ramos C, Gibson GR, Walton GE et al (2022) Systematic review of the effects of exercise and physical activity on the gut microbiome of older adults. Nutrients 14:674. https://doi.org/10.3390/nu14030674

Ren J, Li H, Zeng G et al (2023) Gut microbiome-mediated mechanisms in aging-related diseases: are probiotics ready for prime time? Front Pharmacol 14:1178596. https://doi.org/10.3389/fphar.2023.1178596

Rinninella E, Raoul P, Cintoni M et al (2019) What is the healthy gut microbiota composition? A changing ecosystem across age, environment, diet, and diseases. Microorganisms 7:14. https://doi.org/10.3390/microorganisms7010014

Ros M, Carrascosa JM (2020) Current nutritional and pharmacological anti-aging interventions. Biochim Biophys Acta Mol basis Dis 1866:165612. https://doi.org/10.1016/j.bbadis.2019.165612

Rowland I, Gibson G, Heinken A et al (2018) Gut microbiota functions: metabolism of nutrients and other food components. Eur J Nutr 57:1–24. https://doi.org/10.1007/s00394-017-1445-8

Ruegsegger GN, Booth FW (2018) Health benefits of exercise. Cold Spring Harb Perspect Med 8(7):a029694. https://doi.org/10.1101/cshperspect.a029694

Ryan-Harshman M, Aldoori W (2006) New dietary reference intakes for macronutrients and fibre. Can Fam Physician 52:177–179

Salazar AM, Aparicio R, Clark RI et al (2023) Intestinal barrier dysfunction: an evolutionarily conserved hallmark of aging. Dis Model Mech 16(4):dmm049969. https://doi.org/10.1242/dmm.049969

Salehi B, Martorell M, Arbiser JL et al (2018) Antioxidants: positive or negative actors? Biomol Ther 8:124. https://doi.org/10.3390/biom8040124

Sbierski-Kind J, Grenkowitz S, Schlickeiser S et al (2022) Effects of caloric restriction on the gut microbiome are linked with immune senescence. Microbiome 10:57. https://doi.org/10.1186/s40168-022-01249-4

Sohal RS, Weindruch R (1996) Oxidative stress, caloric restriction, and aging. Science 273:59–63. https://doi.org/10.1126/science.273.5271.59

Thursby E, Juge N (2017) Introduction to the human gut microbiota. Biochem J 474:1823–1836. https://doi.org/10.1042/2FBCJ20160510

Tomova A, Bukovsky I, Rembert E et al (2019) The effects of vegetarian and vegan diets on gut microbiota. Front Nutr 6:47. https://doi.org/10.3389/2Ffnut.2019.00047

Ursell LK, Metcalf JL, Parfrey LW et al (2012) Defining the human microbiome. Nutr Rev 70:38–44. https://doi.org/10.1111/j.1753-4887.2012.00493.x

Vaiserman AM, Koliada AK, Marotta F (2017) Gut microbiota: a player in aging and a target for anti-aging intervention. Ageing Res Rev 35:36–45. https://doi.org/10.1016/j.arr.2017.01.001

Venegas DP, De la Fuente MK, Landskron G et al (2019) Short chain fatty acids (SCFAs)-mediated gut epithelial and immune regulation and its relevance for inflammatory bowel diseases. Front Immunol 10:277. https://doi.org/10.3389/2Ffimmu.2019.00277

Vernocchi P, Del Chierico F, Putignani L (2020) Gut microbiota metabolism and interaction with food components. Int J Mol Sci 21:3688. https://doi.org/10.3390/ijms21103688

Walker AW, Hoyles L (2023) Human microbiome myths and misconceptions. Nat Microbiol 8:1392–1396. https://doi.org/10.1038/s41564-023-01426-7

Wang L, Wang S, Zhang Q et al (2022) The role of the gut microbiota in health and cardiovascular diseases. Mol Biomed 3:30. https://doi.org/10.1186/s43556-022-00091-2

Wu L, Xie X, Li Y et al (2022) Gut microbiota as an antioxidant system in centenarians associated with high antioxidant activities of gut-resident Lactobacillus. NPJ Biofilms Microbiomes 8:102. https://doi.org/10.1038/s41522-022-00366-0
Zeng MY, Inohara N, Nuñez G (2017) Mechanisms of inflammation-driven bacterial dysbiosis in the gut. Mucosal Immunol 10:18–26. https://doi.org/10.1038/mi.2016.75
Zhang P (2022) Influence of foods and nutrition on the gut microbiome and implications for intestinal health. Int J Mol Sci 23:9588. https://doi.org/10.3390/ijms23179588
Zhang F, Fan D, Huang J-l, Zuo T (2022) The gut microbiome: linking dietary fiber to inflammatory diseases. Med Microecol 14:100070. https://doi.org/10.1016/j.medmic.2022.100070
Zheng D, Liwinski T, Elinav E (2020) Interaction between microbiota and immunity in health and disease. Cell Res 30:492–506. https://doi.org/10.1038/s41422-020-0332-7
Zhu X, Li B, Lou P et al (2021) The relationship between the gut microbiome and neurodegenerative diseases. Neurosci Bull 37:1510–1522. https://doi.org/10.1007/s12264-021
Zohoori FV (2019) Chapter 1: Nutrition and diet. Monogr Oral Sci 28:1–13. https://doi.org/10.1159/000455365

SPRINGER NATURE

GPSR Compliance

The European Union's (EU) General Product Safety Regulation (GPSR) is a set of rules that requires consumer products to be safe and our obligations to ensure this.

If you have any concerns about our products, you can contact us on ProductSafety@springernature.com

In case Publisher is established outside the EU, the EU authorized representative is:

Springer Nature Customer Service Center GmbH
Europaplatz 3
69115 Heidelberg, Germany

The manufacturer's authorised representative in the EU is Springer Nature Customer Service Centre GmbH, Europaplatz 3, 69115 Heidelberg, Germany. If you have any concerns regarding our products, please contact ProductSafety@springernature.com

Printed and bound by CPI Group (UK) Ltd, Croydon, CR0 4YY

30/09/2026

02225564-0012